AF411366

Cellular and Molecular Strategies in Cyanobacterial Survival

In Memory of Prof. Dr. Wolfgang Lockau

Cellular and Molecular Strategies in Cyanobacterial Survival

In Memory of Prof. Dr. Wolfgang Lockau

Editors

Iris Maldener

Khaled Selim

MDPI • Basel • Beijing • Wuhan • Barcelona • Belgrade • Manchester • Tokyo • Cluj • Tianjin

Editors
Iris Maldener
Interfaculty Institute of
Microbiology and Infection
Medicine
University of Tübingen
Tübingen
Germany

Khaled Selim
Interfaculty Institute of
Microbiology and Infection
Medicine
University of Tübingen
Tübingen
Germany

Editorial Office
MDPI
St. Alban-Anlage 66
4052 Basel, Switzerland

This is a reprint of articles from the Special Issue published online in the open access journal *Life* (ISSN 2075-1729) (available at: www.mdpi.com/journal/life/special_issues/CMSCS).

For citation purposes, cite each article independently as indicated on the article page online and as indicated below:

LastName, A.A.; LastName, B.B.; LastName, C.C. Article Title. *Journal Name* **Year**, *Volume Number*, Page Range.

ISBN 978-3-0365-1688-2 (Hbk)
ISBN 978-3-0365-1687-5 (PDF)

Contents

Preface to "Cellular and Molecular Strategies in Cyanobacterial Survival"

All species of cyanobacteria are capable of oxygenic photosynthesis. Despite this common mode of metabolism, they exist in different morphologies as single cells or as multicellular filaments, which may even differentiate specialized cells. Furthermore, cyanobacteria occupy almost all illuminated aquatic and terrestrial habitats, including harsh environments of deserts, oceans, and hypersaline, volcanic, and thermal biospheres. Therefore, they represent one of the quantitatively most abundant organisms on earth and can be dated back in evolution for more then 2.4 billion years. In addition to this biodiversity, many species have developed strategies to adapt to various stress conditions, including nutrient starvation, occurring in their own habitat. In recent years, our knowledge on cyanobacterial survival strategies has increased tremendously by applying global studies such as transcriptomics and proteomics as well as advanced microscopic technics. Protein structure and function analysis have revealed the potential of cellular and metabolic adaptation mechanisms of single cells and multicellular filaments towards environmental stress, including cell differentiation, signal transduction, formation of reserve materials, and resuscitation from dormant states, to name a few. In this Special Issue of *Life*, we invited researchers from all over the world to share advances in their understanding of ecological, cellular, and molecular mechanisms of cyanobacterial survival. This includes original work and review articles dealing with signaling pathways, strategies of gene and protein regulation, global studies, and new discoveries related to the differentiation of spore-like akinetes, motile hormogonia, and nitrogen-fixing heterocysts.

Iris Maldener, Khaled Selim
Editors

Editorial

Cellular and Molecular Strategies in Cyanobacterial Survival—"In Memory of Prof. Dr. Wolfgang Lockau"

Khaled A. Selim and Iris Maldener *

Organismic Interactions Department, Interfaculty Institute for Microbiology and Infection Medicine, Cluster of Excellence 'Controlling Microbes to Fight Infections', Tübingen University, Auf der Morgenstelle 28, 72076 Tübingen, Germany
* Correspondence: Khaled.selim@uni-tuebingen.de (K.A.S.); iris.maldener@uni-tuebingen.de (I.M.)

Citation: Selim, K.A., Maldener, I. Cellular and Molecular Strategies in Cyanobacterial Survival—"In Memory of Prof. Dr. Wolfgang Lockau". *Life* **2021**, *11*, 132. https://doi.org/10.3390/life11020132

Academic Editor: Daniela Billi

Received: 3 February 2021
Accepted: 4 February 2021
Published: 9 February 2021

Publisher's Note: MDPI stays neutral with regard to jurisdictional claims in published maps and institutional affiliations.

Aerobic life on Earth evolved about 3.8–2.7 billion years ago with the evolution of oxygenic photosynthesis by cyanobacteria. Approximately 2.4 billion years ago, the ancestors of today's cyanobacteria were the first oxygenic photoautotrophs to release molecular O_2 as a waste product of the oxygenic photosynthesis. Via endosymbiosis, the photosynthetic ability was later transmitted from cyanobacteria to eukaryotes, giving rise to plastids of higher plants. Nowadays, cyanobacteria display diverse cell morphologies and occupy almost all illuminated aquatic and terrestrial habitats, including harsh environments like deserts, oceans, hypersaline, volcanic and thermal biospheres; therefore, they represent one of the quantitatively most abundant organisms on earth.

Cyanobacteria are the most influential photoautotrophic prokaryotes that can perform photosynthesis and fix atmospheric carbon dioxide via the Calvin–Benson cycle for primary metabolism. In addition, several species can fix atmospheric nitrogen (N_2). As primary producers, cyanobacteria play a key role in the global carbon cycle. In addition, the diazotrophic cyanobacteria are the dominant nitrogen fixers in the oceans, introducing combined nitrogen into the global nitrogen cycle. The carbon/nitrogen assimilation reactions require a tight regulation and a constant sensing of the quantity and quality of the carbon and nitrogen availability. Generally, both metabolisms are coordinated by a complex crosstalk between different input signals. The sensing and regulation of the nitrogen/carbon metabolism in cyanobacteria depend on several signal-transduction proteins, which transduce the energy/carbon/nitrogen/day–night status of the cell and include several adaptation strategies. For example, chlorosis in case of nitrogen starvation in unicellular cyanobacteria or a carbon concentrating mechanism to cope with low levels of atmospheric CO_2 and the slow rate of the CO_2 fixing enzyme. Another strategy to cope with nitrogen starvation is realized by filamentous species, which show a division of labor between different cell types, the photosynthetic active vegetative cells and the N_2 fixing heterocysts. These bacteria represent true multicellular organisms which are able to communicate via special cell–cell connections: the septal junctions.

Some of the filamentous heterocyst-forming cyanobacteria differentiate another cell type, the resistant spore-like akinetes, which allow survival under instances of cold and nutrient starvation. When environmental conditions become favorable, these akinetes can germinate and regrow to long filaments. To find and occupy the optimal environment, cyanobacteria deploy yet another strategy: they form small filamentous hormogonia that display pili-based motility. This third differentiated motile cell type is also involved in establishing the plant–cyanobacterium symbiosis.

In this Special Issue of *Life* on cyanobacterial cellular and molecular strategies for survival, we summarize some of the recent research advances in cyanobacterial biology in a wonderful collection of original research and review articles. By mutational studies, Khudyakov et al. [1] investigated the influence of three small proteins (PatS, PatX and HetN) on HetR, the master regulator of heterocyst differentiation in the multicellular cyanobacterium *Nostoc* sp. PCC 7120. PatS, PatX and HetN are negative regulators of HetR with a conserved pentapeptide motif RG(S/T)GR. The paper demonstrates the importance of a tight control of HetR activity for vegetative growth and heterocyst pattern formation. Zulkefli and Hwang [2] investigated the effect of various inorganic nitrogen sources on the growth of *Anabaena variabilis* under nitrogen fixing conditions and on heterocyst differentiation. Álvarez-Escribano et al. [3] analyzed a new example of post-transcriptional regulation by a small regulatory RNA, NsrR1 (nitrogen stress-repressed RNA 1), and identified the gene *all1871*, which is required for the diazotrophic growth of *Nostoc* sp. PCC 7120. The heterocyst specific expression of *all1871* is regulated by direct binding of NsrR1 to the $5'$ untranslated region of the *all1871* mRNA and represents an example of indirect regulation via the global nitrogen regulator NtcA.

Regarding photosynthesis and carbon metabolism, Gollan et al. [4] described the global transcriptional changes in the *Nostoc* sp. PCC 7120 upon the shift from high carbon (3% CO_2) to low carbon (0.04% CO_2) conditions. Selim and Haffner [5] studied the effects of long-term exposure of heavy metal stress on the nitrogen-starved unicellular cyanobacterium *Synechococcus elongatus* PCC 7942. Thurotte et al. [6] showed that the DnaK3 chaperon is required for efficient photosynthesis and for the biogenesis of thylakoid membranes, where the photosynthetic machinery is localized in the unicellular cyanobacterium *Synechocystis* sp. PCC 6803. Margulis et al. [7] demonstrated that overexpression of the gene encoding for Pgr5-like protein, which is required for cyclic photosynthetic electron flow paths in eukaryotes, causes accumulation of chlorophyll, photosystems, and glycogen contents in *Synechocystis* sp. PCC 6803 and is involved in redox regulation. Sun et al. [8] explored the influence of the circadian clock on the dynamic regulation of CO_2 fixation and the carboxysome (CO_2-fixing microcompartment) localization in response to diurnal day–night rhythm in *Synechococcus elongatus* PCC 7942. Koch et al. [9] studied the role of the reserve carbon biopolymer polyhydroxybutyrate (PHB) in *Synechocystis* sp. PCC 6803 physiology. They showed that the intracellular PHB content increases under diurnal day–night cycles.

Besides these original research papers, we received several comprehensive reviews summarizing the updates of some important aspects of cyanobacterial biology. Springstein et al. [10] reviewed the structural determinants and their role in cyanobacterial morphogenesis. Brandenburg and Klähn [11] described the role of small regulatory proteins on cyanobacterial metabolism, while Labella et al. [12] focused on the signaling roles of another small regulatory protein PipX in the coordination of nitrogen metabolism. Rachedi et al. [13] provided a mechanistic overview of stress signaling in cyanobacteria. Conradi et al. [14] and Schirmacher et al. [15] summarized the current advances on the role of pili in motility and natural competence in cyanobacteria. Finally, Lima et al. [16] described the cyanobacterial secretion systems with special focus on extracellular vesicles.

We expect that this collection will provide the reader with a comprehensive overview of some important aspects of cyanobacterial biology. We are confident that this cyanobacterial Special Issue of *Life* will attract the attention of a broad audience, in particular, of any scientist interested in fundamental biological questions, including aspects of cyanobacterial biology.

We would like to dedicate this Special Issue of *Life* to Prof. Dr. Wolfgang Lockau, for his distinguished contribution to the biochemistry and cell biology of cyanobacteria.

Acknowledgments: Khaled A. Selim and Iris Maldener wish to warmly thank all the contributors of the Special Issue of *Life* (ISSN 2075-1729): Cellular and Molecular Strategies in Cyanobacterial Survival. Moreover, we gratefully acknowledge Karl Forchhammer for the continuous support, and we are indebted to the German research foundation (DFG) for continuous funding of the work in our laboratory and to the infrastructural support by the Cluster of Excellence (EXC 2124) of the DFG. Our daily work is dedicated to our beloved family.

Conflicts of Interest: The authors declare no conflict of interest.

References

1. Khudyakov, I.; Gladkov, G.; Elhai, J. Inactivation of Three RG(S/T)GR Pentapeptide-Containing Negative Regulators of HetR Results in Lethal Differentiation of *Anabaena* PCC 7120. *Life* **2020**, *10*, 326. [CrossRef] [PubMed]
2. Zulkefli, N.S.; Hwang, S.-J. Heterocyst Development and Diazotrophic Growth of *Anabaena variabilis* under Different Nitrogen Availability. *Life* **2020**, *10*, 279. [CrossRef] [PubMed]
3. Álvarez-Escribano, I.; Brenes-Álvarez, M.; Olmedo-Verd, E.; Vioque, A.; Muro-Pastor, A.M. The Nitrogen Stress-Repressed sRNA NsrR1 Regulates Expression of *all1871*, a Gene Required for Diazotrophic Growth in *Nostoc* sp. PCC 7120. *Life* **2020**, *10*, 54. [CrossRef] [PubMed]
4. Gollan, P.J.; Muth-Pawlak, D.; Aro, E.-M. Rapid Transcriptional Reprogramming Triggered by Alteration of the Carbon/Nitrogen Balance Has an Impact on Energy Metabolism in *Nostoc* sp. PCC 7120. *Life* **2020**, *10*, 297. [CrossRef]
5. Selim, K.A.; Haffner, M. Heavy Metal Stress Alters the Response of the Unicellular Cyanobacterium *Synechococcus elongatus* PCC 7942 to Nitrogen Starvation. *Life* **2020**, *10*, 275. [CrossRef] [PubMed]
6. Thurotte, A.; Seidel, T.; Jilly, R.; Kahmann, U.; Schneider, D. DnaK3 Is Involved in Biogenesis and/or Maintenance of Thylakoid Membrane Protein Complexes in the Cyanobacterium *Synechocystis* sp. PCC 6803. *Life* **2020**, *10*, 55. [CrossRef] [PubMed]
7. Margulis, K.; Zer, H.; Lis, H.; Schoffman, H.; Murik, O.; Shimakawa, G.; Krieger-Liszkay, A.; Keren, N. Over Expression of the Cyanobacterial Pgr5-Homologue Leads to Pseudoreversion in a Gene Coding for a Putative Esterase in *Synechocystis* 6803. *Life* **2020**, *10*, 174. [CrossRef] [PubMed]
8. Sun, Y.; Huang, F.; Dykes, G.F.; Liu, L.-N. Diurnal Regulation of In Vivo Localization and CO_2-Fixing Activity of Carboxysomes in *Synechococcus elongatus* PCC 7942. *Life* **2020**, *10*, 169. [CrossRef] [PubMed]
9. Koch, M.; Berendzen, K.W.; Forchhammer, K. On the Role and Production of Polyhydroxybutyrate (PHB) in the Cyanobacterium *Synechocystis* sp. PCC 6803. *Life* **2020**, *10*, 47. [CrossRef] [PubMed]
10. Springstein, B.L.; Nürnberg, D.J.; Weiss, G.L.; Pilhofer, M.; Stucken, K. Structural Determinants and Their Role in Cyanobacterial Morphogenesis. *Life* **2020**, *10*, 355. [CrossRef] [PubMed]
11. Brandenburg, F.; Klähn, S. Small but Smart: On the Diverse Role of Small Proteins in the Regulation of Cyanobacterial Metabolism. *Life* **2020**, *10*, 322. [CrossRef] [PubMed]
12. Labella, J.I.; Cantos, R.; Salinas, P.; Espinosa, J.; Contreras, A. Distinctive Features of PipX, a Unique Signaling Protein of Cyanobacteria. *Life* **2020**, *10*, 79. [CrossRef] [PubMed]
13. Rachedi, R.; Foglino, M.; Latifi, A. Stress Signaling in Cyanobacteria: A Mechanistic Overview. *Life* **2020**, *10*, 312. [CrossRef] [PubMed]
14. Conradi, F.D.; Mullineaux, C.W.; Wilde, A. The Role of the Cyanobacterial Type IV Pilus Machinery in Finding and Maintaining a Favourable Environment. *Life* **2020**, *10*, 252. [CrossRef] [PubMed]
15. Schirmacher, A.M.; Hanamghar, S.S.; Zedler, J.A.Z. Function and Benefits of Natural Competence in Cyanobacteria: From Ecology to Targeted Manipulation. *Life* **2020**, *10*, 249. [CrossRef] [PubMed]
16. Lima, S.; Matinha-Cardoso, J.; Tamagnini, P.; Oliveira, P. Extracellular Vesicles: An Overlooked Secretion System in Cyanobacteria. *Life* **2020**, *10*, 129. [CrossRef] [PubMed]

Article

Inactivation of Three RG(S/T)GR Pentapeptide-Containing Negative Regulators of HetR Results in Lethal Differentiation of *Anabaena* PCC 7120

Ivan Khudyakov [1,*], Grigory Gladkov [1] and Jeff Elhai [2]

[1] All-Russia Research Institute for Agricultural Microbiology, 196608 Saint-Petersburg, Russia; ruginodis@gmail.com

[2] Department of Biology, Virginia Commonwealth University, Richmond, VA 23284, USA; elhaij@vcu.edu

* Correspondence: iykhudyakov@yandex.ru

Received: 31 October 2020; Accepted: 1 December 2020; Published: 4 December 2020

Abstract: The filamentous cyanobacterium *Anabaena* sp. PCC 7120 produces, during the differentiation of heterocysts, a short peptide PatS and a protein HetN, both containing an RGSGR pentapeptide essential for activity. Both act on the master regulator HetR to guide heterocyst pattern formation by controlling the binding of HetR to DNA and its turnover. A third small protein, PatX, with an RG(S/T)GR motif is present in all HetR-containing cyanobacteria. In a nitrogen-depleted medium, inactivation of *patX* does not produce a discernible change in phenotype, but its overexpression blocks heterocyst formation. Mutational analysis revealed that PatX is not required for normal intercellular signaling, but it nonetheless is required when PatS is absent to prevent rapid ectopic differentiation. Deprivation of all three negative regulators—PatS, PatX, and HetN—resulted in synchronous differentiation. However, in a nitrogen-containing medium, such deprivation leads to extensive fragmentation, cell lysis, and aberrant differentiation, while either PatX or PatS as the sole HetR regulator can establish and maintain a semiregular heterocyst pattern. These results suggest that tight control over HetR by PatS and PatX is needed to sustain vegetative growth and regulated development. The mutational analysis has been interpreted in light of the opposing roles of negative regulators of HetR and the positive regulator HetL.

Keywords: cyanobacteria; heterocyst; regulation of differentiation

---

1. Introduction

In response to different environmental cues, a subpopulation of vegetative cells from nostocalean cyanobacteria can differentiate into different endpoints: terminally differentiated heterocysts for aerobic nitrogen fixation, suicidal necridia for filament fragmentation to produce motile hormogonia for dissemination, and dormant akinetes for survival in harsh environments [1–5]. The model strain *Anabaena/Nostoc* sp. PCC 7120 (hereafter *Anabaena* PCC 7120) does not produce akinetes and hormogonia in laboratory conditions, although it contains all genes known to be necessary for at least hormogonia production (Figure S5 in [6]). It has been widely used to study different aspects of heterocyst differentiation and nitrogen fixation, including the mechanisms governing what is arguably the simplest and most ancient example of a linear semiregular biological pattern that formed by heterocysts along filaments of vegetative cells. It is evident that despite their apparent simplicity, the regulatory networks underlying heterocyst pattern formation are highly complex and multilayered [3,7].

According to a widely accepted model for the initiation of heterocyst differentiation, the accumulation of 2-oxoglutarate (2-OG) in cyanobacteria under nitrogen deprivation [8] is perceived as a nitrogen starvation signal by the global transcriptional regulator NtcA [9]. Activated by its positive effector 2-OG [10–12], NtcA can act as either an activator or repressor of numerous genes [13]. 2-OG-activated NtcA activates the transcription of *nrrA* [14], which, in turn, activates the transcription of *hetR* [15]. HetR is a master regulator of differentiation specific to filamentous cyanobacteria [16]. The expression of multiple heterocyst-specific genes depends on HetR, but only a few have been shown to be regulated directly, and repression of several promoters in vegetative cells has been reported (see [17] for a recent review).

NtcA and HetR constitute a mutually dependent positive regulatory circuit [18]. Transcriptional regulation of *hetR* is complex—it is transcribed from multiple promoters that are regulated both temporally and spatially [19]. There is more than transcriptional regulation, however, as replacement of normal transcriptional control of the chromosomal *hetR* with ectopic expression from a copper-regulated *petE* promoter resulted in a wild-type-like heterocyst patterning, which depended on post-transcriptional regulation of HetR protein levels [20], of which phosphorylation is a major part [21,22].

Proteins from three distinct genes, a small peptide PatS [23], a small protein PatX [16], and HetN protein [24,25], are known to suppress heterocyst differentiation when overexpressed. The only feature they all have in common is an RGSGR motif (a few PatX alleles, including that in *Anabaena* PCC 7120, contain RGTGR instead), which was shown to be essential for suppression of heterocyst formation [23,26]. Both PatS and HetN have been shown to block the positive autoregulation of *hetR* [27,28]. *patS* and *patX* expression is induced early (6–8 h after combined nitrogen deprivation) in regularly spaced cells that will become heterocysts [16,23] from DIF1 motif (TCCGGA)-containing, NtcA- and HetR-dependent promoters [29,30]. While *patS* expression is downregulated in mature heterocysts, expression of *hetN* first localizes to committed heterocysts [27] and continues also after heterocyst maturation [31,32].

Inactivation of *patS* leads initially to a phenotype of multiple contiguous heterocysts (Mch) during de novo heterocyst differentiation, but eventually, the normal pattern partially reappears [23,33]. In contrast, inactivation of *hetN* results in a normal de novo pattern, but Mch formation during successive rounds of differentiation [27]; thus, PatS appears to be responsible for establishing the spatial heterocyst pattern, while HetN is responsible for its maintenance. Addition to the medium of synthetic penta- or hexapeptides (denoted PatS-5 for RGSGR and PatS-6 for ERGSGR) corresponding to the C-terminal (in PatS) or internal (in HetN) motifs inhibits differentiation, but cannot restore a wild-type heterocyst pattern in a *patS* mutant [23,34]. Analysis of *hetN* deletion variants suggested that the RGSGR motif is required for both inhibitory and patterning activity, and thus PatS and HetN regulate different stages of heterocyst patterning utilizing the same amino acid motif for intercellular signaling [26,35]. The actions of PatS and HetN are affected by a third protein, PatA [36], evidently a response regulator, without which *Anabaena* PCC 7120 produces only terminal heterocysts [17,37].

Besides negative regulators, a positively acting factor, HetL, may be involved in HetR regulation. It is a member of a large family of poorly characterized pentapeptide repeat proteins abundant in cyanobacteria [38] and is composed of 40 pentapeptides (A(D/N)LXX). On a multicopy plasmid *hetL* restored the ability of the PatS-overexpressing strain to differentiate heterocysts, while ectopic overexpression in wild type induced Mch in a nitrate-containing medium [39]. However, inactivation of *hetL* did not impair heterocyst development and diazotrophic growth [33]. A recent publication [40] sheds light on the interplay between HetL, HetR, PatS, and PatX, and provides evidence that HetR interacts with HetL at the same interface as PatS and PatX, but without inhibiting its DNA binding activity, suppressing inhibition of heterocyst differentiation. HetL competes with PatS and PatX for HetR binding, and thus it acts as a competitive activator of HetR, complicating its regulation.

We showed previously that the *hetR* gene arose in filamentous cyanobacteria, likely for the regulation of patterned differentiation of specialized cells, long before they learned how to secure nitrogenase in microoxic heterocysts, and it was invariably accompanied by *patX* and no other RG(S/T)GR-containing regulator protein [16]. It is not alone, however, in *Anabaena* PCC 7120, one of a small a group with three RG(S/T)GR-containing negative regulators of HetR [16]. It is therefore of interest to determine the role PatX plays along with PatS and HetN in regulating heterocyst differentiation. Here, we present the results of mutational analyses of the three RG(S/T)GR-containing proteins in *Anabaena* PCC 7120, alone and in combinations, under conditions that promote heterocyst differentiation, and show that unrestrained HetR activity was lethal under different growth conditions. In a nitrogen-depleted medium, PatX had partially overlapping functions with PatS, but all three negative regulators showed different lesions in intercellular signaling. We attribute these apparently contradictory results to interference of RG[S/T]GR-containing negative regulators with a competitive activator HetL due to a rivalry for HetR binding at the same interface. We also show that in a nitrogen-replete medium, PatX was required in the absence of PatS and PatS was required in the absence of PatX for pattern formation, irrespective of the presence or absence of HetN.

2. Materials and Methods

2.1. Strains, Growth Conditions and Microscopy

Anabaena PCC 7120 was grown in nitrate-replete BG-11 (N+) medium or in BG-11$_0$ medium free of combined nitrogen (N-) [41] and, when appropriate, free of copper (Cu-; for regulation of the *petE* promoter), at 30 °C illuminated by cool white fluorescent light at 2000 lx (ca. 28 µmol photons/m^2/s). For ammonium-containing medium, NH$_4$Cl (2.5 mM) and MOPS buffer (5 mM, pH 8.0) were added to BG-11$_0$. Single and double recombinants and strains carrying replicative plasmids were grown in the presence of appropriate antibiotics at the following final concentrations: neomycin, 25 µg/mL for solid medium and 15 µg/mL for liquid medium; erythromycin, 5 µg/mL; spectinomycin, 5 µg/mL plus streptomycin, 2.5 µg/mL.

To induce heterocyst formation in liquid medium, the filaments from fresh streaks on BG-11 plates were transferred with sterile toothpicks into a 96-well microtiter plate containing liquid BG-11$_0$ medium. Cells were routinely examined by bright-field microscopy with an Axiostar plus microscope (Zeiss International, Jena, Germany) equipped with an EOS 1300D(W) digital camera (Canon Inc., Tokyo, Japan), using 40× objective magnification. To visualize the heterocyst-specific envelope polysaccharide layer, a 0.5% Alcian blue solution in 50% ethanol was used for staining cyanobacterial filaments before microscopic examination [42].

2.2. Plasmid and Strain Construction

The strains and plasmids used in this study are listed in Table 1. *patX* mutant strains RIAM1238 (DR929; *patX*::Ω; Sm^r/Sp^r) and RIAM1239 (DR931; Δ*patX*::Ω; Sm^r/Sp^r) were constructed as follows. A 2.9 kb SalI-XmnI fragment containing the 3′-end of *alr2333*, *asl2332* (*patX*), *alr2331*, *asl2329*, and the 3′-end of *alr2328* from anp03869 was ligated between the SalI and SmaI sites of pK18, producing pRIAM780, and an Ω cassette (a Sm^r/Sp^r determinant flanked by transcriptional terminators) was excised with SmaI from pAM684 and inserted into the internal ScaI site of *patX* in pRIAM780. A construct was selected with the orientation of *aadA* (Sm^r/Sp^r) parallel to *patX*, producing pRIAM796.

Table 1. Strains and plasmids.

Strain or Plasmid Reference	Derivation and/or Relative Characteristics	Source or
Anabaena Strains		
PCC 7120	Wild type	S. Callahan
7120PN	P$_{petE}$-*hetN*	[27]
RIAM1238	DR929; *patX*::Ω; Sm^r/Sp^r	This study
RIAM1239	DR931; Δ*patX*::Ω; Sm^r/Sp^r	This study
RIAM1241	PN DR929; P$_{petE}$-*hetN patX*::Ω; Sm^r/Sp^r	This study
RIAM1242	PN DR931; P$_{petE}$-*hetN* Δ*patX*::Ω; Sm^r/Sp^r	This study
RIAM1243	UHM114 DR931 (pAM1714); Δ*patS* Δ*patX*::Ω(pAM1714); Sm^r/Sp^r Nm^r	This study
RIAM1245	PN DR931 DR1177; P$_{petE}$-*hetN* Δ*patX*::Ω Δ*patS*::C.CE3; Sm^r/Sp^r Em^r	This study

Table 1. *Cont.*

Strain or Plasmid Reference	Derivation and/or Relative Characteristics	Source or
RIAM1248	PN DR931 DR1177; P$_{petE}$-*hetN* Δ*patX*::Ω Δ*patS*::C.CE3; Sm^r/Sp^r Em^r	This study
RIAM1249	PN DR1177; P$_{petE}$-*hetN* Δ*patS*::C.CE3; Em^r	This study
RIAM1250	PN DR1177; P$_{petE}$-*hetN* Δ*patS*::C.CE3; Em^r	This study
UHM114	Δ*patS*	[43]
Plasmids		
anp03226	A *patX* (*asl2332*)-bearing bp 2805907 to 2813409 fragment of *Anabaena* PCC 7120 chromosome in the BamHI site of pUC18; Ap^r	[44]
anp03869	A *patX* (*asl2332*)-bearing bp 2803179 to 2811405 fragment of *Anabaena* PCC 7120 chromosome in the BamHI site of pUC18; Ap^r	[44]
pAM504	Shuttle vector for replication in *E. coli* and *Anabaena*; Km^r Nm^r	[45]
pAM684	A source of the Sp^r Sm^r Ω cassette; Ap^r Sp^r/Sm^r	[46]
pAM1035	*patS* on a 3.3 kb chromosomal fragment in pBluescript II KS(-); Ap^r	[23]
pK18	pBR322-derived cloning vector; Km^r	[47]
pRIAM780	A 2.9 kb SalI-XmnI fragment containing 3′-end of *alr233*, *asl2332* (*patX*), *alr2331* and 3′-end of *alr2330* from anp03869 ligated in SalI-SmaI sites of pK18; Km^r	This study
pRIAM796	Ω cassette inserted into internal ScaI site in *patX* ORF in pRIAM780; Km^r Sp^r/Sm^r	This study
pRIAM860	anp03226 derivative with an AfeI-SmaI fragment deleted; contains *patX* on remaining 4.65 kb insert; Ap^r	This study
pRIAM917	pRIAM780 with ScaI-DraI fragment containing most of *patX* ORF and 3′ UTR replaced with Ω cassette; Km^r Sp^r/Sm^r	This study
pRIAM923	A 4.9 kb chromosomal region (bp 2805907 to 2810809) with ScaI-DraI fragment containing most of *patS* ORF and 3′ UTR replaced with Ω cassette; reconstructed from pRIAM860 and pRIAM917; Km^r Sp^r/Sm^r	This study
pRIAM925	Same as pRIAM917, but the Ω cassette inserted into into internal ScaI site in *patX* ORF; reconstructed from pRIAM860 and pRIAM796; Km^r Sp^r/Sm^r	This study
pRIAM929	Insert from pRIAM925 moved into suicide vector pRL271; Cm^r Em^r Sp^r/Sm^r	This study
pRIAM931	Insert from pRIAM923 moved into suicide vector pRL271; Cm^r Em^r Sp^r/Sm^r	This study
pRIAM1159	Insert from pAM1035 moved in pK18 (probably as BamH-SalI fragment); Km^r	This study
pRIAM1175	C.CE3 Cm^r Em^r cassette excised with EcoICRI from pRL1567 and inserted into EcoRV-ScaI sites of pRIAM1159, replacing *patS*-bearing 0.38 kb chromosomal fragment; Km^r Cm^r Em^r	This study
pRIAM1177	Insert from pRIAM1175 moved into SacI-PstI sites of pRL278; Km^r Cm^r Em^r	This study
pRL271	*sacB*-containing suicide vector; Cm^r Em^r	[48]
pRL278	*sacB*-containing suicide vector; Nm^r/Km^r	[48]
pRL1567	Source of C.CE3 Cm^r Em^r cassette; Ap^r Cm^r Em^r	[49]
pUC19	pBR322-derived cloning vector; Ap^r	[50]

Plasmid pRIAM780 containing the 3′-end of *alr2333*, *asl2332* (*patX*), *alr2331*, *asl2329* and the 3′-end of *alr2328* derived from anp03869 was digested with ScaI and DraI, deleting most of *patX* and its 3′ UTR, replacing the lost sequence with an Ω cassette excised with SmaI from pAM684. A construct was selected with the orientation of *aadA* (Sm^r/Sp^r) parallel to *patX*, producing pRIAM917. Plasmid pRIAM860 is a 4.65-kb derivative of anp03226 with the sequence between AfeI and SmaI sites deleted, leaving all of *all2333*, *asl2332* (*patX*), *alr2331*, *asl2329*, and the 3′-end of *alr2328*. The plasmid was cut with KpnI and partially with BspEI and ligated with pRIAM917 and pRIAM796 digested with KpnI + BspEI, producing pRIAM923 and pRIAM925, respectively. Finally, inserts from pRIAM925 and pRIAM923 were excised with SacI+PstI and placed between the same sites of suicide vector pRL271, producing pRIAM929 and pRIAM931, respectively. All steps of the constructions of pRIAM929 and pRIAM931 are depicted in Supplemental Figure S1.

To delete *patS* from the *Anabaena* PCC 7120 chromosome, an insert from pAM1035 was excised with BamH + SalI and moved to pK18, producing pRIAM1159. A C.CE3 Cm^r Em^r cassette was excised with EcoICRI from pRL1567 and inserted between the EcoRV and ScaI sites of pRIAM1159, replacing a *patS*-bearing 0.38 kb chromosomal fragment with Cm^r Em^r genes parallel to excised *patS*, producing pRIAM1175. The insert from this plasmid was moved as a SacI-PstI fragment between the same sites of suicide vector pRL278, producing pRIAM1177. All steps of the constructions of pRIAM1177 are depicted in Supplemental Figure S2.

The plasmids pRIAM929, pRIAM931, and pRIAM1177 were transformed into *E. coli* strain AM1359 [23] containing a broad host range plasmid pRL443 with conjugal functions and a pRL623 plasmid to provide both methylation and mobilization functions and then conjugated into *Anabaena*

PCC 7120, its derivative strain PN, and other successive mutant derivatives as indicated in Table 1, using standard protocols [51]. The next day, conjugation plates were underlaid with appropriate antibiotics to select for single recombinants. Subsequent selection for double recombinants using the *sacB* gene present on the vectors was performed as described by [52]. After multiple rounds of successive cloning, the genotypes of all constructed mutant strains were confirmed by PCR analysis (Supplemental Figure S3).

3. Results

3.1. Mutational Analysis Suggests That in Nitrogen-Depleted Medium, patX is Impaired in Cell–Cell Signaling and Acts Cell-Autonomously

PatX shares the ability of PatS and HetN to suppress heterocyst differentiation when overexpressed [16]. The phenotypes of *patS* and *hetN* single and double mutants were previously described in detail. The *patS* deletion single mutant strain UHM114 [43] and *patS* replacement mutant [23,33] produce both single heterocysts and Mch in nitrogen-depleted (N-) medium. A conditional P_{petE}-*hetN* mutant transferred into copper-free (Cu-) N- medium initially displays the wild-type pattern, but later, Mch starts to appear [27], while a Δ*hetN* strain CSL7 forms Mch without lag [35]. Unlike the *patS* and *hetN* mutants, inactivation of *patX* did not cause any visibly aberrant phenotype in nitrogen-depleted (N-) medium. The mutant strains RIAM1238 (*patX*::Ω) and RIAM1239 (Δ*patX*::Ω) mutant strains behaved similarly to each other. Qualitatively, both the time course of differentiation and heterocyst pattern were essentially the same as in the wild-type strain (Figure 1a,b). Since the seemingly normal phenotype of *patX* mutants growing in N- medium could be caused by a functional redundancy or impaired function of PatX, we attempted to construct and examine the phenotypes of strains with different combinations of *patX*, *patS*, and *hetN* mutations.

It was previously reported that inactivation of *hetN* results in an unstable Mch phenotype tending to change to a Het$^-$ phenotype upon extended subculturing [24]. To avoid the instability, we exploited an approach used previously by [27], who constructed strain 7120PN in which the coding region of *hetN* was fused to the *petE* promoter controlled by the level of copper. At the concentration of copper in BG-11 medium (0.3 µM), the ectopic expression of *hetN* from this promoter is sufficient to completely suppress heterocyst differentiation in *Anabaena* PCC 7120, while in Cu- medium residual *hetN* expression is low enough to virtually eliminate the HetN-mediated suppression of HetR activity [27,43]. The phenotype of a Δ*patS* derivative of 7120PN was described earlier [43].

In order to assess whether the functional role of PatX is masked by PatS and HetN, we made use of the original conditional mutant strain 7120PN (P_{petE}-*hetN*) [27] with *patX* knocked out: RIAM1241 (P_{petE}-*hetN patX*::Ω) and RIAM1242 (P_{petE}-*hetN* Δ*patX*::Ω). In addition, we looked at the phenotypes of two strains similar to the previously reported 7120PN Δ*patS* mutant UHM100 [43]: RIAM1249 and RIAM1250 (independent isolates of P_{petE}-*hetN* Δ*patS*::C.CE3). Both Δ*patS* mutants showed identical phenotypes, similar to that of UHM100, but different from 7120PN Δ*patX*.

In N$^-$ Cu$^-$ medium, RIAM1241 and RIAM1242 behaved exactly like the parental strain, the *patX*$^+$ *hetN* conditional mutant 7120PN: one day after induction, a single semiregular heterocyst formed (Supplemental Figure 1c,e), but during consecutive rounds of differentiation, Mch started to appear due to the formation of new heterocysts adjacent to existing single or multiple heterocysts (Figure 1d).

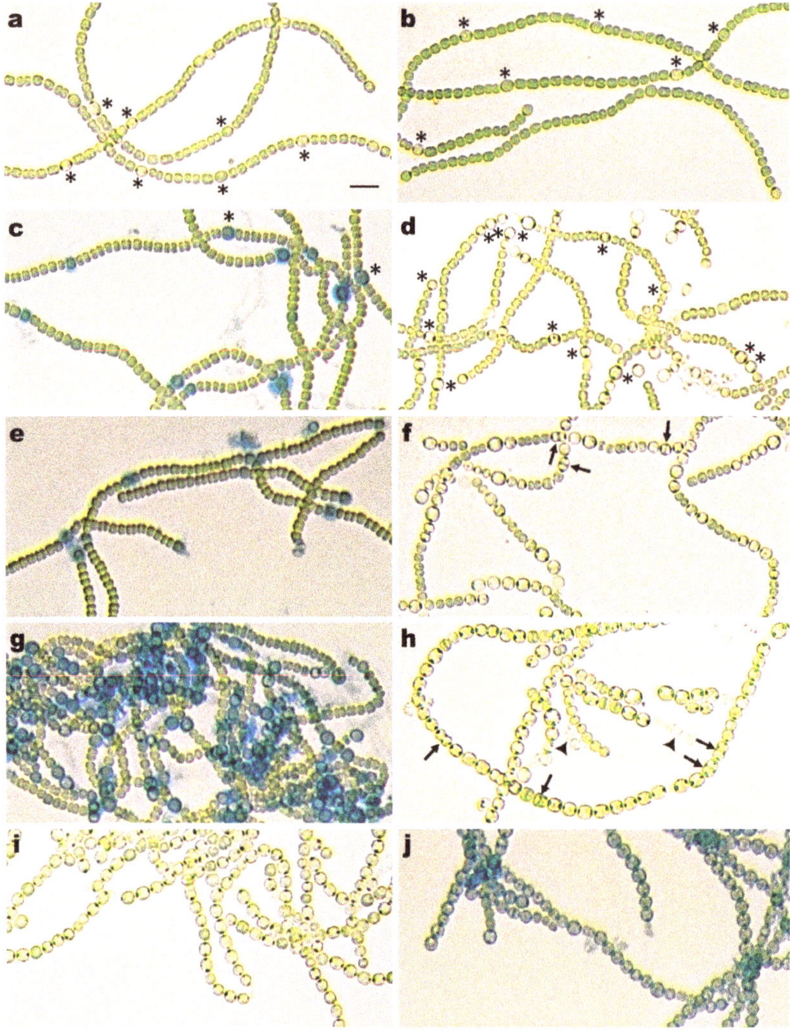

Figure 1. Phenotypes of the wild-type and mutant strains after transfer from a BG-11 plate to liquid BG-11$_0$ copper-free (Cu-) medium. Strains were either unstained (**a**,**b**,**d**,**f**,**h**,**i**) or stained with Alcian Blue (**c**,**e**,**g**,**j**) to identify differentiating cells or mature heterocysts. (**a**) Wild-type *Anabaena* PCC 7120 and (**b**) Δ*patX*::Ω single mutant RIAM1239, 1 day after transfer; (**c**) mutant strain PN P$_{petE}$-*hetN*, 1 day after transfer; (**d**,**e**) P$_{petE}$-*hetN* Δ*patX* conditional double mutant RIAM1242, (**d**) 4 days or (**e**) 1 day after transfer; (**f**,**g**) P$_{petE}$-*hetN* Δ*patS* conditional double mutant RIAM1250, (**f**) 2 days or (**g**) 1 day after transfer; (**h**) Δ*patS* Δ*patX*::Ω(pAM1714) conditional double mutant RIAM1243, 3 days after transfer; (**i**,**j**) *hetN patX patS* conditional triple mutant RIAM1248, (**i**) 2 days or (**j**) 1 day after transfer. Some heterocysts in panels (**a**) through (**d**) are indicated by asterisks. Apparently dividing aberrant heterocysts in panels (**f**) and (**h**) are indicated by arrows, necridia in (**h**) by arrowheads. The bar in panel (**a**) represents 10 μm.

Transfer of RIAM1250 (P*petE*-*hetN* Δ*patS*::C.CE3) filaments in liquid N⁻ Cu⁻ medium resulted in a deteriorated heterocyst pattern with production of a Mch pattern at 24 h (Figure 1g) and progressive ectopic differentiation upon further incubation (Figure 1f), so that nearly complete differentiation occurred after 4–5 days. There was no growth and cell sediment bleached. This behavior was essentially the same as described earlier for a strain UHM100, also lacking *patS* and conditional in *hetN* expression [43]. In the case of our double and triple mutants, the relative abundance of aberrant heterocysts varied in particular mutants and was most prominent in RIAM1243 (Δ*patS* Δ*patX*::Ω(pAM1714)) (ca. 20%, by visual inspection), perhaps due to the presence of a functional copy of *hetN*, which could influence the activity of HetR at late stages and somehow compromise the block of division.

Our initial attempts to construct a Δ*patS* Δ*patX*::Ω double mutant failed: while single recombinants were easily obtained, positive selection for double recombinants [52] on sucrose plates produced only very rare colonies. All retained the Em^r marker of the pRL271 suicide vector, indicating the absence of gene replacement by double recombination. However, after conjugation of a P$_{petE}$-*patS*-containing plasmid pAM1714 marked by Nm resistance [23] into the Δ*patS* SR931/pRL271 meridiploid single recombinants (containing both the WT *patX* and Δ*patX*::Ω genes), which is viable and has a Δ*patS* phenotype, we obtained many Em^s Nm^r colonies on sucrose plates, indicating double recombination. A completely segregated conditional double mutant, RIAM1243 (Δ*patS* Δ*patX*::Ω(pAM1714)), bleached upon transfer to Cu- medium and, apparently, stopped growing, even in nitrate-containing medium. Massive synchronous heterocyst differentiation was visible after 1–2 days in N- medium (Figure 1h), single necridia were rare in some filaments, but formed multiple contiguous necridia in others, and filaments of differentiated cells contained abundant aberrant (pro)heterocysts with mid-cell envelope constrictions and/or division planes.

A triple mutant RIAM1248, defective in *patS*, in *patX*, and conditionally in *hetN*, was constructed, with viability maintained by *hetN* expression from a copper-regulated promoter. As with the *patS patX* double mutant (see above), transfer of the triple mutant into N- Cu- medium resulted in synchronous ectopic differentiation at 24 h (Figure 1j). After 48 h, filaments showed little fragmentation and consisted almost exclusively of proheterocysts and heterocysts (Figure 1i). Aberrant (apparently dividing) heterocysts were comparatively rare in the triple mutant.

3.2. In Nitrogen-Replete Medium, Either patS or patX Alone Can Promote Semiregular Pattern Formation

On BG-11 plates, RIAM1239 (defective in *patX*) did not form heterocysts; however, unlike the wild type, the mutant strain upon transfer from solid to liquid nitrate- or ammonium-containing BG-11 medium produced semiregular heterocysts in many filaments, although at a lower frequency than when transferred from solid BG-11 to liquid N⁻ medium (Figure 2a,b). Differentiation of heterocysts in nitrogen-replete (N+) medium resembles formation of constitutive heterocysts by *patS* and *hetN patS* mutants [23,33,36,43].

Upon transfer to N⁺ Cu⁻ medium, the *hetN* Δ*patX* conditional mutant RIAM1242 formed semiregular predominantly single heterocysts (Figure 2c), and the *hetN* Δ*patS* conditional mutant RIAM1250 in N⁺ Cu⁻ medium behaved similarly to RIAM1242—it also formed semiregular single and some double heterocysts (Figure 2d). However, 5–6 days after the transfer, RIAM1250 started to fragment, first at vegetative cell-heterocyst junctions. Further incubation resulted in progressive fragmentation to very short filaments, single vegetative cells, and detached heterocysts (Figure 2e). Evidently, both PatS and PatX are necessary to restrain HetR sufficiently to prevent differentiation in N+ liquid medium, while only one of them is necessary to produce a semiregular pattern of heterocysts.

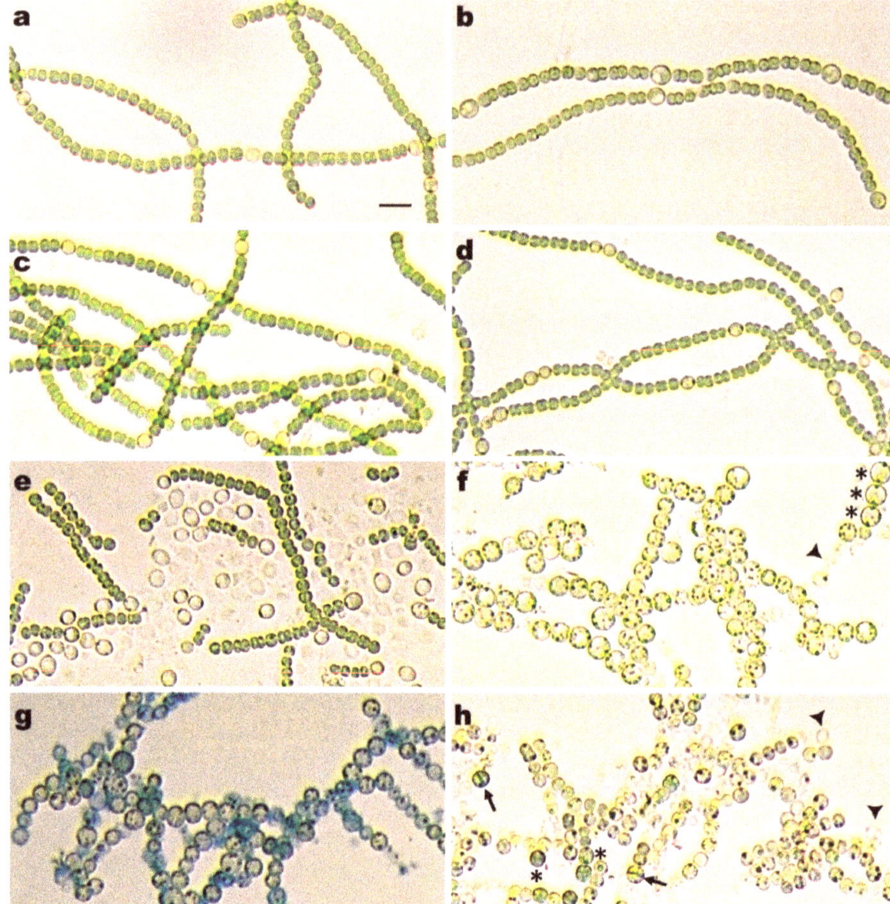

Figure 2. Phenotypes of mutants after transfer from a BG-11 plate into liquid copper-free (**a,c–h**) nitrate
-containing or (**b**) ammonium-containing BG-11 medium. Strains were either unstained (**a-f,h**) or
stained with Alcian Blue (**g**). (**a,b**) Δ*patX*::Ω single mutant RIAM1239, 1 day after transfer; (**c**) P$_{petE}$-*hetN*
Δ*patX* conditional double mutant RIAM1242 and (**d**) P$_{petE}$-*hetN* Δ*patS* conditional double mutant
RIAM1250, (**d**) 2 days or (**e**) 6 days after transfer; (**f,g**) Δ*patS* Δ*patX*::Ω(pAM1714) conditional double
mutant RIAM1243, 3 days after transfer; (**h**) *hetN patX patS* conditional triple mutant RIAM1248, 4 days
after transfer. Some cells resembling heterocysts in panels (**f,h**) are indicated by asterisks and aberrant
round cells with division planes in panel (**h**) by arrows. Some necridia (cell ghosts) are marked in
panels (**f,h**) with arrowheads. The bar in panel (**a**) represents 10 μm.

A double mutant defective in both *patS* and *patX* exhibited a more extreme phenotype under N+
conditions. The Δ*patS* Δ*patX*::Ω(pAM1714) conditional double mutant RIAM1243 grew as filaments of
green vegetative cells on BG-11, but upon transfer into N⁺ Cu⁻ liquid medium, fragmented into chains
of bleached enlarged round cells with small cyanophycin-like granules in the cytoplasm as well as
single and multiple contiguous necridia (Figure 2f). Most of the enlarged round cells morphologically
did not resemble heterocysts, but a few of them had a thicker envelope and polar nodules similar to
those located in the neck regions of heterocysts (Figure 2f, indicated by asterisks). Apparently, in N+
medium, the differentiation process is induced by transfer into liquid BG-11, but is disorganized and
in most cases ceases soon after the initial and early stages. These stages include disintegration of

thylakoids and loss of pigmentation, and deposition of the additional outer polysaccharide layer as indicated by Alcian blue staining (Figure 2g).

The RIAM1248 triple mutant behaved differently from the RIAM1243 double mutant. After transfer to N+ Cu- medium, the induced differentiation process was also disorganized but proceeded more slowly, and during the first two days, many cells retained their shape and size but gradually lost pigmentation, and necridia-like cells appeared. After 4 days, the filaments bleached and fragmented due to massive necridia formation, and some enlarged pale-green round cells with thickened envelopes and division planes were visible but without any sign of envelope constriction. Cells with colorless transparent cytoplasm with large granules were also seen (Figure 2h). Both RIAM1243 and RIAM1248 produced multiple necridia looking like cell ghosts, which increased in number during prolonged incubation (Figure 2f,h).

3.3. Instability of Conditional Mutants Overexpressing patS or hetN

In the course of our experiments with conditional mutants, we observed very high mutation rates of strains overexpressing either *patS* or *hetN*. These strains were constructed to conditionally block differentiation and thus avoid lethality or eliminate a detrimental burden of extra heterocyst production during cultivation. Although this approach permitted us to obtain and document the phenotypes of strains with multiple mutations, working with such mutants was tricky and required constant microscopic control and frequent cloning to prevent otherwise rapid changes in mutant phenotypes.

Several passages on BG-11 plates with or without appropriate antibiotics of fully segregated conditional mutants resulted in a rapid accumulation of secondary mutations. For example, Figure 3 shows a double mutant strain RIAM1243 before and after several subcultures on selective BG-11 + Sp, Sm, Nm plates and subsequent transfer and incubation in N^- Cu^- medium for 2 weeks. While a freshly isolated RIAM1243 clone after prolonged incubation exhibited only rare short stretches of vegetative cells originating from rare single cells that remained undifferentiated (Figure 3a), long filaments of vegetative cells (Het$^-$ phenotype), sometimes with terminal single or multiple heterocysts (PatA-like or PatA Mch phenotypes), accumulated in suspensions of fragmented chains of heterocysts (Figure 3b). Similar instability was exhibited by the conditional P_{petE}-*hetN* Δ*patS* mutant RIAM1250 (Figure 2e) and the triple mutant RIAM1248 (not shown).

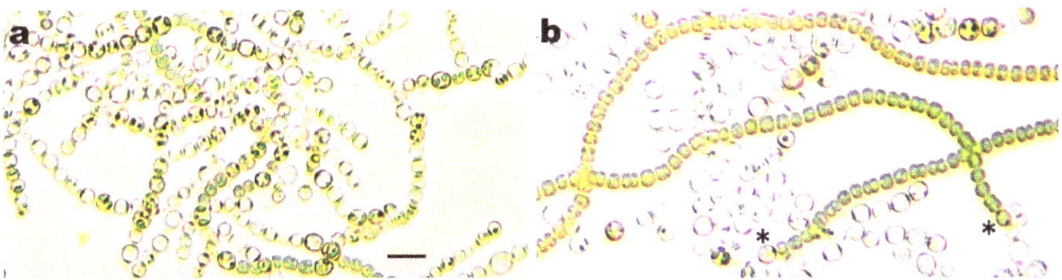

Figure 3. Phenotypes of Δ*patS* Δ*patX*::Ω(pAM1714) conditional double mutant RIAM1243 after transfer from a BG-11 plate into liquid copper-free nitrate-free (Cu- N-) BG-11$_0$ medium and incubation for 14 days. (**a**) A freshly isolated RIAM1243 clone and (**b**) the same mutant after several subcultures on BG-11+Sp, Sm, Nm plate. Asterisks in panel (**b**) indicate terminal heterocysts. The bar in panel (**a**) represents 10 μm.

There was no growth in both cases, and both cultures contained a light yellowish-grey sediment. We used in experiments only freshly isolated clones with all filaments showing similar phenotypic characteristics.

4. Discussion

PatX is the only RG(S/T)GR protein whose association with HetR is more ancient than the heterocyst-forming cyanobacteria [16]; however, its function remains obscure. Prior to this work, all that was known was that ectopic overexpression of PatX blocked heterocyst differentiation in *Anabaena* PCC 7120 [16], and so did PatX orthologs from *Mastigocladus laminosis* [53] (called "alternative PatS") and from *Arthrospira platensis* NIES 39 [54] (called "PatS"), a filamentous strain that does not form heterocysts. To clarify the role of PatX in heterocyst regulation, we disrupted the *patX* gene, but the resulting mutant strain grown in the absence of a nitrogen source showed no heterocyst-related phenotype. This could be a consequence of functional redundancy or a functionality unrelated to heterocyst formation. Loss of PatX also produced no obvious phenotypic change in combination with a conditional *hetN* mutation. Under conditions in which *hetN* was not expressed, two mutant strains, one carrying conditional *hetN* and the other identical but with *patX* knocked out, displayed a similar delayed Mch phenotype (Figure 1d). Both strains were phenotypically different from the conditional *hetN patS* double mutant UHM100 described previously [36,43] and from an analogous double mutant RIAM1250 constructed in this work, which showed gradual asynchronous differentiation so that in a few days, almost all cells became heterocysts (Figure 1f). These results demonstrate that the functionality of *patX* differs from that of *patS*. PatS is required for de novo pattern formation, but is not sufficient to prevent later formation of contiguous heterocysts (Mch). PatX cannot replace PatS in directing de novo pattern under nitrogen-depleted (N-) conditions, nor can it replace HetN in maintaining a pattern.

A residual functionality of *patX* becomes obvious on comparison of a Δ*patS* mutant with a Δ*patS* Δ*patX* double mutant RIAM1243, both retaining a wild-type *hetN* (Figure 1, [23]). The double mutant is inviable unless *patS* is conditionally suppressed by placing the gene under the control of the copper-regulated *petE* promoter. Upon removal of copper from the medium, this double mutant behaved essentially as the conditional *hetN patX patS* triple mutant RIAM1248—both started nearly synchronous differentiation of all vegetative cells into heterocysts when incubated in N⁻ Cu⁻ medium (Figure 1h–j). The intact chromosomal copy of *hetN* in RIAM1243 could not prevent or slow down the differentiation process. Thus, PatX (along with PatS) is required to prevent rapid synchronous differentiation of all vegetative cells.

These results indicate that upon combined nitrogen deprivation, all three negative regulators are instrumental in promoting regular pattern formation and/or maintenance, but in different ways. The roles of PatS and HetN in pattern initiation and maintenance, respectively, are readily discernible, but the role of PatX is masked by the presence of PatS. Our results are in line with those of Risser and Callahan [20], who observed the formation of concentration gradients of HetR in proximity to heterocysts, dependent on either *patS* or *hetN*. When both genes were deleted, no HetR gradients formed, indicating that either *patS* or *hetN* is required for establishing HetR concentration gradients after combined nitrogen removal. Their results demonstrate that *patX* (unknown at the time) is unable to induce HetR gradient formation in the absence of *patS* and *hetN* and thus does not participate in cell–cell signaling in the same way as *patS* and *hetN*.

A role for PatX in regulating heterocyst differentiation is more easily seen when *Anabaena* PCC 7120 is grown in nitrogen-replete (N+) medium. Unlike the wild-type strain, the *patX* mutant RIAM1239 produced morphologically distinct semiregular heterocysts when transferred to liquid nitrate or ammonia-containing media (Figure 2a,b), indicating that its product was needed to prevent unnecessary differentiation. Conditional double mutants RIAM1242 (*hetN patX*) and RIAM1250 (*hetN patS*) also formed semiregular heterocysts in copper-free (Cu-) medium (Figure 2c,d), indicating that in N+, either PatS or PatX is sufficient for rapid pattern formation. At the same time, both conditional double mutant RIAM1243 (Δ*patS* Δ*patX*::Ω(pAM1714)) and conditional triple mutant RIAM1248 (*hetN patX patS*), after removal of copper, started ectopic aberrant differentiation. With both mutants, all cells differentiated into morphologically distinct heterocysts in N- Cu- medium, but in N+ Cu- medium, differentiation into heterocyst-like cells was rare, and aberrant differentiation led to the appearance of enlarged spherical cells that partially retained pigmentation and of necridia, resulting from cell lysis (Figure 2f,h). Apparently, unrestrained HetR in N+ medium promotes differentiation that is misdirected. The low level in N+ medium of NtcA and/or its activator 2-oxoglutarate [55] could exacerbate the situation.

The interactions between the different signaling molecules discussed here can be visualized through the speculative model shown in Figure 4, one of many possible that fit the data. For some other aspects of signaling, see Flores et al., (2019) [17]. Figure 4A introduces the actors considered in the model and their icons, illustrating that a low N/C ratio increases the level of 2-oxoglutarate, inducing both the activity and transcription of NtcA [55]. NtcA and HetR mutually increase the other's transcription, forming a positive feedback loop [18], and a high level of HetR (possibly modified, possibly indirectly) increases the transcription of *patS* [56], *hetL* [40], *patX* [29], and eventually *hetN* [28,29]. Some form of PatS, PatX, and presumably HetN negatively affect the activity of HetR [17,40], while HetL protects HetR from its inhibitors [40]. The model represents the three RG(S/T)GR proteins as five circles, but this is not intended to imply that the pentapeptide is the natural signal within *Anabaena* PCC 7120 [26,35,40,57,58]. Here we refer to the active signal loosely as PatS, PatX, and HetN. Under N+ conditions (Figure 4B), the level of 2-oxoglutarate is reduced, leading to lower levels and activity of NtcA and HetR. PatS is expressed at a low level under these conditions [29,33], and PatX is as well ([29]; Khudyakov, unpublished).

After nitrogen stepdown (-N), cells within wild-type filaments are represented as experiencing different levels of nitrogen starvation [59,60] (Figure 4C, left). N-deprivation increases the level of HetR from a basal level, but the levels of PatS and PatX, whose transcription increases as a result of this change [16,33], are postulated to be sufficient to bind to HetR and inhibit at least part of its activity [17,40]. The level of HetN does not increase during this early period [28]. HetL expression also increases after N-stepdown, probably in differentiating cells [40], and HetL competes with PatS and PatX for binding to HetR [40]. It should be noted that HetL protein is not required for normally spaced heterocysts [39] and the degree to which HetL competes with RG(S/T)GR signals in vivo is speculative [40]. In time, cells in which HetR has escaped inhibition proceed on to form mature heterocysts, while the rest revert back to vegetative cell status (Figure 4C, right). The model shows the presence of HetN and PatX but not PatS in mature heterocysts, reflecting their known time course and levels of transcription and/or protein levels [28,31,61]. The level of HetN in heterocysts and its resulting appearance in adjacent cells is required to maintain the pattern of heterocysts [27]. The state of a wild-type strain is depicted, but phenotypically, a strain defective in PatX behaves the same way in N- medium.

In the absence of PatS, cells that would otherwise differentiate still did, but HetR in adjacent cells that are only mildly starved escaped inhibition, perhaps because of an insufficient RG(S/T)GR signal, leading to MCH (Figure 4D). The inability of the mutation of *patX* (in contrast to *patS*) to affect heterocyst spacing may be due to a difference in the binding to modified HetR in heterocysts of the RGTGR-containing PatX signal relative to RGSGR-containing PatS signal [40]. HetN production restrains more extensive differentiation. In the absence of both PatS and HetN, HetR in cells adjacent to heterocysts is not fully inhibited, and over the course of several days, eventually, nearly all cells differentiate (Figure 4E; [43] and our own observations). In the absence of all three RG(S/T)GR signals (or just PatS and PatX), HetR proteins in both starved and mildly starved cells were not inhibited, and all cells rapidly differentiated (Figure 4F).

In wild-type *Anabaena* PCC 7120, cells grown in nitrate or ammonia did not reach the threshold level of starvation for differentiation (Figure 4G). Even in random cells that are mildly starved, the level of PatS and PatX signal was high enough to titrate the low level of HetR. The presence or absence of HetN did not materially affect the phenotype in N+ medium. In strains lacking PatX (Figure 4H) or PatS (not shown; [23]), the level of RG(S/T)GR signal was no longer high enough to titrate the low level of HetR, and so heterocyst differentiation proceeded even in the presence of nitrate or ammonia. When both PatS and PatX were absent, no early-acting RG(S/T)GR signal was present to restrain HetR, and so, over time, all cells were affected, either by differentiating to aberrant heterocysts or by lysing (Figure 4I).

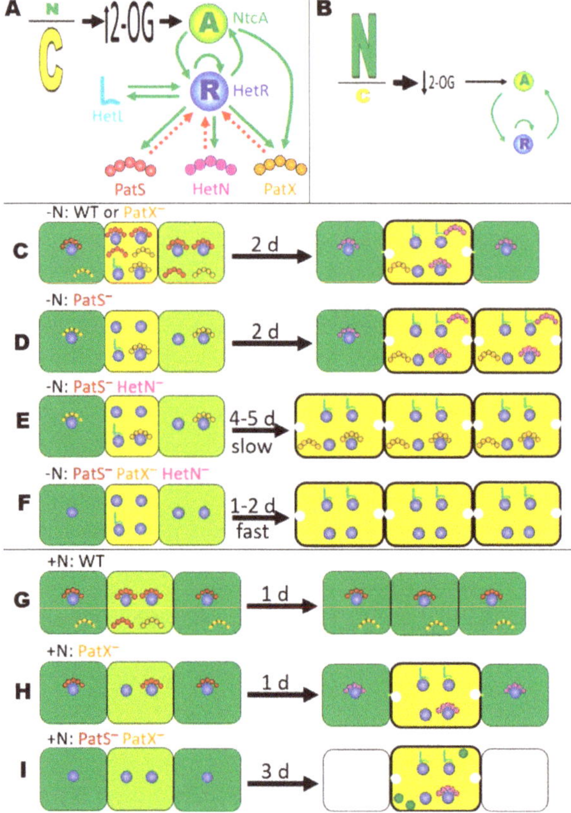

Figure 4. Model integrating roles of different signals in differentiation that are considered in this work. (**A**,**B**) Major actors under conditions of low nitrogen (N) and high carbon (C) or the reverse. Green solid arrows indicate an increase in transcription and/or activity, while red dotted arrows indicate repression of either transcription and/or activity. 2-OG represents 2-oxoglutarate. (**C**–**F**) The state of three adjacent cells after nitrogen stepdown (-N). Cells that are internally nitrogen-replete are represented as dark green, those that are mildly starved as light green, those that have passed the threshold for differentiation as yellow-green, and differentiated cells as brown with a thick envelope and white polar plugs. The level of HetR is crudely represented by the number of blue spheres, in some cases bound to PatS, HetN, PatX, or HetL (see their icons in panel **A**). (**G**–**I**) The state of three adjacent cells under nitrogen-replete (N+) conditions. The conventions are the same as in panels **C**–**F**. In addition, aberrant heterocysts are shown with the green inclusion bodies seen in Figure 2f,h, and necridia are represented as gray cells.

The developmental chaos that results from *patX patS* and *hetN patS* double mutants produces strong selective pressure for the rapid accumulation of suppressor mutations to escape from lethal terminal differentiation. Similar instability has been seen with strains overexpressing HetR [62] or HetZ and HetP [63] and strains with mutated *hetN* [25]. In our experiments, we tried to avoid this problem by using conditional mutants and maintaining them in a medium that blocked heterocyst development completely and thus presumably eliminated negative selection pressure on mutants overproducing heterocysts [43,63]. Nevertheless, a high mutation rate was still observed, resulting in the rapid accumulation of Het$^-$ and PatA-like mutants producing only terminal heterocysts (Figure 3). We know of no studies concerning growth defects in N+ medium in strains affected in HetR function or expression, but HetR may well participate in protein interactions and/or influence expression of genes that are unrelated to heterocyst formation but important for vegetative growth [17]. Its overexpression or inhibition of expression and activity could impose selective pressure for the appearance of suppressor mutations.

In the presence of combined nitrogen, either PatX or PatS as the sole HetR-regulator could establish and maintain a wild-type semiregular heterocyst pattern (Figure 2c,d), and in terms of current models of heterocyst pattern formation, they are fully proficient in cell–cell signaling, a prerequisite for pattern formation. This basic mechanism rests on interactions between two players, HetR and PatX/PatS derivatives, perhaps as modeled by a Turing-like reaction-diffusion mechanism [3], it works irrespective of nitrogen status and is hidden by additional layers of regulation, but can be revealed by mutational analyses. These additional layers may respond to the availability of combined nitrogen or other environmental cues and a set of downstream regulators (which can vary in different strains and complicate regulation) transmitting the patterning signals and converting them into morphogenetic and metabolic changes. We conjecture that this basic mechanism was initially designed by filamentous cyanobacteria for the purpose of dissemination by nonrandom fragmentation of trichomes through necridia formation, and later modified and adopted for patterned heterocyst differentiation.

Supplementary Materials: The following are available online at http://www.mdpi.com/2075-1729/10/12/326/s1, Figure S1: Construction of pRIAM929 and pRIAM931 (Δ*patX*), Figure S2: Construction of pRIAM1177 (Δ*patS*), Figure S3: Segregation of mutations.

Author Contributions: Conceptualization, I.K.; methodology, I.K.; investigation, I.K. and G.G.; visualization, J.E.; writing—original draft preparation, I.K.; writing—review and editing, I.K. J.E. and G.G. All authors have read and agreed to the published version of the manuscript.

Funding: This research was partially funded by the Russian Science Foundation, grant number 19-16-00097.

Conflicts of Interest: The authors declare no conflict of interests.

References

1. Meeks, J.C.; Elhai, J. Regulation of cellular differentiation in filamentous cyanobacteria in free-living and plant-associated symbiotic growth states. *Microbiol. Mol. Biol. Rev.* **2002**, *66*, 94–121. [CrossRef] [PubMed]

2. Tandeau de Marsac, N. Differentiation of hormogonia and relationships with other biological processes. In *The Molecular Biology of Cyanobacteria*; Advances in Photosynthesis and Respiration; Bryant, D.A., Ed.; Springer: Dordrecht, The Netherlands, 1994; Volume 1, pp. 825–842. [CrossRef]

3. Herrero, A.; Stavans, J.; Flores, E. The multicellular nature of filamentous heterocyst-forming cyanobacteria. *FEMS Microbiol. Rev.* **2016**, *40*, 831–854. [CrossRef] [PubMed]

4. Kaplan-Levy, R.N.; Hadas, O.; Summers, M.L.; Rücker, J.; Sukenik, A. Akinetes: Dormant cells of cyanobacteria. In *Dormancy and Resistance in Harsh Environments*; Lubzens, E., Cerda, J., Clark, M., Eds.; Springer: Berlin, Germany, 2010; pp. 5–27.

5. Nürnberg, D.J.; Mariscal, V.; Parker, J.; Mastrolanni, G.; Flores, E.; Mullineaux, C.W. Branching and intercellular communication in the Section V cyanobacterium Mastigocladus laminosus, a complex multicellular prokaryote. *Molec. Microbiol.* **2014**, *91*, 935–949. [CrossRef] [PubMed]

6. Cho, Y.W.; Gonzales, A.; Harwood, T.V.; Huynh, J.; Hwang, Y.; Park, J.S.; Trieu, A.Q.; Italia, P.; Pallipuram, V.K.; Risser, D.D. Dynamic localization of HmpF regulates type IV pilus activity and directional motility in the filamentous cyanobacterium *Nostoc. punctiforme*. *Mol. Microbiol.* **2017**, *106*, 252–265. [CrossRef] [PubMed]

7. Brenes-Álvarez, M.; Olmedo-Verd, E.; Vioque, A.; Muro-Pastor, A.M. Identification of conserved and potentially regulatory small RNAs in heterocystous cyanobacteria. *Front. Microbiol.* **2016**, *7*, 48. [CrossRef] [PubMed]

8. Muro-Pastor, M.I.; Reyes, J.C.; Florencio, F.J. Cyanobacteria perceive nitrogen status by sensing intracellular 2-oxoglutarate levels. *J. Biol. Chem.* **2001**, *276*, 38320–38328. [CrossRef]

9.   Vázquez-Bermúdez, M.F.; Herrero, A.; Flores, E. 2-Oxoglutarate increases the binding affinity of the NtcA (nitrogen control) transcription factor for the *Synechococcus glnA* promoter. *FEBS Lett.* **2002**, *512*, 71–74. [CrossRef]

10. Valladares, A.; Flores, E.; Herrero, A. Transcription activation by NtcA and 2-oxoglutarate of three genes involved in heterocyst differentiation in the cyanobacterium *Anabaena* sp. strain PCC 7120. *J. Bacteriol.* **2008**, *190*, 6126–6133. [CrossRef] [PubMed]

11. Zhao, M.X.; Jiang, Y.L.; He, Y.X.; Chen, Y.F.; Teng, Y.B.; Chen, Y.; Zhang, C.C.; Zhou, C.Z. Structural basis for the allosteric control of the global transcription factor NtcA by the nitrogen starvation signal 2-oxoglutarate. *Proc. Natl. Acad. Sci. USA* **2010**, *107*, 12487–12492. [CrossRef]

12. Laurent, S.; Chen, H.; Bédu, S.; Ziarelli, F.; Peng, L.; Zhang, C.-C. Nonmetabolizable analogue of 2-oxoglutarate elicits heterocyst differentiation under repressive conditions in *Anabaena* sp. PCC 7120. *Proc. Natl. Acad. Sci. USA* **2005**, *102*, 9007–9012. [CrossRef]

13. Picossi, S.; Flores, E.; Herrero, A. ChIP analysis unravels an exceptionally wide distribution of DNA binding sites for the NtcA transcription factor in a heterocyst-forming cyanobacterium. *BMC Genom.* **2014**, *15*, 22. [CrossRef] [PubMed]

14. Muro-Pastor, A.M.; Olmedo-Verd, E.; Flores, E. All4312, an NtcA-regulated two-component response regulator in *Anabaena* sp. strain PCC 7120. *FEMS Microbiol. Lett.* **2006**, *256*, 171–177. [CrossRef] [PubMed]

15. Ehira, S.; Ohmori, M. NrrA directly regulates expression of hetR during heterocyst differentiation in the cyanobacterium *Anabaena* sp. strain PCC 7120. *J. Bacteriol.* **2006**, *188*, 8520–8525. [CrossRef] [PubMed]

16. Elhai, J.; Khudyakov, I. Ancient association of cyanobacterial multicellularity with the regulator HetR and an RGSGR pentapeptide-containing protein (PatX). *Mol. Microbiol.* **2018**, *110*, 931–954. [CrossRef]

17. Flores, E.; Picossi, S.; Valladares, A.; Herrero, A. Transcriptional regulation of development in heterocyst-forming cyanobacteria. *Biochim. Biophys. Acta Gene Regul. Mech.* **2019**, *1862*, 673–684. [CrossRef]

18. Muro-Pastor, A.M.; Valladares, A.; Flores, E.; Herrero, A. Mutual dependence of the expression of the cell differentiation regulatory protein HetR and the global nitrogen regulator NtcA during heterocyst development. *Mol. Microbiol.* **2002**, *44*, 1377–1385. [CrossRef]

19. Rajagopalan, R.; Callahan, S.M. Temporal and spatial regulation of the four transcription start sites of *hetR* from *Anabaena* sp. strain PCC 7120. *J. Bacteriol.* **2010**, *192*, 1088–1096. [CrossRef]

20. Risser, D.D.; Callahan, S.M. Genetic and cytological evidence that heterocyst patterning is regulated by inhibitor gradients that promote activator decay. *Proc. Natl. Acad. Sci. USA* **2009**, *106*, 19884–19888. [CrossRef]

21. Valladares, A.; Flores, E.; Herrero, A. The heterocyst differentiation transcriptional regulator HetR of the filamentous cyanobacterium *Anabaena* forms tetramers and can be regulated by phosphorylation. *Mol. Microbiol.* **2016**, *99*, 808–819. [CrossRef]

22. Roumezi, B.; Xu, X.; Risoul, V.; Fan, Y.; Lebrun, R.; Latifi, A. The Pkn22 kinase of *Nostoc* PCC 7120 is required for cell differentiation via the phosphorylation of HetR on a residue highly conserved in genomes of heterocyst-forming cyanobacteria. *Front. Microbiol.* **2020**, *10*, 3140. [CrossRef]

23. Yoon, H.S.; Golden, J.W. Heterocyst pattern formation controlled by a diffusible peptide. *Science* **1998**, *282*, 935–938. [CrossRef] [PubMed]

24. Black, T.A.; Wolk, C.P. Analysis of a Het- mutation in *Anabaena* sp. strain PCC 7120 implicates a secondary metabolite in the regulation of heterocyst spacing. *J. Bacteriol.* **1994**, *176*, 2282–2292. [CrossRef] [PubMed]

25. Bauer, C.C.; Ramaswamy, K.S.; Endley, S.; Scappino, L.A.; Golden, J.W.; Haselkorn, R. Suppression of heterocyst differentiation in *Anabaena* PCC 7120 by a cosmid carrying wild-type genes encoding enzymes for fatty acid synthesis. *FEMS Microbiol. Lett.* **1997**, *151*, 23–30. [CrossRef] [PubMed]

26. Higa, K.C.; Rajagopalan, R.; Risser, D.D.; Rivers, O.S.; Tom, S.K.; Videau, P.; Callahan, S.M. The RGSGR amino acid motif of the intercellular signalling protein, HetN, is required for patterning of heterocysts in *Anabaena* sp. strain PCC 7120. *Mol. Microbiol.* **2012**, *83*, 682–693. [CrossRef]

27. Callahan, S.M.; Buikema, W.J. The role of HetN in maintenance of the heterocyst pattern in *Anabaena* sp. PCC 7120. *Mol. Microbiol.* **2001**, *40*, 941–950. [CrossRef]

28. Li, B.; Huang, X.; Zhao, J. Expression of *hetN* during heterocyst differentiation and its inhibition of *hetR* up-regulation in the cyanobacterium *Anabaena* sp. PCC 7120. *FEBS Lett.* **2002**, *517*, 87–91. [CrossRef]

29. Mitschke, J.; Vioque, A.; Haas, F.; Hess, W.R.; Muro-Pastor, A.M. Dynamics of transcriptional start site selection during nitrogen stress-induced cell differentiation in *Anabaena* sp. PCC7120. *Proc. Natl. Acad. Sci. USA* **2011**, *108*, 20130–20135. [CrossRef]

30. Muro-Pastor, A.M. The heterocyst-specific NsiR1 small RNA is an early marker of cell differentiation in cyanobacterial filaments. *mBio* **2014**, *5*, e01079-14. [CrossRef]

31. Videau, P.; Oshiro, R.T.; Cozy, L.M.; Callahan, S.M. Transcriptional dynamics of developmental genes assessed with an FMN-dependent fluorophore in mature heterocysts of *Anabaena* sp. strain PCC 7120. *Microbiology* **2014**, *160*, 1874–1881. [CrossRef]

32. Wang, Y.; Xu, X. Regulation by *hetC* of genes required for heterocyst differentiation and cell division in *Anabaena* sp. strain PCC 7120. *J. Bacteriol.* **2005**, *187*, 8489–8493. [CrossRef]

33. Yoon, H.S.; Golden, J.W. PatS and products of nitrogen fixation control heterocyst pattern. *J. Bacteriol.* **2001**, *183*, 2605–2613. [CrossRef] [PubMed]

34. Wu, X.; Liu, D.; Lee, M.H.; Golden, J.W. *patS* minigenes inhibit heterocyst development of *Anabaena* sp. strain PCC 7120. *J. Bacteriol.* **2004**, *186*, 6422–6429. [CrossRef] [PubMed]

35. Corrales-Guerrero, L.; Mariscal, V.; Nürnberg, D.J.; Elhai, J.; Mullineaux, C.W.; Flores, E.; Herrero, A. Subcellular localization and clues for the function of the HetN factor influencing heterocyst distribution in *Anabaena* sp. strain PCC 7120. *J. Bacteriol.* **2014**, *196*, 3452–3460. [CrossRef] [PubMed]

36. Orozco, C.C.; Risser, D.D.; Callahan, S.M. Epistasis analysis of four genes from *Anabaena* sp. strain PCC 7120 suggests a connection between PatA and PatS in heterocyst pattern formation. *J. Bacteriol.* **2006**, *188*, 1808–1816. [CrossRef]

37. Liang, J.; Scappino, L.; Haselkorn, R. The *patA* gene product, which contains a region similar to CheY of *Escherichia coli*, controls heterocyst pattern formation in the cyanobacterium *Anabaena* 7120. *Proc. Natl. Acad. Sci. USA* **1992**, *89*, 5655–5659. [CrossRef]

38. Zhang, R.; Ni, S.; Kennedy, M.A. Crystal structure of Alr1298, a pentapeptide repeat protein from the cyanobacterium *Nostoc* sp. PCC 7120, determined at 2.1 Å resolution. *Proteins* **2020**, *88*, 1143–1153. [CrossRef]

39. Liu, D.; Golden, J.W. *hetL* overexpression stimulates heterocyst formation in *Anabaena* sp. strain PCC 7120. *J. Bacteriol.* **2002**, *184*, 6873–6881. [CrossRef]

40. Xu, X.; Risoul, V.; Byrne, D.; Champ, S.; Douzi, B.; Latifi, A. HetL, HetR and PatS form a reaction-diffusion system to control pattern formation in the cyanobacterium *Nostoc* PCC 7120. *Elife* **2020**, *9*, e59190. [CrossRef]

41. Rippka, R.; Deruelles, J.; Waterbury, J.B.; Herdman, M.; Stanier, R.Y. Generic assignments, strain histories and properties of pure cultures of cyanobacteria. *J. Gen. Microbiol.* **1979**, *111*, 1–61. [CrossRef]

42. Olmedo-Verd, E.; Flores, E.; Herrero, A.; Muro-Pastor, A.M. HetR-dependent and -independent expression of heterocyst-related genes in an *Anabaena* strain overproducing the NtcA transcription factor. *J. Bacteriol.* **2005**, *187*, 1985–1991. [CrossRef]

43. Borthakur, P.B.; Orozco, C.C.; Young-Robbins, S.S.; Haselkorn, R.; Callahan, S.M. Inactivation of *patS* and *hetN* causes lethal levels of heterocyst differentiation in the filamentous cyanobacterium *Anabaena* sp. PCC 7120. *Mol. Microbiol.* **2005**, *57*, 111–123. [CrossRef] [PubMed]

44. Kaneko, T.; Nakamura, Y.; Wolk, C.P.; Kuritz, T.; Sasamoto, S.; Watanabe, A.; Iriguchi, M.; Ishikawa, A.; Kawashima, K.; Kimura, T.; et al. Complete genomic sequence of the filamentous nitrogen-fixing cyanobacterium *Anabaena* sp. strain PCC 7120. *DNA Res.* **2001**, *8*, 205–213. [CrossRef] [PubMed]

45. Wei, T.F.; Ramasubramanian, T.S.; Golden, J.W. *Anabaena* sp. strain PCC 7120 *ntcA* gene required for growth on nitrate and heterocyst development. *J. Bacteriol.* **1994**, *176*, 4473–4482. [CrossRef] [PubMed]

46. Ramaswamy, K.S.; Carrasco, C.D.; Fatma, T.; Golden, J.W. Cell-type specificity of the *Anabaena fdxN*-element rearrangement requires *xisH* and *xisI*. *Mol. Microbiol.* **1997**, *23*, 1241–1249. [CrossRef] [PubMed]

47. Pridmore, R.D. New and versatile cloning vectors with kanamycin-resistance marker. *Gene* **1987**, *56*, 309–312. [CrossRef]

48. Black, T.A.; Cai, Y.; Wolk, C.P. Spatial expression and autoregulation of *hetR*, a gene involved in the control of heterocyst development in *Anabaena*. *Mol. Microbiol.* **1993**, *9*, 77–84. [CrossRef]

49. Khudyakov, I.; Wolk, C.P. Evidence that the *hanA* gene coding for HU protein is essential for heterocyst differentiation in, and cyanophage A-4(L) sensitivity of, *Anabaena* sp. strain PCC 7120. *J. Bacteriol.* **1996**, *178*, 3572–3577. [CrossRef]

50. Vieira, J.; Messing, J. The pUC plasmids, an M13mp7-derived system for insertion mutagenesis and sequencing with synthetic universal primers. *Gene* **1982**, *19*, 259–268. [CrossRef]

51. Elhai, J.; Vepritskiy, A.; Muro-Pastor, A.M.; Flores, E.; Wolk, C.P. Reduction of conjugal transfer efficiency by three restriction activities of *Anabaena* sp. strain PCC 7120. *J. Bacteriol.* **1997**, *179*, 1998–2005. [CrossRef]

52. Cai, Y.P.; Wolk, C.P. Use of a conditionally lethal gene in *Anabaena* sp. strain PCC 7120 to select for double recombinants and to entrap insertion sequences. *J. Bacteriol.* **1990**, *172*, 3138–3145. [CrossRef]

53. Antonaru, L.A.; Nürnberg, D.J. Role of PatS and cell type on the heterocyst spacing pattern in a filamentous branching cyanobacterium. *FEMS Microbiol. Lett.* **2017**, *364*, fnx154. [CrossRef] [PubMed]

54. Zhang, J.Y.; Chen, W.L.; Zhang, C.C. *hetR* and *patS*, two genes necessary for heterocyst pattern formation, are widespread in filamentous nonheterocyst-forming cyanobacteria. *Microbiology* **2009**, *155*, 1418–1426. [CrossRef] [PubMed]

55. Herrero, A.; Flores, E. Genetic responses to carbon and nitrogen availability in *Anabaena*. *Environ. Microbiol.* **2019**, *21*, 1–17. [CrossRef]

56. Huang, X.; Dong, Y.; Zhao, J. HetR homodimer is a DNA-binding protein required for heterocyst differentiation, and the DNA-binding activity is inhibited by PatS. *Proc. Natl. Acad. Sci. USA* **2004**, *101*, 4848–4853. [CrossRef]

57. Corrales-Guerrero, L.; Mariscal, V.; Flores, E.; Herrero, A. Functional dissection and evidence for intercellular transfer of the heterocyst-differentiation PatS morphogen. *Mol. Microbiol.* **2013**, *88*, 1093–1105. [CrossRef] [PubMed]

58. Zhang, L.; Zhou, F.; Wang, S.; Xu, X. Processing of PatS, a morphogen precursor, in cell extracts of Anabaena sp. PCC 7120. *FEBS Lett.* **2017**, *591*, 751–759. [CrossRef]

59. Popa, R.; Weber, P.K.; Pett-Ridge, J.; Finzi, J.A.; Fallon, S.J.; Hutcheon, I.D.; Nealson, K.H.; Capone, D.G. Carbon and nitrogen fixation and metabolite exchange in and between individual cells of Anabaena oscillarioides. *ISME J.* **2007**, *1*, 354–360. [CrossRef] [PubMed]

60. Brown, A.I.; Rutenberg, A.D. A storage-based model of heterocyst commitment and patterning in cyanobacteria. *Phys. Biol.* **2014**, *11*, 16001. [CrossRef]

61. Park, J.-J.; Lechno-Yossef, S.; Wolk, C.P.; Vieille, C. Cell-specific gene expression in Anabaena variabilis grown phototrophically, mixotrophically, and heterotrophically. *BMC Genom.* **2013**, *14*, 759. [CrossRef]

62. Buikema, W.J.; Haselkorn, R. Characterization of a gene controlling heterocyst differentiation in the cyanobacterium *Anabaena* 7120. *Genes Dev.* **1991**, *5*, 321–330. [CrossRef]

63. Zhang, H.; Xu, X. Manipulation of pattern of cell differentiation in a *hetR* mutant of *Anabaena* sp. PCC 7120 by overexpressing *hetZ* alone or with *hetP*. *Life* **2018**, *8*, 60. [CrossRef] [PubMed]

Publisher's Note: MDPI stays neutral with regard to jurisdictional claims in published maps and institutional affiliations.

Article

Heterocyst Development and Diazotrophic Growth of *Anabaena variabilis* under Different Nitrogen Availability

Nur Syahidah Zulkefli [1] **and Soon-Jin Hwang** [2,*]

[1] Department of Environmental Health Science, Konkuk University, Seoul 05029, Korea; nsyahidah.zulkefli@gmail.com

[2] Department of Environmental Health Science and Human and Eco-care Center, Konkuk University, Seoul 05029, Korea

* Correspondence: sjhwang@konkuk.ac.kr; Tel.: +82-2-450-3748; Fax: +82-2-456-5062

Received: 31 August 2020; Accepted: 11 November 2020; Published: 13 November 2020

Abstract: Nitrogen is globally limiting primary production in the ocean, but some species of cyanobacteria can carry out nitrogen (N) fixation using specialized cells known as heterocysts. However, the effect of N sources and their availability on heterocyst development is not yet fully understood. This study aimed to evaluate the effect of various inorganic N sources on the heterocyst development and cellular growth in an N-fixing cyanobacterium, *Anabaena variabilis*. Growth rate, heterocyst development, and cellular N content of the cyanobacteria were examined under varying nitrate and ammonium concentrations. *A. variabilis* exhibited high growth rate both in the presence and absence of N sources regardless of their concentration. Ammonium was the primary source of N in *A. variabilis*. Even the highest concentrations of both nitrate (1.5 g L^{-1} as $NaNO_3$) and ammonium (0.006 g L^{-1} as Fe-NH_4-citrate) did not exhibit an inhibitory effect on heterocyst development. Heterocyst production positively correlated with the cell N quota and negatively correlated with vegetative cell growth, indicating that both of the processes were interdependent. Taken together, N deprivation triggers heterocyst production for N fixation. This study outlines the difference in heterocyst development and growth in *A. variabilis* under different N sources.

Keywords: *Anabaena variabilis*; heterocyst; diazotrophic growth; nitrogen-fixation; cell quota; nitrogen availability

1. Introduction

Cyanobacteria can survive in harsh environmental conditions such as darkness, extreme temperatures, and high salinity. They also have the ability to grow under nutrient limitations [1,2]. They often pose a threat to various freshwater ecosystems worldwide as they cause cyanobacterial blooms and produce toxins and unpleasant odor [3–5]. Cyanobacterial blooms could be controlled by reducing nutrient input, especially nitrogen (N) and phosphorus (P), which in turn leads to the reduction in their growth and development [6,7]. P deprivation is more effective in controlling cyanobacterial blooms than N deprivation. N deprivation exhibits inhibitory effect only on non-N-fixing taxa, leaving N-fixing taxa less or unaffected owing to their ability to fix N for survival [8].

Certain bloom-forming cyanobacteria develop an adaptation mechanism to cope with N deprivation by converting atmospheric N into ammonium through N fixation. These cyanobacteria use ammonium as a major N source for N assimilation followed by nitrate, nitrite, and nitrogen fixation [9,10]. Nitrogen-fixing cyanobacteria possess nitrogenase enzyme, which fixes atmospheric N into ammonium under N-deprived condition, thereby introducing a novel N source to the N cycle in aquatic ecosystems for their survival [11,12]. This enzyme consists of two component proteins, dinitrogenase (molybdenum-iron (MoFe) protein, which in turn is composed of two identical subunits encoded by *nif*D and *nif*K genes) and dinitrogenase reductase (an iron (Fe)-containing protein composed of two identical units encoded by *nif*H gene) [13]. Nitrogenase is highly sensitive to the presence of oxygen, which suppresses the N-fixation process. Therefore, successful N fixation by diazotrophic cyanobacteria that depend on oxygenic photosynthesis for energy, faces a constant challenge in optimizing N fixation, as the process can only occur under anaerobic conditions [14,15].

Nostocales, a group of filamentous N-fixing cyanobacteria, comprises the majority of cyanobacteria that respond to N deprivation by fixing N in specialized cells called heterocysts [14,16,17]. Heterocysts consist of a thick multilayered wall and lack oxygenic photosystem II. This structure creates an anaerobic condition suitable for N fixation to take place efficiently both in light and dark periods, utilizing ATP generated via photosystem-I of vegetative cells [18,19]. Heterocysts derived from N-deprived trichomes appear paler and are easily distinguished from vegetative cells [17,20]. Heterocysts are unevenly distributed among the vegetative cells with zero to more than 20 vegetative cells between two heterocysts. Heterocyst frequency can be measured as heterocyst per filament length or as heterocyst to vegetative cell ratio [21–24]. In natural ecosystems, heterocyst frequency of *Anabaena* spp. was found to be up to 9.8 heterocyst mm^{-1}, while *Aphanizomenon flos-aquae* was shown to have 1 or 2 heterocysts per filament [21,25]. Heterocyst frequency (heterocyst to vegetative cell ratio) in a cyanobacterial community composed of *Dolichospermum crassum*, *Aphanizomenon gracile*, and *Cuspidothrix issatschenkoi* in Laguna del Sauce Lake was in the range of 0.006–0.018% [22]. The maturation and fixation activities in heterocysts begin approximately 18–24 h after extracellular N deprivation [17,26].

Previous studies have attempted to demonstrate the effect of N limitation on N fixation and heterocyst development in cyanobacteria via continuous culture by regulating the levels of combined N or nitrate alone. [8,27–29]. Enhanced development of heterocyst and nitrogenase activity under low levels of combined N or nitrate alone indicated that N fixation by heterocystous cyanobacteria was critical when available N source was limited or absent. Heterocyst development and the increase in nitrogenase activity that occurred as early as a few hours to days after incubation were evident in cyanobacteria species such as *Anabaena variabilis*, *Nostoc sp.* and *Cylindrospermopsis raciborskii* [27,30,31]. However, heterocyst development and N fixation also occurred in the presence of N [32,33]. Although the presence of ammonium may inhibit N fixation, it can occur in the presence of N uptake if its concentration falls below cellular N requirement [27,33]. A few other studies indicated that heterocyst differentiation and N fixation were also dependent on the type of N source and the minimum N required in response to N deprivation [10,34].

Taken together, N fixation and heterocyst differentiation in cyanobacteria depend on the availability of the N source and concentration of N required to meet the demands for growth and other physiological mechanisms [8]. However, the factors that influence N fixation in terms of heterocyst formation and contribute to cyanobacterial growth are highly species-specific and vary among cyanobacterial strains [8,12,22,35]. Particularly, among Nostocales strains, various physiological conditions could affect heterocyst formation, N-fixation rate, and filament growth, which in turn, is responsible for their distinguished fitness and dominance among phytoplankton assemblages with respect to N depletion [21,36]. Among bloom-forming Nostocales, *Anabaena* has been suggested as a good model to study N fixation including the mechanism of heterocyst differentiation owing to its filamentous property and cellular differentiation capability in severe conditions [27,35,37,38]. The *Anabaena* species is known to cause harmful blooms in freshwater systems. However, the extent of N fixation and heterocyst development under a wide range of N availability is seldom explored.

There are confounding results on *Anabaena* growth and heterocyst development under the influence of N source [35,39–41]. Therefore, a wide range of N availability is necessary to understand the extent to which *Anabaena* can survive and produce heterocysts. The aim of this study was to evaluate the effects of different N sources (nitrate and ammonium) and their concentrations on cellular growth and heterocyst production with respect to the change in cellular N quota. We tested two different sources of inorganic N with varying concentrations. Furthermore, cyanobacteria can utilize other sources of nitrogen if there is sufficient external N source or ammonium, which is the most preferred N source. Therefore, we hypothesized that N uptake and growth are higher in the presence of ammonium than nitrate, and that a low concentration of nitrate could trigger heterocyst production at a higher frequency than that of ammonium [27,42,43].

2. Materials and Methods

2.1. Preparation of Stock Culture

Anabaena variabilis-AG40092 was obtained from KCTC (Korea Collection for Type Cultures, Daejeon, Korea). The test strain was maintained in BG-11 medium [44] in 200-mL culture flasks at 25 °C with a light intensity of 30 μmol m^{-2} s^{-1} [45,46] maintained using a cool white ring-shaped fluorescent light (FCL-32EXD/30, Kumho Electric Inc., Seoul, Korea) with an alternating light–dark cycle of 14:10 h in a temperature-controlled incubator (Vision Scientific, VS-1203P4S, Daejeon, Korea). The stock culture was sustained by seeding the subcultures into a fresh medium. The final subculture was prepared three weeks before the experiment. Prior to the experiment, the subculture kept in the stationary phase was nutrient starved for 3 d to minimize the effect of previous original BG-11 medium on the growth of our tested cyanobacterium. *A. variabilis* cells were centrifuged at 1000 rpm for 10 min (Table Top Centrifuge VS-5000i, Vision Scientific Co., Daejeon, Korea) and the pellets were resuspended in fresh N&P-free BG-11 medium to remove the residual N and P present in the cell surface [29,47]. The process was repeated twice, and the cells were finally resuspended in N&P-free BG-11 medium. Nutrient-starved cells were incubated under the same temperature and light intensity, as mentioned above. The cell density was measured using an SR chamber (Graticules S52 Sedgewick Rafter Counting Chamber, Structure Probe, Inc., West Chester, USA) under a light microscope (Axiostar plus, Carl ZEISS, Oberkochen, Germany) at 200× magnification to ensure that an appropriate cell density is used to observe the growth and heterocyst production for further experiments.

2.2. Preparation of Culture Medium

A normal BG-11 medium used for the stock cultures was prepared. To induce nutrient starvation in the test cyanobacterium for three days, all N and P sources (sodium nitrate (NaNO$_3$), ferric ammonium citrate (C$_6$H$_8$FeNO$_7$), and dipotassium phosphate (K$_2$HPO$_4$)) were omitted from the original BG-11 medium. The BG-11 medium was also used to test the effect of various N sources on cyanobacterial growth and heterocyst development for the 14-day experiment. However, it was modified by completely omitting the ammonium source in nitrate supplement and vice versa. The P source was kept at a constant concentration in all the media as shown in Table 1. Sodium nitrate (NaNO$_3$) and ferric ammonium citrate (C$_6$H$_8$FeNO$_7$) were used to regulate the concentrations of nitrate and ammonium as N source. A control medium was also prepared by removing all N sources in the BG-11 medium. All experimental media were autoclaved at 121 °C for 15 min and prepared a day before the experiment was carried out.

Table 1. Nutrient composition of BG-11 medium used for cyanobacteria culture in this study.

Component	Concentration (g L^{-1}. dH_2O)
$NaNO_3$	1.5
K_2HPO_4	0.04
$MgSO_4.7H_2O$	0.075
$CaCl_2.2H_2O$	0.036
Citric acid	0.006
Ferric ammonium citrate (Fe-NH_4-citrate)	0.006
EDTA Na_2	0.001
Na_2CO_3	0.02
H_3BO_3	2.86
$MnCl_2.4H_2O$	1.81
$ZnSO_4.7H_20$	0.22
$CuSO_4.5H_2O$	0.08
$Na_2MoO_4.2H_2O$	0.39
$Co(NO_3)_2.6H_2O$	0.05

2.3. Experimental Preparation

A. variabilis was grown in N&P-free BG-11 medium for 3 d, prior to the experiment. Nitrate treatment was established by omitting ammonium source (ammonium ferric citrate, ($C_6H_8FeNO_7$)) from BG-11 medium to ensure that only nitrate source (sodium nitrate, ($NaNO_3$)) was available in the medium, while ammonium treatment was prepared by omitting nitrate source (sodium nitrate, ($NaNO_3$)) from BG-11 medium and only ammonium source (ammonium ferric citrate, ($C_6H_8FeNO_7$)) was retained. Three different concentrations of both nitrate and ammonium treatments that represented a range from N-rich to N-stressed condition were prepared. They were prepared by serial dilution (10^0, 10^{-2}, 10^{-5}) of the original nitrate concentration, 1.5 g L^{-1} (17.6 mM), and ammonium concentration, 0.006 g L^{-1} (0.0229 mM), in the unmodified BG-11. Control was prepared by omitting both N sources (nitrate and ammonium) from the original BG-11 medium (Table 1).

Twenty milliliter aliquots of nutrient-starved cells were transferred to 500-mL cell culture flasks containing 480 mL of modified nitrate, ammonium, and control medium, comprising seven experimental treatments (Table 2). All treatments were prepared in triplicates. Average inoculating cell concentration at the beginning of the experiment was 4.20 ± 1.16 × 10^5 cells mL^{-1}. Experimental cultures were incubated at 25 °C with a light intensity of 30 μmol photon m^{-2} s^{-1} using a cool white fluorescent light with an alternating 14:10 light:dark photoperiod for 14 d. We had 7 different treatments in triplicates equal to a total of 21 flasks. Subsamples were obtained under sterile conditions on a clean bench (Model: CB-1600C, Solution Lab, Korea) at every 2-d intervals for microscopy, nutrient, and cell quota (N) analyses. On day 0, subsamples were obtained after 3 h of incubation in culture medium under the light cycle period. All treatments were conducted separately in different sets of flasks in the same manner.

Table 2. The concentrations of nitrogen (N) sources used in each treatment.

Treatment	N Source	Concentration (g L^{-1})
A	Absent	-
B		1.5
C	NaNO$_3$ (nitrate)	1.5×10^{-2}
D		1.5×10^{-5}
E		0.006
F	Fe-NH$_4$-citrate (ammonium)	0.006×10^{-2}
G		0.006×10^{-5}

2.4. Microscopic Count of Vegetative Cells and Heterocysts

At each sampling time, 5-mL aliquots from each treatment were obtained and fixed with 1% (final conc. v/v) Lugol's solution. One milliliter of Lugol-fixed sample from each treatment was placed onto a gridded SR chamber (Graticules S52 Sedgewick Rafter Counting Chamber, Structure Probe, Inc., West Chester, PA, USA) and the cells were measured using an inverted light microscope (Axiostar plus, Carl ZEISS, Oberkochen, Germany) at ×200 or ×400 magnification. Heterocysts were identified and distinguished from the vegetative cells by their thickened cell wall, pale appearance and pole formation compared to the adjacent cells [20,31,39]. Heterocyst frequency was calculated as the percentage of heterocyst density to the total vegetative cell density [48]. Heterocysts were counted in a random manner at 10 fields per sample per treatment and the cell density was calculated according to following equation:

$$C\ [\text{cells mL}^{-1}] = \frac{N}{F} \times \frac{1000\ \text{fields}}{1\ \text{mL}} \times D$$

N = number of cells or units counted; D = dilution factor; F = number of fields counted.

2.5. Chlorophyll-a Analysis

To analyze chlorophyll-*a*, 5 mL aliquots obtained from each treatment were filtered through a 1.2-μm pore-sized GF/C filter (Whatman, Product No. 1823-047). The filters were frozen in the dark until chlorophyll-*a* was extracted and measured using spectrophotometry. The chlorophyll-*a* was extracted by placing the filter in 10 mL of 90% acetone for 24 h in the dark and was measured using a spectrophotometer (Model: Optizen 3220UV, Mecasys Co., Ltd., Daejeon, Korea) at the wavelengths of 630 nm, 645 nm, 663 nm, and 750 nm. Further, chlorophyll-*a* concentration was calculated using the following equation [49]:

$$\text{Chl} - a(\mu\text{g L}^{-1}) = \frac{Y \times v}{V}$$

v = volume of 90% acetone used (mL); V = volume of sample filtered (mL);
Y = 11.64 X_1 − 2.16 X_2 + 0.10 X_3;
X_1 = A(663 nm) − A(750 nm); X_2 = A(645 nm) − A(750 nm); X_3 = A(630 nm) − A(750 nm).

2.6. Growth Rate Analysis

Cyanobacterial growth rate was determined by evaluating the changes in cell density at 2-d intervals of sampling time for 14-d incubation. Cell density was used to calculate the specific growth rate (μ) at each sampling interval by using the following equation [49]:

$$\mu\left(\text{d}^{-1}\right) = \ln(N_1 - N_0)/(t_1 - t_0)$$

N_0: Cell density (cells mL^{-1}) at the beginning of the sampling interval, t_0;
N_1: Cell density (cells mL^{-1}) at the end of the sampling interval, t_1;

($t_1 - t_0$): time interval of sampling (d).

2.7. Cellular Nitrogen Measurement

A ten-microliter aliquot was filtered using a 0.2 μm pore-sized polycarbonate membrane filter (Product code: 1223036, GVS Filter Technology, Sanford, FL, USA) and processed following the persulfate digestion method [50] to measure cellular N content. After filtration, the filters were inserted into 25-mL PTFE (polytetrafluoroethylene) Teflon cylinders with lined screw caps and were frozen at −20 °C until digestion. Digestion stock solution was prepared freshly beforehand, which consisted of potassium persulfate ($K_2S_2O_8$) and sodium hydroxide (NaOH). Prior to the digestion process, 10 mL of distilled water was added to each Teflon cylinder tube containing the filtered samples including the negative control tube containing only the blank filter paper. The same volume also was used to prepare the standard solutions covering the analytical range (0, 2, 4, 6, 8, 10 mg L^{-1}) using potassium nitrate (KNO_3). Further, 2.5 mL of digestion reagent was added to each tube and the tubes were closed tightly, inverted twice to mix, and autoclaved at 120 °C for 30 min. Digested samples and standards were allowed to cool at room temperature (22–25 °C). To each tube, 1 mL of borate buffer solution (made from boric acid (H_3BO_3), 1 mL of sodium hydroxide (NaOH)) and 0.2 mL of 1N hydrochloric acid (HCl) were added, and then measured using a spectrophotometer (Model: Optizen 3220UV, Mecasys Co., Ltd., Daejeon, Korea) at 220 nm and 275 nm of wavelength. Standard curves for nitrate or ammonium were prepared by plotting absorbance readings of the standards against their concentration. All calculations were carried out using the standard curve as reference.

2.8. Dissolved Inorganic Nitrogen Measurement

Dissolved inorganic N concentration in each treatment was measured to determine the N remaining in the medium at each sampling time. The filtrate obtained using a 0.2 μm pore-sized polycarbonate membrane filter was processed to determine the dissolved nitrate and ammonium concentration using standard methods. Nitrate concentration was determined by the UV spectrophotometric screening method (Standard Method-4500-NO_3, American Public Health Association (APHA), 2017). Prior to the spectrophotometric measurement, 0.1 mL of 1N HCl were added to 5 mL filtrate in a 15-mL conical tube (Model: Optizen 3220UV, Mecasys Co., Ltd., Daejeon, Korea) and the measurements were performed at 220 nm and 275 nm wavelength. Nitrate standards (0, 2, 4, 6, 8, 10 mg L^{-1}) were prepared using potassium nitrate (KNO_3) and spectrophotometric measurements were conducted as mentioned above. Ammonium concentration was measured using the phenate method (Standard Method-4500-NH_3, APHA, 2017), in which 0.2 mL of phenol, 0.2 mL of sodium nitroprusside, and 0.5 mL of oxidizing solution (alkaline citrate solution and sodium hypochlorite mixture) were added into 5 mL of sample filtrates and standards, with thorough mixing via vortexing (Vortex Mixer KMC-1300V, Vision Scientific Co., Ltd., Daejeon, Korea) after the addition of each component. Conical tubes containing samples were covered and incubated at room temperature in subdued light for at least 1 h to develop color. Absorbance at 640 nm was measured using a spectrophotometer. Standard curves were prepared for nitrate or ammonium by plotting absorbance readings of standards against their concentrations. All calculations were carried out using the standard curve as reference.

2.9. Statistical Analyses

All data were analyzed to estimate any significant relationship between analytical factors and the treatments. A normal distribution test (Shapiro–Wilk test) was performed to determine the suitability of parametric tests to be applied to the data. Log (x + 1) transformation was applied to non-normal data to achieve normality. Spearman and Pearson's correlations were used to evaluate the relationship between analytical parameters and treatments. One-way ANOVA with Post-hoc Tukey HSD multiple comparison analysis was used to check for the differences in chl-*a* concentration, cell density, heterocyst density, heterocyst frequency, growth rate, cell quota (N) and residual nitrate or ammonium concentration between various treatments of nitrate and ammonium treatment on

A. variabilis. Mean difference of residual nitrate concentrations in the medium was analyzed using Kruskal–Wallis test (mean rank ANOVA) since the data did not pass the normality test even after log transformation. Two-way ANOVA with Post-hoc Scheffe multiple comparison was used to determine the significant difference in mean chl-*a* concentration, cell density, heterocyst density, heterocyst frequency, growth rate, and cell quota (N) between the various types of N sources and N concentration. The difference between samples was considered significant at $p < 0.05$. All data obtained were presented using SigmaPlot® v. 10.0 (Systat Software Inc. (SSI), San Jose, California) and statistical analyses were carried out using PASW® Statistic v. 18 (SPSS Inc., Chicago, IL, USA).

3. Results

3.1. Changes in A. variabilis under Nitrate Treatment

3.1.1. Chl-a, Cell Density, and Growth Rate

A. variabilis exposed to three different nitrate treatments ($n = 3$) and control showed a similar pattern with a relatively slow growth during the 14-d incubation period (Figure 1a). The highest growth rate was observed in the highest concentration of nitrate (1.5 g L^{-1}). However, the significant difference in average growth rates was found between the moderate treatment of nitrate (1.5×10^{-2} g L^{-1}) and the control ($F_{(3,8)} = 5.969$, $p = 0.034$). The cell growth rate in the presence of 1.5×10^{-2} g L^{-1} nitrate treatment was also significantly different from that in the lowest nitrate treatment, 1.5×10^{-5} g L^{-1} ($p = 0.022$) (Table 3). We observed a gradual increase in chl-*a* levels in nitrate treatments over time. Throughout the experimental period, chl-*a* concentration started to exhibit a significant decrease since day 6 in the case of 1.5×10^{-2} g L^{-1} nitrate treatment, as compared to that in the 1.5 g L^{-1} treatment ($F_{(3,8)} = 7.673$, $p = 0.007$). On day 12, 1.5 g L^{-1} nitrate treatment had significantly higher chl-*a* concentration than that in N-free treatment ($F_{(3,8)} = 17.819$, $p = 0.25$), 1.5×10^{-2} g L^{-1} nitrate treatment ($p < 0.001$), and 1.5×10^{-5} g L^{-1} nitrate treatment ($p = 0.009$). At the end of experiment, chl-*a* concentration in the highest nitrate treatment was significantly higher than those in the other treatments ($F_{(3,8)} = 17.819$, $p = 0.001$) (Figure 1a).

Table 3. Average growth rate (μ) of *Anabaena variabilis* calculated from the cell density over 14-d incubation under different nitrate and ammonium treatments. N-free treatment was set as the control. The experiments were carried out in triplicates and the average of all the values was obtained for each treatment.

Treatment	N Source	Concentration (g L^{-1})	Growth Rate (μ)
N-free	Absent	0	0.103 ± 0.024
		1.5	0.121 ± 0.023
Nitrate	NaNO$_3$	1.5×10^{-2}	0.063 ± 0.023
		1.5×10^{-5}	0.127 ± 0.006
		0.006	0.156 ± 0.013
Ammonium	Fe-NH$_4$-citrate	0.006×10^{-2}	0.126 ± 0.001
		0.006×10^{-5}	0.125 ± 0.012

The initial cell densities of *A. variabilis* inoculated on the first day showed no significant difference among all the treatments ($n = 4$), in which the average cell density was $4.29 \pm 1.44 \times 10^5$ cells mL^{-1} (Figure 1c). Cell density increased in all treatments including the control during the 14-d incubation, and significantly correlated with chl-*a* concentration ($r(96) = 0.829$, $p < 0.001$). However, changes in cell density showed a different trend from that of chl-*a* concentration, particularly towards the end of the experiment, in which cell density was either maintained at the same level or decreased (Figure 1a,c). There was a significant difference in cell densities between the highest nitrate treatment and other treatments from day 4 to day 8 ($F_{(3,8)} = 12.635$, $p = 0.002$), in which the average cell density on day 4 doubled from $8.15 \pm 1.25 \times 10^5$ cells mL^{-1} to $16.96 \pm 2.25 \times 10^5$ cells mL^{-1} on day 8.

However, on day 8, only cell density in 1.5 g L^{-1} nitrate treatment became significantly lower than that in N-free treatment (F$_{(3,8)}$ = 2.474, p = 0.032). Cell density of N-free treatment showed an abrupt peak on day 10, reaching almost similar density to those of other treatments, and declined gradually until the experiment was terminated.

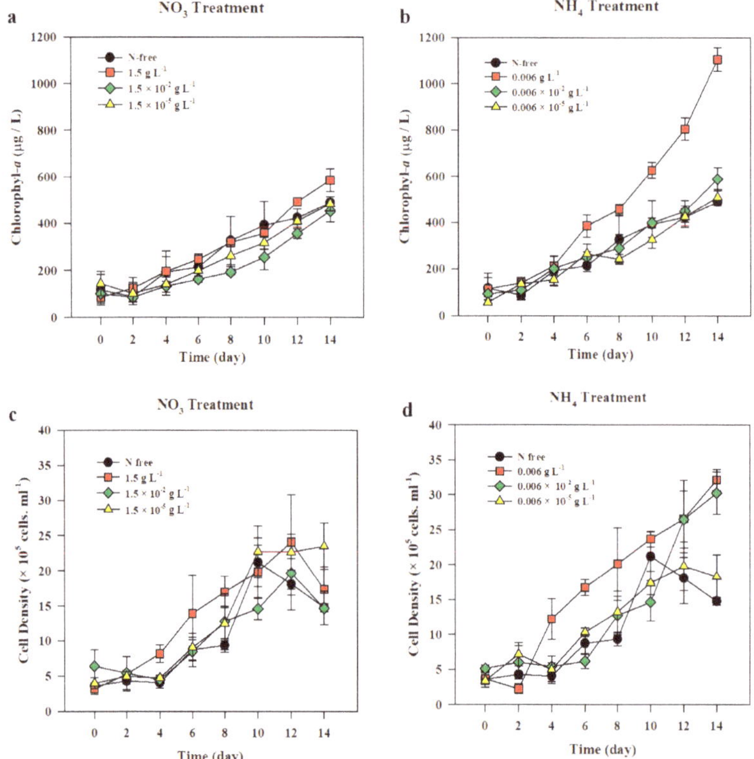

Figure 1. The changes in the concentration of chl-*a* (**a**,**b**) and cell density (**c**,**d**) of *Anabaena variabilis* under various nitrate (**a**,**c**) and ammonium (**b**,**d**) treatments over 14 d of incubation. N-free treatment was set as control. Measurements on day 0 was made 3 h after inoculation. The experiments were carried out in triplicates and the average of all the values was calculated. The error bars represent the standard deviation among replicates.

3.1.2. Heterocyst Development

Heterocysts appeared to be paler than the neighboring vegetative cells and thereby could be easily distinguished under the light microscope (Figure 2). Under 400× magnification, both heterocyst poles next to their adjacent cells were clearly observed. The average number of vegetative cells between heterocysts in N-free treatment ranged from 8 to 20 cells during the 14-d observation, which was within the range found in previous studies [17,20].

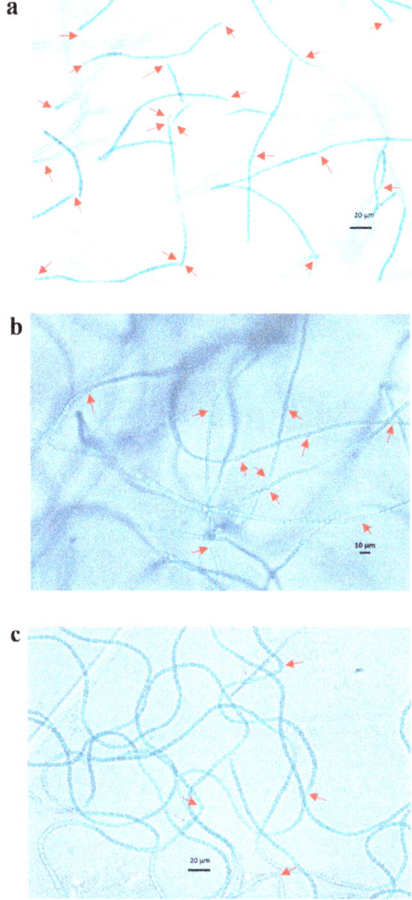

Figure 2. Microscopic photos of *A. variabilis* trichomes under 400× magnification. Heterocysts shown in the photos were formed during the experiment with N-free medium (**a**), low nitrate (1.5×10^{-5} g L^{-1} of NaNO$_3$) (**b**) and ammonium (0.006×10^{-2} g L^{-1} of Fe-NH$_4$-citrate) supplements (**c**) on day 4 when the highest heterocyst frequencies were measured. Heterocysts were able to be distinguished with adjacent vegetative cells by roundish shape and pale color. Arrows in figures indicate heterocysts.

The initial average heterocyst density in all nitrate treatments was $8.3 \pm 1.97 \times 10^3$ cells mL^{-1} and it increased by 1.42–6.83×10^4 cells mL^{-1} after 2 d of incubation, following which heterocyst accounted for 3–16 % of the total cells counted (Figure 3a,c). During the 14-d incubation, N-starved *A. variabilis* produced heterocysts in all nitrate treatments, and the highest heterocyst density was observed in 1.5×10^{-5} g L^{-1} nitrate treatment, followed by N-free treatment (Figure 4). Heterocyst number rapidly increased within 2 d after incubation both in 1.5×10^{-5} g L^{-1} nitrate treatment and N-free treatment. The heterocyst densities of these treatments were significantly higher than those of other nitrate treatments ($F_{(3,8)} = 7.645$, $p = 0.010$). On day 4, they peaked at the highest densities of $1.54 \pm 0.52 \times 10^5$ cells mL^{-1} and $1.29 \pm 0.34 \times 10^5$ cells mL^{-1}, respectively, before a steep decrease on day 6, and then reached an equilibrium upon reaching 0.5×10^5 cells mL^{-1} in most treatments (Figure 3a). Heterocyst density in 1.5×10^{-2} g L^{-1} nitrate treatment showed no significant difference compared to that in 1.5 g L^{-1} nitrate treatment, in which heterocysts produced were the lowest among all treatments (Figure 3a). On average, incubation under 1.5×10^{-5} g L^{-1} nitrate treatment produced the highest number of heterocysts but had no significant difference with that of the control, N-free condition.

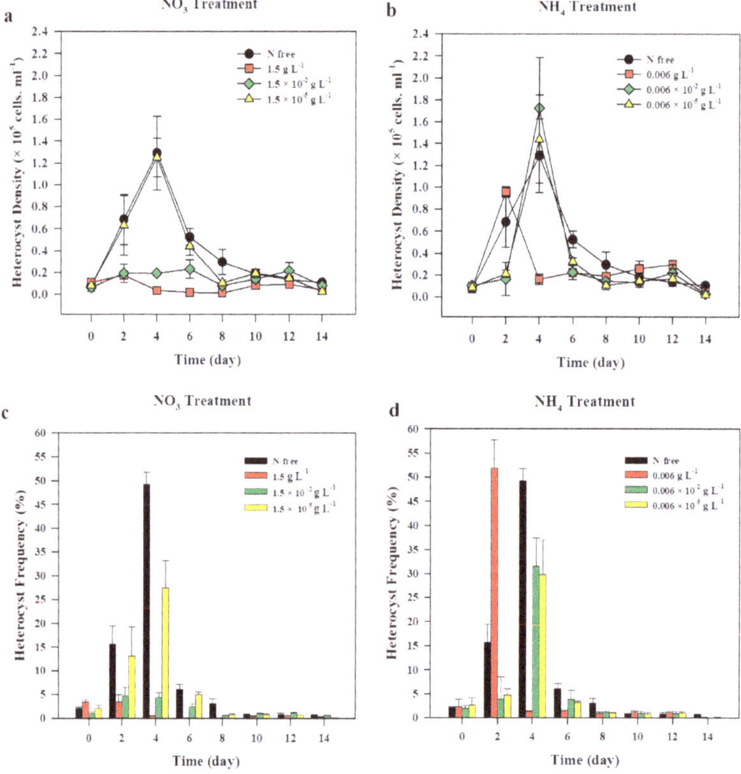

Figure 3. The changes in the heterocyst density (**a**,**b**) and heterocyst frequency (**c**, **d**) of *Anabaena variabilis* under various nitrate (**a**,**c**) and ammonium (**b**,**d**) treatments over 14 d of incubation. N-free treatment was set as the control. Measurements on day 0 were made 3 h after inoculation. The experiments were carried out in triplicates and the average of all the values was obtained. The error bars represent the standard deviation among the replicates.

Heterocyst frequencies significantly decreased over time with the decrease in the heterocyst density and the increase in the vegetative cells ($r(96) = -0.227$, $p = 0.026$) (Figure 4). The Pearson correlation test revealed that heterocyst density and frequency had significantly negative correlation with vegetative cell density ($r(96) = -0.349$, $p = 000$ and $r(96) = -0.685$, $p < 0.001$, respectively). The results of ANOVA showed that there was a significant difference in heterocyst frequencies on day 2 ($F_{(3,8)} = 7.762$, $p = 0.009$) and on day 4 ($F_{(3,8)} = 12.212$, $p = 0.004$) among all nitrate treatments. On day 4, the highest frequency observed was in N-free treatment followed by 1.5×10^{-5} g L^{-1} nitrate treatment, where heterocysts accounted for about 49% and 27%, respectively, of total cell density (Figures 3 and 4). On day 6, heterocyst frequency in both of N-free and 1.5×10^{-5} g L^{-1} treatments abruptly dropped below 7% of the total cell density and gradually declined until day 8, when heterocyst frequency remained significantly high in N-free treatment compared to all the other treatments ($F_{(3,8)} = 51.831$, $p < 0.001$).

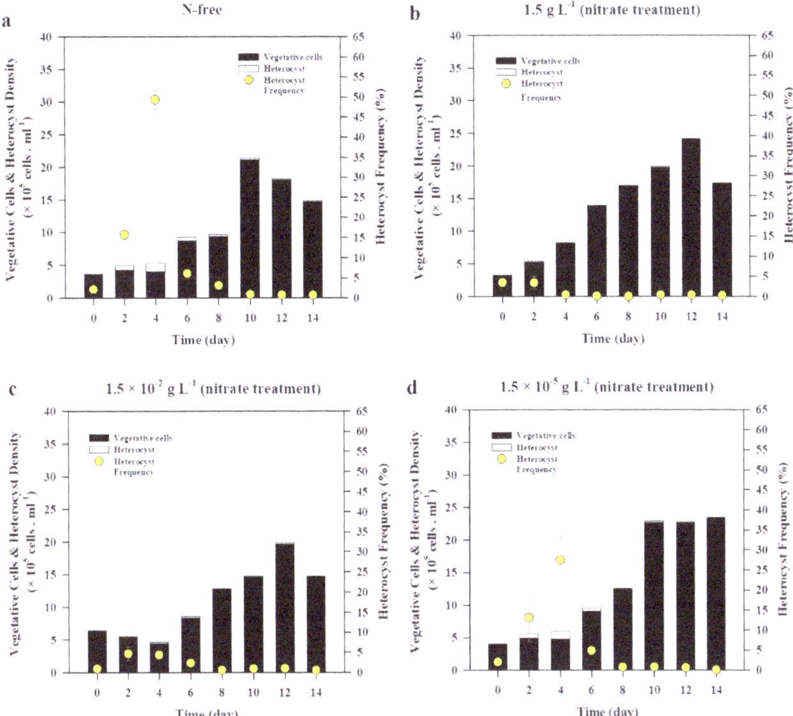

Figure 4. The changes in vegetative cell density, heterocyst density, and heterocyst frequency of *Anabaena variabilis* under various nitrate concentrations over 14 d of incubation. Measurements on day 0 were made 3 h after inoculation. These experiments were carried out in triplicates and the average of all the values was obtained. The error bars represent the standard deviation among the replicates.

3.1.3. Nitrogen Cell Quota

The initial N cell quota exhibited no significant difference among all the treatments, with an exception of the 1.5 g L^{-1} nitrate treatment (Figure 5a). N-starved cells showed significantly higher N uptake after 3-h of incubation in the highest nitrate treatment. The N cell quota for this treatment reached 1.77 ± 0.29 ng N cell^{-1} ($F_{(3,8)}$ = 53.884, p < 0.001), while that in other treatments was as low as 0.46 ± 0.14 ng N cell^{-1} (Figure 5a). However, on day 4, all nitrate treatments, with the exception of the highest treatment, reached the maximum value of the N cell quota, resulting in no significant difference between all the nitrate treatments. Further, the values of the N cell quota of *A. variabilis* in all nitrate treatments gradually decreased towards the end of the experiment. The average N cell quota in nitrate treatments on day 4 was 0.81 ± 0.13 ng N cell^{-1} (Figure 5a). The N cell quotas on day 6 dropped by 60–70% as compared to those on day 4 in all nitrate treatments, with the exception of 1.5 g L^{-1} treatment, in which it decreased only by 30%, and remained significantly higher than those in all the other treatments ($F_{(3,8)}$ = 7.559, p = 0.010). On the last day of the experiment, the 1.5 g L^{-1} nitrate treatment maintained the highest value of the N cell quota ($F_{(3,8)}$ = 16.006, p = 0.002) (Figure 5a).

The N cell quota exhibited significant positive correlation with heterocyst frequency (r(96) = 0.359, p = 0.000) and negative correlation with cell density (r(96) = −0.742, p = 0.000) in nitrate treatments. However, there was no significant correlation between N cell quota and heterocyst density (r(96) = 0.044, p = 0.667).

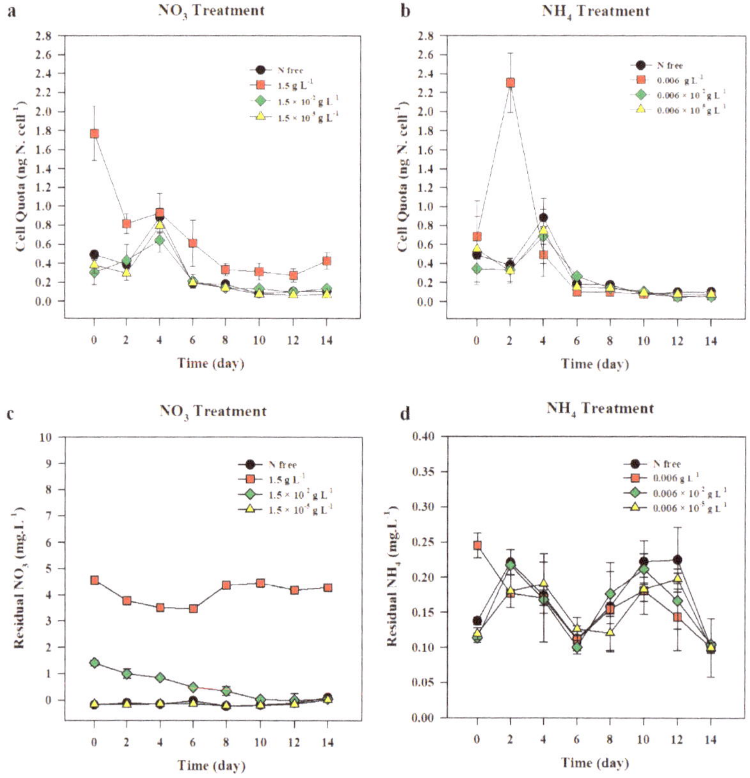

Figure 5. Nitrogen cell quota (**a,b**) of *Anabaena variabilis* and the residual nitrate (**c**) and ammonium (**d**) concentrations in the medium over 14 d of incubation under various nitrate (**a,c**) and ammonium (**b,d**) treatments. N-free treatment was set as the control. Measurements on day 0 were made 3 h after inoculation. All values were averaged from the triplicate samples and the error bars represent the standard deviation among replicates.

3.1.4. Residual Nitrate in the Medium

The changes in the residual nitrate concentrations in the medium varied among different nitrate treatments (Figure 5c). The residual nitrate concentration in 1.5×10^{-2} g L^{-1} nitrate treatment alone exhibited a moderate decline until the concentration reached zero on day 12. The concentration of the residual nitrate in the highest nitrate treatment (1.5 g L^{-1}) gradually decreased until day 6, and then subsequently increased until it reached a stable state by the end of the experiment. However, a Kruskal–Wallis test showed that it remained significantly higher than all the other nitrate treatments throughout the experimental period (X^2 (3) = 9.409, $p < 0.001$).

The residual nitrate concentration in the medium of all the nitrate treatments showed a significant negative correlation with both heterocyst density (r(92) = −0.561, $p < 0.001$) and heterocyst frequency (r(92) = −0.457, $p < 0.001$), and it showed a positive correlation with the N cell quota (r(91) = 0.515, $p < 0.001$).

3.2. Changes in A. variabilis under Ammonium Treatment

3.2.1. Chl-a, Cell Density, and Growth Rate

A. variabilis in all ammonium treatments ($n = 3$) and N-free treatment exhibited a linear growth with a relative steady rate, with the exception of the highest ammonium treatment (0.006 g L^{-1}), which exhibited highest growth rate during the 14-d experiment (Figure 1b, Table 3). The growth rate of *A. variabilis* in 0.006 g L^{-1} ammonium treatment was significantly different from that of the control ($F_{(3,8)}$ = 4.386, p = 0.042). The initial chl-*a* concentration in all ammonium treatments was averaged and was 95.27 ± 25.05 µg L^{-1} and increased to 190.45 ± 25.21 µg L^{-1} on day 4. On day 6, chl-*a* concentration in the highest ammonium treatment markedly increased and was significantly different from those of all the other treatments ($F_{(3,8)}$ = 12.601, p = 0.002). The final chl-*a* concentration on day 14 maintained the highest concentration among all the treatments, with an average concentration of 1105.51 ± 52.31 µg L^{-1} (Figure 1b). ANOVA test revealed that the chl-*a* concentration in 0.006 g L^{-1} ammonium treatment was significantly higher than those of other treatments ($F_{(3,8)}$ = 156.280, p < 0.001) on the final day of the experiment.

Cell density of *A. variabilis* in ammonium treatment was positively correlated with chl-*a* concentration ($r(96)$ = 0.869, p < 0.05). On average, the highest density was 17.18 ± 9.96 × 10^5 cell mL^{-1}, which was observed in the highest ammonium treatment (0.006 g L^{-1}). The initial cell densities in all ammonium treatments were not significantly different from each other ($F_{(3,8)}$ = 3.247, p = 0.081) and the average initial cell density was 3.96 ± 0.80 × 10^5 cells mL^{-1}. Cell density increased rapidly under 0.006 g L^{-1} ammonium treatment from day 4, after it reached at 12.20 ± 2.94 × 10^5 cell mL^{-1}, and was significantly higher than those in all the other ammonium treatments ($F_{(3,8)}$ = 12.458, p = 0.002) (Figure 1d). On day 10, there was a significant difference in cell density between 0.006 × 10^{-2} g L^{-1} and 0.006 g L^{-1} ammonium treatments ($F_{(3,8)}$ = 8.231, p = 0.008). The cell density in 0.006 ×10^{-2} g L^{-1} ammonium treatment was also significantly higher than that of the control (p = 0.042). The final cell density on day 14 ranged from 14.78 ± 0.55 × 10^5 cell mL^{-1} to 32.10 ± 1.51 × 10^5 cell mL^{-1} and was significantly higher than that of the control ($F_{(3,8)}$ = 41.155, p < 0.001).

3.2.2. Heterocyst Development

A. variabilis produced heterocysts regardless of the concentration of the ammonium treatment and they significantly decreased over time ($r(96)$ = −0.342, p = 0.001) after reaching the peak (Figure 3b). Average heterocyst density at the beginning of experiment was 8.75 ± 1.13 × 10^3 cells mL^{-1}. The highest treatment of ammonium (0.006 g L^{-1}) reached the peak of heterocyst production on day 2, and the heterocyst frequency was 52% (Figures 3d and 6). On day 2, a significant difference in heterocyst density was observed between 0.006 g L^{-1} ammonium treatment and 0.006 × 10^{-2} g L^{-1} treatment ($F_{(3,8)}$ = 23.467, p = 0.009). Heterocyst density in 0.006 g L^{-1} ammonium treatment was also significantly higher than that in 0.006 × 10^{-5} g L^{-1} ammonium treatment (p = 0.016) (Figure 3b). On day 4, heterocyst produced by *A. variabilis* reached its highest density under all treatments, with the exception of the 0.006 g L^{-1} ammonium treatment. Heterocyst density developed under 0.006 g L^{-1} treatment abruptly dropped to 80% of its initial value on day 2, reaching 1.63 × 10^4 cells mL^{-1}. Heterocyst density recorded from the highest ammonium treatment on day 4 was significantly lower than those of the rest of ammonium treatments ($F_{(3,7)}$ = 11.196, p = 0.005). After reaching the highest peak on day 4, heterocyst density in 0.006 × 10^{-2} g L^{-1}, 0.006 × 10^{-5} g L^{-1}, and N-free treatments rapidly declined on day 6 when heterocyst densities in all ammonium treatments were significantly lower than that in the control ($F_{(3,8)}$ = 19.547, p < 0.001). Heterocyst density was maintained below 3.0 × 10^4 cells mL^{-1} in all treatments including control until the end of the experiment. The average final heterocyst density in all ammonium treatments was 5.1 × 10^3 cells mL^{-1}, which was slightly lower than the initial heterocyst density. Under ammonium treatment, the Pearson correlation test showed that heterocyst density had a significant negative correlation with chl-*a* concentration ($r(96)$ = −0.250, p = 0.014), cell density ($r(96)$ = −0.400, p = 0.000), and growth rate ($r(96)$= −0.239, p = 0.019).

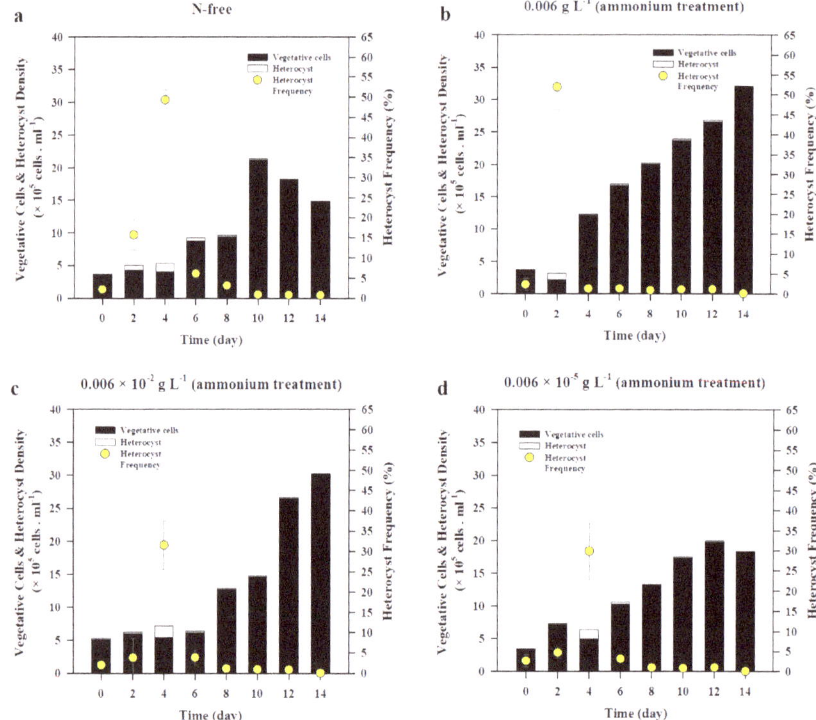

Figure 6. The changes in vegetative cell density, heterocyst density, and heterocyst frequency of *Anabaena variabilis* under various ammonium concentrations over 14 d of incubation. Measurements on day 0 were made 3 h after inoculation. All values were averaged from triplicate samples and the error bars represent the standard deviation among replicates.

The average initial heterocyst frequency was 2.3% in all ammonium treatments (Figure 3d). Until day 2, *A. variabilis* produced heterocysts at a frequency of 52% in the highest ammonium treatment (0.006 g L^{-1}) (Figure 6), which was significantly higher than those in other ammonium treatments ($F_{(3,8)}$ = 13.861, p = 0.002). The frequency dropped to 1.3% on day 4 and remained at around 1% until the end of experiment. In contrast, low ammonium treatments (0.006 × 10^{-2} g L^{-1} and 0.006 × 10^{-5} g L^{-1}) showed similar heterocyst frequency to the control (N-free treatment) on day 4, where the heterocyst frequencies ranged from 29.9 to 34.9%, and then rapidly decreased to a value below 6% until the end of experiment (Figures 3d and 6). On day 4, heterocysts in 0.006 g L^{-1} ammonium treatment showed the lowest frequency among all treatments ($F_{(3,7)}$ = 9.789, p = 0.007). The heterocyst frequency slowly decreased after the peak, especially after day 8, and remained below 10% in all the treatments. On average, the frequency on the last day of the experiment was below 1%, being much lower than the initial frequency.

3.2.3. Nitrogen Cell Quota

The average N cell quota of *A. variabilis* on day 0 was 0.52 ± 0.14 ng N cell^{-1} (Figure 5b). It sharply increased in the highest ammonium treatment (0.006 g L^{-1}) on day 2, resulting in a significantly higher N cell quota than that in the other ammonium treatments, where the value was 2.31 ± 1.0 ng N cell^{-1} ($F_{(3,8)}$ = 8.516, p = 0.002). The peak of the N cell quota on day 2 in the highest ammonium treatment sharply decreased on day 4, resulting in the lowest cell quota, but it was not significantly different from values in the other treatments ($F_{(3,8)}$ = 1.673, p = 0.249). The lowest value of N cell quota in the highest ammonium treatment was maintained by day 10. On day 4, the maximal N cell quota appeared for the rest of the treatments in which N-free treatment exhibited the highest value. N cell quotas in all treatments decreased further on day 6 and ranged from 0.10 ± 0.0 ng N cell^{-1} to 0.27 ± 0.03 ng N cell^{-1}, and the value was then maintained low until the end of the experiment. The N cell quota showed no significant difference among the different treatments until day 12 ($F_{(3,6)} = 5.765$, $p = 0.034$). However, on the last day of the experiment (day 14), the cell quota in the N-free treatment was significantly higher than that in the 0.006 g L^{-1} treatment ($F_{(3,7)} = 11.001$, $p = 0.019$) and the 0.006×10^{-2} g L^{-1} treatment ($p = 0.004$).

The N cell quota showed a significant positive correlation with heterocyst density ($r(93) = 0.479$, $p = 0.000$) and heterocyst frequency ($r(93) = 0.772$, $p < 0.001$) and a negative correlation with vegetative cell density ($r(96) = -0.924$, $p < 0.001$) and cell growth rate ($r(93) = -0.299$, $p = 0.004$) in the ammonium treatment.

3.2.4. Residual Ammonium in the Medium

The residual ammonium in the highest ammonium concentration (0.006 g L^{-1}) alone was significantly different from those in other treatments on day 0, with a concentration of 0.245 mg L^{-1} ($F_{(3,8)} = 99.137$, $p < 0.001$) (Figure 5d). On day 2, there was no significant difference observed in the residual ammonium concentrations between all the ammonium treatments ($F_{(3,6)} = 3.974$, $p = 0.071$), and it showed no statistical difference by the end of the experiment. Pearson's correlation test showed that the residual ammonium concentration across all ammonium treatments had a positive correlation with the heterocyst density ($r(92) = 0.219$, $p = 0.036$).

4. Discussion

This study evaluated the changes in cell growth, heterocyst development, and N cell quota in a cyanobacterial species, *A. variabilis*, under various concentrations of nitrate and ammonium over 14-d laboratory incubation. The test cyanobacteria possessed a diazotrophic growth due to its ability to develop heterocysts. Two N sources (nitrate and ammonium) with different concentrations exhibited different patterns of growth and heterocyst development. Both the growth rate and N cell quota of *A. variabilis* were higher under ammonium treatment than under nitrate treatment. These results were consistent with our first hypothesis stating that ammonium causes relatively high N uptake and growth; however, its heterocyst production was contrary to what we had predicted. Low concentration of nitrate produced a high number of heterocysts within nitrate treatment; however, the presence of ammonium did not suppress heterocyst production in *A. variabilis* and instead increased heterocyst production compared to that of the nitrate treatments. This result suggests that we probably used too low ammonium concentration to see the suppression of heterocyst production through a serial dilution. Only the highest ammonium concentration showed significant difference of heterocyst production from the other two concentrations. Hence, we suspect that our different ammonium treatments were no longer working for comparison after day 2, thereby making our second hypothesis difficult to prove.

The development of heterocyst in cyanobacteria is a response to a certain type of N source and its concentrations in the environment [14,16,17]. Since cyanobacteria prioritize ammonium over the other types of N sources for uptake and assimilation [34,50,51], the presence of ammonium can repress the initiation of heterocyst development. Lindell and Post (2001) [43] observed that ammonium concentrations greater than 1 µM (1.8×10^{-5} g L^{-1}) inhibited the uptake of other forms of N by repressing ntcA gene expression. Moreover, another study reported that ammonium at a concentration of 1 mM (0.018 g L^{-1}) reduced heterocyst density, while 20 mM (1.24 g L^{-1}) nitrate had no significant effect on heterocyst production, indicating that nitrate is not essentially utilized to meet the N requirement in cyanobacteria [52]. In contrast to our expectation, in this study, heterocyst density in *A. variabilis* was higher under ammonium treatments than under nitrate treatments. The highest concentration of ammonium supplement (0.006 g L^{-1}) exhibited the highest production of heterocyst. In contrast, the residual nitrate concentration in the highest nitrate treatment (1.5 g L^{-1}) did not show any significant changes throughout the experimental period. This result indicated that the *A. variabilis* strain used in our study produces heterocyst at a low frequency in the presence of high nitrate concentration to meet the remaining N demand. On the other hand, prior studies showed that a marine cyanobacterial species, *Nodularia spumigena*, showed different intraspecific strategies between strains in terms of the C:N cellular ratio and the specific N fixation under the same environmental conditions [53,54]. Hence, we also assume that it is possible that different strains of *A.variabilis* might possess different strategies in heterocyst development to meet the N requirement. The activation of N fixation and heterocyst differentiation in cyanobacteria may depend on the N content of cells [10]. In this study, we observed that high N cell quota was associated with higher production of heterocyst, indicating that N fixation occurs to meet the N requirement of *A. variabilis* and, subsequently, N might be assimilated for cellular growth. It has been suggested that N assimilation in cyanobacteria is dependent on signals from an internal mechanism by which the filament senses N sufficiency or insufficiency within the cells based on the external N availability [35]. Even in the presence of ammonium or nitrate at any concentration, it would be an added advantage to possess both systems of N assimilation from an external source and N fixation if the proteins allow both processes to take place at the same time [35].

A. variabilis, in this study, exhibited a linear growth both in the presence and absence of the N sources (nitrate or ammonium), irrespective of their concentrations. *Anabaena* species are likely to survive in unfavorable conditions such as N-depleted environments [27,35,48]. Past studies have observed that a few Nostocales including *A. variabilis* were capable of growing even without N source in the presence of increased heterocyst cells [27,55]. Even under critically low N:P ratio and high light conditions, *Anabaena flos-aquae* dominated over non-N-fixing species, indicating that N fixation could support its growth under harsh conditions [56]. The result of the current study showed that chl-*a* concentrations of *A. variabilis* both in nitrate and ammonium treatments exhibited a similarly increasing trend, but ammonium, particularly in the highest concentration of 0.006 g L^{-1} (0.0229 mM), caused an acceleration in the cell growth, while actively producing a large number of heterocysts at the same time (Figures 1 and 3), indicating that a high concentration of ammonium supports biomass development of cyanobacteria. Some studies also demonstrated that ammonium enrichment led to an increase in most cyanobacteria and their domination in the phytoplankton community [57–59]. High N cell quota was observed in the highest ammonium treatment after 2 d of incubation and it corresponded with the rapid decline in the residual ammonium in the medium (Figure 5b,d). The rapid ammonium assimilation exhibited by *A. variabilis* in this study is also supported by the results of other studies, where cyanobacteria could take up ammonium at a higher rate than other N sources, such as nitrate and urea, because it is the most reduced form [9,10,42]. In contrast, the N cell quota of *A. variabilis* in the highest nitrate supplement, 1.5 g L^{-1} (17.6 mM), used in this study decreased over time and the residual nitrate concentration in the medium remained high, indicating that cell growth in *A. variabilis* is not triggered instantly upon N uptake, but instead, might rely on N fixation, which in turn depends on heterocyst production (Figures 4b and 5a,c). The absence of N, or nitrate concentrations as low as 1.5×10^{-5} g L^{-1}, did not inhibit its growth, suggesting that *A. variabilis* is capable of growing under N-limiting conditions in the presence of high frequency of heterocyst production.

We used lab-grown culture kept in stationary phase to minimize the effect of original BG-11 medium on the cyanobacterial growth during the 14-day experiment. We also included N starvation before assigning the culture into each nitrate-depleted or ammonium-depleted treatment, so that only growth after the introduction of single N source could be observed. In addition to this, we removed P from the medium for starvation to minimize the formation of heterocyst before the start of the experiment, as P is also an important nutrient in the heterocyst formation [60–62]. It is important to impose nutrient starvation prior to the growth experiment to exclude N sources other than the ones included in the experiment. This is carried out by thorough washing of the residual N and P present on the cell surface and subsequent incubation of the cells in an N-free medium for a certain period [29,47]. Upon Fe-NH$_4$-citrate removal in nitrate-supplemented medium in the current study, Fe might be limited in the nitrate treatment. However, our results showed that Fe limitation was not noticeable in heterocyst production and growth of *A. variabilis*. Nevertheless, we do not deny a potential role of iron ion in the regulation of heterocyst production in cyanobacteria [31].

A low frequency of heterocyst was detected in *A. variabilis* prior to the beginning of the experiment during nutrient starvation. Since *A. variabilis* rapidly adapts to the nutrient-depleted condition through a unique mechanism called N fixation, it is assumed that initiation of heterocyst-mediated N fixation occurs during the starvation period [20]. However, the low density of heterocyst at day 0 did not affect the results of this study since we observed a specific pattern of heterocyst formation during the 14-d-long experiment. Across the various N conditions involved in this study, there was a short period during which *A. variabilis* gradually developed high heterocyst density, followed by a sudden decline, and then maintained the heterocyst production at a low frequency until the end of experiment (Figure 3). We observed heterocyst formation after a 24-h exposure to nitrate-free medium. However, the thickening of the cell walls and appearance of nodules were observed only on day 2 [27], even though the regulation in the expression of related genes such as HetR, NtcA, and PatS was observed within few hours after deprivation [43,63]. In this study, a significant increase (3–52%) in heterocyst formation has been observed in *A. variabilis* after 2 d of exposure to either nitrate or ammonium treatment, compared to its initial density (Figure 3c,d). The subsequent decline in heterocyst density after reaching its peak on day 4 was observed with a consistent increase in vegetative cells, suggesting that *A. variabilis* preferred vegetative growth over heterocyst development in the presence of an optimal concentration of fixed N within the cells. Previous studies have reported the presence of an intracellular transfer of ATP and fixed N between vegetative cells and heterocysts [64]. Therefore, cyanobacteria meet their high energy requirements in the form of ATP via cellular interactions within the filament, indicating that the process of heterocyst differentiation and vegetative growth are interdependent [17,65].

The diazotrophic growth demonstrated by the high growth rate and heterocyst formation of *A. variabilis* under various N conditions included in this study indicates its ecological fitness under harsh environmental conditions in freshwater ecosystems. Nutrient fluxes in the ecosystem may influence the dominance of some cyanobacterial populations, especially those with N-fixation ability. A prior study observed that a significant increase in N fixation after a rapid decline in N availability stimulated an alternation in the dominance of toxic *Aphanizomenon* and *Microcystis* in an eutrophic lake [66]. In a multispecies community with both N fixers and non-N fixers, the dominance of a heterocystous N-fixer species called *Anabaena flos-aquae* was observed under high light intensity and low N:P ratio conditions. These findings suggest that the occurrence of a bloom with developed heterocysts under those conditions could potentially outcompete the non-N-fixer species [56]. Ecological fitness due to the N fixation ability also varies among N-fixing species [22,67]. Certain N-fixing cyanobacteria produce high biomass with increased heterocyst production only under low N conditions, while some of the other species produce bloom with heterocyst under both N-limiting and N-sufficient conditions [22,67]. This suggests that differences in fitness due to the N-fixation ability among N-fixing species play a major role in determining the biomass distribution and the dominance of N-fixing species within the cyanobacterial community.

Although this study did not consider the effect of phosphorus on the growth and heterocyst development, prior studies showed its important role in heterocyst expression as well as accelerating cyanobacterial growth [60–62]. Since heterocysts are differentiated from existing vegetative cells, there is evidence of phosphorus involvement in the oxidative phosphate pathway of the N fixation process and also the accumulation of intracellular phosphorus at low levels in the heterocyst cells compared to vegetative cells [39,68,69]. Under limiting nitrogen, *Dolichospermum flos-aquae* produced heterocysts with significantly higher nitrogen fixation in high phosphorus treatment than low phosphorus treatment, supporting the assertion that nitrogen fixation is promoted by high phosphorus availability [36,70,71]. Another study also showed that cyanobacterial community composition was shifted to community dominating N-fixers under the condition of low N:P ratio [56].

A previous study has suggested that heterocyst-to-vegetative cell ratio could be an adequate indicator of N fixation in cyanobacteria [36]. An increased heterocyst frequency exhibited by *A. variabilis* in this study indicated that N fixation occurred to meet the N demand after its adaptation to various N-limiting conditions. However, other studies using a few marine cyanobacterial strains such as *Nodularia spumigena*, *Aphanizomenon* sp., and *Dolichospermum* spp. demonstrated that the genes related to heterocyst formation and genes related to N fixation were not activated simultaneously under the same N-limited condition, indicating that heterocyst frequency alone does not adequately represent the measurement of N fixation [53,72–74]. Nevertheless, heterocyst frequency could indicate the potential of N fixation [35,75,76]. Adaptation to N-stressed conditions through N fixing ability potentially introduces a novel source of N into the ecosystem and also provides N to the other non-N-fixing microorganism communities in aquatic ecosystems [77,78]. A recent study demonstrated that *Aphanizomenon* spp., dominant colony-forming, N-fixing cyanobacterium found in the N-limiting Baltic Sea during summer, transfer about 50% of its newly fixed N in the form of ammonium to the microbial and classical food web within the plankton community, indicating that N-fixing cyanobacterial blooms can stimulate ecosystem productivity and biogeochemical processes within a short period of time [12]. Another study showed that the N-fixation activity of an *Anabaena* sp. disrupted the N removal efficiency of the constructed wetlands [77]. Fixed N introduced by *Anabaena* spp. was suspected to be used by non-N-fixing *Microcystis* sp. that caused blooms in an adjacent lake, indicating that the N-fixing ability of cyanobacteria can also benefit the phytoplankton community [77]. The findings of this study indicated that the N-fixing ability of *A. variabilis*, through the mechanism of heterocyst differentiation, plays an important role in the N cycle in freshwater ecosystems, and its abundance potentially influences the advancement of the phytoplankton community.

5. Conclusions

This study demonstrated that a filamentous cyanobacteria, *Anabaena variabilis*, developed heterocysts in response to N availability. Heterocyst production was inversely correlated to cell growth in *A. variabilis*. It maintained a slow growth during high production of heterocysts, and exhibited a linear cell growth when heterocysts were produced at a low frequency. Ammonium was the primary source of N that contributed to a higher growth rate and heterocyst production than those of nitrate. The cell N quota was fulfilled by active uptake of ammonium rather than nitrate for growth. However, when the external N supply was no longer able to support cyanobacterial N demand, potentially fixed N in heterocysts majorly contributed to the cell N quota. Even nitrate supplement as high as 1.5 g L^{-1} could not affect the formation of heterocysts, indicating that nitrate is not crucial for N assimilation. The results of this study provide basic information on how heterocyst development and diazotrophic growth of *A. variabilis* occur with respect to N availability. Moreover, the difference exhibited by *A. variabilis* contributed to the understanding of its ecological fitness under N-limiting environment, indicating its potential dominance and its role in the N cycle in freshwater ecosystems. Further studies are required to investigate the role of other essential nutrients, such as P, in the development of heterocysts and cell growth, particularly with regard to heterocyst expression and N fixation.

Author Contributions: Laboratory experiments, analyses, writing, and original draft preparation, N.S.Z.; Research and review, S.-J.H. All authors have read and agreed to the published version of the manuscript.

Funding: This paper was supported by Konkuk University in 2019.

Acknowledgments: We thank Keon-Hee Kim for his advice and assistance in the research. We also thank the crew of The Limnology Lab, Department of Environmental Health Science, Konkuk University, for the technical support and discussion. Finally, we are grateful to anonymous reviewers who provided critical comments and constructive suggestions to improve the earlier version of this manuscript.

Conflicts of Interest: The authors declare no conflict of interest.

References

1. Schlichting, J.; Harold, E. Survival of some fresh-water algae under extreme environmental conditions. *Trans. Am. Microsc. Soc.* **1974**, *93*, 610–613. [CrossRef] [PubMed]
2. Carey, C.C.; Ibelings, B.W.; Hoffmann, E.P.; Hamilton, D.P.; Brookes, J.D. Eco-physiological adaptations that favour freshwater cyanobacteria in a changing climate. *Water Res.* **2012**, *46*, 1394–1407. [CrossRef] [PubMed]
3. Landsberg, J.H. The effects of harmful algal blooms on aquatic organisms. *Rev. Fish. Sci.* **2002**, *10*, 113–390. [CrossRef]
4. Ibelings, B.W.; Havens, K.E. Cyanobacterial toxins: A qualitative meta–analysis of concentrations, dosage and effects in freshwater, estuarine and marine biota. In *Cyanobacterial Harmful Algal Blooms: State of the Science and Research Needs*; Springer: New York, NY, USA, 2008; pp. 675–732.
5. Hamilton, D.P.; Wood, S.A.; Dietrich, D.R.; Puddick, J. Costs of harmful blooms of freshwater cyanobacteria. In *Cyanobacteria: An Economic Perspective*; John Wiley Sons: New York, NY, USA, 2014; pp. 247–256.
6. Xu, H.; Paerl, H.; Qin, B.; Zhu, G.; Hall, N.; Wu, Y. Determining critical nutrient thresholds needed to control harmful cyanobacterial blooms in eutrophic Lake Taihu, China. *Environ. Sci. Technol.* **2015**, *49*, 1051–1059. [CrossRef]
7. Paerl, H.W. Controlling cyanobacterial harmful blooms in freshwater ecosystems. *Microb. Biotechnol.* **2017**, *10*, 1106–1110. [CrossRef]
8. Kolzau, S.; Dolman, A.M.; Voss, M.; Wiedner, C. The response of nitrogen fixing cyanobacteria to a reduction in nitrogen loading. *Int. Rev. Hydrobiol.* **2018**, *103*, 5–14. [CrossRef]
9. Dortch, Q. The interaction between ammonium and nitrate uptake in phytoplankton. *Mar. Ecol. Prog. Series Oldend.* **1990**, *61*, 183–201. [CrossRef]
10. Spröber, P.; Shafik, H.M.; Présing, M.; Kovács, A.W.; Herodek, S. Nitrogen uptake and fixation in the cyanobacterium Cylindrospermopsis raciborskii under different nitrogen conditions. *Hydrobiologia* **2003**, *506*, 169–174. [CrossRef]
11. Howarth, R.W.; Marino, R.; Cole, J.J. Nitrogen fixation in freshwater, estuarine, and marine ecosystems. 2. Biogeochemical controls1. *Limnol. Oceanogr.* **1988**, *33*, 688–701. [CrossRef]
12. Adam, B.; Klawonn, I.; Svedén, J.B.; Bergkvist, J.; Nahar, N.; Walve, J.; Littmann, S.; Whitehouse, M.J.; Lavik, G.; Kuypers, M.M. N 2-fixation, ammonium release and N-transfer to the microbial and classical food web within a plankton community. *ISME J.* **2016**, *10*, 450–459. [CrossRef]
13. Ben-Porath, J.; Zehr, J. Detection and characterization of cyanobacterial nifH genes. *Appl. Environ. Microbiol.* **1994**, *60*, 880–887. [CrossRef] [PubMed]
14. Fay, P. Oxygen relations of nitrogen fixation in cyanobacteria. *Microbiol. Mol. Biol. Rev.* **1992**, *56*, 340–373. [CrossRef]
15. Paerl, H. The cyanobacterial nitrogen fixation paradox in natural waters. *F1000Research* **2017**, *6*, 244. [CrossRef] [PubMed]
16. Stal, L.J. The effect of oxygen concentration and temperature on nitrogenase activity in the heterocystous cyanobacterium Fischerella sp. *Sci. Rep.* **2017**, *7*, 1–10. [CrossRef]
17. Kumar, K.; Mella-Herrera, R.A.; Golden, J.W. Cyanobacterial heterocysts. *Cold Spring Harb. Perspect. Biol.* **2010**, *2*, a000315. [CrossRef]
18. Wolk, C.P. Movement of carbon from vegetative cells to heterocysts in Anabaena cylindrica. *J. Bacteriol.* **1968**, *96*, 2138–2143. [CrossRef]
19. Thomas, J. Absence of the Pigments of Photosystem II of Photosynthesis in Heterocysts of a Blue–Green Alga. *Nature* **1970**, *228*, 181–183. [CrossRef]
20. Herrero, A.; Stavans, J.; Flores, E. The multicellular nature of filamentous heterocyst-forming cyanobacteria. *Fems Microbiol. Rev.* **2016**, *40*, 831–854. [CrossRef]
21. Laamanen, M.; Kuosa, H. Annual variability of biomass and heterocysts of the N. *Boreal Environ. Res.* **2005**, *10*, 19–30.
22. González-Madina, L.; Pacheco, J.P.; Yema, L.; de Tezanos, P.; Levrini, P.; Clemente, J.; Crisci, C.; Lagomarsino, J.J.; Méndez, G.; Fosalba, C. Drivers of cyanobacteria dominance, composition and nitrogen fixing behavior in a shallow lake with alternative regimes in time and space, Laguna del Sauce (Maldonado, Uruguay). *Hydrobiologia* **2019**, *829*, 61–76. [CrossRef]

23. J, K. Cyanoprokaryota 3. Teil/3rd Part: Heterocytous genera. In *Süswasserflora von Mitteleuropa–Freshwater Flora of Central Europe*; Springer: Berlin/Heidelberg, Germany, 2013.

24. Kangatharalingam, N.; Priscu, J.C.; Paerl, H.W. Heterocyst envelope thickness, heterocyst frequency and nitrogenase activity in Anabaena flos-aquae: Influence of exogenous oxygen tension. *Microbiology* **1992**, *138*, 2673–2678. [CrossRef]

25. Komárek, J.; Kováčik, L.U. Trichome structure of fourAphanizomenon taxa (Cyanophyceae) from Czechoslovakia, with notes on the taxonomy and delimitation of the genus. *Plant Syst. Evol.* **1989**, *164*, 47–64. [CrossRef]

26. Yoon, H.-S.; Golden, J.W. Heterocyst pattern formation controlled by a diffusible peptide. *Science* **1998**, *282*, 935–938. [CrossRef] [PubMed]

27. Ogawa, R.E.; Carr, J.F. The Influence of Nitrogen on Heterocyst Production in Blue-Green Algae 1. *Limnol. Oceanogr.* **1969**, *14*, 342–351. [CrossRef]

28. Willis, A.; Chuang, A.W.; Burford, M.A. Nitrogen fixation by the diazotroph Cylindrospermopsis raciborskii (Cyanophyceae). *J. Phycol.* **2016**, *52*, 854–862. [CrossRef] [PubMed]

29. Wang, S.; Xiao, J.; Wan, L.; Zhou, Z.; Wang, Z.; Song, C.; Zhou, Y.; Cao, X. Mutual dependence of nitrogen and phosphorus as key nutrient elements: One facilitates Dolichospermum flos-aquae to overcome the limitations of the other. *Environ. Sci. Technol.* **2018**, *52*, 5653–5661. [CrossRef] [PubMed]

30. Willis, A.; Adams, M.P.; Chuang, A.W.; Orr, P.T.; O'Brien, K.R.; Burford, M.A. Constitutive toxin production under various nitrogen and phosphorus regimes of three ecotypes of Cylindrospermopsis raciborskii ((Wołoszyńska) Seenayya et Subba Raju). *Harmful Algae* **2015**, *47*, 27–34. [CrossRef]

31. Aly, W.; Andrews, S. Iron regulation of growth and heterocyst formation in the nitrogen fixing cyanobacterium Nostoc sp. PCC 7120. *J. Ecol. Health Environ.* **2016**, *4*, 103–109. [CrossRef]

32. Chaffin, J.D.; Bridgeman, T.B. Organic and inorganic nitrogen utilization by nitrogen-stressed cyanobacteria during bloom conditions. *J. Appl. Phycol.* **2014**, *26*, 299–309. [CrossRef]

33. Masuda, T.; Furuya, K.; Kodama, T.; Takeda, S.; Harrison, P.J. Ammonium uptake and dinitrogen fixation by the unicellular nanocyanobacterium Crocosphaera watsonii in nitrogen-limited continuous cultures. *Limnol. Oceanogr.* **2013**, *58*, 2029–2036. [CrossRef]

34. Mickelson, J.C.; Davis, E.B.; Tischer, R. The effect of various nitrogen sources upon heterocyst formation in Anabaena flos-aquae A–37. *J. Exp. Bot.* **1967**, *18*, 397–405. [CrossRef]

35. Thiel, T.; Pratte, B. Effect on Heterocyst Differentiation of Nitrogen Fixation in Vegetative Cells of the Cyanobacterium Anabaena variabilis ATCC 29413. *J. Bacteriol.* **2001**, *183*, 280–286. [CrossRef] [PubMed]

36. Lilen, Y.; Elena, L.; Paula, d.T.P. The role of heterocytes in the physiology and ecology of bloom-forming harmful cyanobacteria. *Harmful Algae* **2016**, *60*, 131–138. [CrossRef]

37. Videau, P.; Cozy, L.M. Anabaena sp. strain PCC 7120: Laboratory Maintenance, Cultivation, and Heterocyst Induction. *Curr. Protoc. Microbiol.* **2019**, *52*, e71. [CrossRef] [PubMed]

38. Wolk, C.P. Heterocyst formation in Anabaena. In *Prokaryotic Development*; American Society of Microbiology: Washington, DC, USA, 2000; pp. 83–104.

39. Bradley, S.; Carr, N. Heterocyst and nitrogenase development in Anabaena cylindrica. *Microbiology* **1976**, *96*, 175–184. [CrossRef]

40. Golden, J.W.; Yoon, H.-S. Heterocyst development in Anabaena. *Curr. Opin. Microbiol.* **2003**, *6*, 557–563. [CrossRef]

41. Wilcox, M.; Mitchison, G.; Smith, R. Pattern formation in the blue-green alga, Anabaena: I. Basic mechanisms. *J. Cell Sci.* **1973**, *12*, 707–723.

42. McCarthy, J. The kinetics of nutrient utilization. *Can. Bull. Fish. Aquat. Sci.* **1981**, *210*, 211–233.

43. Lindell, D.; Post, A.F. Ecological Aspects of ntcA Gene Expression and Its Use as an Indicator of the Nitrogen Status of Marine Synechococcus spp. *Appl. Environ. Microbiol.* **2001**, *67*, 3340–3349. [CrossRef]

44. Stanier, R.; Kunisawa, R.; Mandel, M.; Cohen-Bazire, G. Purification and properties of unicellular blue-green algae (order Chroococcales). *Bacteriol. Rev.* **1971**, *35*, 171. [CrossRef]

45. Sabour, B.; Sbiyyaa, B.; Loudiki, M.; Oudra, B.; Belkoura, M.; Vasconcelos, V. Effect of light and temperature on the population dynamics of two toxic bloom forming Cyanobacteria–Microcystis ichthyoblabe and Anabaena aphanizomenoides. *Chem. Ecol.* **2009**, *25*, 277–284. [CrossRef]

46. Islam, M.A.; Beardall, J. Growth and photosynthetic characteristics of toxic and non-toxic strains of the cyanobacteria Microcystis aeruginosa and Anabaena circinalis in relation to light. *Microorganisms* **2017**, *5*, 45. [CrossRef] [PubMed]

47. Guimarães, P.; Yunes, J.S.; Cretoiu, M.S.; Stal, L.J. Growth Characteristics of an Estuarine Heterocystous Cyanobacterium. *Front. Microbiol.* **2017**, *8*, 1132. [CrossRef] [PubMed]

48. Fogg, G. Growth and Heterocyst Production in Anabaena cylindrica Lemm. *New Phytol.* **1944**, *43*, 164–175. [CrossRef]

49. Rice, E.W.; Baird, R.B.; Eaton, A.D. *Standard Methods for the Examination of Water and Wastewater*; American Public Health Association (APHA): Washington, DC, USA, 2017.

50. Muro-Pastor, M.I.; Reyes, J.C.; Florencio, F.J. Cyanobacteria perceive nitrogen status by sensing intracellular 2-oxoglutarate levels. *J. Biol. Chem.* **2001**, *276*, 38320–38328. [PubMed]

51. Herrero, A.; Muro-Pastor, A.M.; Flores, E. Nitrogen control in cyanobacteria. *J. Bacteriol.* **2001**, *183*, 411–425. [CrossRef]

52. Sanz-Alférez, S.; del Campo, F.F. Relationship between nitrogen fixation and nitrate metabolism in the Nodularia strains M1 and M2. *Planta* **1994**, *194*, 339–345. [CrossRef]

53. Olofsson, M.; Egardt, J.; Singh, A.; Ploug, H. Inorganic phosphorus enrichments in Baltic Sea water have large effects on growth, carbon fixation, and N_2 fixation by Nodularia spumigena. *Aquat. Microb. Ecol.* **2016**, *77*, 111–123. [CrossRef]

54. Wulff, A.; Mohlin, M.; Sundbäck, K. Intraspecific variation in the response of the cyanobacterium Nodularia spumigena to moderate UV-B radiation. *Harmful Algae* **2007**, *6*, 388–399. [CrossRef]

55. Fogg, G. Growth and heterocyst production in Anabaena cylindrica lemm.: II. in relation to carbon and nitrogen metabolism. *Ann. Bot.* **1949**, *13*, 241–259. [CrossRef]

56. De Tezanos Pinto, P.; Litchman, E. Interactive effects of N: P ratios and light on nitrogen-fixer abundance. *Oikos* **2010**, *119*, 567–575. [CrossRef]

57. Glibert, P.M.; Wilkerson, F.P.; Dugdale, R.C.; Raven, J.A.; Dupont, C.L.; Leavitt, P.R.; Parker, A.E.; Burkholder, J.M.; Kana, T.M. Pluses and minuses of ammonium and nitrate uptake and assimilation by phytoplankton and implications for productivity and community composition, with emphasis on nitrogen-enriched conditions. *Limnol. Oceanogr.* **2016**, *61*, 165–197. [CrossRef]

58. Donald, D.B.; Bogard, M.J.; Finlay, K.; Leavitt, P.R. Comparative effects of urea, ammonium, and nitrate on phytoplankton abundance, community composition, and toxicity in hypereutrophic freshwaters. *Limnol. Oceanogr.* **2011**, *56*, 2161–2175. [CrossRef]

59. Glibert, P.M.; Berg, G.M. Nitrogen form, fate and phytoplankton composition. In *Experimental Ecosystems and Scale: Tools for Understanding and Managing Coastal Ecosystems*; Kennedy, V.S., Kemp, W.M., Peterson, J.E., Dennison, W.C., Eds.; Springer: New York, NY, USA, 2009; pp. 183–189.

60. Wolk, C.P.; Ernst, A.; Elhai, J. Heterocyst metabolism and development. In *The Molecular Biology of Cyanobacteria*; Springer: Dordrecht, The Netherlands, 1994; pp. 769–823.

61. Walve, J.; Larsson, U. Blooms of Baltic Sea Aphanizomenon sp. (Cyanobacteria) collapse after internal phosphorus depletion. *Aquat. Microb. Ecol.* **2007**, *49*, 57–69. [CrossRef]

62. Xu, H.; Paerl, H.W.; Qin, B.; Zhu, G.; Gaoa, G. Nitrogen and phosphorus inputs control phytoplankton growth in eutrophic Lake Taihu, China. *Limnol. Oceanogr.* **2010**, *55*, 420–432. [CrossRef]

63. Hu, H.-X.; Jiang, Y.-L.; Zhao, M.-X.; Cai, K.; Liu, S.; Wen, B.; Lv, P.; Zhang, Y.; Peng, J.; Zhong, H. Structural insights into HetR– PatS interaction involved in cyanobacterial pattern formation. *Sci. Rep.* **2015**, *5*, 1–11. [CrossRef]

64. Plominsky, Á.M.; Delherbe, N.; Mandakovic, D.; Riquelme, B.; González, K.; Bergman, B.; Mariscal, V.; Vásquez, M. Intercellular transfer along the trichomes of the invasive terminal heterocyst forming cyanobacterium Cylindrospermopsis raciborskii CS-505. *FEMS Microbiol. Lett.* **2015**, *362*, fnu009. [CrossRef]

65. Horne, A.J.; Goldman, C.R. Nitrogen fixation in Clear Lake, California. I. Seasonal variation and the role of heterocysts 1. *Limnol. Oceanogr.* **1972**, *17*, 678–692. [CrossRef]

66. Beversdorf, L.J.; Miller, T.R.; McMahon, K.D. The role of nitrogen fixation in cyanobacterial bloom toxicity in a temperate, eutrophic lake. *PLoS ONE* **2013**, *8*, e56103. [CrossRef]

67. Moisander, P.H.; Cheshire, L.A.; Braddy, J.; Calandrino, E.S.; Hoffman, M.; Piehler, M.F.; Paerl, H.W. Facultative diazotrophy increases Cylindrospermopsis raciborskii competitiveness under fluctuating nitrogen availability. *Fems Microbiol. Ecol.* **2012**, *79*, 800–811. [CrossRef]

68. Kruger, N.J.; von Schaewen, A. The oxidative pentose phosphate pathway: Structure and organisation. *Curr. Opin. Plant Biol.* **2003**, *6*, 236–246. [CrossRef]

69. Braun, P.D.; Schulz-Vogt, H.N.; Vogts, A.; Nausch, M. Differences in the accumulation of phosphorus between vegetative cells and heterocysts in the cyanobacterium Nodularia spumigena. *Sci. Rep.* **2018**, *8*, 1–6. [CrossRef] [PubMed]

70. Kenesi, G.; Shafik, H.M.; Kovács, A.W.; Herodek, S.; Présing, M. Effect of nitrogen forms on growth, cell composition and N 2 fixation of Cylindrospermopsis raciborskii in phosphorus-limited chemostat cultures. *Hydrobiologia* **2009**, *623*, 191–202. [CrossRef]

71. Stewart, W.; Alexander, G. Phosphorus availability and nitrogenase activity in aquatic blue-green algae. *Freshw. Biol.* **1971**, *1*, 389–404. [CrossRef]

72. Mohlin, M.; Wulff, A. Interaction effects of ambient UV radiation and nutrient limitation on the toxic cyanobacterium Nodularia spumigena. *Microb. Ecol.* **2009**, *57*, 675–686. [CrossRef] [PubMed]

73. Vintila, S.; El-Shehawy, R. Ammonium ions inhibit nitrogen fixation but do not affect heterocyst frequency in the bloom-forming cyanobacterium Nodularia spumigena strain AV1. *Microbiology* **2007**, *153*, 3704–3712. [CrossRef]

74. Klawonn, I.; Nahar, N.; Walve, J.; Andersson, B.; Olofsson, M.; Svedén, J.; Littmann, S.; Whitehouse, M.J.; Kuypers, M.; Ploug, H. Cell-specific nitrogen-and carbon-fixation of cyanobacteria in a temperate marine system (Baltic Sea). *Environ. Microbiol.* **2016**, *18*, 4596–4609. [CrossRef]

75. Jewell, W.J.; Kulasooriya, S. The relation of acetylene reduction to heterocyst frequency in blue-green algae. *J. Exp. Bot.* **1970**, *21*, 874–880. [CrossRef]

76. Chan, F.; Pace, M.L.; Howarth, R.W.; Marino, R.M. Bloom formation in heterocystic nitrogen-fixing cyanobacteria: The dependence on colony size and zooplankton grazing. *Limnol. Oceanogr.* **2004**, *49*, 2171–2178. [CrossRef]

77. Zhang, X.; Jia, X.; Yan, L.; Wang, J.; Kang, X.; Cui, L. Cyanobacterial nitrogen fixation influences the nitrogen removal efficiency in a constructed wetland. *Water* **2017**, *9*, 865. [CrossRef]

78. Karlson, A.M.; Duberg, J.; Motwani, N.H.; Hogfors, H.; Klawonn, I.; Ploug, H.; Svedén, J.B.; Garbaras, A.; Sundelin, B.; Hajdu, S. Nitrogen fixation by cyanobacteria stimulates production in Baltic food webs. *Ambio* **2015**, *44*, 413–426. [CrossRef] [PubMed]

Publisher's Note: MDPI stays neutral with regard to jurisdictional claims in published maps and institutional affiliations.

Article

The Nitrogen Stress-Repressed sRNA NsrR1 Regulates Expression of *all1871*, a Gene Required for Diazotrophic Growth in *Nostoc* sp. PCC 7120

Isidro Álvarez-Escribano, Manuel Brenes-Álvarez, Elvira Olmedo-Verd, Agustín Vioque and Alicia M. Muro-Pastor

Instituto de Bioquímica Vegetal y Fotosíntesis, Consejo Superior de Investigaciones Científicas and Universidad de Sevilla, 41092 Sevilla, Spain; isidroae9@hotmail.com (I.Á.-E.); mabreal92@gmail.com (M.B.-Á.); eolmedoverd@hotmail.com (E.O.-V.); vioque@us.es (A.V.)
* Correspondence: alicia@ibvf.csic.es; Tel.: +34-954489521

Received: 25 March 2020; Accepted: 27 April 2020; Published: 29 April 2020

Abstract: Small regulatory RNAs (sRNAs) are post-transcriptional regulators of bacterial gene expression. In cyanobacteria, the responses to nitrogen availability, that are mostly controlled at the transcriptional level by NtcA, involve also at least two small RNAs, namely NsiR4 (nitrogen stress-induced RNA 4) and NsrR1 (nitrogen stress-repressed RNA 1). Prediction of possible mRNA targets regulated by NsrR1 in *Nostoc* sp. PCC 7120 allowed, in addition to previously described *nblA*, the identification of *all1871*, a nitrogen-regulated gene encoding a protein of unknown function that we describe here as required for growth at the expense of atmospheric nitrogen (N_2). We show that transcription of *all1871* is induced upon nitrogen step-down independently of NtcA. All1871 accumulation is repressed by NsrR1 and its expression is stronger in heterocysts, specialized cells devoted to N_2 fixation. We demonstrate specific interaction between NsrR1 and the 5′ untranslated region (UTR) of the *all1871* mRNA, that leads to decreased expression of *all1871*. Because transcription of NsrR1 is partially repressed by NtcA, post-transcriptional regulation by NsrR1 would constitute an indirect way of NtcA-mediated regulation of *all1871*.

Keywords: regulatory RNA; cyanobacteria; post-transcriptional regulation; heterocyst

---

1. Introduction

Small non-coding RNAs (sRNAs) are relevant players in regulatory circuits affecting essentially every aspect of bacterial physiology. These types of molecules are usually post-transcriptional regulators fine-tuning the responses to different environmental conditions [1]. In fact, the interplay between the regulation exerted by transcription factors and that exerted by small RNAs produces complex regulatory circuits in the form of feed-forward loops (coherent or incoherent) involving a transcription factor, an sRNA and their regulated target(s) [2,3].

Global nitrogen regulation is controlled in cyanobacteria by NtcA, a protein that belongs to the CRP/FNR family of transcriptional regulators [4]. Direct binding of NtcA to the corresponding promoters accounts for regulation of many nitrogen-regulated genes [5–7]. However, the mechanisms involved in expression of some nitrogen-regulated genes whose promoters do not contain NtcA binding sites might involve the participation of other NtcA-regulated factor(s). NtcA has been shown to regulate expression of several sRNAs, some of them with a wide distribution among phylogenetically distant cyanobacteria [5,8,9]. Among these sRNAs, NsiR4 (nitrogen-stress inducible RNA 4) is involved in nitrogen assimilation control via regulation of IF7, the inactivating factor of the key enzyme glutamine synthetase [8]. NsrR1 (nitrogen-stress repressed RNA 1) modulates translation of NblA [10], a protein required for the degradation of phycobilisomes that provide amino acids as a source of nitrogen under nitrogen deficiency [11]. Transcription of both *gifA* (encoding IF7) and *nblA* is directly regulated by NtcA [5,12], therefore the post-transcriptional regulation exerted on these transcripts by NsiR4 and NsrR1, respectively, constitutes a second level of NtcA-mediated, indirect regulation.

In the absence of combined nitrogen, filamentous cyanobacteria such as model strain *Nostoc* sp. PCC 7120 differentiate heterocysts, specialized cells devoted to fixation of atmospheric nitrogen [13,14]. Differentiation of functional heterocysts is ultimately under control of NtcA, but also requires HetR, a regulator specifically involved in cellular differentiation. The nitrogen-regulated, HetR-dependent transcriptome includes transcripts for genes involved in specific aspects of heterocyst physiology, such as the sequential deposition of specialized envelopes or the fixation of nitrogen by the enzyme nitrogenase. The HetR-dependent transcriptome also includes non-coding transcripts, both antisense and small RNAs, that would participate in the metabolic reprogramming that takes place in heterocysts [15,16], again pointing to the relevance of post-transcriptional regulation on cyanobacterial physiology.

In this work, we identify *all1871* as a gene required for heterocyst function and describe its regulation by NsrR1. Expression of *all1871* is induced upon nitrogen step down, but its induction does not require NtcA or HetR. We verify that NsrR1 regulates accumulation of All1871 at the post-transcriptional level by its interaction with the 5'-UTR of *all1871*.

2. Materials and Methods

2.1. Strains and Growth Conditions

Cultures of wild-type and the different mutant derivatives of *Nostoc* sp. PCC 7120 (Table S1) were bubbled with an air/CO_2 mixture (1% *v/v*) and grown photoautotrophically at 30 °C in BG11 medium [17] containing ferric citrate instead of ammonium ferric citrate, lacking $NaNO_3$ and containing 6 mM NH_4Cl, 10 mM $NaHCO_3$, and 12 mM *N*-tris(hydroxymethyl)methyl-2-aminoethanesulfonic acid-NaOH (TES) buffer (pH 7.5). Nitrogen deficiency was induced by removal of combined nitrogen. Occasionally, 17.6 mM $NaNO_3$ was used as nitrogen source. Solid media were solidified with 1% Difco Agar. Mutant strains were grown in the presence of appropriate antibiotics at the following concentrations: streptomycin (Sm) and spectinomycin (Sp), 2–3 µg/mL each (liquid medium) or 5 µg/mL each (solid medium), neomycin (Nm), 5 µg/mL (liquid medium) or 25 µg/mL (solid medium). *Escherichia coli* strains (Table S1) were grown in Luria-Bertani (LB) medium, supplemented with appropriate antibiotics.

2.2. Reporter Assays for In Vivo Verification of Targets

For the experimental target verification in *E. coli*, we used a previously described reporter system [18] and the superfolder green fluorescent protein (sfGFP) plasmid pXG10-SF [19]. The 5'-UTR of *all1871*, from the transcriptional start site (TSS) at position −137 with respect to the initiation codon (coordinate 2234072) to 60 nucleotides within the coding region, containing the predicted NsrR1 interaction sequence, was amplified from genomic DNA using oligonucleotides 247 and 248 (see Table S2 for oligonucleotide sequences and description). The information about the TSS was taken from [5]. The corresponding polymerase chain reaction (PCR) product was digested with NsiI and XbaI and cloned into the vector pXG10-SF digested with NsiI and NheI, resulting in plasmid pIAE9, bearing a translational fusion of a truncated All1871 protein with sfGFP (see Table S3 for plasmid descriptions). For NsrR1 expression in *E. coli*, plasmid pAVN1 [10] was used (Table S3).

Mutation U51G (Mut-51) was introduced in NsrR1 by overlapping PCR using primer pairs 197 and 296, and 198 and 295, and cloned as described for the wild type version [10] generating plasmid pIAE20 (Table S3). The compensatory mutation in the 5'-UTR of *all1871* (Comp-51) was generated in the same way with primer pairs 247 and 304, and 303 and 248, and cloned as described above for the wild-type version resulting in plasmid pIAE22 (Table S3). The sequences of inserts in plasmids containing NsrR1 and *all1871*-sf*gfp* fusions are shown in Tables S4 and S5, respectively.

For testing various combinations of both plasmids, these were introduced into *E. coli* DH5α. Plasmid pJV300 [20] was used as a control expressing an unrelated RNA. Plasmid pXG0 [18] was used as control for background fluorescence. Fluorescence measurements were done with a microplate reader (Varioskan) using liquid cultures from eight individual colonies bearing each combination of plasmids, and normalized to the OD_{600} of each culture as described previously [21]. Fluorescence was also visualized in *E. coli* cells plated on solid LB medium by excitation with a 302-nm wavelength lamp.

2.3. RNA Isolation, Northern Blot and Primer Extension Analysis

RNA samples were isolated from cells collected at different times after removing combined nitrogen (ammonium) from the media. Alternatively, cells were grown in media lacking combined nitrogen and RNA was isolated from cells collected at different times after the addition of 10 mM NH_4Cl and 20 mM TES buffer. Total RNA was isolated using hot phenol as described [22] with modifications [9]. Northern blot hybridization was performed as previously described [23,24]. Strand-specific ^{32}P-labelled probes for Northern blot were prepared with Taq DNA polymerase using a PCR fragment as template in a reaction with $[\alpha\text{-}^{32}P]$dCTP and one single oligonucleotide as primer (corresponding to the complementary strand of the sRNA or mRNA to be detected). Hybridization to *rnpB* [25] was used as loading and transfer control. Hybridization signals were quantified on a Cyclone Storage Phosphor System with Optiquant software (PerkinElmer). Primer extension analysis of 5′ ends of *all1871* was carried out as previously described [23] using 5 µg of total RNA and oligonucleotide 161 labeled with $[\gamma\text{-}^{32}P]$ATP.

2.4. In Vitro Synthesis and Labelling of RNA

The DNA templates for the in vitro transcription of NsrR1 and *all1871* 5′-UTR RNA were generated by PCR with a forward primer that includes a T7 promoter sequence and three extra Gs upstream the 5′-end of the coded RNA, and a reverse primer corresponding to the 3′ end of the RNA (see Tables S2 and S6). The *all1871* 5′-UTR fragment extends from the TSS at position −137 to 60 nucleotides downstream the translational start. RNA transcripts were generated with the MEGAscript High-Yield Transcription Kit (AM1333, Ambion). After transcription, RNAs were treated with DNase I and purified by phenol and chloroform extraction, ethanol-precipitated at −20 °C, and washed with 70% ethanol. In vitro transcribed RNAs were 5′-labelled and purified as described [10].

2.5. In Vitro Structure Probing and Footprinting

We mixed 0.1 pmol (about 50,000 cpm) of labeled NsrR1 RNA in 7 µL with 2 pmol of unlabelled *all1871* 5′-UTR RNA, denatured for 1 min at 95 °C and chilled on ice for 5 min, followed by the addition of 1 µL of 1 mg/mL yeast RNA (Ambion AM7118) and 1 µL of 10× structure buffer (Ambion). The samples were incubated further for 15 min at 37 °C. Treatment with RNase T1, RNase A or lead(II) acetate were performed as described [10].

An alkaline ladder was obtained by incubating 0.2 pmol of 5′-labelled RNA at 95 °C for 3 min in 7.5 µL of alkaline hydrolysis buffer (Ambion) containing 1.5 µg of yeast RNA (Ambion AM7118). Reactions were stopped by the addition of 15 µL of denaturing formamide loading buffer.

RNase T1 G ladders were obtained by incubating 0.1 pmol of 5′-labelled RNA and 1 µL of 1 mg/mL yeast RNA (Ambion AM7118) in 9 µL sequencing buffer (Ambion) for 10 min at 50 °C, followed by the addition of 1 µL of 0.1 U/mL RNase T1 (Ambion AM2283) and incubation at room temperature for 15 min. Reactions were stopped by the addition of 20 µL of Inactivation/Precipitation buffer (Ambion) and incubation at −20 °C for 15 min. The precipitate was washed with 70% ethanol and resuspended in 3–7 µL of denaturing formamide loading buffer.

All samples were run on 10% polyacrylamide, 7 M urea gels and bands visualized with a Cyclone Storage Phosphor System (PerkinElmer).

2.6. Expression and Purification of Protein All1871 and Western Blot

To produce His-tagged All1871 protein, the *all1871* gene was amplified using *Nostoc* DNA as template and primers 335 and 336, and the PCR product was cloned in vector pET-28a (+) (Novagen) using NcoI and XhoI, producing plasmid pIAE30. Plasmid pIAE30, containing downstream a T7 polymerase-dependent promoter the *all1871* gene fused at the 3′ end to a sequence encoding a His_6-tag, was transferred by electroporation to *E. coli* BL21-(DE3)-RIL, in which the gene encoding T7 RNA polymerase is under the control of an isopropyl-β-D-1-thiogalactopyranoside (IPTG)-regulated promoter. A 25-mL pre-inoculum of this strain was grown overnight in LB medium supplemented with chloramphenicol and kanamycin and used to inoculate 275 mL of the same medium. The culture was incubated at 37 °C until OD_{600} = 0.6. Recombinant All1871 expression was induced by the addition of 1 mM IPTG. After 4 h at 37 °C, cells were collected and resuspended in 20 mM sodium phosphate buffer (pH 7.2) containing 6 M urea, 500 mM NaCl, 5 mM imidazole and 1 mM phenylmethylsulfonyl fluoride (6 mL/g of cells). Cells were broken by sonication and after centrifugation at 15,000× *g* (15 min, 4 °C), and the His_6-All1871 protein was purified from the supernatant by chromatography through a 1-mL HisTrap HP column (GE Healthcare), using an imidazole gradient in the same buffer to elute the retained proteins. An additional purification step was carried out by size-exclusion chromatography in a HiLoad 16/60 Superdex 75 column (Pharmacia) using 20 mM sodium phosphate buffer (pH 7.2) containing 2 M urea and 150 mM NaCl. A total amount of 3.5 mg of purified protein was used in seven subcutaneous injections of a rabbit to produce antibodies in the Animal Production and Experimentation Center, Universidad de Sevilla (Seville, Spain). Antiserum was recovered several times up to five months after the first injection and stored at −80 °C until used. Antibodies specific for All1871 were purified from the serum by affinity chromatography on immobilized His-tagged All1871 using the AminoLink® Plus Immobilization Kit (ThermoFisher, Waltham, MA, USA) following the manufacturer's instructions.

For Western blot analysis, *E. coli* cells were resuspended in sodium dodecyl sulphate-polyacrylamide gel electrophoresis (SDS-PAGE) loading buffer and the proteins fractionated on 15% SDS-PAGE. Antibodies against All1871 (see above), GFP (Roche) and *E. coli* GroEL (SIGMA-ALDRICH) were used. The ECL Plus immunoblotting system (GE Healthcare) was used to detect the different antibodies using anti-rabbit (SIGMA-ALDRICH) or anti-mouse (Bio-Rad) horse radish peroxidase conjugated secondary antibodies.

For Western blot analysis of *Nostoc* proteins, soluble fractions of cell-free extracts were used. Cells from 25 mL of culture were resuspended in 500 µL of Tris-HCl buffer (pH 8) containing 2 mM β-mercaptoethanol and protease inhibitor cocktail (cOmplete™ ultra tablets, EDTA-free, Roche) in an Eppendorf 1.5 mL tube. A volume of approximately 75 µL of glass beads (0.25–0.3 mm diameter) was added and the suspension subjected to seven cycles of 1 min vortex followed by 1 min on ice. The resulting extract was centrifuged 3 min at 3000× *g* at 4 °C and the supernatant further centrifuged for 30 min at 16,000× *g* at 4 °C. The supernatant of the last centrifugation constitutes the soluble fraction.

2.7. Generation and Complementation of all1871 Mutant Strain

To generate a strain lacking *all1871*, two overlapping fragments were amplified by PCR using as template genomic DNA with oligonucleotides 162 and 163 and oligonucleotides 164 and 165, respectively (Table S2). The resulting products were then used as templates for a third PCR with oligonucleotides 162 and 165 resulting in deletion of sequences corresponding to *all1871* and the generation of a unique XhoI site between the two amplified fragments. The fragment was cloned into pSpark (Canvax Biotech), rendering pSAM318 and its sequence was verified (Eurofins Genomics). After digestion with BamHI at the sites provided by oligonucleotides 162 and 165, the fragment was cloned into BamHI-digested *sacB*-containing Sm^RSp^R vector pCSRO [26], rendering pSAM324. A Nm^R gene was excised from pRL278 [27] as a SalI-XhoI fragment and cloned into the XhoI site in pSAM324, rendering pSAM326, which was transferred to *Nostoc* sp. strain PCC 7120 by conjugation [28], with selection for resistance to Nm. Cultures of the exconjugants obtained were used to select for clones resistant to 5% sucrose [29], and individual sucrose resistant colonies were checked by PCR using flanking oligonucleotides 162 and 180. Clones bearing the *all1871* region interrupted by the Nm^R gene were named *all1871*::Nm.

Plasmid pIAE65 was constructed to express *all1871* from the *trc* promoter. A PCR fragment was amplified using oligonucleotides 247 and 735, digested with NsiI and XhoI and cloned into NsiI and XhoI-digested pMBA37 [15]. pIAE65 was transferred to strain *all1871*::Nm by conjugation with selection of Sm^R Sp^R cells.

2.8. Construction of a Strain Bearing a Translational Fusion all1871-gfpmut2

In order to analyze spatial expression of *all1871* along *Nostoc* filaments, plasmid pELV75 (Table S3) was constructed. A fragment containing the sequence from position −200 with respect to the TSS plus the entire *all1871* gene was amplified using oligonucleotides 451 and 596. The fragment was digested with ClaI and EcoRV and cloned in pCSAM147 [30], in frame with the *gpfmut2* gene, rendering pELV73. The EcoRI fragment from pELV73 containing the *all1871-gfpmut2* translational fusion was cloned into pCSV3 [31] producing pELV75, that was transferred to *Nostoc* by conjugation with selection for Sm^{R}Sp^R cells.

2.9. Microscopy

Fluorescence of *Nostoc* sp. PCC 7120 filaments carrying plasmid pELV75 growing on top of solidified nitrogen-free medium, was analyzed by confocal microscopy and quantified as described [32] using a Leica HCX PLAN-APO 63× 1.4 NA oil immersion objective attached to a Leica TCS SP2 laser-scanning confocal microscope. Samples were excited at 488 nm by an argon ion laser and the fluorescent emission was monitored by collection across windows of 500–538 nm (GFP) and 630–700 nm (cyanobacterial autofluorescence). Filaments were stained with Alcian blue and visualized at the microscope as described [33].

3. Results

3.1. NsrR1 Interacts with all1871 mRNA 5′-UTR

NsrR1 (nitrogen stress-repressed RNA 1) is a small RNA previously described to post-transcriptionally regulate expression of *nblA* [10]. A prediction of mRNAs possibly regulated by NsrR1 also identified the 5′-UTR of gene *all1871* as likely interacting with NsrR1 [10]. *all1871* would encode a conserved protein of unknown function. The high score of the prediction (*p*-value = 6.11 × 10^{-9}), together with the conservation of predicted interactions between NsrR1 homologs and *all1871* homologs in most cyanobacteria encoding NsrR1 (Figure S1), prompted us to further analyze a possible post-transcriptional regulation of *all1871* by NsrR1.

Figure 1 shows the predicted interaction between NsrR1 from *Nostoc* sp. PCC 7120 and the 5′-UTR of *all1871*, that extends from position −40 to −3, overlaps the translation initiation region and therefore is expected to affect translation of the mRNA (Figure 1A). To verify the interaction between NsrR1 and the mRNA of *all1871*, we used a heterologous reporter assay in *E. coli* [18], in which the 5′-UTR of *all1871*, from the TSS at position −137 with respect to the initiation codon (Mitschke et al., 2011), plus 60 nucleotides of the coding sequence of *all1871* were fused to the sf*gfp* gene and co-expressed in *E. coli* with NsrR1 or with a control RNA. Accumulation of the GFP protein was measured in *E. coli* cells bearing different combinations of plasmids. The translation initiation region of *all1871* was functional in *E. coli* resulting in significant GFP protein accumulation and GFP fluorescence (Figure 1B–D). The GFP fluorescence of cells bearing the *all1871*::sf*gfp* fusion (and the amount of GFP protein) decreased to less than 20% of the control when NsrR1 was co-expressed, indicating a direct interaction between NsrR1 and the 5′-UTR of *all1871*, which affects translation (Figure 1B–C).

To verify the interaction at the predicted site, a point mutation was introduced in NsrR1 affecting the predicted helix between NsrR1 and the 5′-UTR of *all1871* (nucleotide 51, U to G). A compensatory mutation was introduced in the corresponding positions of the 5′-UTR of *all1871* (nucleotide −13 with respect to the start codon, G to C) and different combinations of the wild-type and mutated versions of both NsrR1 and 5′-UTR of *all1871* were co-expressed. Mutation of nucleotide 51 in NsrR1 reduced the interaction between NsrR1 and the 5′-UTR of *all1871*, as suggested by the lower degree of fluorescence reduction with respect to the control (Figure 1C). When the mutated version of NsrR1 was combined with the mutated version of the mRNA containing the compensatory change, base pairing was restored resulting in a stronger fluorescence reduction than with wild-type NsrR1. Because of the long region of interaction, mutation of one single nucleotide produces a small effect, but differences observed are statistically significant. These data support a direct interaction of NsrR1 with the 5′-UTR of the *all1871* mRNA that affects translation of All1871.

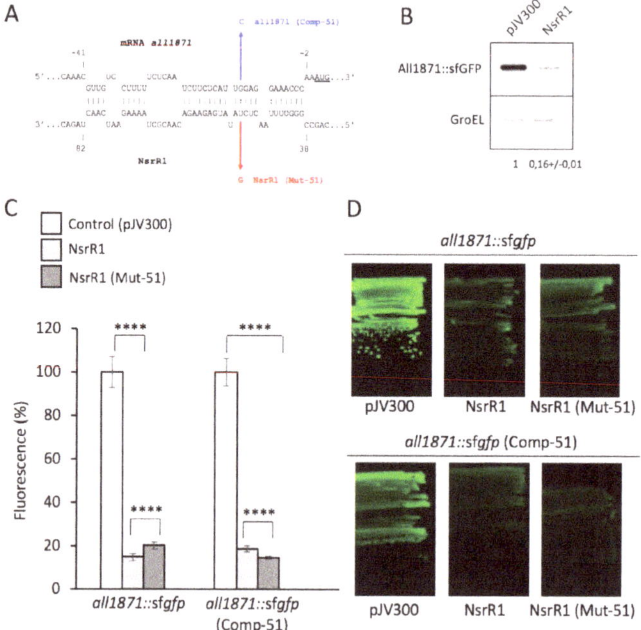

Figure 1. Verification of NsrR1 interaction with the 5′-UTR of *all1871* using an in vivo reporter system. (**A**) Predicted interaction between NsrR1 and the 5′-UTR of *all1871* according to IntaRNA [34]. *all1871* nucleotides are numbered with respect to the start of the coding sequence (AUG start codon is underlined). The mutation introduced in NsrR1 at position 51 (U to G, Mut-51) and the corresponding compensatory mutation in *all1871* 5′-UTR position −13 (G to C, Comp-51) are indicated in red and blue, respectively. (**B**) Accumulation of GFP protein in *E. coli* DH5α cells bearing an *all1871::sfgfp* fusion combined with plasmid pJV300 (encoding a control RNA) or with a plasmid encoding NsrR1. Western blots were carried out using antibodies against GFP or GroEL. Numbers at the bottom of the image indicate relative GFP levels with respect to control after normalization with GroEL (average of two experiments). (**C**) Fluorescence measurements of *E. coli* DH5α cultures bearing combinations of plasmids expressing different versions of NsrR1 (wild type or Mut-51) and *all1871::sfgfp* fusions (wild type or Comp-51). Plasmid pJV300 (encoding a control RNA) was used as control. The data are presented as the mean +/− standard deviation of cultures from eight independent colonies after subtraction of fluorescence in cells bearing pXG0. Fluorescence is normalized to the OD$_{600}$ of each culture. *T*-test *p*-value < 0.0001****. (**D**) Fluorescence intensities of *E. coli* cells bearing *all1871::sfgfp* fusions (wild type or Comp-51) combined with different versions of NsrR1 (wild type or Mut-51) or with the control plasmid pJV300. Strains growing on LB agar plates were photographed under ultraviolet (UV) light.

In addition, we have studied the interaction between NsrR1 and the *all1871* mRNA by in vitro footprinting analysis. ^{32}P-labelled NsrR1 was incubated with unlabeled *all1871* mRNA (a fragment extending from positions −137 to +60 with respect to the start of the coding sequence) and probed with RNase T1, RNase A or lead(II) acetate (Figure 2A). A clear footprint was detected between positions 45 and 62 of NsrR1 (highlighted in the secondary structure model, Figure 2B), in good agreement with the bioinformatic prediction (Figure 1A). These in vitro results, therefore, confirm the in vivo results obtained from the analysis in *E. coli* using the sfGFP fusion system.

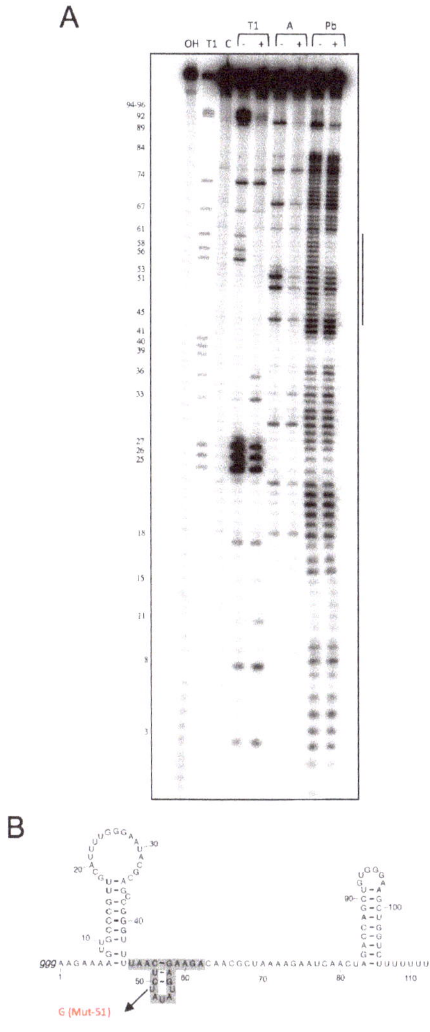

Figure 2. In vitro footprinting assay of the interaction between NsrR1 and *all1871* 5′-UTR. (**A**) RNase T1, RNase A or Pb(II) footprinting of the interaction between NsrR1 and the 5′-UTR of *all1871*. -/+ above lanes indicate absence/presence of *all1871* 5′ UTR. The protected area is indicated by a vertical bar. NsrR1 was 5′ end-labelled. C, untreated control; OH, alkaline ladder; T1, RNase T1 ladder. Nucleotide positions of NsrR1 are shown on the left. (**B**) The nucleotides in NsrR1 involved in the interaction with the *all1871* 5′-UTR are indicated in grey on the previously described secondary structure model of NsrR1 [10]. The nucleotide changed (U to G) in version Mut-51 of NsrR1 is also indicated.

3.2. Expression of all1871 Is Regulated by Nitrogen Availability But Does Not Require NtcA or HetR

According to global transcriptomic analyses, accumulation of the *all1871* transcript is induced in response to nitrogen deficiency [16,35]. We have analyzed transcription from the TSS identified at position 2234072r [5] in the wild type strain and in two mutant derivatives, *ntcA* mutant strain CSE2 [36] and *hetR* mutant strain DR884a [27]. Although in all strains there is significant expression in the presence of ammonium, transcription is similarly induced upon nitrogen step down in the wild type and in both mutant strains, suggesting the induction of transcription does not require NtcA or HetR (Figure 3). According to primer extension analysis, induction is transient both in the wild type and the *ntcA* strain, with a decreased amount of transcript at 24 h *vs.* 8 h after nitrogen removal, but in the *hetR* mutant the accumulation of transcript continues to be strong even at 24 h after nitrogen removal.

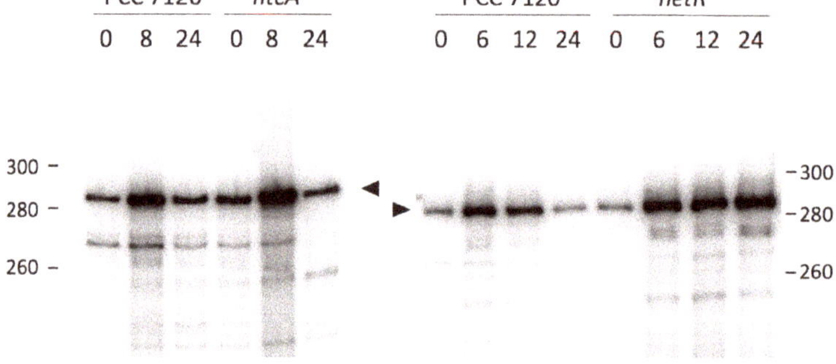

Figure 3. Nitrogen-regulated expression of *all1871*. Primer extension analysis of *all1871* transcripts in *Nostoc* sp. PCC 7120 and its *ntcA* mutant derivative CSE2 (left panel) or in *Nostoc* sp. PCC 7120 and its *hetR* mutant derivative DR884a (right panel). Expression was analyzed in cells grown in the presence of ammonium and transferred to medium containing no source of combined nitrogen for the number of hours indicated. Size markers (nucleotides) are indicated. Triangles point to the products corresponding to the 5′ end at position 2234072r [5].

3.3. Expression of all1871 Is Reduced by NsrR1 in Nostoc sp. PCC 7120

In order to study in vivo in *Nostoc* the possible effect of NsrR1 on *all1871* expression, we used a mutant strain that lacks NsrR1 (Δ*nsrR1*) [10]. Cells of strain Δ*nsrR1* have no apparent difference in growth with respect to wild-type cells in media containing nitrate, ammonia, or lacking a source of combined nitrogen [10].

We have previously shown that transcription of NsrR1 is only partially (and transiently) repressed upon nitrogen step down, whereas maximal expression of NsrR1 is achieved upon ammonium addition to cells growing in the absence of combined nitrogen [10]. Therefore, to maximize the difference between wild-type cells expressing NsrR1 and cells lacking NsrR1 (Δ*nsrR1* strain), we chose to analyze cells grown in the absence of combined nitrogen (N$_2$) and compared them with cells grown in the absence of combined nitrogen to which ammonium was added and incubation continued for 4, 8 or 24 h (Figure 4).

As previously described, expression of NsrR1 was induced and reached its highest level 8 h after ammonium addition (Figure 4A, middle panel). Concomitantly, expression of *all1871* was repressed in the wild type, with a minimum in the sample corresponding to 8 h after ammonium addition (Figure 4A, upper panel). Furthermore, repression of *all1871* was weaker in the strain lacking NsrR1, suggesting NsrR1 has a significant negative effect on the accumulation of *all1871* mRNA in *Nostoc*, consistent with the interaction observed in the *E. coli* assay between NsrR1 and the *all1871* mRNA. The effects of NsrR1 on the accumulation of All1871 protein were also analyzed by Western blot using antibodies we have generated against purified recombinant All1871 protein. The levels of All1871 protein observed in extracts of strain Δ*nsrR1* were about five-fold higher than those in the wild-type strain in the absence of combined nitrogen (Figure 4B). Consistent with the mRNA levels (Figure 4A), the reduction of the amount of All1871 protein in response to ammonium addition was stronger in the wild-type strain than in the Δ*nsrR1* strain lacking NsrR1.

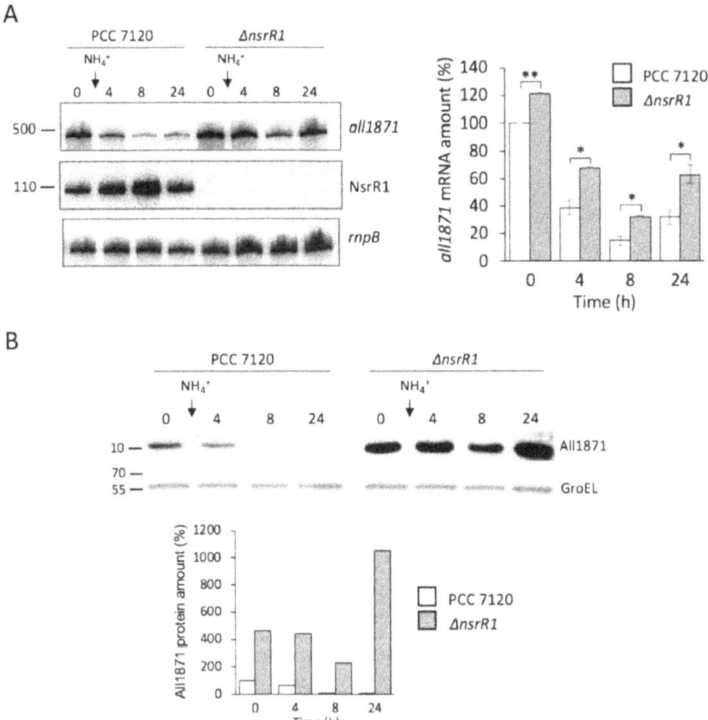

Figure 4. NsrR1 affects nitrogen-regulated expression of *all1871*. (**A**) Expression of *all1871* was analyzed by Northern blot in *Nostoc* sp. PCC 7120 and in a mutant strain lacking NsrR1 (Δ*nsrR1*). Cells were grown in nitrogen-free liquid medium for one week (steady-state N_2-fixing cultures) and 10 mM ammonium chloride was added to increase expression of NsrR1. Samples were taken before ammonium addition (0) and at the number of hours indicated after ammonium addition. The upper panel shows hybridization to the *all1871* probe. The middle panel shows hybridization to the probe for NsrR1. The lower panel shows hybridization to a probe for *rnpB* used as loading and transfer control. Quantification of *all1871* mRNA accumulation is indicated on the right, relative to the amount present in the wild type strain in N_2 using the amount of *rnpB* for normalization. Results from two technical replicates were averaged. *T*-test *p*-value < 0.05*; < 0.01**. (**B**) Accumulation of the All1871 protein was determined by Western blot in samples containing 40 µg of soluble fraction from cells analyzed in (**A**). Upper panels show detection of All1871. Lower panels show detection of GroEL, used as loading and transfer control. Quantification of All1871 protein is shown at the bottom, relative to the amount present in the wild type strain in N_2. Quantification was performed with ImageLab software (Bio-Rad) using the amount of GroEL for normalization. One representative experiment is shown.

3.4. All1871 Is Differentially Expressed in Heterocysts and Required for Diazotrophic Growth But Not for Heterocyst Differentiation

In order to obtain an insight into a possible function of All1871, we analyzed expression of *all1871* along filaments of *Nostoc*. We prepared plasmid pELV75 containing a segment from position −200 with respect to the TSS of *all1871* plus a translational fusion between the entire *all1871* gene and the *gfpmut2* gene. This plasmid was introduced in *Nostoc* by conjugation. Integration of the plasmid in the region encoding *all1871* leads to a partial duplication of this region in which the translational fusion is preceded by the natural context of *all1871* in the wild type strain (Figure 5A).

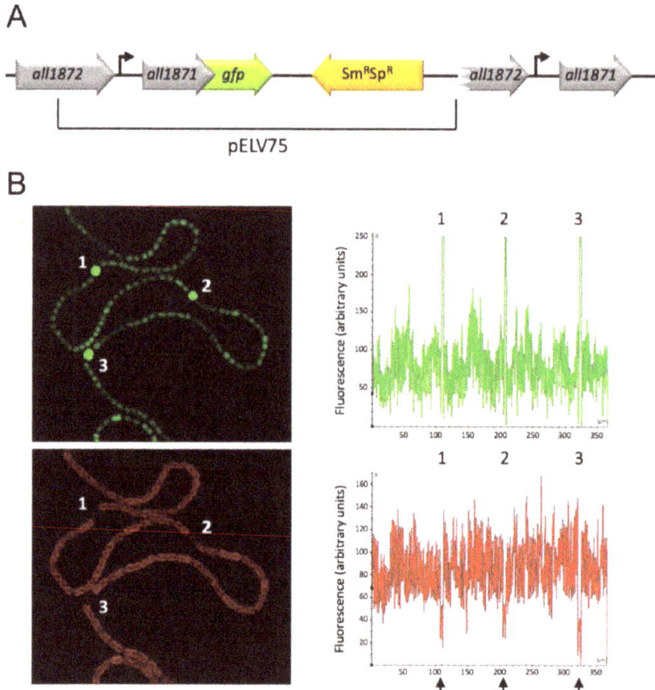

Figure 5. Expression of P$_{all1871}$-*all1871*-*gfp* along nitrogen-fixing filaments of *Nostoc* sp. PCC 7120 bearing pELV75. (**A**) Schematic representation of plasmid pELV75 integrated by single recombination in the chromosomal region encoding *all1871*. The location of the transcriptional start site of *all1871* (bent arrows) and the segment (from -336 with respect to the start codon of *all1871*) fused to *gfp* in plasmid pELV75 are depicted. Not drawn to scale. (**B**) Confocal fluorescence image of a filament growing on top of nitrogen-free medium is shown for the green channel (upper panel, GFP fluorescence) and red channel (lower panel, autofluorescence). Quantification of the green and red signals along the filament is shown on the right of each image. Mature heterocysts are indicated with numbers. The positions of lowest autofluorescence, corresponding to heterocysts, are indicated with black arrows.

GFP fluorescence was analyzed by confocal fluorescence microscopy of filaments growing on top of medium lacking combined nitrogen (Figure 5B). Quantification of the green signal (GPF) and the red autofluorescence along the filaments showed that although all cells showed green fluorescence, fluorescence peaks were associated with cells that had differentiated as heterocysts, as indicated by their larger size and reduced red autofluorescence.

We then prepared an *all1871* null mutant by interrupting the *all1871* gene with a Nm^R gene (Figure S2A). Complete segregation of mutant chromosomes was verified by PCR amplification and Western blot with antibodies against All1871 (Figure S2B–C). Cells of the *all1871*::Nm mutant were unable to grow on plates in the absence of combined nitrogen but showed no growth defect in the presence of combined nitrogen (nitrate) (Figure 6A). The mutation was complemented by introduction of a plasmid bearing *all1871* expressed from the *trc* promoter from *E. coli*, that provides constitutive expression in *Nostoc* (see e.g., [37]), leading to partial recovery of the wild-type phenotype, as indicated by the greenish growth of complemented cells vs. the blue-green growth of wild-type cells (Figure 6A).

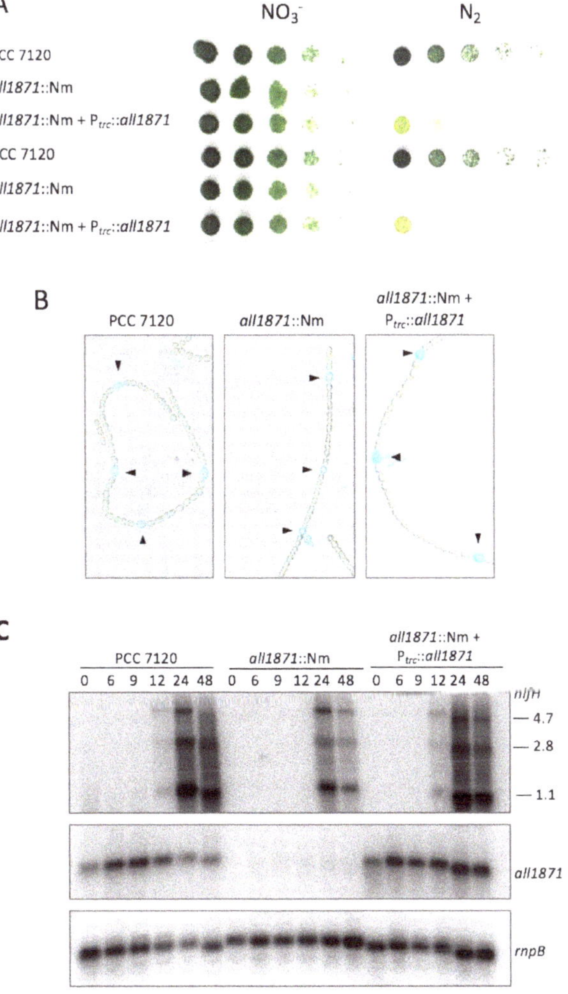

Figure 6. *all1871* mutants are defective in diazotrophic growth. (**A**) Cells were grown in the presence of nitrate, washed and resuspended in BG11$_0$ at an OD$_{750}$ = 0.3. Five-fold serial dilutions of liquid cultures of wild type, insertional mutant *all1871*::Nm and complemented mutant strain (*all1871*::Nm + P$_{trc}$::*all1871*) were prepared and 10 µL of each dilution plated on BG11$_0$ plates lacking nitrogen (N$_2$) or containing nitrate (NO$_3^-$). Two different clones of each strain were analyzed. Pictures were taken after 10 days (NO$_3^-$) or 13 days (N$_2$) of incubation at 30 °C. (**B**) Alcian blue staining of heterocyst polysaccharides in filaments of *Nostoc* sp. PCC 7120, insertional mutant *all1871*::Nm and complemented mutant strain (*all1871*::Nm + P$_{trc}$::*all1871*), 26 h after combined nitrogen removal. (**C**) Northern blots with RNA isolated from the indicated strains at different time points (indicated in hours) after nitrogen removal and hybridized with probes for *nifH* (upper panel), *all1871* (middle panel) and *rnpB* (lower panel) as loading control. Sizes of the *nifH* transcripts are indicated on the right in kb.

We wondered whether the observed defects in diazotrophic growth of the *all1871*::Nm strain were due to lack of heterocyst differentiation or heterocyst function. Because steady-state nitrogen-fixing cultures could not be obtained, we analyzed morphological differentiation of heterocysts 26 h after combined nitrogen removal by staining the cells with Alcian blue, a molecule that binds to the external polysaccharide layer of heterocysts. Figure 6B shows the presence of Alcian blue-stained cells in a regular pattern similar to the wild type in the mutant strain *all1871*::Nm as well as in the complemented strain *all1871*::Nm + P_{trc}-*all1871*, indicating differentiation of heterocysts took place in both strains, at least to the relatively initial stage in which polysaccharides are deposited outside the outer membrane of the vegetative cells undergoing differentiation into heterocysts. As an indication of heterocyst maturity, transcription of the *nifHDK* genes encoding nitrogenase was also analyzed by Northern blot (Figure 6C). Transcription of *nifHDK* seems delayed and reduced in the *all1871*::Nm mutant. Again, complementation with a plasmid bearing *all1871* expressed from the *trc* promoter from *E. coli* leads to recovery of the wild-type timing of *nifHDK* expression.

Finally, we have measured the nitrogenase activity of the different strains (Table 1). Strain *all1871*::Nm had no detectable nitrogenase activity in either oxic or anoxic conditions. This result excludes the possibility that the absence of activity in strain *all1871*::Nm was due to a defect in the maturation of heterocyst envelopes that results in the presence of inactivating O_2 amounts inside the heterocyst.

Table 1. Nitrogenase activity of different *Nostoc* strains.

Strain	Nitrogenase Activity [1] (μmol Ethylene·h^{-1}·mg Chl^{-1})	
	Oxic Conditions	Anoxic Conditions
Nostoc sp. PCC 7120	10.50 ± 3.49	10.64 ± 2.05
all1871::Nm	0.00 ± 0.00	0.00 ± 0.00
all1871::Nm + P_{trc}::*all1871*	3.66 ± 1.67	5.05 ± 3.41

[1] Nitrogenase activity was measured in cultures grown with nitrate and transferred for 24 h to nitrogen-free medium. Data are the average and standard deviation of assays performed with two independent cultures of *Nostoc* sp. PCC 7120 and strain *all1871*::Nm + P_{trc}::*all1871* or eight independent cultures of strain *all1871*::Nm.

4. Discussion

Small RNAs are important components in regulatory networks that involve classical transcription factors. For instance, in *E. coli*, members of the CRP/FNR family of transcriptional regulators are known to control the expression of several small RNAs, all of them contributing to the regulatory effects exerted by these two major transcription factors [2]. Cyanobacterial transcription factors also regulate the expression of small RNAs that exert regulatory functions involved in the adaptation to different environmental situations [38]. For instance, the transcriptional regulator RpaB and small RNA PsrR1 constitute a feed-forward loop controlling acclimation to different light intensities [39]. Accumulating evidence concerning nitrogen-regulated non-coding RNAs indicates this type of molecules is involved in NtcA-mediated post-transcriptional regulation [5,9]. One case analyzed in detail is NsiR4, whose transcription is induced in response to nitrogen deficiency and is involved in post-transcriptional regulation of glutamine synthetase [8]. Expression of another sRNA, NsrR1, is repressed by NtcA in response to nitrogen deficiency. Among predicted targets of NsrR1 are *nblA* [10], and *all1871*, whose interaction with NsrR1 we describe in this work.

Interaction between homologs of NsrR1 and homologs of *all1871* is conserved in many cyanobacteria encoding NsrR1 (Figure S1), consistent with the observation that *all1871* has the highest score in CopraRNA predictions of candidates to be regulated by NsrR1 [10]. Using an in vivo reporter system established in *E. coli* we demonstrate that NsrR1 represses translation, leading to reduced expression of an *all1871*::sfGFP fusion in the presence of NsrR1. A direct interaction of NsrR1 with the predicted region in the 5′-UTR of *all1871* was also verified by a point mutation in the region involved (Figure 1) and by in vitro footprinting experiments (Figure 2).

We have demonstrated higher accumulation of *all1871* transcripts in a strain of *Nostoc* sp. PCC 7120 that lacks NsrR1 (Δ*nsrR1*) than in wild-type cells cultured in the absence of combined nitrogen (Figure 4A). Upon addition of ammonium, that induces NsrR1 expression, the amount of *all1871* mRNA is strongly reduced in the wild type but not in the Δ*nsrR1* strain. Similarly, the amount of All1871 protein, detected with a specific antibody (Figure 4B), is much higher in cells lacking NsrR1 than in the wild-type strain. Upon the addition of ammonium, the amount of All1871 protein is reduced in the wild type, becoming barely detectable 8 h after ammonium addition, but only slightly reduced in the Δ*nsrR1* strain. The magnitude of the repression upon ammonium addition is higher at the protein level than at the transcript level, in agreement with NsrR1 inhibiting translation through its interaction with the 5′-UTR of *all1871*. The reduced accumulation of *all1871* mRNA in the presence of NsrR1 can be attributed to destabilization of the mRNA when translation is inhibited.

Upregulation of *all1871* upon nitrogen deprivation would be mediated, at least in part, by alleviating the repression exerted by NsrR1, whose transcription is repressed upon nitrogen stress. NsrR1 repression is only partially dependent on NtcA, with an additional factor possibly involved [10], explaining that up-regulation of *all1871* is independent of NtcA (Figure 3). The observation that there is partial repression of *all1871* expression upon ammonium addition even in the Δ*nsrR1* mutant strain (Figure 4) points to some additional factor(s) being involved in regulation of *all1871*. In any case, because the interaction between NsrR1 and *all1871* takes place at the translation initiation region, even if another regulatory mechanism influences transcription of *all1871*, the accumulation of the All1871 protein would be ultimately regulated by nitrogen availability through NsrR1 (Figure 7).

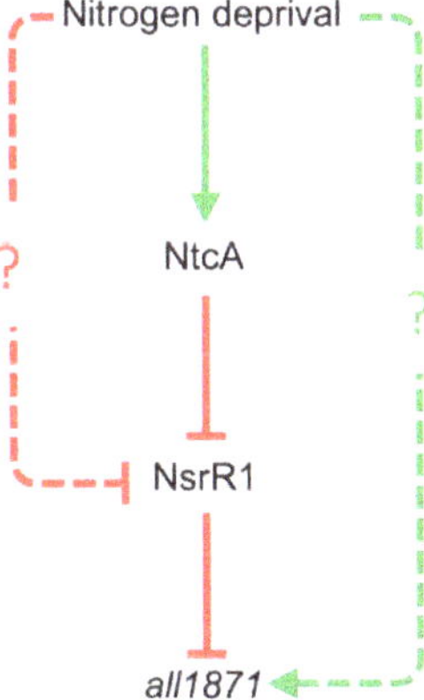

Figure 7. Schematic representation of elements involved in the regulation of expression of *all1871*. Green arrows represent positive effects. Red arrows represent negative effects. Dashed lines represent mechanisms operated by currently unknown factors (question marks).

By means of a translational fusion to the *gfp* gene we have shown higher expression of *all1871* in heterocysts than in vegetative cells (Figure 5). We have also shown that expression of *all1871* does not require HetR (Figure 3) suggesting it is not heterocyst-specific. Taken together these observations point to the operation of post-transcriptional regulatory mechanisms leading to differential accumulation of *all1871* transcripts in heterocysts *vs.* vegetative cells. Because NtcA is subjected to differential accumulation in heterocysts [31,40], the amount of NsrR1 is also expected to be different. A higher amount of NtcA in heterocysts would result in lower levels of NsrR1, and consequently increased translation of All1871 in these specialized cells. NsrR1 would thus contribute to differential expression in heterocysts of a transcript found to be required for diazotrophic growth. Whether the stronger accumulation of All1871-GFP protein in heterocysts (Figure 5) is only a consequence of reduced post-transcriptional regulation by NsrR1 or involves additional differential regulation in these specialized cells, is currently unknown.

All1871 is conserved in many unicellular and filamentous strains of cyanobacteria. Its function is unknown and it does not contain domains of known function. Strong accumulation in heterocysts together with the observation that its expression is regulated by NsrR1, a nitrogen-repressed sRNA, suggest a possible role in the response to nitrogen stress and/or in heterocyst function. A mutant lacking All1871 (*all1871*::Nm) could not grow in media lacking combined nitrogen (Figure 6A). The inability to grow in media lacking combined nitrogen of the *all1871*::Nm strain is only partially complemented by constitutive expression of *all1871* from P$_{trc}$ (Figure 6A). Such phenotype could be explained by toxicity due to unregulated expression of *all1871* in this strain, in contrast to the transient induction observed in the wild type upon nitrogen step-down. It can also be speculated that constitutive expression of *all1871* in the complemented strain could lead to sequestration of NsrR1 affecting accumulation of other mRNAs also regulated by NsrR1, such as *nblA* [10]. However, it has been previously shown that cells lacking NsrR1 have no apparent phenotypic differences with respect to wild-type cells [10]. Despite its ability to develop Alcian blue-stained cells (Figure 6B), suggestive of heterocyst differentiation, and the induction of nitrogenase transcription (Figure 6C), *all1871*::Nm filaments lacked nitrogenase activity (Table 1), indicating that All1871 is required for heterocyst function through currently unknown mechanisms. We have observed that in the absence of combined nitrogen, filaments of the *all1871*::Nm strain are extensively fragmented in liquid media (not shown). Because exchange of metabolites between vegetative cells and heterocysts is a requisite for sustained nitrogen fixation, filament fragmentation would preclude diazotrophic growth of the *all1871*::Nm strain.

In summary, in this work we describe a new gene required for diazotrophic growth, although not for heterocyst differentiation, and characterize its regulation by NsrR1, a nitrogen-regulated small RNA involved in acclimation to nitrogen deficiency.

Supplementary Materials: The following are available online at http://www.mdpi.com/2075-1729/10/5/54/s1: Figure S1: Conservation in cyanobacteria of the predicted interaction between the mRNA of *all1871* and NsrR1; Figure S2: Construction of mutant *all1871*::Nm; Table S1: Strains; Table S2: Oligonucleotides; Table S3: Plasmids; Table S4: Sequences of inserts in plasmids containing NsrR1 used for verification in *E. coli*; Table S5: Sequences of inserts in the *all1871-sfgfp* fusion plasmids; Table S6: Sequences of templates used for in vitro transcription.

Author Contributions: Conceptualization, A.V. and A.M.M.-P.; Formal analysis, A.V. and A.M.M.-P.; Funding acquisition, A.M.M.-P.; Investigation, I.Á.-E., M.B.-Á., E.O.-V., A.V. and A.M.M.-P.; Methodology, A.V. and A.M.M.-P.; Supervision, A.V. and A.M.M.-P.; Writing—original draft, A.M.M.-P.; Writing—review and editing, I.Á.-E., M.B.-Á., A.V., E.O.-V. and A.M.M.-P. All authors have read and agreed to the published version of the manuscript.

Funding: This research was funded by Ministerio de Economía y Competitividad, grant number BFU2013-48282-C2-1-P, and by Agencia Estatal de Investigación (AEI), Ministerio de Economía, Industria y Competitividad, grant number BFU2016-74943-C2-1-P, both cofinanced by Fondo Europeo de Desarrollo Regional (FEDER). IA-E is the recipient of a predoctoral contract from Ministerio de Economía y Competitividad, Spain (BES-2014-068488). MB-A is the recipient of a predoctoral contract from Ministerio de Educación, Cultura y Deporte, Spain (FPU014/05123 and EST16-00088).

Acknowledgments: We thank M. Isabel Muro-Pastor (IBVF, CSIC-Universidad de Sevilla) for help with protein purification, and Alicia Orea (IBVF, CSIC- Universidad de Sevilla) for technical assistance with microscopy.

Conflicts of Interest: The authors declare no conflict of interest.

References

1. Wagner, E.G.; Romby, P. Small RNAs in bacteria and archaea: Who they are, what they do, and how they do it. *Adv. Genet.* **2015**, *90*, 133–208. [CrossRef]
2. Brosse, A.; Guillier, M. Bacterial small RNAs in mixed regulatory networks. *Microbiol. Spectr.* **2018**, *6*, RWR-0014-2017. [CrossRef]
3. Nitzan, M.; Rehani, R.; Margalit, H. Integration of bacterial small RNAs in regulatory networks. *Annu. Rev. Biophys.* **2017**, *46*, 131–148. [CrossRef]

4. Herrero, A.; Muro-Pastor, A.M.; Flores, E. Nitrogen control in cyanobacteria. *J. Bacteriol.* **2001**, *183*, 411–425. [CrossRef]

5. Mitschke, J.; Vioque, A.; Haas, F.; Hess, W.R.; Muro-Pastor, A.M. Dynamics of transcriptional start site selection during nitrogen stress-induced cell differentiation in *Anabaena* sp. PCC7120. *Proc. Natl. Acad. Sci. USA* **2011**, *108*, 20130–20135. [CrossRef]

6. Picossi, S.; Flores, E.; Herrero, A. ChIP analysis unravels an exceptionally wide distribution of DNA binding sites for the NtcA transcription factor in a heterocyst-forming cyanobacterium. *BMC Genom.* **2014**, *15*, 22. [CrossRef]

7. Giner-Lamia, J.; Robles-Rengel, R.; Hernández-Prieto, M.A.; Muro-Pastor, M.I.; Florencio, F.J.; Futschik, M.E. Identification of the direct regulon of NtcA during early acclimation to nitrogen starvation in the cyanobacterium *Synechocystis* sp. PCC 6803. *Nucleic Acids Res.* **2017**, *45*, 11800–11820. [CrossRef]

8. Klähn, S.; Schaal, C.; Georg, J.; Baumgartner, D.; Knippen, G.; Hagemann, M.; Muro-Pastor, A.M.; Hess, W.R. The sRNA NsiR4 is involved in nitrogen assimilation control in cyanobacteria by targeting glutamine synthetase inactivating factor IF7. *Proc. Natl. Acad. Sci. USA* **2015**, *112*, E6243–E6252. [CrossRef]

9. Brenes-Álvarez, M.; Olmedo-Verd, E.; Vioque, A.; Muro-Pastor, A.M. Identification of conserved and potentially regulatory small RNAs in heterocystous cyanobacteria. *Front. Microbiol.* **2016**, *7*, 48. [CrossRef]

10. Álvarez-Escribano, I.; Vioque, A.; Muro-Pastor, A.M. NsrR1, a nitrogen stress-repressed sRNA, contributes to the regulation of *nblA* in *Nostoc* sp. PCC 7120. *Front. Microbiol.* **2018**, *9*, 2267. [CrossRef]

11. Collier, J.L.; Grossman, A.R. A small polypeptide triggers complete degradation of light-harvesting phycobiliproteins in nutrient-deprived cyanobacteria. *EMBO J.* **1994**, *13*, 1039–1047. [CrossRef]

12. García-Domínguez, M.; Reyes, J.C.; Florencio, F.J. NtcA represses transcription of *gifA* and *gifB*, genes that encode inhibitors of glutamine synthetase type I from *Synechocystis* sp. PCC 6803. *Mol. Microbiol.* **2000**, *35*, 1192–1201. [CrossRef]

13. Flores, E.; Herrero, A. Compartmentalized function through cell differentiation in filamentous cyanobacteria. *Nature Rev. Microbiol.* **2010**, *8*, 39–50. [CrossRef]

14. Muro-Pastor, A.M.; Hess, W.R. Heterocyst differentiation: From single mutants to global approaches. *Trends Microbiol.* **2012**, *20*, 548–557. [CrossRef]

15. Olmedo-Verd, E.; Brenes-Álvarez, M.; Vioque, A.; Muro-Pastor, A.M. A heterocyst-specific antisense RNA contributes to metabolic reprogramming in *Nostoc* sp. PCC 7120. *Plant Cell Physiol.* **2019**, *60*, 1646–1655. [CrossRef]

16. Brenes-Álvarez, M.; Mitschke, J.; Olmedo-Verd, E.; Georg, J.; Hess, W.R.; Vioque, A.; Muro-Pastor, A.M. Elements of the heterocyst-specific transcriptome unravelled by co-expression analysis in *Nostoc* sp. PCC 7120. *Environ. Microbiol.* **2019**, *21*, 2544–2558. [CrossRef]

17. Rippka, R.; Deruelles, J.; Waterbury, J.B.; Herdman, M.; Stanier, R.Y. Generic assignments, strain stories and properties of pure cultures of cyanobacteria. *J. Gen. Microbiol.* **1979**, *111*, 1–61.

18. Urban, J.H.; Vogel, J. Translational control and target recognition by *Escherichia coli* small RNAs in vivo. *Nucleic Acids Res.* **2007**, *35*, 1018–1037. [CrossRef]

19. Corcoran, C.P.; Podkaminski, D.; Papenfort, K.; Urban, J.H.; Hinton, J.C.; Vogel, J. Superfolder GFP reporters validate diverse new mRNA targets of the classic porin regulator, MicF RNA. *Mol. Microbiol.* **2012**, *84*, 428–445. [CrossRef]

20. Sittka, A.; Pfeiffer, V.; Tedin, K.; Vogel, J. The RNA chaperone Hfq is essential for the virulence of *Salmonella typhimurium*. *Mol. Microbiol.* **2007**, *63*, 193–217. [CrossRef]

21. Wright, P.R.; Richter, A.S.; Papenfort, K.; Mann, M.; Vogel, J.; Hess, W.R.; Backofen, R.; Georg, J. Comparative genomics boosts target prediction for bacterial small RNAs. *Proc. Natl. Acad. Sci. USA* **2013**, *110*, E3487–E3496. [CrossRef]

22. Mohamed, A.; Jansson, C. Influence of light on accumulation of photosynthesis-specific transcripts in the cyanobacterium *Synechocystis* 6803. *Plant Mol. Biol.* **1989**, *13*, 693–700. [CrossRef]

23. Muro-Pastor, A.M.; Valladares, A.; Flores, E.; Herrero, A. The *hetC* gene is a direct target of the NtcA transcriptional regulator in cyanobacterial heterocyst development. *J. Bacteriol.* **1999**, *181*, 6664–6669. [CrossRef]

24. Steglich, C.; Futschik, M.E.; Lindell, D.; Voß, B.; Chisholm, S.W.; Hess, W.R. The challenge of regulation in a minimal photoautotroph: Non-coding RNAs in *Prochlorococcus*. *PLoS Genet.* **2008**, *4*, e1000173. [CrossRef]

25. Vioque, A. Analysis of the gene encoding the RNA subunit of ribonuclease P from cyanobacteria. *Nucleic Acids Res.* **1992**, *20*, 6331–6337. [CrossRef]

26. Merino-Puerto, V.; Mariscal, V.; Mullineaux, C.W.; Herrero, A.; Flores, E. Fra proteins influencing filament integrity, diazotrophy and localization of septal protein SepJ in the heterocyst-forming cyanobacterium *Anabaena* sp. *Mol. Microbiol.* **2010**, *75*, 1159–1170. [CrossRef]

27. Black, T.A.; Cai, Y.; Wolk, C.P. Spatial expression and autoregulation of *hetR*, a gene involved in the control of heterocyst development in *Anabaena*. *Mol. Microbiol.* **1993**, *9*, 77–84. [CrossRef]

28. Elhai, J.; Wolk, C.P. Conjugal transfer of DNA to cyanobacteria. *Methods Enzymol.* **1988**, *167*, 747–754.

29. Cai, Y.; Wolk, C.P. Use of a conditionally lethal gene in *Anabaena* sp. strain PCC 7120 to select for double recombinants and to entrap insertion sequences. *J. Bacteriol.* **1990**, *172*, 3138–3145. [CrossRef]

30. Muro-Pastor, A.M.; Flores, E.; Herrero, A. NtcA-regulated heterocyst differentiation genes *hetC* and *devB* from *Anabaena* sp. strain PCC 7120 exhibit a similar tandem promoter arrangement. *J. Bacteriol.* **2009**, *191*, 5765–5774. [CrossRef]

31. Olmedo-Verd, E.; Muro-Pastor, A.M.; Flores, E.; Herrero, A. Localized induction of the *ntcA* regulatory gene in developing heterocysts of *Anabaena* sp. strain PCC 7120. *J. Bacteriol.* **2006**, *188*, 6694–6699. [CrossRef]

32. Muro-Pastor, A.M. The heterocyst-specific NsiR1 small RNA is an early marker of cell differentiation in cyanobacterial filaments. *mBio* **2014**, *5*, e01079-14. [CrossRef]

33. Olmedo-Verd, E.; Flores, E.; Herrero, A.; Muro-Pastor, A.M. HetR-dependent and -independent expression of heterocyst-related genes in an *Anabaena* strain overproducing the NtcA transcription factor. *J. Bacteriol.* **2005**, *187*, 1985–1991. [CrossRef]

34. Mann, M.; Wright, P.R.; Backofen, R. IntaRNA 2.0: Enhanced and customizable prediction of RNA-RNA interactions. *Nucleic Acids Res.* **2017**, *45*, W435–W439. [CrossRef]

35. Flaherty, B.L.; Van Nieuwerburgh, F.; Head, S.R.; Golden, J.W. Directional RNA deep sequencing sheds new light on the transcriptional response of *Anabaena* sp. strain PCC 7120 to combined-nitrogen deprivation. *BMC Genom.* **2011**, *12*, 332. [CrossRef]

36. Frías, J.E.; Flores, E.; Herrero, A. Requirement of the regulatory protein NtcA for the expression of nitrogen assimilation and heterocyst development genes in the cyanobacterium *Anabaena* sp. PCC 7120. *Mol. Microbiol.* **1994**, *14*, 823–832. [CrossRef]

37. Brenes-Álvarez, M.; Vioque, A.; Muro-Pastor, A.M. The integrity of the cell wall and its remodeling during heterocyst differentiation are regulated by phylogenetically conserved small RNA Yfr1 in *Nostoc* sp. strain PCC 7120. *mBio* **2020**, *11*, e02599-19. [CrossRef]

38. Muro-Pastor, A.M.; Hess, W.R. Regulatory RNA at the crossroads of carbon and nitrogen metabolism in photosynthetic cyanobacteria. *Biochim. Biophys. Acta Gene Regul. Mech.* **2020**, *1863*, 194477. [CrossRef]

39. Kadowaki, T.; Nagayama, R.; Georg, J.; Nishiyama, Y.; Wilde, A.; Hess, W.R.; Hihara, Y. A feed-forward loop consisting of the response regulator RpaB and the small RNA PsrR1 controls light acclimation of photosystem I gene expression in the cyanobacterium *Synechocystis* sp. PCC 6803. *Plant Cell Physiol.* **2016**, *57*, 813–823. [CrossRef]

40. Sandh, G.; Ramstrom, M.; Stensjö, K. Analysis of the early heterocyst Cys-proteome in the multicellular cyanobacterium *Nostoc punctiforme* reveals novel insights into the division of labor within diazotrophic filaments. *BMC Genom.* **2014**, *15*, 1064. [CrossRef]

Article

Rapid Transcriptional Reprogramming Triggered by Alteration of the Carbon/Nitrogen Balance Has an Impact on Energy Metabolism in *Nostoc* sp. PCC 7120

Peter J. Gollan *, Dorota Muth-Pawlak and Eva-Mari Aro

Department of Biochemistry, Molecular Plant Biology, University of Turku, Tykistökatu 6A, 20520 Turku, Finland; dokrmu@utu.fi (D.M.-P.); evaaro@utu.fi (E.-M.A.)
* Correspondence: petgol@utu.fi

Received: 29 October 2020; Accepted: 18 November 2020; Published: 20 November 2020

Abstract: *Nostoc* (*Anabaena*) sp. PCC 7120 is a filamentous cyanobacterial species that fixes N_2 to nitrogenous compounds using specialised heterocyst cells. Changes in the intracellular ratio of carbon to nitrogen (C/N balance) is known to trigger major transcriptional reprogramming of the cell, including initiating the differentiation of vegetative cells to heterocysts. Substantial transcriptional analysis has been performed on *Nostoc* sp. PCC 7120 during N stepdown (low to high C/N), but not during C stepdown (high to low C/N). In the current study, we shifted the metabolic balance of *Nostoc* sp. PCC 7120 cultures grown at 3% CO_2 by introducing them to atmospheric conditions containing 0.04% CO_2 for 1 h, after which the changes in gene expression were measured using RNAseq transcriptomics. This analysis revealed strong upregulation of carbon uptake, while nitrogen uptake and metabolism and early stages of heterocyst development were downregulated in response to the shift to low CO_2. Furthermore, gene expression changes revealed a decrease in photosynthetic electron transport and increased photoprotection and reactive oxygen metabolism, as well a decrease in iron uptake and metabolism. Differential gene expression was largely attributed to change in the abundances of the metabolites 2-phosphoglycolate and 2-oxoglutarate, which signal a rapid shift from fluent photoassimilation to glycolytic metabolism of carbon after transition to low CO_2. This work shows that the C/N balance in *Nostoc* sp. PCC 7120 rapidly adjusts the metabolic strategy through transcriptional reprogramming, enabling survival in the fluctuating environment.

Keywords: cyanobacteria; *Nostoc* sp. PCC 7120; transcriptomics; photosynthesis; carbon/nitrogen

1. Introduction

Cyanobacteria use light energy to fix inorganic carbon (C_i) and nitrogen (N), harvested from their aquatic environment, into the metabolic components required for growth and propagation. Environmental sources of C_i include dissolved CO_2 and bicarbonate (HCO_3^-), while N can be supplied by nitrate (NO_3^-), nitrite (NO_2^-), ammonium (NH_4^+), urea or N_2 (in diazotrophic cyanobacteria; reviewed in [1]). The tight coupling of the concentrations of C_i and N taken up from the environment prevents metabolic imbalance within the cell, which allows cyanobacteria to thrive amidst varying nutritional conditions. This is achieved in large part by transcriptional modifications that are triggered by fluctuations in the cellular homeostasis of organic carbon (C) and N, which are represented by changes in the relative abundances of key metabolite signals (reviewed in [2–4]). One such metabolite is 2-oxoglutarate (2OG), also known as α-ketoglutarate (αKG), which is a product of the tricarboxylic acid (TCA) cycle. The metabolite 2OG provides the inorganic carbohydrate skeleton for glutamate synthesis that occurs by the incorporation of NH_4^+ in the glutamine synthetase/glutamine-oxoglutarate aminotransferase (GS/GOGAT) cycle. Cellular 2OG levels therefore represent the abundances of both C and N (reviewed in [5,6]), making 2OG a central signalling metabolite that triggers transcriptional adjustments to restore C/N balance [7–9]. 2-phosphoglycolate (2PG) is another metabolite that controls transcriptional reprogramming in response to C/N balance [10], and 2PG is formed when Rubisco catalyses the oxygenation of RuBP (photorespiration), instead of the favoured carboxylation reaction between RuBP and CO_2 (reviewed in [4,11]). An increase in 2PG concentration therefore represents CO_2 deficiency, triggering transcriptional reprogramming designed to upregulate C_i import into the cell [12–15].

Nostoc (*Anabaena*) sp. PCC 7120 is a filamentous, diazotrophic cyanobacterium wherein C/N balance controls the formation of heterocyst cells specialised for fixing N_2 into NH_4^+, while photosynthetic CO_2 fixation is restricted to vegetative cells (reviewed in [16,17]). In *Nostoc* sp. PCC 7120 and other heterocystous cyanobacteria, differentiation in cell structure and function is triggered by changes in C/N homeostasis and enacted by massive transcriptional reprogramming [1]. Emphasis on the impact of N concentration on heterocyst differentiation has revealed the central roles of 2OG and several regulatory proteins in instigating transcriptional and physiological responses to N deficit [6]. However, the transcriptional response of heterocystous cyanobacteria to C_i availability has drawn only little attention [18], in sharp contrast to that in non-diazotrophic species [12,13,15,19–22]. In the current study of the global transcriptome of *Nostoc* sp. PCC 7120, we found that a shift from 3% CO_2 to 0.04% CO_2 for 1 h can be largely attributed to changes in the levels of the metabolites 2PG and 2OG, which has a strong effect on genes involved in the import and metabolism of C_i and N. This study also identified that genes encoding factors involved in photosynthetic electron transport, glycolysis and iron homeostasis are regulated by C/N homeostasis, which is suggested to trigger a transition from efficient photoautotrophic growth and energy storage to photoinhibition and glycolysis.

2. Materials and Methods

2.1. Growth and CO_2 Stepdown

Nostoc sp. PCC 7120 cultures were grown in BG11 medium [23] buffered with 10 mM TES-KOH (pH 8.0) at 30 °C under constant illumination of 50 μmol photons m^{-2} s^{-1} with 120 rpm agitation, in air enriched with 3% (*v/v*) CO_2. During the exponential growth phase (OD_{750} = 1.0), 2 mL samples were taken from three individual replicate cultures and frozen for RNA isolation. For CO_2 stepdown, the cultures were pelleted and the pellets washed once with fresh BG11, before resuspension in fresh BG11 and growth in air containing 0.04% (*v/v*) CO_2. After 1 h, 2 mL samples were collected from three replicates and frozen for RNA isolation.

2.2. RNA Isolation and Transcriptomics

Total RNA was isolated as described in [24]. Total RNA samples were submitted to the Beijing Genomics Institute (China) for library construction and RNA sequencing using Illumina HiSeq2500. RNA reads were aligned using Strand NGS 2.7 software (Avadis) using the *Nostoc* sp. PCC 7120 reference genome and annotations downloaded from Ensembl (EBI). Aligned reads were normalised and quantified using the DESeq package (R). Significantly differentially expressed genes were identified using a 2-way ANOVA. *p*-Values were adjusted for false discovery rate (FDR) using the Benjamini–Hochberg procedure.

3. Results

The transcriptome of *Nostoc* sp. PCC 7120 grown in BG11 under 3% CO_2-enriched air was compared with that of the same strain shifted from 3% CO_2 to 0.04% CO_2 for 1 h, revealing 230 genes to be upregulated >2-fold and 211 genes to be downregulated >2-fold, in the low CO_2 condition. The RNAseq data are available at the NCBI Sequence Read Archive (submission SUB8244772). As expected, given the short period under a new metabolic condition, no differences in the growth rates, lengths of filaments or frequencies of heterocysts (approximately 4% of all cells under 3% CO_2)

were observed between the cultures exposed to 0.04% CO_2 conditions, compared to those grown at 3% CO_2. Therefore, statistical tests of these parameters in the different cultures were not performed.

3.1. Uptake and Metabolism of Carbon and Nitrogen Are Inversely Responsive to Low CO_2 Conditions

The operons encoding three plasma membrane-localised HCO_3^- uptake systems were among the most strongly upregulated entities in *Nostoc* sp. PCC 7120 following CO_2 stepdown (Table 1). In cyanobacteria, the Cmp (BCT1) system is powered by ATP hydrolysis [25], while the SbtA and BicA systems depend on Na^+ ions for HCO_3^- symport [21,26]. The upregulation of the operon encoding the Mrp Na^+:H^+ antiporter upon CO_2 stepdown may also be linked to HCO_3^- uptake through the extrusion of Na^+ to support SbtA and BicA activity [27,28]. HCO_3^- uptake in cyanobacteria forms a major part of the carbon concentration mechanism (CCM), which also involves the concentration of cellular CO_2 into HCO_3^- by a customised NAD(P)H dehydrogenase (NDH-1) complex [29]. Genes *ndhF3*, *ndhD3* and *cupA*, which encode subunits that specialise the NDH1–MS complex for inducible CO_2 uptake, were upregulated after CO_2 stepdown (Table 1), as were several other genes encoding the core NDH–1M complex (see below). Notably, a putative *cupS* orthologue (*alr1320*), which is encoded separately from the *ndhF3*/*ndhD3*/*cupA* cluster in the *Nostoc* sp. PCC 7120 genome [30], was not differentially expressed (DE) in the current work. Genes encoding Rubisco and some carboxysome subunits were mildly upregulated in low CO_2 (1.3 to 1.7 fold change (FC); not shown), while other CCM components were not DE, suggesting that these components were already in sufficient abundance before CO_2 stepdown, whereas C_i uptake from the environment, especially HCO_3^- uptake, was apparently a primary concern for survival after 1 h under low CO_2.

The decrease in CO_2 was found to cause downregulation of the *nir* operon, which encodes subunits of an ATP-dependent nitrate (NO_3^-) transporter, as well as NO_3^- and nitrite (NO_2^-) reductases [31,32]. The *nir* operon is responsible for NO_3^- uptake and reduction to ammonia (NH_3), and is broadly conserved across cyanobacteria (reviewed in [5]). Unlike the *nir* operon, the majority of *nif* genes that encode subunits for the assembly and function of nitrogenase, which reduces atmospheric N_2 to NH_3, were not DE in the current work. Exceptions were *nifB* and *nifH2*, which were downregulated (Table 1). A gene that encodes a protein similar to the C terminus of Mo-like nitrogenase (*alr1713*) was strongly downregulated, along with its neighbour (*asr1714*); however, the function of the encoded proteins is not known. Since heterocyst development is upregulated by a high C/N ratio (reviewed in [33]), it was not surprising to see many genes involved in the structural development of heterocysts repressed by the shift to low CO_2. In particular, the *hpd*, *hgl* and *dev* gene clusters that encode many components for the synthesis and export of glycolipids, which form the oxygen-impermeable heterocyst envelope (reviewed in [34,35]), and were downregulated in the current data. Notably, the *hep* genes encoding heterocyst outer layer polysaccharides were only moderately downregulated by the shift to low CO_2 (average FC −1.5, data not shown).

3.2. Expression of Genes Encoding Photosynthetic and Respiratory Components Responds to Low CO_2 Conditions

Expression of genes encoding photosystem II (PSII) core proteins D1 and D2 was upregulated after the shift to low CO_2 (Table 2), indicating an increase in the damage and turnover of PSII reaction centres [36,37]. In keeping with this, the genes encoding the two FtsH proteases involved in the degradation and turnover of damaged D1 protein were also upregulated [38,39]. We also found strongly induced expression of the *flv2–flv4* operon, as well as several genes encoding orange carotenoid proteins (OCPs) and early light-inducible proteins (ELIPs), all of which are associated with PSII photoprotection [40–44]. These expression data suggest that the shift to low CO_2 led to the over-reduction, damage and repair of PSII. Given this evidence for PSII over-reduction, it was surprising to find the gene encoding the "plastid" terminal oxidase (PTOX) as one of the most strongly downregulated in the current study (Table 2), being highly expressed under 3% CO_2 and strongly repressed in 0.04% CO_2. PTOX is part of a water–water cycle that moves electrons from reduced plastoquinone (PQ) to O_2, and is thought to be an electron valve for balancing the photosynthetic redox state [45], which would presumably be important under low CO_2 (discussed below).

Virtually all genes encoding subunits of photosystem I (PSI) were substantially downregulated in the current study, which is in contrast to their increased expression in *Synechocystis* sp. PCC 6803 and unchanged expression *Synechococcus elongatus* PCC 7942 in low CO_2 [13,20]. PSI downregulation in *Nostoc* sp. PCC 7120 points towards a decrease in PSI electron transport that may be related to the diminution of the terminal electron acceptor CO_2. These conditions would also be expected to upregulate O_2 reduction and the subsequent formation of toxic superoxide radicals ($O_2^{\bullet-}$); indeed, genes encoding a superoxide dismutase (SodB; *alr2938*) and peroxiredoxin (*all2375*), involved in reactive oxygen species (ROS) scavenging, were upregulated 3.2-fold and 2.3-fold, respectively (not shown). The expression of *isiB* (*alr2405*), which encodes a flavodoxin (Fld) that accepts electrons from PSI via a flavin mononucleotide cofactor [46], was also upregulated after low CO_2 treatment (Table 2), suggesting a shortage of oxidised ferredoxin (Fd) acceptors [47]. The upregulated expression of genes encoding F_0-F_1 ATP synthase points to an increased demand for energy in low CO_2, required for HCO_3^- import and CO_2 hydration (reviewed in [2]). Notably, the gene *alr1004* encoding an enzyme that converts glyoxylate to glycine for the detoxification of 2PG [18] was found to be downregulated after CO_2 stepdown, while other enzymes in the glycolate metabolism pathway were not DE.

Table 1. Differentially Expressed (DE) Genes Involved in Carbon and Nitrogen Metabolism.

Name [1]	Gene ID [1]	Description	Process	Fold Change [2]	*p*-Value [3]
sbt operon	*all2133–all2134*	Na^+-dependent bicarbonate permease, P_{II}-like regulatory protein	Bicarbonate import	78.8	<0.001
cmp operon	*alr2877–alr2880*	ATP-dependent bicarbonate uptake subunits		36.4	<0.001
bicA operon	*all1303–all1304*	Na^+-dependent bicarbonate permease, Na^+:H^+ antiporter		8.7	<0.001
mrp operon	*all1837–all1843*	Na^+:H^+ antiporter subunits	Na^+ extrusion, pH regulation	5.4	<0.001
ndhF3	*alr4156*	NDH–1MS subunit 5		2.5	0.002
ndhD3	*alr4157*	NDH–1MS subunit 4		2.0	<0.001
cupA	*alr4158*	NDH–1MS CO_2 uptake subunit	CO_2 uptake	5.3	<0.001
nir operon	*alr0607–alr0612*	Nitrate/nitrite reductase, ATP-dependent nitrate permease	Nitrate/nitrite import and metabolism	−3.5	0.017
nirB	*all0605*	Nitrate-dependent expression of *nir* cluster		−2.9	<0.001
nifB	*all1517*	Fe–Mo cofactor biosynthesis subunit		−2.0	0.002
nifH2	*alr0874*	Fe–S cluster-binding nitrogenase reductase	N_2 fixation, heterocyst development and function	−3.2	0.024
hgd, hgl clusters	*all5341–alr5359*	Heterocyst glycolipid layer biosynthesis		−2.6	0.004
dev operon	*alr3710–alr3712*	ATP-binding subunit, membrane transport subunits		−2.1	0.004
alr1713	*alr1713*	Similar to Mo-dependent nitrogenase, C-terminus	Unknown	−5.6	0.002
asr1714	*asr1714*	Uncharacterised protein		−5.8	<0.001

[1] Shaded cells represent operons or clusters of neighbouring genes; [2] Fold changes of genes upregulated or downregulated in low CO_2, compared to high CO_2, are coloured orange or green, respectively. In cases of multiple genes, average fold changes are shown; [3] *p*-values determined by moderated *t*-test. In cases of multiple genes, largest *p*-value is shown.

Table 2. Differentially Expressed (DE) Genes Involved in Photosynthesis and Respiration.

Name [1]	Gene ID [1]	Description	Process	Fold Change [2]	*p* Value [3]
psbAII	*alr3727*	Photosystem II D1 protein	PSII electron transport	6.9	<0.001
psbAIII	*alr4592*	Photosystem II D1 protein		1.9	0.003
psbAIV	*all3572*	Photosystem II D1 protein		3.7	<0.001
psbD	*alr4548*	Photosystem II D2 protein		3.4	<0.001
psaA	*alr5154*	Photosystem I core protein A1		−1.9	0.035
psaB1	*alr5155*	Photosystem I core protein A2		−1.9	0.038
psaB2	*alr5314*	Photosystem I core protein A2		−2.2	0.031
psaC	*asr3463*	Photosystem I Fe-S subunit	PSI electron transport	−2.2	<0.001
psaD	*all0329*	Photosystem I reaction centre subunit 2		−2.7	<0.001
psaE	*asr4319*	Photosystem I subunit E		−2.2	0.001
psaI	*asl3849*	Photosystem I subunit I		−2.5	0.012
pasK	*asr4775*	Photosystem I subunit K		−3.1	0.004
psaM	*asr4657*	Photosystem I subunit M		−2.3	0.005
flv2	*all4444*	Flavodiiron protein		23.7	<0.001
all4445	*all4445*	Unknown protein		31.8	<0.001
flv4	*all4446*	Flavodiiron protein	Other photosynthetic/respiratory electron transport	16.7	<0.001
isiB	*alr2405*	Flavodoxin		2.9	0.003
cytA	*alr4251*	Cytochrome c_6		−2.2	0.011
flv1B-flv3B	*all0177–all0178*	Flavodiiron protein (heterocyst-specific)		−2.0	0.004
ptox	*all2096*	Alternative plastoquinone oxidase		−65.0	<0.001
ftsH	*alr1261*	FtsH protease	PSII turnover	2.3	<0.001
ftsH2	*all3642*	FtsH protease		2.4	<0.001
pec operon	*alr0523–alr0527*	Phycoerythrocyanin synthesis		−3.1	0.015
chlL, chlN operon	*all5076–all5078*	Protochlorophyllide reductase, ATP-binding protein		−3.3	<0.001
chlG	*all4480*	Chlorophyll synthase 33 kDa subunit	Metabolism/binding of light-harvesting pigments	2.2	0.003
hemH	*alr4616*	Ferrochelatase		8.6	<0.001
ocp	*all3149*	Orange carotenoid-binding protein		23.9	<0.001
ocp-like	*all4941*	Orange carotenoid protein-like		3.1	0.010
asl3726	*asl3726*	CAB/ELIP/HLIP superfamily		9.8	<0.001
asr5262	*asr5262*	CAB/ELIP/HLIP superfamily		8.6	0.005
atpase cluster	*all0004–all0010*	ATP synthase subunits	ATP synthesis	2.7	<0.001
ndh-1 operon	*alr0223–alr0227*	NDH-1 complex subunits	Electron and proton transport	2.3	0.005
ndh-1 operon	*all3840–all3842*			2.1	0.001
ndhB	*all4883*			2.5	0.003
ndhN	*alr4216*			1.9	0.001
alr1004	*alr1004*	Alanine-glyoxylate transaminase	Glycolate metabolim	−2.7	<0.001
ndbA	*all1553*	NDH-2 NAD(P)H:PQ reductase	Respiration	−2.1	<0.001
hupS	*all0688*	Uptake hydrogenase, small subunit		−2.5	0.002
nifJ/PFOR	*alr1911*	Pyruvate-ferredoxin/ flavodoxin oxidoreductase	H_2 uptake/evolution	−42.4	<0.001
hox clusters	*alr0750–all0752* *alr0760–alr0766*	Bidirectional hydrogenase subunits, assembly and regulation		−73.9	0.002
ppsA	*all0635*	Phophoenolpyruvate synthase	Glycolysis	−107.4	<0.001

[1] Shaded cells represent operons or clusters of neighbouring genes; [2] Fold changes of genes upregulated or downregulated in low CO_2, compared to high CO_2, are coloured orange or green, respectively. In cases of multiple genes, average fold changes are shown; [3] *p*-values determined by moderated *t*-test. In cases of multiple genes, largest *p*-value is shown.

The downregulation of PSI abundance in response to low CO_2 can partially clarify the apparent PSII over-reduction discussed above. Lower PSI levels may also be linked to a decrease in the number of heterocysts, which contain a higher PSI:PSII ratio than in vegetative cells [16]. Although such a decrease in heterocysts was not observed after 1 h in low CO_2, suppressed heterocyst development was evident in the downregulation of *hgl* and *dev* clusters (Table 1), and this can also explain the suppression of genes encoding cytochrome c6, Flv1B and Flv3B (Table 2) that operate predominately [48] or exclusively [49] in heterocysts. The gene encoding the small subunit of the heterocyst-specific uptake hydrogenase (HupS) was also downregulated here, reflecting the downregulation of the heterocystous nitrogenase activity under low CO_2.

We observed downregulation of the *pec* cluster that encodes the phycoerythrocyanin (PEC) parts of the light-harvesting phycobilisome (PBS) complex [50,51], while genes encoding the other components of the PBS were not DE, indicating that the light-harvesting cross-section of PBS in *Nostoc* sp. PCC 7120 is altered in response to low CO_2. An operon encoding a subunit of the light-independent protochlorophyllide reductase (DPOR) was also downregulated after 1 h in low CO_2, while the expression of *chlG* and *hemH* genes, involved in later stages of chlorophyll and haem synthesis, respectively, were upregulated (Table 2).

Most genes encoding subunits of the NDH-1 complex were upregulated under low CO_2 (Table 2), probably to fulfil their role in CO_2 uptake as part of the NDH–1MS complex described above. In contrast, *ndbA* encoding NDH-2 was downregulated in the current study, suggesting a decrease in NDH-2-mediated respiration after the shift to low CO_2. In addition, the gene encoding phosphoenolpyruvate (PEP) synthase, which converts pyruvate to PEP that is consumed in the TCA cycle, was strongly downregulated (Table 2). Many genes encoding subunits of the bidirectional hydrogenase (Hox) were among the most strongly downregulated in response to low CO_2 conditions (Table 2), being both highly expressed in 3% CO_2 and strongly repressed in low CO_2. Hox reversibly catalyses the reduction of H^+ to form H_2, powered by reduced Fd/Fld with electrons derived from either PSI or pyruvate, the latter route by way of pyruvate ferredoxin/flavodoxin oxidoreductase (PFOR), which converts pyruvate to acetyl-CoA (reviewed in [52]). The *nifJ* gene encoding PFOR followed a similar expression profile to Hox subunits, being another of the most strongly downregulated genes in the current study. The physiological role of the Hox enzyme is not known, but has been described as an electron valve that can maintain redox balance and store reducing power as H_2 during excess photosynthesis or fermentation [52–55]. The expression profile of Hox and PFOR genes suggests that pyruvate-powered hydrogen production is active under 3% CO_2 and inactivated by the shift to low CO_2.

3.3. Expression of Transcription Regulators Responds to Changes in CO_2 Conditions

The current study revealed substantial changes in the expression of several genes encoding transcription regulators, providing evidence of an ongoing cascade of transcriptional reprogramming after 1 h under low CO_2 (Table 3). Upregulated expression of the LysR-type regulator (LTTR) *cmpR* corresponds to the strong upregulation of its target, the *cmp* cluster encoding the BCT1 HCO_3^- transporter (Table 1), as previously shown in *Synechocystis* sp. PCC 6803, *Synechococcus* sp. PCC 7942 [56] and *Nostoc* sp. PCC 7120 [57]. Two sigB-type group 2 sigma factors, which have roles in the transcriptional response to low CO_2 and C/N balance [58–60], were upregulated after the shift to low CO_2 (Table 3). SigB has also been implicated in response to environmental stress and resistance to photoinhibition in *Synechocystis* sp. PCC 6803 [61,62], which is in line with the upregulation in this study of *groES* and *groEL* genes (4.3 to 5.5 FC; not shown) and photoprotective factors such as OCPs and the *flv2–flv4* operon (Table 2). Two homologous two-component response regulator clusters, which each comprise a histidine kinase and a DNA-binding regulator, were found to be upregulated by low CO_2. Of the two, the chromosomal *hik31* (C-*hik31*) operon was more highly upregulated, and has been found to be involved in the responses to oxygen concentration, light and metabolism [63,64]. A TetR-family transcription regulator with unknown function was also upregulated by low CO_2 (Table 3).

Another LTTR gene that was highly expressed under high CO_2 and was strongly downregulated after low CO_2 transition (Table 3) shared substantial sequence homology with the *ndhR* transcription repressor (also called *ccmR*) of *Synechocystis* sp. PCC 6803 [12,27]. In other cyanobacteria, NdhR represses the expression of CCM genes, including the *sbt* and *bicA* HCO_3^- importers, the *mrp* cluster and the *ndhF3/ndhD3/cupA* cluster [12,15,20,27,65]. The rapid downregulation of a putative *ndhR* orthologue in *Nostoc* sp. PCC 7120 may reveal the mechanism behind the strong upregulation of CCM genes after CO_2 stepdown in the current study (Table 1). Downregulation of the transcription enhancer *ntcB*, which increases N metabolism through upregulation of the *nir* operon [66], also correlates with downregulation of other N-related genes in the current work (Table 1). Expression of *ntcB* is controlled by NtcA [1]. As *ntcA* was not DE in the current study, downregulation of *ntcB* and other NtcA regulons may be due to the low CO_2-induced inactivation of NtcA (discussed below). Similarly, downregulated transcriptional regulator genes *patB* (also called *cnfR*), *devH* and *nrrA* can also be attributed to inhibited NtcA activity [67–71]. These genes are expressed in heterocysts, where PatB upregulates the expression of *nifB* [72], DevH regulates heterocyst glycolipids [48,68,73] and NrrA induces expression of both the heterocyst regulator *hetR* and *fraF* encoding a filament integrity protein [70,74]. Both *nifB* and the *hgl* cluster were downregulated in this study (Table 1), while *hetR* and *fraF* were not DE (not shown).

Table 3. Differentially Expressed (DE) Genes Encoding Transcription Regulators.

Name	Gene ID	Description	Process	Fold Change [1]	p Value [2]
cmpR	*all0862*	LysR-type transcriptional regulator	Regulates *cmp* cluster	3.1	<0.001
sigB	*all7615*	Group 2 sigma factor	Response to stress	4.6	<0.001
sigB3	*all7608*	Group 2 sigma factor		3.8	<0.001
C-*hik31* operon	*all7583–all7584*	Two-component sensor His kinase, response regulator	Regulation of central metabolism in response to glucose, low O_2	2.8	<0.001
P-*hik31* operon	*alr1170–alr1171*			1.6	0.001
all7523	*all7523*	TetR family regulator	Unknown	2.4	0.046
putative *ndhR* orthologue	*all4986*	LysR-type transcriptional regulator	Repression of CCM expression	−146.2	<0.001
ntcB	*all0602*	LysR-type transcriptional regulator	Co-activation of *nir* operon	−2.4	0.002
devH	*alr3952*	CRP family transcriptional regulator	Het glycolipid biosynthesis	−2.2	0.006
patB/cnfR	*all2512*	Heterocyst patterning	Heterocyst development	−2.4	0.001
nrrA	*all4312*	OmpR family regulator		−2.1	<0.001

[1] Fold changes of genes upregulated or downregulated in low CO_2, compared to high CO_2, are coloured orange or green, respectively. In cases of multiple genes, average fold changes are shown; [2] p-values determined by moderated *t*-test. In cases of multiple genes, largest p-value is shown.

3.4. Low CO_2 Conditions Influence the Expression of Metal Homeostasis Genes

A number of genes and gene clusters related to cellular homeostasis of iron (Fe) and other metals were found to be DE after CO_2 stepdown (Table 4). A gene cluster encoding subunits of a periplasmic ferrous Fe (Fe(II)) transporter [75,76] was strongly downregulated in the current study (Table 4), indicating a lower uptake of Fe from the environment under low CO_2. The *suf* cluster, encoding proteins involved in Fe mobilization and Fe–S cluster assembly, was also downregulated. The expression of both the Fe(II) transporter and the *suf* cluster is upregulated by Fe deprivation [77–79], suggesting a surplus of cellular Fe after low CO_2 treatment. Several neighbouring clusters of genes encoding subunits of metal cation efflux systems such as copper, nickel, zinc, cadmium and cobalt, were upregulated after low CO_2 treatment (Table 4). These genes have been implicated in heavy metal resistance [80,81], although the link to the current conditions is not clear. Taken together, low CO_2 appears to induce an active decrease in cellular metal content, which may be a strategy to avoid oxidative stress during the reducing conditions induced by an insufficient availability of photosynthetic electron acceptors.

Table 4. Differentially Expressed (DE) Genes Involved in the Transport and Metabolism of Metals.

Name	Gene ID	Description	Process	Fold Change [1]	p Value [2]
Fe(II) transport operon	*alr2118–asr2120*	Ferrous iron transporter subunits	Periplasmic iron import	−42.6	<0.001
suf operon	*alr2492–alr2496*	ATPase, iron and sulphur transfer	Fe–S cluster assembly, transfer	−3.8	<0.001
Metal efflux cluster	*all7606–all7611*	Proton extrusion, cation efflux		3.1	<0.001
Metal efflux cluster	*all7616–all7619*	Cadmium/nickel/zinc/cobalt efflux system	Metal cation efflux, cellular metal homeostasis	3.9	<0.001
Metal efflux cluster	*all7629–all7633*	Cadmium/nickel/zinc/cobalt efflux system		5.0	<0.001
Cu^{2+} efflux cluster	*alr7634–alr7636*	Putative copper efflux		5.1	0.002

[1] Fold changes of genes upregulated or downregulated in low CO_2, compared to high CO_2, are coloured orange or green, respectively. In cases of multiple genes, average fold changes are shown; [2] p-values determined by moderated *t*-test. In cases of multiple genes, largest p-value is shown.

4. Discussion

4.1. Transcriptional Regulation in Response to CO_2 Stepdown is Triggered by Metabolites

Induction of the most strongly upregulated genes, the HCO_3^- transporters (Table 1), after CO_2 stepdown, suggests a rapid transcriptional response that is highly sensitive to cellular C_i levels. In some unicellular cyanobacteria, repression of the CCM genes by NdhR (also called CcmR) can be modulated by both 2OG and 2PG [12,15,65,82]. Increased cellular 2PG concentration caused by increased photorespiration in low CO_2 leads to increased abundance of the NdhR–2PG complex that is unable to bind DNA to repress expression [4]. Additionally, declining 2OG levels due to lower CO_2 fixation and potentially lower TCA cycle activity decrease the abundance of the NdhR-2OG repressor complex, although it is unclear whether 2OG levels would actually decrease after only 1 h in low CO_2 due to the mobilisation of stored glycogen into the TCA cycle [2,15,82,83]. $NADP^+$, another co-repressor of NdhR [82], is also theoretically far less abundant after CO_2 stepdown, due to decreased CO_2 fixation and lower NADPH consumption despite constant light conditions. Although a putative NdhR in *Nostoc* sp. PCC 7120 has not been studied, it appears that the LTTR encoded by *all4986* represents such an orthologue, and that the strong downregulation of *all4986* after CO_2 stepdown led to de-repression of the NdhR regulon, which includes the *ndhR* gene itself [84]. Previous transcriptomics studies have shown *ndhR* expression to be upregulated in *Synechocystis* sp. PCC 6803 after >3 h in low C_i conditions [12,13], which is in contrast to the strong downregulation of *all4986* seen here after 1 h (Table 3). This suggests that NdhR de-repression in response to CO_2 stepdown may be transient, and/or that expression of the NdhR regulon is also controlled by other transcription factors [18]. In the current study, de-repression by the putative NdhR is proposed to have caused a rapid and strong increase in HCO_3^- and CO_2 uptake under C limitation, with the upregulated *cmp* operon (Table 1) inducing further increase in HCO_3^- uptake. The *cmp* inducer CmpR is activated by 2PG or RuBP [82,85], both of which are in higher concentrations after CO_2 stepdown due to decreased CO_2 fixation. Furthermore, *cmpR* expression is also auto-upregulated (Table 3) [57]. In *Synechococcus* sp. PCC 7942 CmpR additionally upregulates the expression of PSII core subunits [86,87], found upregulated also in the current study alongside factors for PSII photoprotection and turnover, and downregulation of most PSI genes (Table 2). The overlap between cellular responses to either low CO_2 or high light stress is well documented [12,20,86,88], highlighting insufficient electron sinks similarly created by both high light and low CO_2, and resulting in the over-reduction of photosynthetic electron carriers [2]. Notably, several PSII photoprotection factors encoded by genes that were DE in the current study, including *psbAIII*, *flv4* and *sodB*, were shown to be regulated together with Rubisco and other CCM genes by another LTTR in *Nostoc* sp. PCC 7120 called PacR [18], suggesting the likely involvement of PacR in the transcriptional reprogramming seen here.

In addition to the LTTR transcription factors, the transcriptional response to low CO_2 is also regulated by LexA and the cyAbrB paralogues [89–92]. The vast change in expression of the *hox* operon after CO_2 stepdown (Table 2) may be related to the activity of LexA [93] and/or cyAbrB [90,94–96]. In *Synechocystis* sp. PCC 6803, cyAbrB2 controls the expression of many CCM components that were likewise found to be upregulated in this study (Table 1) [90], while the cyAbrB2 orthologue in *Nostoc* sp. PCC 7120 regulates the expression of FeSOD [96], also upregulated here. *Nostoc* sp. PCC 7120 cyAbrB1 has been recently implicated in transcriptional regulation of heterocyst differentiation [97], demonstrating a role close to the interface of C_i and N availability that suggests cyAbrB1/2 were likely active in the transcription regulation observed in the current study.

Given the transcriptional activation of NtcA, the master regulator of N metabolism, by high levels of 2OG upon a shift to low N (high C/N ratio) [7,98,99], it is widely assumed that a decrease in CO_2 leads to a decline in NtcA activity by decreasing the abundance of the 2OG–NtcA–PipX activator complex [100]. In the current study, lower NtcA activity was indeed evident in the downregulation of *nir* genes encoding NO_3^- uptake and metabolism (Table 1), as well as downregulation of the *nir* co-activator *ntcB* (Table 3; reviewed in [101]). This transcriptional regulation would effectively bring N metabolism into alignment with decreased C metabolism after CO_2 stepdown, despite the presence of N sources in the BG11 media. Overall, nearly 50% of genes downregulated in the current study have NtcA-binding promoters [71], although many NtcA-regulated genes, such as those involved in the uptake of NH_3 and urea, regulation of GS-GOGAT enzymes, as well as NtcA itself [71,102–104], were not DE after 1 h in low CO_2. The current work may therefore include only the early NtcA regulon. The NtcA-activated differentiation of vegetative cells to heterocysts occurs over approximately 24 h, via a cascade of transcriptional regulation that includes early upregulation of the co-activator *nrrA* [33]. The rapid downregulation in the current work of some heterocyst regulators, including *nrrA*, may block the commencement of heterocyst differentiation in response to relative N excess over C after CO_2 stepdown, and may signal an eventual decrease in the small number of heterocysts that are known to occur under high C/N [105]. Downregulation of NtcA activity in the current study appears to indicate a decline in 2OG levels after only 1 h in low CO_2, although an increased concentration of NH_3 derived from 2PG metabolism can also explain suppression the NtcA regulon under low CO_2 (low C/N ratio) [83]. N excess under low CO_2 is also evident in the −2.2 FC downregulated expression of cyanophycinase chb2 (*all0571*; not shown), which is regulated by NrrA, suggesting a decrease in the mobilisation of stored N under low CO_2 (reviewed in [106]). It can also be argued that an increase in the ADP/ATP ratio under CO_2 stepdown, caused by decreased photosynthetic electron transport and rapid changes in metabolism, increases the abundance of both the ADP–PII–PipX complex and the inactive form of NtcA [6].

The strong downregulation of operons involved in ferrous Fe import and Fe–S cluster assembly in the current work (Table 4) suggests a connection between cellular Fe homeostasis and C/N balance in *Nostoc* sp. PCC 7120, which has been explored [107]. The current results indicate a CO_2 stepdown-induced cellular excess of Fe and other transition metals, which may be due to downregulation of Fe-rich PSI complexes (Table 2) [108] and/or an excess of reductant caused by insufficient electron sinks. Both conditions pose the danger of ROS formation, evidenced by upregulated expression of SOD, peroxiredoxin and protein chaperones.

4.2. Altered C/N Balance Modulates the Energetic Strategy of Nostoc sp. PCC 7120

The current study shows that a stepdown from 3% CO_2 in enriched air to 0.04% CO_2 (atmospheric) for just 1 h led to substantial reprogramming of global gene expression in *Nostoc* sp. PCC 7120 cultures. As discussed above, most of the transcriptional changes observed here can be directly attributed to metabolite signalling instigated by the alteration of the cellular C/N balance (Figure 1), initiated by the decrease in CO_2 concentration. Furthermore, these results highlight rapid transcriptional reprogramming of photosynthesis and energy metabolism in *Nostoc* sp. PCC 7120 in response to CO_2 levels (Figure 2). High CO_2 in light promotes a high rate of photosynthesis and the storage of photosynthate in the form of glycogen [13,109–111]. Strong expression of PEP synthase and PFOR under these conditions indicate glycolysis/gluconeogenesis through the metabolism of pyruvate, which is consumed in the TCA cycle to facilitate respiratory electron transport and to provide 2OG for amino acid synthesis (reviewed in [112,113]). Therefore, growth under 3% CO_2 somewhat resembles photomixotrophy, with photosynthetic and glycolytic metabolisms occurring concomitantly, even though glucose was not externally provided to cultures. Under these conditions, the cells experience a high C/N, which is evident in the relatively high expression of N metabolic genes, reflecting a cellular excess of 2OG [7,98,99]. In high CO_2, strong expression of Hox, which is important under mixotrophy and N deprivation [55], and PTOX, may provide a system to maintain redox poise [45], and in the case of Hox, also store surplus energy as H_2 [53]. The transfer of *Synechocystis* sp. PCC 6803 and *Synechococcus elongatus* PCC 7942 cultures from high to low CO_2 showed that the toxic effects of 2PG transiently block Calvin–Benson–Bassham (CBB) activity [15,20,83] and this is also evident here in the upregulated expression of PSII repair, photoprotection and ROS scavenging enzymes in *Nostoc* sp. PCC 7120, which indicate over-reduction of the photosynthetic electron transport chain. Interestingly, detoxification of 2PG appeared to be downregulated through repression of *alr1004* during CO_2 stepdown (Table 2), perhaps highlighting the importance of the metabolite for signalling during the early stages of C_i deprivation. We also observed downregulation of the terminal proteins in the phycobilisomes, suggesting modified harvesting of light energy to alleviate excitation pressure on the photosynthetic system. Under these conditions, inhibition of photoassimilation is compensated by glycolysis of stored glucose and CBB intermediates, providing an important supply of substrates for anaplerosis of the CBB and TCA cycles during acclimation to the transition [83,114,115]. In the current work, enhanced glycolytic activity is indicated by the upregulation of NDH-1, suggesting an increased reliance on respiratory electron transport for ATP generation, while strong downregulation of PEP synthase and PFOR after CO_2 stepdown may prevent diversion of pyruvate away from the TCA cycle. Notably, PEP abundance increased substantially in *Synechocystis* sp. PCC 6803 and *Synechococcus elongatus* PCC 7942 after a shift from high to low CO_2 [20,83], while genes encoding both pyruvate kinase and PFOR were highly expressed in *Synechococcus elongatus* PCC 7942 after long-term acclimation to low CO_2, but not directly after the transition [20]. These findings support the results of this study and suggest that the metabolism of PEP and pyruvate are tightly regulated after the transition to low CO_2. This may be linked to the role of PEP and pyruvate as substrates to anaplerotic carbon fixation to produce TCA cycle intermediates oxaloacetate and malate (reviewed in [116]). During the adjustment to a low C/N ratio, TCA cycle activity generates 2OG for amino acid synthesis, utilising excess NH_4^+ generated through 2PG detoxification [11,83].

This work has revealed rapid transcriptional reprogramming in *Nostoc* sp. PCC 7120 in response to a decrease in C_i availability, namely strong upregulation of CCM components and photoprotection, and downregulation of N uptake and early stages of heterocyst differentiation. Despite the vast increase in HCO_3^- uptake, glycolysis of stored C apparently plays an important role in energy metabolism at low CO_2, likely due to 2PG-induced inhibition of the CBB cycle. A majority of the transcriptional effects induced by low CO_2 in *Nostoc* sp. PCC 7120 can be attributed to 2PG modulation of CmpR and a putative NdhR homologue; however, the effects of changing abundance of 2OG and NtcA activity after 1 h in 0.04% CO_2, as well as the roles of other transcriptional regulators cannot be discounted. Finally, this work highlights the sensitivity of *Nostoc* sp. PCC 7120 to factors that influence the cellular C/N balance and demonstrates the speed at which genetic and metabolic reprogramming can take place, allowing rapid acclimation for surviving and thriving in the fluctuating environment.

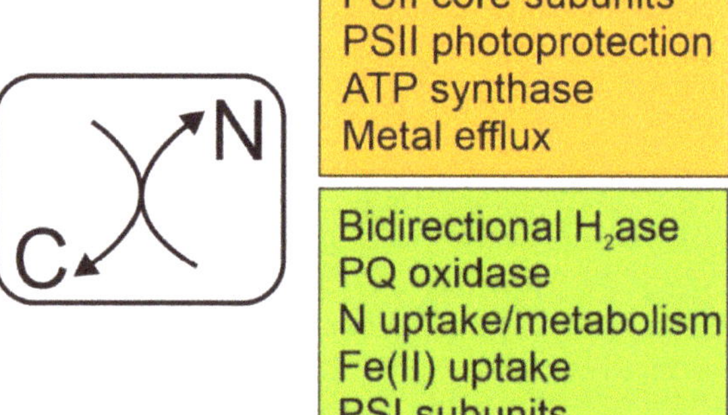

Figure 1. The transcriptional response of *Nostoc* sp. PCC 7120 cells to a change in the cellular concentrations of carbon and nitrogen (C/N balance). Decrease in the external CO_2 concentration of cultures causes a decline in the cellular C/N balance signalled by an increased production of 2PG and decreased 2OG levels relative to N. This metabolic change leads to upregulation of genes encoding processes depicted in orange, and downregulation of genes encoding processes depicted in green.

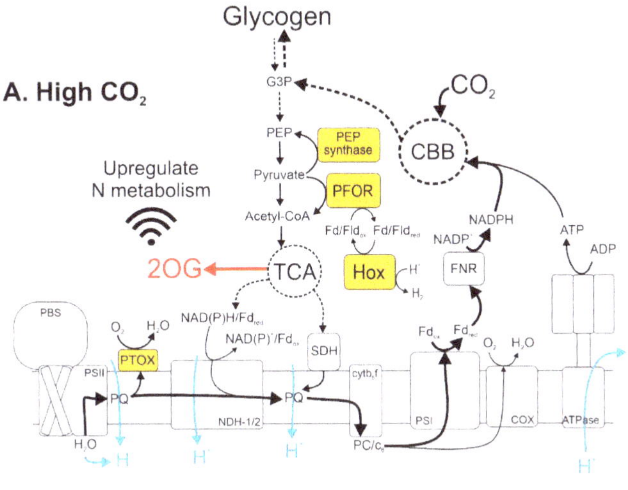

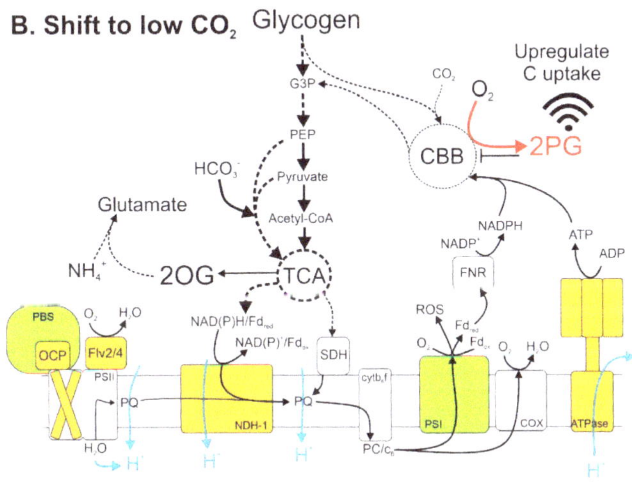

Figure 2. Schematic representation of the interactions between photosynthetic/respiratory electron transport and accumulation of 2OG and 2PG for signalling the C/N balance in *Nostoc* sp. PCC 7120 cells. Scheme is based on global transcriptomic profiling of *Nostoc* sp. PCC 7120 cultures under high CO_2 conditions and after CO_2 stepdown. (**A**) Under 3% CO_2, efficient photosynthetic electron transport and Calvin–Benson–Bassham (CBB) cycle activity enables gluconeogenesis of glyceraldehyde-3-phosphate (G3P), leading to accumulation of carbohydrate storage in the form of glycogen. Glycolysis of glycogen stores and/or photosynthate supplies pyruvate to the incomplete tricarboxylic acid (TCA) cycle, which produces reductant, ATP and succinate to drive respiratory electron transport through NAD(P)H-dehydrogenase (NDH) and succinate dehydrogenase (SDH), as well as other cellular processes. The TCA cycle also generates 2-oxoglutarate (2OG), which accumulates under high CO_2 due to a relative shortage of NH_4^+ and glutamine. 2OG operates as a signal for upregulating genes involved in N uptake and metabolism. Strong expression of phosphoenolpyruvate (PEP) synthase converts pyruvate to PEP. Pyruvate:ferredoxin/flavodoxin oxidoreductase (PFOR) converts pyruvate to acetyl-CoA, reducing oxidised ferredoxin/flavodoxin (Fd/Flv_{ox}) that is consumed by bidirectional hydrogenase (Hox) for storage of excess energy as H_2. Strong expression of plastoquinone terminal oxidase (PTOX) maintains redox homeostasis of the plastoquinone (PQ) pool during high photosynthetic electron transport, while cytochrome c6 oxidase (COX) maintains the redox poise of lumenal electron

carriers cytochrome c_6 (c_6) and plastocyanin (PC). Genes encoding factors coloured orange are highly expressed under 3% CO_2 and strongly downregulated by the shift to 0.04% CO_2. (**B**) After 1 h in 0.04% CO_2, a deficiency of CO_2 electron acceptors leads to oxygenation of Rubisco, causing photorespiration that produces 2-phosphoglycolate (2PG). The CBB cycle and other metabolic pathways are inhibited by 2PG, which also signals upregulation of the transcriptomic response to low CO_2. Low CBB activity causes over-reduction of the photosynthetic electron transport chain, leading to the production of reactive oxygen species (ROS) at photosystem I (PSI) and downregulation of PSI subunits. Increased reducing pressure on photosystem II (PSII) also causes upregulation of the PSII repair cycle and upregulation of PSII photoprotection by flavoproteins (Flv2/4) and orange carotenoid proteins (OCPs), as well as downregulation of phycoerythrocyanin in the phycobilisome (PBS). Glycolysis triggered by decreased CO_2 assimilation provides substrates for anaplerotic supplementation of the CBB and TCA cycles, including the production of oxaloacetate from PEP and bicarbonate (HCO_3^-). TCA cycle activity also produces 2OG for glutamate production from excess ammonium (NH_4^+) resulting from the shift from 3% to 0.04% CO_2. Genes encoding factors coloured orange or green are upregulated or downregulated, respectively, 1 h after the shift from 3% to 0.04% CO_2. Black arrows indicate the movement of electrons or ATP, dashed arrows summarise multiple enzymatic reactions in carbohydrate metabolism, blue arrows indicate the movement of protons. The red arrow in (**A**) shows the production of 2OG by TCA cycle activity, the red arrow in (**B**) shows the production of 2PG by Rubisco oxygenation in the CBB cycle and the black bar in (**B**) indicates CBB inhibition by 2PG.

Author Contributions: Conceptualization, P.J.G., E.-M.A.; investigation and analysis, P.J.G.; original draft preparation, P.J.G.; writing—review and editing, P.J.G., D.M.-P., E.-M.A.; funding acquisition, P.J.G., E.-M.A. All authors have read and agreed to the published version of the manuscript.

Funding: This research was funded by Academy of Finland, projects 26080341 (P.J.G.) and 307335 (E.-M.A.), and the Jane and Aatos Erkko Foundation (E.-M.A.).

Acknowledgments: The authors acknowledge Julia Walter for the growth of cultures and the isolation of RNA used in this study.

Conflicts of Interest: The authors declare no conflict of interest. The funders had no role in the design of the study; in the collection, analyses, or interpretation of data; in the writing of the manuscript, or in the decision to publish the results.

References

1. Herrero, A.; Flores, E. Genetic responses to carbon and nitrogen availability in Anabaena. *Environ. Microbiol.* **2019**, *21*, 1–17. [CrossRef] [PubMed]

2. Burnap, R.L.; Hagemann, M.; Kaplan, A. Regulation of CO_2 Concentrating Mechanism in Cyanobacteria. *Life* **2015**, *5*, 348–371. [CrossRef] [PubMed]

3. Huergo, L.F.; Dixon, R. The Emergence of 2-Oxoglutarate as a Master Regulator Metabolite. *Microbiol. Mol. Biol. Rev.* **2015**, *79*, 419–435. [CrossRef]

4. Zhang, C.-C.; Zhou, C.-Z.; Burnap, R.L.; Peng, L. Carbon/Nitrogen Metabolic Balance: Lessons from Cyanobacteria. *Trends Plant Sci.* **2018**, *23*, 1116–1130. [CrossRef]

5. Flores, E.; Frías, J.E.; Rubio, L.M.; Herrero, A. Photosynthetic nitrate assimilation in cyanobacteria. *Photosynth. Res.* **2005**, *83*, 117–133. [CrossRef]

6. Forchhammer, K.; Selim, K.A. Carbon/nitrogen homeostasis control in cyanobacteria. *FEMS Microbiol. Rev.* **2019**, *44*, 33–53. [CrossRef]

7. Muro-Pastor, M.I.; Reyes, J.C.; Florencio, F.J. Cyanobacteria perceive nitrogen status by sensing intracellular 2-oxoglutarate levels. *J. Biol. Chem.* **2001**, *276*, 38320–38328. [PubMed]

8. Vázquez-Bermúdez, M.F.; Herrero, A.; Flores, E. 2-Oxoglutarate increases the binding affinity of the NtcA (nitrogen control) transcription factor for the *Synechococcus glnA* promoter. *FEBS Lett.* **2002**, *512*, 71–74. [CrossRef]

9. Li, J.-H.; Laurent, S.; Konde, V.; Bédu, S.; Zhang, C.-C. An increase in the level of 2-oxoglutarate promotes heterocyst development in the cyanobacterium *Anabaena* sp. strain PCC 7120. *Microbiology* **2003**, *149*, 3257–3263. [CrossRef]

10. Haimovich-Dayan, M.; Lieman-Hurwitz, J.; Orf, I.; Hagemann, M.; Kaplan, A. Does 2-phosphoglycolate serve as an internal signal molecule of inorganic carbon deprivation in the cyanobacterium *Synechocystis* sp. PCC 6803? *Environ. Microbiol.* **2015**, *17*, 1794–1804. [CrossRef]

11. Hagemann, M.; Bauwe, H. Photorespiration and the potential to improve photosynthesis. *Curr. Opin. Chem. Biol.* **2016**, *35*, 109–116. [CrossRef] [PubMed]

12. Wang, H.-L.; Postier, B.L.; Burnap, R.L. Alterations in Global Patterns of Gene Expression in *Synechocystis* sp. PCC 6803 in Response to Inorganic Carbon Limitation and the Inactivation of ndhR, a LysR Family Regulator. *J. Biol. Chem.* **2004**, *279*, 5739–5751. [CrossRef] [PubMed]

13. Eisenhut, M.; Von Wobeser, E.A.; Jonas, L.; Schubert, H.; Ibelings, B.W.; Bauwe, H.; Matthijs, H.C.; Hagemann, M. Long-Term Response toward Inorganic Carbon Limitation in Wild Type and Glycolate Turnover Mutants of the Cyanobacterium *Synechocystis* sp. Strain PCC 6803. *Plant Physiol.* **2007**, *144*, 1946–1959. [CrossRef] [PubMed]

14. Kaplan, A.; Hagemann, M.; Bauwe, H.; Kahlon, S.; Ogawa, T. Carbon acquisition by cyanobacteria: Mechanisms, comparative genomics and evolution. In *The Cyanobacteria: Molecular Biology, Genomics and Evolution*; Herrero, A., Flores, E., Eds.; Caister Academic Press: Norwich, UK, 2008; pp. 305–323.

15. Klähn, S.; Orf, I.; Schwarz, D.; Matthiessen, J.K.; Kopka, J.; Hess, W.R.; Hagemann, M. Integrated Transcriptomic and Metabolomic Characterization of the Low-Carbon Response Using an *ndhR* Mutant of *Synechocystis* sp. PCC 68031. *Plant Physiol.* **2015**, *169*, 1540–1556. [CrossRef]

16. Wolk, C.P.; Ernst, A.; Elhai, J. Heterocyst Metabolism and Development. In *The Molecular Biology of Cyanobacteri*; Springer: Dordrecht, The Netherlands, 1994; pp. 769–823. [CrossRef]

17. Herrero, A.; Muro-Pastor, A.M.; Valladares, A.; Flores, E. Cellular differentiation and the NtcA transcription factor in filamentous cyanobacteria. *FEMS Microbiol. Rev.* **2004**, *28*, 469–487. [CrossRef]

18. Picossi, S.; Flores, E.; Herrero, A. The LysR-type transcription factor PacR is a global regulator of photosynthetic carbon assimilation in Anabaena. *Environ. Microbiol.* **2015**, *17*, 3341–3351. [CrossRef]

19. Allahverdiyeva, Y.; Vainonen, J.P.; Vorontsova, N.; Keränen, M.; Carmel, D.; Aro, E.-M. Dynamic Changes in the Proteome of *Synechocystis* 6803 in Response to CO_2 Limitation Revealed by Quantitative Proteomics. *J. Proteome Res.* **2010**, *9*, 5896–5912. [CrossRef]

20. Schwarz, D.; Nodop, A.; Hüge, J.; Purfürst, S.; Forchhammer, K.; Michel, K.-P.; Bauwe, H.; Kopka, J.; Hagemann, M. Metabolic and Transcriptomic Phenotyping of Inorganic Carbon Acclimation in the Cyanobacterium *Synechococcus elongatus* PCC 7942. *Plant Physiol.* **2011**, *155*, 1640–1655. [CrossRef]

21. Price, G.D.; Woodger, F.J.; Badger, M.R.; Howitt, S.M.; Tucker, L. Identification of a SulP-type bicarbonate transporter in marine cyanobacteria. *Proc. Natl. Acad. Sci. USA* **2004**, *101*, 18228–18233. [CrossRef]

22. McGinn, P.J.; Price, G.D.; Maleszka, R.; Badger, M.R. Inorganic Carbon Limitation and Light Control the Expression of Transcripts Related to the CO_2-Concentrating Mechanism in the Cyanobacterium *Synechocystis* sp. Strain PCC6803. *Plant Physiol.* **2003**, *132*, 218–229. [CrossRef]

23. Rippka, R.; Stanier, R.Y.; Deruelles, J.; Herdman, M.; Waterbury, J.B. Generic Assignments, Strain Histories and Properties of Pure Cultures of Cyanobacteria. *Microbiology* **1979**, *111*, 1–61. [CrossRef]

24. Walter, J.; Lynch, F.; Battchikova, N.; Aro, E.-M.; Gollan, P.J. Calcium impacts carbon and nitrogen balance in the filamentous cyanobacterium *Anabaena* sp. PCC 7120. *J. Exp. Bot.* **2016**, *67*, 3997–4008. [CrossRef] [PubMed]

25. Omata, T.; Price, G.D.; Badger, M.R.; Okamura, M.; Gohta, S.; Ogawa, T. Identification of an ATP-binding cassette transporter involved in bicarbonate uptake in the cyanobacterium *Synechococcus* sp. strain PCC 7942. *Proc. Natl. Acad. Sci. USA* **1999**, *96*, 13571–13576. [CrossRef] [PubMed]

26. Shibata, M.; Katoh, H.; Sonoda, M.; Ohkawa, H.; Shimoyama, M.; Fukuzawa, H.; Kaplan, A.; Ogawa, T. Genes Essential to Sodium-dependent Bicarbonate Transport in Cyanobacteria: Function and phylogenetic analysis. *J. Biol. Chem.* **2002**, *277*, 18658–18664. [CrossRef] [PubMed]

27. Blanco-Rivero, A.; Leganés, F.; Fernández-Valiente, E.; Calle, P.; Fernández-Piñas, F. *mrpA*, a gene with roles in resistance to Na^+ and adaptation to alkaline pH in the cyanobacterium *Anabaena* sp. PCC7120. *Microbiology* **2005**, *151*, 1671–1682. [CrossRef] [PubMed]

28. Fukaya, F.; Promden, W.; Hibino, T.; Tanaka, Y.; Nakamura, T.; Takabe, T. An Mrp-like cluster in the halotolerant cyanobacterium *Aphanothece halophytica* functions as a Na^+/H^+ antiporter. *Appl. Environ. Microbiol.* **2009**, *75*, 6626–6629. [CrossRef]

29. Allahverdiyeva, Y.; Aro, E.-M.; Nixon, P.J. Structure and Physiological Function of NDH-1 Complexes in Cyanobacteria. In *Bioenergetic Processes of Cyanobacteria*; Springer: Dordrecht, The Netherlands, 2011; pp. 445–467. [CrossRef]

30. Kaneko, T.; Nakamura, Y.; Wolk, C.P.; Kuritz, T.; Sasamoto, S.; Watanabe, A.; Iriguchi, M.; Ishikawa, A.; Kawashima, K.; Kimura, T.; et al. Complete Genomic Sequence of the Filamentous Nitrogen-fixing Cyanobacterium *Anabaena* sp. Strain PCC 7120. *DNA Res.* **2001**, *8*, 205–213. [CrossRef]

31. Frías, J.E.; Flores, E.; Herrero, A. Nitrate assimilation gene cluster from the heterocyst-forming cyanobacterium *Anabaena* sp. strain PCC 7120. *J. Bacteriol.* **1997**, *179*, 477–486. [CrossRef]

32. Frías, J.E.; Flores, E. Negative Regulation of Expression of the Nitrate Assimilation nirA Operon in the Heterocyst-Forming Cyanobacterium *Anabaena* sp. Strain PCC 7120. *J. Bacteriol.* **2010**, *192*, 2769–2778. [CrossRef]

33. Flores, E.; Picossi, S.; Valladares, A.; Herrero, A. Transcriptional regulation of development in heterocyst-forming cyanobacteria. *Biochim. Biophys. Acta Bioenergy* **2019**, *1862*, 673–684. [CrossRef]

34. Kumar, K.; Mella-Herrera, R.A.; Golden, J.W. Cyanobacterial Heterocysts. *Cold Spring Harb. Perspect. Biol.* **2010**, *2*, a000315. [CrossRef] [PubMed]

35. Maldener, I.; Muro-Pastor, A.M. Cyanobacterial Heterocysts. In *Encyclopedia of Life Sciences*; John Wiley & Sons, Ltd.: Chichester, UK, 2010. [CrossRef]

36. Aro, E.-M.; Virgin, I.; Andersson, B. Photoinhibition of Photosystem II. Inactivation, protein damage and turnover. *Biochim. Biophys. Acta Bioenergy* **1993**, *1143*, 113–134. [CrossRef]

37. Komenda, J.; Hassan, H.A.; Diner, B.A.; Debus, R.J.; Barber, J.; Nixon, P.J. Degradation of the Photosystem II D_1 and D_2 proteins in different strains of the cyanobacterium *Synechocystis* PCC 6803 varying with respect to the type and level of psbA transcript. *Plant Mol. Biol.* **2000**, *42*, 635–645. [CrossRef]

38. Silva, P.; Thompson, E.P.; Bailey, S.; Kruse, O.; Mullineaux, C.W.; Robinson, C.; Mann, N.H.; Nixon, P.J. FtsH Is Involved in the Early Stages of Repair of Photosystem II in *Synechocystis* sp PCC 6803. *Plant Cell* **2003**, *15*, 2152–2164. [CrossRef] [PubMed]

39. Komenda, J.; Barker, M.; Kuviková, S.; De Vries, R.; Mullineaux, C.W.; Tichý, M.; Nixon, P.J. The FtsH Protease slr0228 Is Important for Quality Control of Photosystem II in the Thylakoid Membrane of *Synechocystis* sp. PCC 6803. *J. Biol. Chem.* **2006**, *281*, 1145–1151. [CrossRef] [PubMed]

40. Eisenhut, M.; Georg, J.; Klähn, S.; Sakurai, I.; Mustila, H.; Zhang, P.; Hess, W.R.; Aro, E.-M. The Antisense RNA As1_flv4 in the Cyanobacterium *Synechocystis* sp. PCC 6803 Prevents Premature Expression of the *flv4-2* Operon upon Shift in Inorganic Carbon Supply. *J. Biol. Chem.* **2012**, *287*, 33153–33162. [CrossRef]

41. Bersanini, L.; Battchikova, N.; Jokel, M.; Rehman, A.; Vass, I.; Allahverdiyeva, Y.; Aro, E.M. Flavodiiron protein Flv2/Flv4-related photoprotective mechanism dissipates excitation pressure of PSII in cooperation with phycobilisomes in cyanobacteria. *Plant Physiol.* **2014**, *164*, 805–818. [CrossRef]

42. Gwizdala, M.; Wilson, A.; Kirilovsky, D. In Vitro Reconstitution of the Cyanobacterial Photoprotective Mechanism Mediated by the Orange Carotenoid Protein in *Synechocystis* PCC 6803. *Plant Cell* **2011**, *23*, 2631–2643. [CrossRef]

43. López-Igual, R.; Wilson, A.; Leverenz, R.L.; Melnicki, M.R.; De Carbon, C.B.; Sutter, M.; Turmo, A.; Perreau, F.; Kerfeld, C.A.; Kirilovsky, D. Different Functions of the Paralogs to the N-Terminal Domain of the Orange Carotenoid Protein in the Cyanobacterium *Anabaena* sp. PCC 7120. *Plant Physiol.* **2016**, *171*, 1852–1866. [CrossRef]

44. Kufryk, G.; Hernandez-Prieto, M.A.; Kieselbach, T.; Miranda, H.; Vermaas, W.; Funk, C. Association of small CAB-like proteins (SCPs) of *Synechocystis* sp. PCC 6803 with Photosystem II. *Photosynth. Res.* **2008**, *95*, 135–145. [CrossRef]

45. Lea-smith, D.J.; Bombelli, P.; Vasudevan, R.; Howe, C.J. Photosynthetic, Respiratory and extracellular electron transport pathways in cyanobacteria. *Biochim. Biophys. Acta Bioenergy* **2016**, *1857*, 247–255. [CrossRef] [PubMed]

46. González, A.; Fillat, M.F. Overexpression, immunodetection, and site-directed mutagenesis of *Anabaena* sp. PCC 7120 flavodoxin: A comprehensive laboratory practice on molecular biology. *Biochem. Mol. Biol. Educ.* **2018**, *46*, 493–501. [CrossRef]

47. Laudenbach, D.E.; Straus, N.A. Characterization of a cyanobacterial iron stress-induced gene similar to psbC. *J. Bacteriol.* **1988**, *170*, 5018–5026. [CrossRef] [PubMed]

48. Torrado, A.; Ramírez-Moncayo, C.; Navarro, J.A.; Mariscal, V.; Molina-Heredia, F.P. Cytochrome c6 is the main respiratory and photosynthetic soluble electron donor in heterocysts of the cyanobacterium *Anabaena* sp. PCC 7120. *Biochim. Biophys. Acta Bioenergy* **2019**, *1860*, 60–68. [CrossRef] [PubMed]

49. Ermakova, M.; Allahverdiyeva, Y.; Allahverdiyeva, Y.; Aro, E.-M.; Allahverdiyeva, Y. Novel heterocyst-specific flavodiiron proteins in *Anabaena* sp. PCC 7120. *FEBS Lett.* **2013**, *587*, 82–87. [CrossRef]

50. Swanson, R.V.; De Lorimier, R.; Glazer, A.N. Genes encoding the phycobilisome rod substructure are clustered on the Anabaena chromosome: Characterization of the phycoerythrocyanin operon. *J. Bacteriol.* **1992**, *174*, 2640–2647. [CrossRef]

51. Ducret, A.; Sidler, W.; Wehrli, E.; Frank, G.; Zuber, H. Isolation, Characterization and Electron Microscopy Analysis of A Hemidiscoidal Phycobilisome Type from the Cyanobacterium *Anabaena* sp. PCC 7120. *JBIC J. Biol. Inorg. Chem.* **1996**, *236*, 1010–1024. [CrossRef]

52. Khanna, N.; Lindblad, P. Cyanobacterial Hydrogenases and Hydrogen Metabolism Revisited: Recent Progress and Future Prospects. *Int. J. Mol. Sci.* **2015**, *16*, 10537–10561. [CrossRef]

53. Appel, J.; Phunpruch, S.; Steinmüller, K.; Schulz, R. The bidirectional hydrogenase of *Synechocystis* sp. PCC 6803 works as an electron valve during photosynthesis. *Arch. Microbiol.* **2000**, *173*, 333–338. [CrossRef]

54. Carrieri, D.; Wawrousek, K.; Eckert, C.; Yu, J.; Maness, P. The role of the bidirectional hydrogenase in cyanobacteria. *Bioresour. Technol.* **2011**, *102*, 8368–8377. [CrossRef]

55. Gutekunst, K.; Chen, X.; Schreiber, K.; Kaspar, U.; Makam, S.; Appel, J. The Bidirectional NiFe-hydrogenase in *Synechocystis* sp. PCC 6803 Is Reduced by Flavodoxin and Ferredoxin and Is Essential under Mixotrophic, Nitrate-limiting Conditions. *J. Biol. Chem.* **2014**, *289*, 1930–1937. [CrossRef] [PubMed]

56. Omata, T.; Gohta, S.; Takahashi, Y.; Harano, Y.; Maeda, S.-I. Involvement of a CbbR Homolog in Low CO_2-Induced Activation of the Bicarbonate Transporter Operon in Cyanobacteria. *J. Bacteriol.* **2001**, *183*, 1891–1898. [CrossRef] [PubMed]

57. López-Igual, R.; Picossi, S.; López-Garrido, J.; Flores, E.; Herrero, A. N and C control of ABC-type bicarbonate transporter Cmp and its LysR-type transcriptional regulator CmpR in a heterocyst-forming cyanobacterium, *Anabaena* sp. *Environ. Microbiol.* **2012**, *14*, 1035–1048. [CrossRef] [PubMed]

58. Brahamsha, B.; Haselkorn, R. Identification of multiple RNA polymerase sigma factor homologs in the cyanobacterium *Anabaena* sp. strain PCC 7120: Cloning, expression, and inactivation of the sigB and sigC genes. *J. Bacteriol.* **1992**, *174*, 7273–7282. [CrossRef]

59. Caslake, L.F.; Gruber, T.M.; Bryant, N.A. Expression of two alternative sigma factors of *Synechococcus* sp. strain PCC 7002 is modulated by carbon and nitrogen stress. *Microbiology* **1997**, *143*, 3807–3818. [CrossRef]

60. Muro-Pastor, A.M.; Herrero, A.; Flores, E. Nitrogen-Regulated Group 2 Sigma Factor from *Synechocystis* sp. Strain PCC 6803 Involved in Survival under Nitrogen Stress. *J. Bacteriol.* **2001**, *183*, 1090–1095. [CrossRef]

61. Tuominen, I.; Pollari, M.; Tyystjärvi, E.; Tyystjärvi, T. The SigB σ factor mediates high-temperature responses in the cyanobacterium *Synechocystis* sp. PCC6803. *FEBS Lett.* **2006**, *580*, 319–323. [CrossRef]

62. Hakkila, K.; Antal, T.; Gunnelius, L.; Kurkela, J.; Matthijs, H.C.; Tyystjärvi, E.; Tyystjärvi, T. Group 2 Sigma Factor Mutant sigCDE of the Cyanobacterium *Synechocystis* sp. PCC 6803 Reveals Functionality of Both Carotenoids and Flavodiiron Proteins in Photoprotection of Photosystem II. *Plant Cell Physiol.* **2013**, *54*, 1780–1790. [CrossRef]

63. Summerfield, T.C.; Nagarajan, S.; Sherman, L.A. Gene expression under low-oxygen conditions in the cyanobacterium *Synechocystis* sp. PCC 6803 demonstrates Hik31-dependent and -independent responses. *Microbiology* **2011**, *157*, 301–312. [CrossRef]

64. Nagarajan, S.; Sherman, D.M.; Shaw, I.; Sherman, L.A. Functions of the Duplicated hik31 Operons in Central Metabolism and Responses to Light, Dark, and Carbon Sources in *Synechocystis* sp. Strain PCC 6803. *J. Bacteriol.* **2011**, *194*, 448–459. [CrossRef]

65. Jiang, Y.-L.; Wang, X.-P.; Sun, H.; Han, S.-J.; Li, W.-F.; Cui, N.; Lin, G.-M.; Zhang, J.-Y.; Cheng, W.; Cao, D.-D.; et al. Coordinating carbon and nitrogen metabolic signaling through the cyanobacterial global repressor NdhR. *Proc. Natl. Acad. Sci. USA* **2017**, *115*, 403–408. [CrossRef]

66. Frías, J.E.; Flores, E.; Herrero, A. Activation of the Anabaena nir operon promoter requires both NtcA (CAP family) and NtcB (LysR family) transcription factors. *Mol. Microbiol.* **2000**, *38*, 613–625. [CrossRef] [PubMed]

67. Liang, J.; Scappino, L.; Haselkorn, R. The patB gene product, required for growth of the cyanobacterium *Anabaena* sp. strain PCC 7120 under nitrogen-limiting conditions, contains ferredoxin and helix-turn-helix domains. *J. Bacteriol.* **1993**, *175*, 1697–1704. [CrossRef] [PubMed]

68. Hebbar, P.B.; Curtis, S.E. Characterization of devH, a Gene Encoding a Putative DNA Binding Protein Required for Heterocyst Function in *Anabaena* sp. Strain PCC 7120. *J. Bacteriol.* **2000**, *182*, 3572–3581. [CrossRef] [PubMed]

69. Jones, K.M.; Buikema, W.J.; Haselkorn, R. Heterocyst-Specific Expression of patB, a Gene Required for Nitrogen Fixation in *Anabaena* sp. Strain PCC 7120. *J. Bacteriol.* **2003**, *185*, 2306–2314. [CrossRef]

70. Ehira, S.; Ohmori, M. NrrA, a nitrogen-responsive response regulator facilitates heterocyst development in the cyanobacterium Anabaena sp. strain PCC 7120. *Mol. Microbiol.* **2006**, *59*, 1692–1703. [CrossRef]

71. Picossi, S.; Flores, E.; Herrero, A. ChIP analysis unravels an exceptionally wide distribution of DNA binding sites for the NtcA transcription factor in a heterocyst-forming cyanobacterium. *BMC Genom.* **2014**, *15*, 22. [CrossRef]

72. Tsujimoto, R.; Kamiya, N.; Fujita, Y. Identification of a *cis*-acting element in nitrogen fixation genes recognized by CnfR in the nonheterocystous nitrogen-fixing cyanobacterium *Leptolyngbya boryana*. *Mol. Microbiol.* **2016**, *101*, 411–424. [CrossRef]

73. Ramírez, M.E.; Hebbar, P.B.; Zhou, R.; Wolk, C.P.; Curtis, S.E. Anabaena sp. Strain PCC 7120 Gene devH Is Required for Synthesis of the Heterocyst Glycolipid Layer. *J. Bacteriol.* **2005**, *187*, 2326–2331. [CrossRef]

74. Ehira, S.; Ohmori, M. NrrA directly regulates expression of the fraF gene and antisense RNAs for fraE in the heterocyst-forming cyanobacterium *Anabaena* sp. strain PCC 7120. *Microbiology* **2014**, *160*, 844–850. [CrossRef]

75. Katoh, H.; Hagino, N.; Grossman, A.R.; Ogawa, T. Genes Essential to Iron Transport in the Cyanobacterium *Synechocystis* sp. Strain PCC 6803. *J. Bacteriol.* **2001**, *183*, 2779–2784. [CrossRef] [PubMed]

76. Fresenborg, L.S.; Graf, J.; Schätzle, H.; Schleiff, E. Iron homeostasis of cyanobacteria: Advancements in siderophores and metal transporters. In *Advances in Cyanobacterial Biology*; Academic Press: Cambridge, MA, USA, 2020. [CrossRef]

77. Shen, G.; Balasubramanian, R.; Wang, T.; Wu, Y.; Hoffart, L.M.; Krebs, C.; Bryant, N.A.; Golbeck, J.H. SufR Coordinates Two [4Fe-4S]$^{2+, 1+}$ Clusters and Functions as a Transcriptional Repressor of the *sufBCDS* Operon and an Autoregulator of *sufR* in Cyanobacteria. *J. Biol. Chem.* **2007**, *282*, 31909–31919. [CrossRef] [PubMed]

78. Shcolnick, S.; Summerfield, T.C.; Reytman, L.; Sherman, L.A.; Keren, N. The Mechanism of Iron Homeostasis in the Unicellular Cyanobacterium *Synechocystis* sp. PCC 6803 and Its Relationship to Oxidative Stress. *Plant Physiol.* **2009**, *150*, 2045–2056. [CrossRef]

79. Vuorijoki, L.; Tiwari, A.; Kallio, P.T.; Aro, E.-M. Inactivation of iron-sulfur cluster biogenesis regulator SufR in *Synechocystis* sp. PCC 6803 induces unique iron-dependent protein-level responses. *Biochim. Biophys. Acta Gen. Subj.* **2017**, *1861*, 1085–1098. [CrossRef]

80. Garcia-Dominguez, M.; López-Maury, L.; Florencio, F.J.; Reyes, J.C.; Braunstein, M.; Brown, A.M.; Kurtz, S.; Jacobs, W.R. A Gene Cluster Involved in Metal Homeostasis in the Cyanobacterium *Synechocystis* sp. Strain PCC 6803. *J. Bacteriol.* **2000**, *182*, 1507–1514. [CrossRef]

81. Huertas, M.; López-Maury, L.; Giner-Lamia, J.; Riego, A.M.S.; Florencio, F.J. Metals in Cyanobacteria: Analysis of the Copper, Nickel, Cobalt and Arsenic Homeostasis Mechanisms. *Life* **2014**, *4*, 865–886. [CrossRef]

82. Daley, S.M.E.; Kappell, A.D.; Carrick, M.J.; Burnap, R.L. Regulation of the Cyanobacterial CO_2-Concentrating Mechanism Involves Internal Sensing of NADP+ and α-Ketogutarate Levels by Transcription Factor CcmR. *PLoS ONE* **2012**, *7*, e41286. [CrossRef]

83. Eisenhut, M.; Huege, J.; Schwarz, D.; Bauwe, H.; Kopka, J.; Hagemann, M. Metabolome Phenotyping of Inorganic Carbon Limitation in Cells of the Wild Type and Photorespiratory Mutants of the Cyanobacterium *Synechocystis* sp. Strain PCC 6803. *Plant Physiol.* **2008**, *148*, 2109–2120. [CrossRef]

84. Figge, R.M.; Cassier-Chauvat, C.; Chauvat, F.; Cerff, R. Characterization and analysis of an NAD(P)H dehydrogenase transcriptional regulator critical for the survival of cyanobacteria facing inorganic carbon starvation and osmotic stress. *Mol. Microbiol.* **2001**, *39*, 455–469. [CrossRef]

85. Nishimura, T.; Takahashi, Y.; Yamaguchi, O.; Suzuki, H.; Maeda, S.-I.; Omata, T. Mechanism of low CO_2-induced activation of the cmp bicarbonate transporter operon by a LysR family protein in the cyanobacterium *Synechococcus elongatus* strain PCC 7942. *Mol. Microbiol.* **2008**, *68*, 98–109. [CrossRef]

86. Takahashi, Y.; Yamaguchi, O.; Omata, T. Roles of CmpR, a LysR family transcriptional regulator, in acclimation of the cyanobacterium *Synechococcus* sp. strain PCC 7942 to low-CO_2 and high-light conditions. *Mol. Microbiol.* **2004**, *52*, 837–845. [CrossRef] [PubMed]

87. Tanaka, H.; Kitamura, M.; Nakano, Y.; Katayama, M.; Takahashi, Y.; Kondo, T.; Manabe, K.; Omata, T.; Kutsuna, S. CmpR is Important for Circadian Phasing and Cell Growth. *Plant Cell Physiol.* **2012**, *53*, 1561–1569. [CrossRef] [PubMed]

88. Zhang, P.; Sicora, C.I.; Vorontsova, N.; Allahverdiyeva, Y.; Allahverdiyeva, Y.; Nixon, P.J.; Aro, E.-M. FtsH protease is required for induction of inorganic carbon acquisition complexes in *Synechocystis* sp. PCC 6803. *Mol. Microbiol.* **2007**, *65*, 728–740. [CrossRef] [PubMed]

89. Domain, F.; Houot, L.; Chauvat, F.; Cassier-Chauvat, C. Function and regulation of the cyanobacterial genes lexA, recA and ruvB: LexA is critical to the survival of cells facing inorganic carbon starvation. *Mol. Microbiol.* **2004**, *53*, 65–80. [CrossRef] [PubMed]

90. Lieman-Hurwitz, J.; Haimovich, M.; Shalev-Malul, G.; Ishii, A.; Hihara, Y.; Gaathon, A.; Lebendiker, M.; Kaplan, A. A cyanobacterial AbrB-like protein affects the apparent photosynthetic affinity for CO_2 by modulating low-CO_2-induced gene expression. *Environ. Microbiol.* **2009**, *11*, 927–936. [CrossRef]

91. Kaniya, Y.; Kizawa, A.; Miyagi, A.; Kawai-Yamada, M.; Uchimiya, H.; Kaneko, Y.; Nishiyama, Y.; Hihara, Y. Deletion of the Transcriptional Regulator cyAbrB2 Deregulates Primary Carbon Metabolism in *Synechocystis* sp. PCC 68031[W]. *Plant Physiol.* **2013**, *162*, 1153–1163. [CrossRef]

92. Orf, I.; Schwarz, D.; Kaplan, A.; Kopka, J.; Hess, W.R.; Hagemann, M.; Klähn, S. CyAbrB2 Contributes to the Transcriptional Regulation of Low CO_2 Acclimation in *Synechocystis* sp. PCC 6803. *Plant Cell Physiol.* **2016**, *57*, 2232–2243. [CrossRef]

93. Sjöholm, J.; Oliveira, P.; Lindblad, P. Transcription and Regulation of the Bidirectional Hydrogenase in the Cyanobacterium *Nostoc* sp. Strain PCC 7120. *Appl. Environ. Microbiol.* **2007**, *73*, 5435–5446. [CrossRef]

94. Oliveira, P.; Lindblad, P. Transcriptional regulation of the cyanobacterial bidirectional Hox-hydrogenase. *Dalton Trans.* **2009**, 9990–9996. [CrossRef]

95. Ishii, A.; Hihara, Y. An AbrB-Like Transcriptional Regulator, Sll0822, Is Essential for the Activation of Nitrogen-Regulated Genes in *Synechocystis* sp. PCC 6803. *Plant Physiol.* **2008**, *148*, 660–670. [CrossRef]

96. Agervald, Å.; Baebprasert, W.; Zhang, X.; Incharoensakdi, A.; Lindblad, P.; Stensjö, K. The CyAbrB transcription factor CalA regulates the iron superoxide dismutase in *Nostoc* sp. strain PCC 7120. *Environ. Microbiol.* **2010**, *12*, 2826–2837. [CrossRef] [PubMed]

97. Higo, A.; Nishiyama, E.; Nakamura, K.; Hihara, Y.; Ehira, S. cyAbrB Transcriptional Regulators as Safety Devices To Inhibit Heterocyst Differentiation in *Anabaena* sp. Strain PCC 7120. *J. Bacteriol.* **2019**, *201*. [CrossRef] [PubMed]

98. Tanigawa, R.; Shirokane, M.; Maeda, S.-I.; Omata, T.; Tanaka, K.; Takahashi, H. Transcriptional activation of NtcA-dependent promoters of *Synechococcus* sp. PCC 7942 by 2-oxoglutarate in vitro. *Proc. Natl. Acad. Sci. USA* **2002**, *99*, 4251–4255. [CrossRef]

99. Zhao, M.-X.; Jiang, Y.-L.; He, Y.-X.; Chen, Y.-F.; Teng, Y.-B.; Chen, Y.; Zhang, C.-C.; Zhou, C.-Z. Structural basis for the allosteric control of the global transcription factor NtcA by the nitrogen starvation signal 2-oxoglutarate. *Proc. Natl. Acad. Sci. USA* **2010**, *107*, 12487–12492. [CrossRef]

100. Espinosa, J.; Forchhammer, K.; Burillo, S.; Contreras, A. Interaction network in cyanobacterial nitrogen regulation: PipX, a protein that interacts in a 2-oxoglutarate dependent manner with PII and NtcA. *Mol. Microbiol.* **2006**, *61*, 457–469. [CrossRef] [PubMed]

101. Ohashi, Y.; Shi, W.; Takatani, N.; Aichi, M.; Maeda, S.-I.; Watanabe, S.; Yoshikawa, H.; Omata, T. Regulation of nitrate assimilation in cyanobacteria. *J. Exp. Bot.* **2011**, *62*, 1411–1424. [CrossRef]

102. Su, Z.; Olman, V.; Mao, F.; Xu, Y. Comparative genomics analysis of NtcA regulons in cyanobacteria: Regulation of nitrogen assimilation and its coupling to photosynthesis. *Nucleic Acids Res.* **2005**, *33*, 5156–5171. [CrossRef]

103. Mitschke, J.; Vioque, A.; Haas, F.; Hess, W.R.; Muro-Pastor, A.M. Dynamics of transcriptional start site selection during nitrogen stress-induced cell differentiation in *Anabaena* sp. PCC7120. *Proc. Natl. Acad. Sci. USA* **2011**, *108*, 20130–20135. [CrossRef]

104. Galmozzi, C.V.; Saelices, L.; Florencio, F.J.; Muro-Pastor, M.I. Posttranscriptional Regulation of Glutamine Synthetase in the Filamentous Cyanobacterium *Anabaena* sp. PCC 7120: Differential Expression between Vegetative Cells and Heterocysts. *J. Bacteriol.* **2010**, *192*, 4701–4711. [CrossRef]

105. Kang, R.-J.; Shi, D.-J.; Cong, W.; Cai, Z.-L.; Ouyang, F. Regulation of CO_2 on heterocyst differentiation and nitrate uptake in the cyanobacterium Anabaena sp. PCC 7120. *J. Appl. Microbiol.* **2005**, *98*, 693–698. [CrossRef]

106. Flores, E.; Arévalo, S.; Burnat, M. Cyanophycin and arginine metabolism in cyanobacteria. *Algal Res.* **2019**, *42*, 101577. [CrossRef]

107. Lopez-Gomollon, S.; Hernández, J.A.; Pellicer, S.; Angarica, V.E.; Peleato, M.L.; Fillat, M.F. Cross-talk Between Iron and Nitrogen Regulatory Networks in *Anabaena* (*Nostoc*) sp. PCC 7120: Identification of Overlapping Genes in FurA and NtcA Regulons. *J. Mol. Biol.* **2007**, *374*, 267–281. [CrossRef] [PubMed]

108. Raven, J.A.; Evans, M.C.W.; Korb, R.E. The role of trace metals in photosynthetic electron transport in O_2-evolving organisms. *Photosynth. Res.* **1999**, *60*, 111–150. [CrossRef]

109. Gupta, J.K.; Rai, P.; Jain, K.K.; Srivastava, S. Overexpression of bicarbonate transporters in the marine cyanobacterium *Synechococcus* sp. PCC 7002 increases growth rate and glycogen accumulation. *Biotechnol. Biofuels* **2020**, *13*, 17. [CrossRef] [PubMed]

110. Cano, M.; Holland, S.C.; Artier, J.; Burnap, R.L.; Ghirardi, M.; Morgan, J.A.; Yu, J. Glycogen Synthesis and Metabolite Overflow Contribute to Energy Balancing in Cyanobacteria. *Cell Rep.* **2018**, *23*, 667–672. [CrossRef]

111. Schwarz, D.; Orf, I.; Kopka, J.; Hagemann, M. Effects of Inorganic Carbon Limitation on the Metabolome of the *Synechocystis* sp. PCC 6803 Mutant Defective in glnB Encoding the Central Regulator PII of Cyanobacterial C/N Acclimation. *Metabolites* **2014**, *4*, 232–247. [CrossRef]

112. Vermaas, W.F. Photosynthesis and Respiration in Cyanobacteria. In *Encyclopedia of Life Sciences*; John Wiley & Sons, Ltd.: Chichester, UK, 2001. [CrossRef]

113. Welkie, D.G.; Rubin, B.E.; Diamond, S.; Hood, R.D.; Savage, D.F.; Golden, S.S. A Hard Day's Night: Cyanobacteria in Diel Cycles. *Trends Microbiol.* **2019**, *27*, 231–242. [CrossRef]

114. Makowka, A.; Nichelmann, L.; Schulze, D.; Spengler, K.; Wittmann, C.; Forchhammer, K.; Gutekunst, K. Glycolytic Shunts Replenish the Calvin–Benson–Bassham Cycle as Anaplerotic Reactions in Cyanobacteria. *Mol. Plant* **2020**, *13*, 471–482. [CrossRef]

115. Huege, J.; Goetze, J.; Schwarz, D.; Bauwe, H.; Hagemann, M.; Kopka, J. Modulation of the Major Paths of Carbon in Photorespiratory Mutants of Synechocystis. *PLoS ONE* **2011**, *6*, e16278. [CrossRef]

116. Tang, J.K.-H.; Tang, Y.J.; Blankenship, R.E. Carbon Metabolic Pathways in Phototrophic Bacteria and Their Broader Evolutionary Implications. *Front. Microbiol.* **2011**, *2*, 165. [CrossRef]

Publisher's Note: MDPI stays neutral with regard to jurisdictional claims in published maps and institutional affiliations.

Communication

Heavy Metal Stress Alters the Response of the Unicellular Cyanobacterium *Synechococcus elongatus* PCC 7942 to Nitrogen Starvation

Khaled A. Selim * and Michael Haffner

Organismic Interactions Department, Interfaculty Institute for Microbiology and Infection Medicine, Cluster of Excellence 'Controlling Microbes to Fight Infections', Tübingen University, Auf der Morgenstelle 28, 72076 Tübingen, Germany; michael.haffner@student.uni-tuebingen.de
* Correspondence: Khaled.selim@uni-tuebingen.de

Received: 19 October 2020; Accepted: 5 November 2020; Published: 7 November 2020

Abstract: Non-diazotrophic cyanobacteria are unable to fix atmospheric nitrogen and rely on combined nitrogen for growth and development. In the absence of combined nitrogen sources, most non-diazotrophic cyanobacteria, e.g., *Synechocystis* sp. PCC 6803 or *Synechococcus elongatus* PCC 7942, enter a dormant stage called chlorosis. The chlorosis process involves switching off photosynthetic activities and downregulating protein biosynthesis. Addition of a combined nitrogen source induces the regeneration of chlorotic cells in a process called resuscitation. As heavy metals are ubiquitous in the cyanobacterial biosphere, their influence on the vegetative growth of cyanobacterial cells has been extensively studied. However, the effect of heavy metal stress on chlorotic cyanobacterial cells remains elusive. To simulate the natural conditions, we investigated the effects of long-term exposure of *S. elongatus* PCC 7942 cells to both heavy metal stress and nitrogen starvation. We were able to show that elevated heavy metal concentrations, especially for Ni^{2+}, Cd^{2+}, Cu^{2+} and Zn^{2+}, are highly toxic to nitrogen starved cells. In particular, cells exposed to elevated concentrations of Cd^{2+} or Ni^{2+} were not able to properly enter chlorosis as they failed to degrade phycobiliproteins and chlorophyll a and remained greenish. In resuscitation assays, these cells were unable to recover from the simultaneous nitrogen starvation and Cd^{2+} or Ni^{2+} stress. The elevated toxicity of Cd^{2+} or Ni^{2+} presumably occurs due to their interference with the onset of chlorosis in nitrogen-starved cells, eventually leading to cell death.

Keywords: heavy metal stress; chlorosis; resuscitation; non-diazotrophic cyanobacteria; *Synechococcus elongatus* PCC 7942

---

1. Introduction

Cyanobacteria occupy a privileged position in the Earth's history as a key player in global C/N cycles [1] and as inventors of oxygenic photosynthesis by evolving two coupled photosystems. Thereby, they enriched the Earth's atmosphere with oxygen around 2.4 billion years ago [2]. Physiologically, with respect to nitrogen demand, cyanobacteria can be classified into two distinct groups: (1) diazotrophic cyanobacteria, which can fix atmospheric gaseous nitrogen via an enzyme called nitrogenase (e.g., *Anabaena variabilis* PCC 7937 and *Nostoc* sp. PCC 7120 of heterocyst-forming cyanobacteria and *Cyanothece* sp. 51142 of unicellular non-heterocystous cyanobacteria) [3,4], and (2) non-diazotrophic cyanobacteria, which are unable to fix atmospheric nitrogen and rely on the availability of a combined nitrogen source, such as nitrate or ammonia, for growth and intracellular catabolic and anabolic reactions (e.g., *Synechocystis* sp. PCC 6803 and *Synechococcus elongatus* PCC 7942) [1,5].

In nature, nitrogen availability is highly variable. Under environmental depletion of combined nitrogen, most non-diazotrophic cyanobacteria respond by degrading their photosynthetic pigments, in a process termed chlorosis [6]. Upon prolonged starvation, they switch to a maintenance lifestyle by tuning down their anabolic processes. Later on, they enter a dormant state, where the chlorotic cells survive the prolonged periods of nitrogen starvation [7]. The chlorosis process involves degradation of phycobilisomes (light-harvesting complex), chlorophyll a, the bulk of cellular proteins, and thylakoid membranes; downregulation of protein biosynthesis and energy-consuming reactions; accumulation of reserve polymers of glycogen and poly-β-hydroxybutyrate (PHB) [5,8]. A hallmark of chlorosis is the cell cycle arrest and the turn from a blue-green color to a yellow color due to the degradation of the photosynthetic apparatus machinery [8,9]. A minimal residual photosynthesis [8,10] (~0.1%) is, however, maintained to retain cell viability over prolonged periods up to 6 months.

Upon re-availability of combined nitrogen, the chlorotic dominant cyanobacteria rapidly awake and return greenish to the vegetative growth in a process called resuscitation [9]. Therefore, resuscitation can be considered as the reverse process of chlorosis and takes place through two phases [8,9]. Within the first 12–16 h (the first phase of resuscitation), cells reactivate their protein biosynthesis and machinery for nitrate assimilation (*narB*; *nir* operon; *moa* gene cluster for the molybdenum co-factor of nitrate reductase) and rebuild the entire F-type ATPase machinery [5]. The required energy within the first phase of resuscitation is generated by glycogen degradation [11]. In the second phase of resuscitation, a metabolic rewiring towards the photosynthetic machinery takes place, including re-pigmentation, re-building of thylakoid membranes and re-generation of photosystem II and oxygen evolution activities [8,9]. The reconstitution and recovery of the photosynthetic apparatus is recognizable after 24 h of resuscitation, as evidenced by re-greening of the cells and increased photosystem II activity, while proper cell growth is first detectable after 48 h [8].

Cyanobacteria have specific needs for metals, which frequently differ from those of other bacteria—Mn^{2+} is required for chlorophyll biosynthesis and for water-splitting oxygen-evolving complex, Cu^{2+} for plastocyanin and cytochrome oxidase, Zn^{2+} for DNA and RNA polymerases as well as for the carboxysome-localized carbonic anhydrase (for carbon assimilation), Co^{2+} for vitamin B_{12} biosynthesis and Mo^{2+} for nitrogenase in heterocysts [12–14]. Those metals are considered as essential elements and required in trace amounts, while non-essential metals (e.g., Cd^{2+}, Pb^{2+}, Cr^{2+} and Hg^{2+}) are toxic to cyanobacteria [13,14]. In addition, the essential metals are more generally required to provide defense against oxidative stresses, transfer electrons in photosynthetic reaction centers and serve as cofactors for metalloproteins. However, high concentrations of heavy metals, irrespective of whether they are essential or not, negatively influence cyanobacterial metabolism. Mostly affected is the photosynthetic machinery with alterations in the photosynthetic electron transport chain, several photosynthetic enzymes, dark reactions, ATP synthesis and photosynthetic pigments, leading to photoinhibition. For example, Pb^{2+}, Mn^{2+} and Cd^{2+} can inhibit chlorophyll a and b biosynthesis [13,14].

Based on mutational analysis in *Escherichia coli*, the CutA protein (**Cu**$^{2+}$ tolerance protein **A**) has been proposed to be implicated in bacterial divalent ion tolerance. However, we recently were unable to link CutA to heavy metal tolerance in cyanobacteria under vegetative growth [15]. As heavy metals are ubiquitous in the cyanobacterial biosphere, a proper control of metal homeostasis is, therefore, essential for the photosynthetic lifestyle of cyanobacteria. While nitrogen remains the growth-limiting factor, other environmental factors, such as heavy metals, could be toxic to the vegetative cells owing to their interference with vital intracellular processes [13,14]. However, it is unclear how they affect the cyanobacteria when they are in the dormant chlorotic state. Since nitrogen limitation is a frequent environmental stress, we aimed to reveal how heavy metal stress and nitrogen limitation could affect wild-type and $\Delta cutA$ cells in the cyanobacterium model organism, *S. elongatus* PCC 7942.

2. Materials and Methods

2.1. Microbiology Biology Methods

Exponentially growing wild-types of either *Synechococcus elongatus* PCC 7942 and Δ*cutA* mutant cells of OD_{750} 0.3–0.5 were subjected to heavy metals stress by supplementing unbuffered BG11 media with one of the following heavy metals: $PbCl_2$, $CrCl_2$, $MnCl_2$, $ZnCl_2$, $CuSO_4$, $NiCl_2$ and $CdCl_2$ in concentrations from 2.5 to 50 µM in a 24-well plate. For chlorosis experiments, the heavy metal treatments were added to $BG11_0$ (BG11 media without nitrate) to induce metal stress and chlorosis [8] for *S. elongatus* PCC 7942 and Δ*cutA* mutant of an initial OD_{750} 0.5. The survival of nitrogen-starved/heavy-metal-stressed cells was checked by drop assay [15,16], where 5 µl cells from each treatment was spotted on nitrate-supplemented BG11 agar plates in the absence of heavy metals and incubated at 28 °C under a light intensity of 30–50 µmol photons m^{-2} s^{-1} for one week.

2.2. Pulse-Amplitude-Modulation (PAM) Measurements

The photosynthetic fitness for the recovery of the nitrogen-starved/heavy-metal-treated cells was estimated by measuring photosystem II (PSII) activity under a 50 µE light intensity using WATER-PAM chlorophyll fluorescence (Walz GmbH), as described previously [16,17]. The maximal PSII quantum yield was determined with the saturation pulse method using the *F0−Fm/Fm* ratio [17].

2.3. Photosynthetic Oxygen Evolution Measurements

In vivo photosynthetic oxygen evolution was estimated using an oxygen electrode of the Clark-type (Hansatech DW1) [16,17]. Oxygen evolution of 2 mL of recovering cultures normalized to an OD_{750} of 0.4 was measured at room temperature at 50 µE. The 50 µE light intensity was provided using a high-intensity white-light source (Hansatech L2).

2.4. Molecular Biology Methods

Transformation of *S. elongatus* PCC 7942 to create a knockout mutant in the open reading frame (ORF *Synpcc7942_2261*) encoding for CutA was achieved by replacement of the ORF *Synpcc7942_2261* by kanamycin resistance cassette, as described previously for *Synechocystis* sp. PCC 6803 [16]. The Δ*cutA* mutant was selected on BG11 plates supplemented with 50 µg/ml kanamycin and verified with PCR, as shown previously elsewhere [15].

3. Results

To simulate the natural situation and to study the influence of long-time exposure of chlorotic cells to heavy metals, we combined heavy metal stress and nitrogen starvation on *S. elongatus* wild-type (WT) and *cutA* null mutant (Δ*Se*CutA) cells, which we previously showed not to be involved in heavy metal tolerance in vegetative growing cyanobacteria [15]. The Δ*Se*CutA mutant was generated by replacement of the *cutA* encoding gene *synpcc7942_2261* with a kanamycin-resistant gene in *S. elongatus* PCC 7942 [15].

When exposed to low concentrations of heavy metals (2.5 µM) and nitrogen starvation, the WT and Δ*cutA* cells were able to resuscitate on nitrate-supplemented BG11 plates lacking heavy metals (Figure 1A; first row). These results were confirmed further by measuring the recovery of the photosynthetic apparatus (photosystem II; PSII) using pulse-amplitude-modulation (PAM) fluorometry for WT and Δ*cutA* cells, resuscitating from a long period (28 days) of chlorosis in the presence of 2.5 µM heavy metals ($PbCl_2$, $CuSO_4$ or $CdCl_2$). Over 50 h of resuscitation, the efficiency of PSII regeneration for WT and Δ*cutA* cells was comparable (Figure 1B). Additionally, in agreement with the increase in PSII quantum yield in Figure 1B, photosynthetic oxygen evolution of the same cells was also recognizable after 50 h of resuscitation (Figure 1C). However, the net oxygen evolution of heavy-metal-treated cells was reduced by 30 to 40% in comparison to untreated cells, implying that the cells were severely stressed by the presence of all the heavy metals tested but were still able to recover. A similar phenotype was observed for the Δ*cutA* mutant. This observation motivated us to check further for the influence of higher concentrations of heavy metals on chlorotic cells.

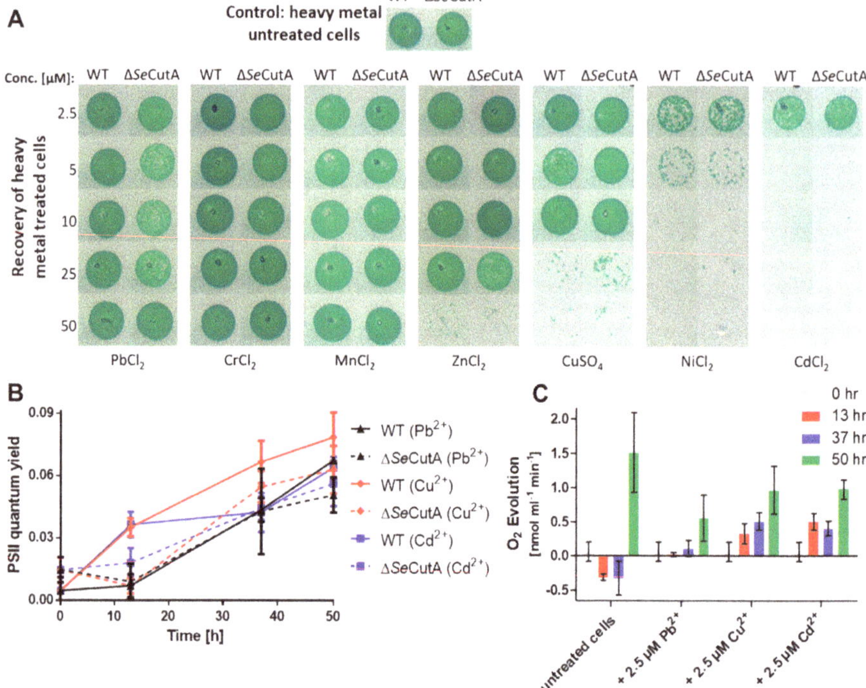

Figure 1. Influence of heavy metal treatments on the chlorotic *S. elongatus* WT and Δ*Se*CutA cells. (**A**) The survival of nitrogen-starved/heavy-metal-treated cells (as indicated) was evaluated by drop assay on BG11 (nitrate-supplemented, in absence of heavy metals) media. (**B**,**C**) Shown are the PSII quantum yield (**B**) and photosynthetic oxygen evolution (**C**) for the recovery of *S. elongatus* cells over 50 h of resuscitation from a long period (28 days) of chlorosis and 2.5 µM heavy metal treatment (as indicated). Both PSII quantum yield in (**B**) and oxygen evolution in (**C**) were determined under a light intensity of 50 µE, using pulse-amplitude-modulation (PAM) fluorometry and an oxygen electrode, respectively.

Interestingly, when *S. elongatus* cells were nitrogen-starved for 5 days and simultaneously exposed to different heavy metals (Pb^{2+}, Cr^{3+}, Mn^{2+}, Zn^{2+}, Cu^{2+}, Ni^{2+} and Cd^{2+}) at concentrations higher than 2.5 µM, different toxicities were observed (Figure 1A). Recovery of treated cells on BG11 plates (nitrate-supplemented, in absence of heavy metals) revealed a high toxicity of Ni^{2+} and Cd^{2+} ions to chlorotic cells, already apparent at the concentration of 5 µM, whereas Cu^{2+} and Zn^{2+} ions showed intermediate toxicity and the rest of the tested ions showed no toxicity (Figure 1A). A similar pattern of toxicity was observed for the Δ*cutA* mutant (Figure 1A).

During normal nitrogen starvation-induced chlorosis, the cells turn from a deep blue-green to a yellowish color due to NblA-induced pigment degradation [5,8]. Surprisingly, we observed that under nitrogen starvation conditions and in the presence of elevated levels of the heavy metals Ni^{2+}, Cd^{2+}, Cu^{2+} and Zn^{2+}, the cells remained green due to impaired degradation of phycobiliproteins and chlorophyll a (visible in absorbance changes at 625 and 680 nm) and were not able to enter chlorosis properly (Figure 2A). These non-bleaching cells died after several days of heavy metal stress and nitrogen starvation as indicated by the inability of the cells to recover on BG11 (Figure 1A). To confirm these results, PAM fluorometry was used to monitor (over 104 h) the PSII activity of cells recovering from a 92-hour-long treatment with high concentrations (25 and 50 µM) of either medium toxic (Cu^{2+}) or high toxic (Ni^{2+}) ions during nitrogen starvation (Figure 2B,C). This analysis revealed that PSII activity resumed in cells that were recovering from Cu^{2+} treatment during chlorosis (Figure 2B), confirming the low toxicity of Cu^{2+} ions to the chlorotic cells. By contrast, cells that had been exposed to elevated concentrations of Ni^{2+} during chlorosis were not able to recover after the shift to nitrogen-rich and Ni^{2+}-free media (Figure 2C), confirming the high toxicity of Ni^{2+} ions and the death of the chlorotic cells after $NiCl_2$ treatment.

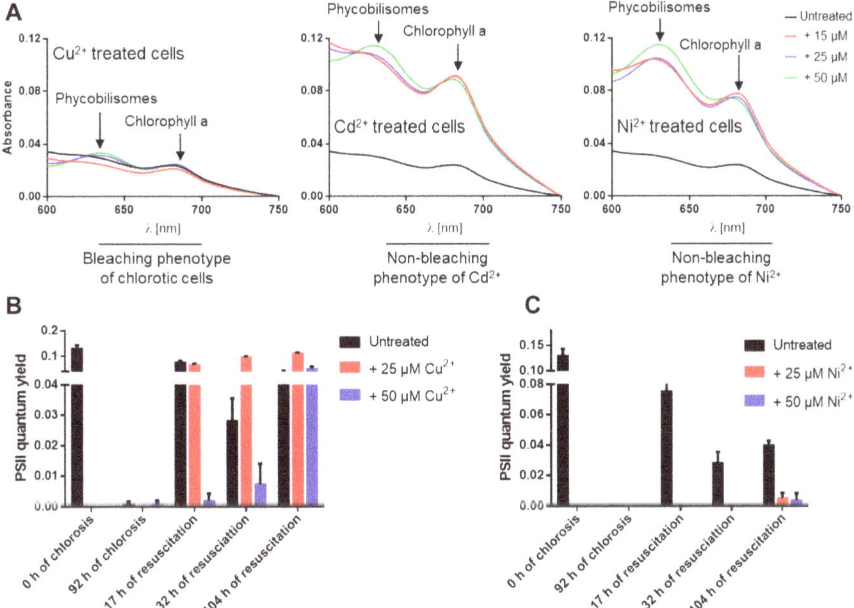

Figure 2. The non-bleaching phenotype of nitrogen-starved and heavy-metal-treated *S. elongatus* cells leads to cell death. (**A**) Absorption spectra of nitrogen-starved and heavy-metal-treated cells as indicated, revealing impair degradation of phycobiliproteins and chlorophyll a under Cd^{2+} and Ni^{2+} treatment. (**B**,**C**) The PSII quantum yield for the recovery of resuscitated cells from chlorosis and high metal concentrations (25 or 50 µM) of Cu^{2+} (**B**) or Ni^{2+} (**C**) by PAM fluorometry under a light intensity of 50 µE.

4. Discussion

Nitrogen-starved WT and Δ*cutA* cells were exposed to different concentrations of heavy metals to check the toxicity of heavy metals on chlorotic cells (Figure 1A,B). Our results, presented here, clearly reveal that elevated heavy metal concentrations, especially for Ni^{2+}, Cd^{2+}, Cu^{2+} and Zn^{2+}, are severely toxic to nitrogen-starved cells. Our data suggest that this is most probably due to dysregulation of the bacterium metabolism and the highly oxidative stress imposed on the cells. In particular, cells exposed to elevated concentrations of Cd^{2+} or Ni^{2+} failed to enter chlorosis properly as they were not able to degrade phycobilisomes and chlorophyll a (Figure 2A). In resuscitation assays, these cells were able to recover from chlorosis coupled with Cu^{2+} or Zn^{2+} stress but not Cd^{2+} or Ni^{2+} stress (Figures 1A and 2B,C). Therefore, we concluded that the elevated toxicity of Cd^{2+} or Ni^{2+} occurs due to their interference with the onset of chlorosis in nitrogen-starved cells, which finally leads to cell death, most likely due to accumulation of reactive oxygen species.

The Δ*cutA* mutant was shown, previously, not to be involved in heavy metal stress in cyanobacteria [15]. In agreement with our previous results [15], we were unable to link the *S. elongatus* Δ*cutA* mutant to heavy metal tolerance even under severe stress conditions by combining heavy metal and nitrogen stress (Figure 1A,B). Indeed, we found no difference between the *S. elongatus cutA* mutant and wild-type strains recovering from heavy metal stress either in the presence [15] or absence of combined nitrogen sources, confirming that CutA plays a different role in cyanobacteria rather than participating in heavy metal resistance.

As photoautotroph microbes, cyanobacteria are considered promising microbial factories for sunlight- and CO_2-fueled production of a vast array of high-value chemicals (e.g., biopolymers) that are of remarkable interest for industry and human health. Synthetic biology strategies to engineer cyanobacteria rely on robust promoters with predictable input–output responses to tightly control the desired gene expression for sustainable biotechnological applications [18]. Among the promoters used to engineer cyanobacteria, several are metal-inducible promoters (e.g., Cu^{2+}-dependent P_{petJ} and P_{petE}, Ni^{2+}-dependent P_{nrsB}, Zn^{2+}-dependent P_{zia} and P_{smt}, Co^{2+}-dependent P_{coaT}, and P_{copM} which is induced by Ni^{2+}, Cd^{2+}, Cu^{2+} and Zn^{2+}) [19]. The Ni^{2+}-inducible P_{nrsB} promotor was shown to be among the most strong and versatile promoters in the unicellular cyanobacterium *Synechocystis* sp. PCC 6803, with expression levels nearly up to the activity of the strongest *Synechocystis* promotor, P_{psbA2} [20,21]. However, the bottleneck in using metal-dependent promoters in cyanobacterial synthetic biology applications is the metal toxicity at higher concentrations, especially for Cu^{2+}, Ni^{2+}, Zn^{2+} and Co^{2+} [12,18–20].

Some of useful cyanobacterial products accumulate under nitrogen starvation conditions—for example, PHB [22]. As a biopolymer, PHB receives, nowadays, a lot of attention for biotechnological applications as a replacement of conventional plastic. Consistent with our results, attempts to overexpress the PHB synthase-encoding gene in *Synechocystis* sp. PCC 6803 under the control of the strong Ni^{2+}-inducible promotor in nitrogen-depleted conditions were unsuccessful [23]. Our data suggest that this is likely due to the high toxicity of Ni^{2+} under nitrogen starvation. Furthermore, recent studies revealed a limitation in the use of the P_{nrsB}, P_{petJ} and P_{coaT} promoters for biotechnological applications in cyanobacteria at high metal concentrations [18,20,24]. All of these observations hinder the use of metal-dependent promoters for synthetic biology applications in cyanobacteria, especially under nitrogen starvation conditions.

Collectively, this study sheds light on an unexplored area of cyanobacteria physiology and serves as a valuable guide for synthetic biology approaches in cyanobacteria by defining the clear limitation of using heavy metal-inducible promoters under nitrogen starvation. A recent proteomic study defined the global proteomic response of *Synechocystis* sp. PCC 6803 to Cu^{2+} applied to regulate the Cu^{2+}-inducible promoter P_{petJ} [24]. Notably, they identified clear irreversible proteomic changes due to Cu^{2+} stress, including significant alternations in the protein amounts of the outer and inner membranes and of the cell surface and the downregulation of ribosomal proteins.

Author Contributions: K.A.S. conceived and initiated the project. M.H. performed the experiments with inputs from K.A.S. K.A.S. generated the mutant, analyzed and interpreted the data with help from M.H., generated the final figures, and wrote the manuscript. All authors have read and agreed to the published version of the manuscript.

Funding: This research work was supported by DFG Graduiertenkolleg Grant 1708-2, and Deutscher Akademischer Austauschdienst (DAAD) to K.A.S.

Acknowledgments: The authors gratefully acknowledge Karl Forchhammer for continued support. Furthermore, we would like to acknowledge Libera Lo-Presti for critical scientific and linguistic editing of the manuscript, the infrastructural support by the Cluster of Excellence *'Controlling Microbes to Fight Infections'* (EXC 2124) of the German research foundation (DFG), and the support by Open Access Publishing of Tübingen University.

Conflicts of Interest: The authors declare no conflict of interest.

References

1. Forchhammer, K.; Selim, K.A. Carbon/nitrogen homeostasis control in cyanobacteria. *FEMS Microbiol. Rev.* **2020**, *44*, 33–53. [CrossRef] [PubMed]

2. Schirrmeister, B.E.; de Vos, J.M.; Antonelli, A.; Bagheri, H.C. Evolution of multicellularity coincided with increased diversification of cyanobacteria and the Great Oxidation Event. *Proc. Natl. Acad. Sci. USA* **2013**, *110*, 1791–1796. [CrossRef] [PubMed]

3. Flores, E.; Herrero, A. Compartmentalized function through cell differentiation in filamentous cyanobacteria. *Nat. Rev. Microbiol.* **2010**, *8*, 39–50. [CrossRef] [PubMed]

4. Welsh, E.A.; Liberton, M.; Stöckel, J.; Loh, T.; Elvitigala, T.; Wang, C.; Wollam, A.; Fulton, R.S.; Clifton, S.W.; Jacobs, J.M.; et al. The genome of *Cyanothece* 51142, a unicellular diazotrophic cyanobacterium important in the marine nitrogen cycle. *Proc. Natl. Acad. Sci. USA* **2008**, *105*, 15094–15099. [CrossRef] [PubMed]

5. Forchhammer, K.; Schwarz, R. Nitrogen chlorosis in unicellular cyanobacteria—A developmental program for surviving nitrogen deprivation. *Environ. Microbiol.* **2019**, *21*, 1173–1184. [CrossRef] [PubMed]

6. Allen, M.M.; Smith, A.J. Nitrogen chlorosis in blue-green algae. *Arch. Mikrobiol.* **1969**, *69*, 114–120. [CrossRef]

7. Görl, M.; Sauer, J.; Baier, T.; Forchhammer, K. Nitrogen-starvation-induced chlorosis in Synechococcus PCC 7942: Adaptation to long-term survival. *Microbiology (Reading)* **1998**, *144*, 2449–2458. [CrossRef]

8. Klotz, A.; Georg, J.; Bučinská, L.; Watanabe, S.; Reimann, V.; Januszewski, W.; Sobotka, R.; Jendrossek, D.; Hess, W.R.; Forchhammer, K. Awakening of a Dormant Cyanobacterium from Nitrogen Chlorosis Reveals a Genetically Determined Program. *Curr. Biol.* **2016**, *26*, 2862–2872. [CrossRef]

9. Spät, P.; Klotz, A.; Rexroth, S.; Maček, B.; Forchhammer, K. Chlorosis as a Developmental Program in Cyanobacteria: The Proteomic Fundament for Survival and Awakening. *Mol. Cell. Proteom.* **2018**, *17*, 1650–1669.

10. Sauer, J.; Schreiber, U.; Schmid, R.; Völker, U.; Forchhammer, K. Nitrogen starvation-induced chlorosis in Synechococcus PCC 7942. Low-level photosynthesis as a mechanism of long-term survival. *Plant Physiol.* **2001**, *126*, 233–243. [CrossRef]

11. Doello, S.; Klotz, A.; Makowka, A.; Gutekunst, K.; Forchhammer, K. A Specific Glycogen Mobilization Strategy Enables Rapid Awakening of Dormant Cyanobacteria from Chlorosis. *Plant Physiol.* **2018**, *177*, 594–603. [CrossRef]

12. Cavet, J.S.; Borrelly, G.P.; Robinson, N.J. Zn, Cu and Co in cyanobacteria: Selective control of metal availability. *FEMS Microbiol. Rev.* **2003**, *27*, 165–181. [CrossRef]

13. Pfeiffer, T.Ž.; čamagajevac, I.Š.; Maronić, D.Š.; Maksimović, I. Regulation of photosynthesis in algae under metal stress. In *Environment and Photosynthesis: A Future Prospect*; Singh, V., Singh, S., Singh, R., Prasad, S., Eds.; New Delhi-Stadium Press: Delhi, India, 2018; pp. 261–286.

14. Khan, M.; Nawaz, N.; Ali, I.; Azam, M.; Rizwan, M.; Ahmad, P.; Ali, S. Regulation of Photosynthesis Under Metal Stress. In *Photosynthesis, Productivity, and Environmental Stress*, 1st ed.; Parvaiz Ahmad, P., Ahanger, M.A., Alyemeni, M.N., Alam, P., Eds.; John Wiley & Sons Ltd.: Hoboken, NJ, USA, 2019; pp. 95–105.

15. Selim, K.A.; Tremiño, L.; Marco-Marín, C.; Alva, V.; Espinosa, J.; Contreras, A.; Hartmann, M.D.; Forchhammer, K.; Rubio, V. Functional and structural characterization of PII-like protein CutA does not support involvement in heavy metal tolerance and hints at a small-molecule carrying/signaling role. *FEBS J.* **2020**, *2020*. [CrossRef]

16. Selim, K.A.; Haase, F.; Hartmann, M.D.; Hagemann, M.; Forchhammer, K. P$_{II}$-like signaling protein SbtB links cAMP sensing with cyanobacterial inorganic carbon response. *Proc. Natl. Acad. Sci. USA* **2018**, *115*, E4861–E4869. [CrossRef]

17. Dai, G.Z.; Qiu, B.S.; Forchhammer, K. Ammonium tolerance in the cyanobacterium *Synechocystis* sp. strain PCC 6803 and the role of the psbA multigene family. *Plant Cell Environ.* **2014**, *37*, 840–851. [CrossRef]

18. Behle, A.; Saake, P.; Germann, A.T.; Dienst, D.; Axmann, I.M. Comparative Dose–Response Analysis of Inducible Promoters in Cyanobacteria. *ACS Synth. Biol.* **2020**, *9*, 843–855. [CrossRef] [PubMed]

19. Till, P.; Toepel, J.; Bühler, B.; Mach, R.L.; Mach-Aigner, A.R. Regulatory systems for gene expression control in cyanobacteria. *Appl. Microbiol. Biotechnol.* **2020**, *104*, 1977–1991. [CrossRef]

20. Englund, E.; Liang, F.; Lindberg, P. Evaluation of promoters and ribosome binding sites for biotechnological applications in the unicellular cyanobacterium *Synechocystis* sp. PCC 6803. *Sci. Rep.* **2016**, *6*, 36640. [CrossRef] [PubMed]

21. Blasi, B.; Peca, L.; Vass, I.; Kós, P.B. Characterization of stress responses of heavy metal and metalloid inducible promoters in synechocystis PCC6803. *J. Microbiol. Biotechnol.* **2012**, *22*, 166–169. [CrossRef]

22. Koch, M.; Doello, S.; Gutekunst, K.; Forchhammer, K. PHB is Produced from Glycogen Turn-over during Nitrogen Starvation in *Synechocystis* sp. PCC 6803. *Int. J. Mol. Sci.* **2019**, *20*, 1942. [CrossRef]

23. Alford, J. Etablierung Eines Neuen Kultivierungssystems und Genetische Modifikationen zur Steigerung der PHB-Produktion in *Synechocystis* sp. PCC 6803. Bachelor's Thesis, University of Tübingen, Tübingen, Germany, 2018.

24. Angeleri, M.; Muth-Pawlak, D.; Wilde, A.; Aro, E.M.; Battchikova, N. Global proteome response of Synechocystis 6803 to extreme copper environments applied to control the activity of the inducible petJ promoter. *J. Appl. Microbiol.* **2019**, *126*, 826–841. [CrossRef] [PubMed]

Publisher's Note: MDPI stays neutral with regard to jurisdictional claims in published maps and institutional affiliations.

Article

DnaK3 Is Involved in Biogenesis and/or Maintenance of Thylakoid Membrane Protein Complexes in the Cyanobacterium *Synechocystis* sp. PCC 6803

Adrien Thurotte [1,2,†], **Tobias Seidel** [1,†], **Ruven Jilly** [1], **Uwe Kahmann** [3] **and Dirk Schneider** [1,*]

[1] Department of Chemistry, Biochemistry, Johannes Gutenberg University Mainz, 55128 Mainz, Germany; adrienthurotte@netcourrier.com (A.T.); TobiasSeidel@gmx.net (T.S.); ruvenjilly@gmail.com (R.J.)

[2] Institute of Molecular Biosciences, Goethe University Frankfurt, Max-von-Laue Straße 9, 60438 Frankfurt, Germany

[3] Department of Molecular Cell Biology, Bielefeld University, 33615 Bielefeld, Germany; ZUD@gmx.de

* Correspondence: dirk.schneider@uni-mainz.de; Tel.: +49-6131-39-25833

† These authors contributed equally.

Received: 8 April 2020; Accepted: 28 April 2020; Published: 30 April 2020

Abstract: DnaK3, a highly conserved cyanobacterial chaperone of the Hsp70 family, binds to cyanobacterial thylakoid membranes, and an involvement of DnaK3 in the biogenesis of thylakoid membranes has been suggested. As shown here, light triggers synthesis of DnaK3 in the cyanobacterium *Synechocystis* sp. PCC 6803, which links DnaK3 to the biogenesis of thylakoid membranes and to photosynthetic processes. In a DnaK3 depleted strain, the photosystem content is reduced and the photosystem II activity is impaired, whereas photosystem I is regular active. An impact of DnaK3 on the activity of other thylakoid membrane complexes involved in electron transfer is indicated. In conclusion, DnaK3 is a versatile chaperone required for biogenesis and/or maintenance of thylakoid membrane-localized protein complexes involved in electron transfer reactions. As mentioned above, Hsp70 proteins are involved in photoprotection and repair of PS II in chloroplasts.

Keywords: chaperone; Hsp70; photosynthesis; thylakoid membrane biogenesis; photosystem maintenance; *Synechocystis* sp. PCC6803

---

1. Introduction

In plants and cyanobacteria, the biogenesis and dynamics of thylakoid membranes (TMs) is light-controlled [1,2]. In plants, proplastids develop into chloroplasts, involving the de novo formation of an internal TM network [3], and a developed TM network dynamically reorganizes in the light [4]. When the cyanobacterium *Synechocystis* sp. PCC 6803 (from here on: *Synechocystis*) is grown in the dark under light-activated heterotrophic growth (LAHG) conditions, where glucose is the only available energy source, *Synechocystis* cells exhibit reduced or even just rudimentary TMs [5,6]. However, after shifting dark-adapted cells into the light, the *Synechocystis* cells quickly rebuild a TM network and recover photosynthetic activity [5,7]. While dark-adapted *Synechocystis* cells do not harbor active photosystem II (PS II) complexes, complete photosynthetic activity is regained within 24 h after transferring dark-adapted cells into the light, and reappearance of photosynthetic electron transfer processes is coupled to the formation of internal TMs [7]. However, it is still enigmatic how the formation of internal TMs is controlled, both in chloroplasts and cyanobacteria, although some proteins that might be involved in this process have already been described previously [8]. These proteins include the inner membrane-associated protein of 30 kDa (IM30, also known as Vipp1: The vesicle-inducing protein in plastids 1), Hsp70 (Heat shock protein 70) chaperones, dynamin-like proteins, a prohibitin-like protein, as well as YidC, a membrane protein integrase [9–16]. Nevertheless, while some proteins are probably more directly involved in TM formation, the structure and stability of TMs are also affected more indirectly by pathways, which control the biogenesis of lipids and/or cofactors, and, e.g., mutants defective in synthesis of chlorophyll or of the membrane lipid phosphatidylglycerol (PG) have severely reduced TM systems [17–20].

Molecular chaperones of the Hsp70 family are involved in multiple cellular processes, such as folding of newly synthesized proteins, protein disaggregation, prevention of protein misfolding, protein transport, or the control of regulatory protein functions [21]. The thus far best characterized Hsp70 chaperone is the DnaK protein of the bacterium *Escherichia coli* [22]. In cyanobacteria, at least two DnaK proteins, DnaK2 and DnaK3, are highly conserved, and most cyanobacteria contain an additional DnaK1 protein as well as further DnaK-like proteins [15,23,24]. While cyanobacterial genomes typically encode several DnaK chaperones together with multiple DnaJ (Hsp40) proteins, which serve as DnaK co-chaperones, the physiological function of this DnaK-DnaJ network in cyanobacteria is essentially not understood. In recent years, the physiological roles of individual DnaK and DnaJ proteins have been analyzed to some extent in the cyanobacteria *Synechococcus* sp. PCC 7942 and *Synechocystis* [16,24–26]. In *Synechocystis*, three DnaK proteins are expressed together with at least seven DnaJ proteins [15,25]. The two *dnaK* genes *dnaK2* and *dnaK3* are essential in *Synechocystis*, but not *dnaK1* [15]. The DnaK2 protein has been classified as the canonical DnaK protein involved in cellular stress responses, and DnaK2 most likely functions together with Sll0897, the only type I DnaJ protein expressed in *Synechocystis* [24,25]. In line with this, deletion of the *sll0897* gene resulted in a heat-sensitive phenotype [25]. However, interactions with other DnaJ proteins cannot be excluded, and in fact, the DnaK2 protein interacts and cooperates with the type II J protein DnaJ2 in *Synechococcus* sp. PCC 7942 [27].

In contrast to the remaining *dnaJ* genes, the *dnaJ* gene *sll1933* (*dnaJ3*) could not be deleted in *Synechocystis*, indicating that the encoded DnaJ3 protein is essential [25]. The *dnaK3* and *dnaJ3* genes are organized in a conserved gene cluster in cyanobacteria, and a functional interaction of DnaK3 with DnaJ3 is assumed [28] DnaK3- and DnaJ3-homologs are encoded in essentially all cyanobacterial genomes, except in *Gloeobacter violaceus* PCC 4721, a cyanobacterium that lacks TMs [29,30]. Based on this observation it has been suggested that the physiological function of both proteins might be linked to TMs, and consequently, DnaK3 and DnaJ3 were suggested to be involved in the biogenesis and/or maintenance of TMs [16,25,31]. The DnaK3s of both *Synechococcus* and *Synechocystis* co-purify with membranes, and the unique DnaK3 C-terminus has been implicated to mediate tight membrane binding of DnaK3 in *Synechocystis* [15,31]. However, what might be the function of DnaK3 at TMs?

The function of a cyanobacterial DnaK3 has recently been linked to the PS II reaction center protein D1 [16], the main target of stress-induced damage in the photosynthetic electron transport chain, which is constantly degraded and replaced by newly synthesized proteins in a PS II repair cycle [32,33]. Furthermore, a Hsp70 chaperone is involved in the biogenesis, protection and/or repair of PS II complexes in chloroplasts [34,35]. Based on these observations we hypothesized that the physiological functions of DnaK proteins might have diverged in cyanobacteria, and DnaK3 potentially is specifically involved in biosynthesis/maintenance of TM complexes involved in photosynthesis.

In the present study, we have analyzed the role of the Hsp70 protein DnaK3 in TM maintenance in the cyanobacterium *Synechocystis* sp. PCC 6803. Expression of DnaK3 is light-regulated. Reduction of the cellular DnaK3 content resulted in decreased PS and phycobilisome (PBS) contents, a lowered PS I-to-PS II ratio, a generally reduced photosynthetic activity as well as disturbed PS II activity at elevated light conditions. The observation that the PS II activity is affected after photoinhibition in a mutant strain, where the cellular DnaK3 content is reduced, and the comparison of the mutant strain with *Synechocystis* wt suggests a specific function of DnaK3 in PS II protection and/or repair. However, based on the here presented data its activity must be wider. Thus, our findings support the assumption that DnaK3 is involved in biogenesis and/or maintenance of TM-localized electron transfer complexes in cyanobacteria.

2. Materials and Methods

2.1. Growth Conditions

A glucose-tolerant *Synechocystis* sp. PCC 6803 wild type (wt) and the merodiploid *dnaK3* (*sll1932*) knock-down (KD) strain [15] were cultivated photomixotrophically at 30 °C in liquid BG11 medium [36] supplemented with 5 mM glucose. Kanamycin (80 µg/mL) was added in case of the *dnaK3KD* strain. The cultures were aerated with air enriched with 2% CO_2 and grown under fluorescent white light at a light intensity of 20 (LL, low light) or 120 (HL, high light) µmol/m^2 s, respectively. To determine growth rates, the strains were initially adjusted to an OD_{750} of 0.05 in BG11 medium, containing 5 mM glucose, and growth was followed by monitoring OD_{750}. For LAHG cultures, *Synechocystis* cells were grown in a dark cabinet for at least two weeks, during which the cultures were diluted at least five times in fresh medium, as described previously (Barthel et al., 2013).

2.2. SDS-PAGE and Immunoblot Analysis

Synechocystis cells were harvested in the exponential growth phase at an OD_{750} below 2.0. Cell pellets were resuspended in buffer (50 mM HEPES, pH 7.0, 25 mM $CaCl_2$, 5 mM $MgCl_2$, 10% (v/v) glycerol) and a proteinase inhibitor mix (Sigma Aldrich) was added at a 1:1000 dilution. Cells were broken with glass beads (0.25–0.5 mm diameter) in a beadbeater. Unbroken cells and glass beads were removed by centrifugation at 1600 g and the respective protein concentrations were determined by three independent Bradford assays. After addition of SDS sample buffer and heating at 65 °C for 15 min, cell extracts were loaded on an 8% polyacrylamide gel and proteins were separated by SDS gel electrophoresis. Subsequently, proteins were transferred to a polyvinylidene difluoride membrane, using a wet electroblotting system from Bio-Rad. The rabbit primary antibodies were used at 1:2000 (anti-L23 directed against the large ribosomal subunit protein L23 encoded by *sll1801*, Gramsch laboratories, Schwabhausen, Germany), 1:1000 (anti-DnaK1, anti-DnaK2 and anti-DnaK3 [15], anti-PsaA/PsaB [37]) or 1:100 (anti-PsbA [38]) dilutions, respectively, whereas the goat anti-rabbit secondary antibody (Sigma Aldrich) was diluted 1:10,000. PsbA/D1-HRP antibodies were obtained from Agrisera and used in 1:15,000 dilution. To visualize the protein bands, membranes were incubated with the enhanced chemiluminescence kit from Pierce. Each immunoblot analysis has been repeated at least three times.

2.3. Complete Deletion of DnaK3 in Synechocystis Cells Grown under LAHG Conditions

To test whether DnaK3 is dispensable in the dark, the *dnaK3KD* strain [15] was cultivated in liquid BG11 medium under LAHG conditions and diluted if necessary. During each dilution step, the concentration of kanamycin was enhanced in the growth medium from 80 to 275 µg mL^{-1}. To check whether the strain was completely segregated, genomic DNA was isolated and analyzed by PCR using the primers NtdnaK3check (5′-gtttttagaagcggagaaagtgg-3′) and CtdnaK3check (5′-cctttgggttggaaaccattgg-3′).

2.4. Cell Number and Chlorophyll Concentration Determination

Cell numbers were counted with a light microscope using a Thoma counting chamber. Chlorophyll concentrations were determined photometrically after methanol extraction [39].

2.5. Electron Microscopy

To study the cell morphology of the different *Synechocystis* strains, cell pellets obtained from a 10 mL cell suspension were washed and resuspended in buffer (50 mM KH_2PO_4, pH 7). Ultrastructural investigations were performed as described previously [37]. The number of thylakoid layers per cell was determined, evaluating more than 200 individual cells of wt and the DnaK3 depleted *Synechocystis* strain, respectively.

2.6. Absorbance and Low Temperature (77K) Fluorescence Spectra

Absorbance spectra of whole cells were recorded using a Perkin-Elmer Lambda 25 spectrophotometer equipped with an integrating sphere. Cell suspensions were adjusted to a constant value of 300,000 cells mL^{-1}. Ratios of cyanobacterial chromophores were determined using the absorption ratio at 625/680 (phycocyanin/chlorophyll) or at 490/440 (carotenoids/chlorophyll).

Low-temperature (77 K) fluorescence emission spectra were recorded using an Aminco Bowman Series 2 spectrofluorimeter. Cultures were adjusted to a chlorophyll concentration of 3 µg·mL^{-1} in BG11 medium and frozen in liquid nitrogen. Chlorophylls were excited at 435 nm and phycobilisomes (PBs) at 580 nm. Fluorescence emission was recorded from 630 to 760 nm.

2.7. Oxygen Evolution

Oxygen production of the cell suspensions was determined in the presence of 500 µM phenyl-p-benzoquinone (PPBQ) using a fiber-optic oxygen meter (PreSens) under actinic light (600 µmol photons m^{-2}·s^{-1}). Prior to the measurement, the cultures were adjusted to a chlorophyll concentration of 3 µg·mL^{-1} in BG11 medium. For experiments in presence of a protein synthesis inhibitor, 100 µg·mL^{-1} lincomycin was added prior to illumination (1500 µmol photons m^{-2}·s^{-1}).

2.8. Chlorophyll Fluorescence Induction Curves

Cultures were adjusted to a chlorophyll concentration of 3 µg·mL^{-1} in BG11 medium, and subsequently fluorescence induction curves were recorded at room temperature, using a Dual-PAM-100 measuring system equipped with Dual-E and DUAL-DR modules (Heinz Walz GmbH). During the initial dark phase, background fluorescence was probed by weak measuring light (0.024 µmol photons m^{-2}·s^{-1}) and after 40 s fluorescence was induced by switching on red actinic light (95 µmol photons m^{-2}·s^{-1}). Saturating pulses (600 ms, 10.000 µmol photons m^{-2}·s^{-1}) were applied once during the dark phase and at 30 s intervals during the light phase, to obtain minimal (F_0) and maximal (F_m and $F_{m'}$) fluorescence values [40,41]. The coefficient of photochemical quenching of the PS II Chl fluorescence (qP) was calculated using the software routine for light induction measurements (qP = (Fm-Fm')/(Fm-Fo')) after 250 s illumination with red actinic light.

2.9. P_{700} Re-Reduction Kinetics

Re-reduction kinetics were recorded using a Dual-PAM-100 measuring system. P_{700} was first reduced by 10 sec far-red and then oxidized by a 20 ms saturation light pulse (10.000 µmol photons m^{-2}·s^{-1}). 15 individual re-reduction curves were recorded, averaged, and fitted with single exponential functions to determine decay halftimes ($t_{1/2}$). Prior to the measurement, the different cultures were adjusted to a chlorophyll concentration of 3 µg·mL^{-1} in BG11 medium.

3. Results

3.1. DnaK3 Synthesis is Light-Induced and Essential in the Dark

The *Synechocystis dnaK2* and *dnaK3* genes are essential in the light [15], and the DnaK1-3 proteins were detected by Western blot analyses in *Synechocystis* cells grown under constant illumination [15]. However, when *Synechocystis* cells were grown in the dark under LAHG conditions, the DnaK2 protein, but not DnaK1 and DnaK3, were detectable (Figure 1, 0 h). Yet, when dark-adapted cells were shifted into the light, the DnaK2 level did not substantially alter, whereas the DnaK1 level quickly increased until two hours after shifting the cells into the light. DnaK3 was detectable already after one hour, and its cellular content increased steadily. Thus, the synthesis of DnaK1 and DnaK3 clearly is triggered by light in *Synechocystis*.

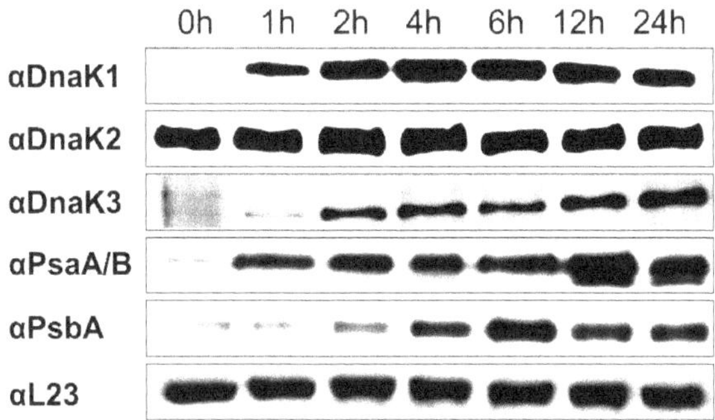

Figure 1. Light-dependent accumulation of DnaK1, 2, and 3. Dark-adapted *Synechocystis* cultures were shifted into the light (0–24 h). Cell extracts (20 µg protein) were analyzed at different time points via immunodetection, using anti-DnaK1, 2, or 3 antibodies as well as antibodies directed against PS I (PsaA/B) and PS II (PsbA) core subunits or the ribosomal protein L23 (loading control).

Since DnaK1 is not essential for the viability of *Synechocystis* cells [15], we focused our subsequent analyses on DnaK3.

As DnaK3 is essential in the light [15], the observation of a light-induced DnaK3 synthesis indicated that DnaK3 might be dispensable in the dark. Therefore, we next attempted to completely delete the *Synechocystis dnaK3* gene in cells grown in the dark under LAHG conditions. Yet, even after more than half a year of cultivation under LAHG conditions and increasing the kanamycin concentration in the growth medium up to 275 µg·mL^{-1}, a fragment corresponding in size to the wild type (wt) *dnaK3* gene was always detected via PCR in the *dnaK3* knock-down (*KD)* strain in addition to the *dnaK3* gene disrupted by the kanamycin resistance (*aphA*) cassette (Figure 2A,B). As *Synechocystis* contains multiple identical genome copies, this result indicates that some, but not all, of the genomic *dnaK3* copies were deleted in the mutant strain. Thus, DnaK3 likely is essential not only in the light but also in the dark under LAHG conditions.

Yet, we recently showed that expression of *dnaJ3* [25], which is organized in a gene cluster together with *dnaK3*, is essential in *Synechocystis*, and thus deletion of *dnaK3* might have affected the expression of *dnaJ3*. To assess this potential polar effect, we also quantified the amount of the DnaJ3 protein in the *dnaK3KD* strain (Figure 2A). Since the DnaJ3 level was not decreased compared to the wt, we concluded that insertion of the *aphA* cassette into the *dnaK3* gene locus did not dramatically affect the expression of *dnaJ3*. Nevertheless, a polar effect on expression of *dnaJ3* cannot be completely excluded.

To quantify the relative cellular DnaK3 content in the *dnaK3KD* strain, total cellular extracts of the wt and the KD strain were analyzed via Western blots (Figure 2C). The intensity of each band was quantified using the Image J software and divided by the quantity of cellular extract loaded. Based on this analysis, the DnaK3 content was decreased by about 60% ± 10% in the *dnaK3KD* strain compared to the wt.

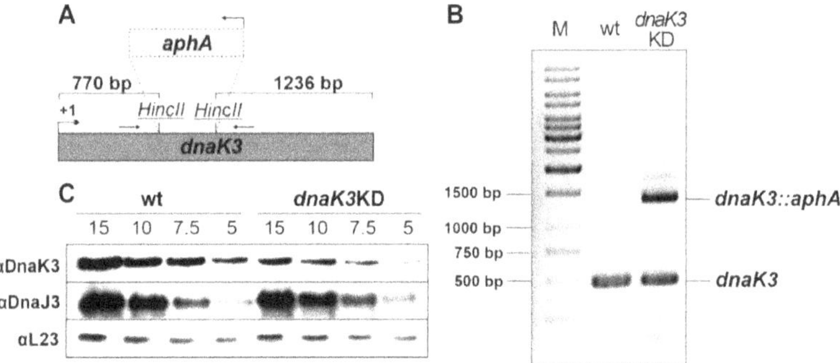

Figure 2. Deletion of *dnaK3* in the dark and the DnaK3 content in the *Synechocystis dnaK3KD* strain. (**A**) In the *Synechocystis dnaK3KD* strain [15], the *dnaK3* gene was disrupted by insertion of a kanamycin resistance cassette (*aphA* gene). (**B**) The *dnaK3* gene locus of wt and *dnaK3KD* cells grown in the dark was analyzed via PCR using genomic DNA as a template, and the PCR products were loaded on a 1.5% agarose gel together with a molecular size marker (M). Fragments of about 500 bp and 1500 bp represent the wt and the *dnaK3* gene interrupted by a kanamycin resistance cassette (*aphA*), respectively. (**C**) The relative DnaK3 content in the *dnaK3KD* strain was determined by immunoblot analysis. Cell extracts prepared from the wt and *dnaK3KD* strains, respectively, were loaded on a SDS-polyacrylamide gel in descending protein concentrations (15 µg to 5 µg) followed by a Western blot analysis using α-DnaK3, α-DnaJ3 and α-L23 (loading control) antibodies.

3.2. Reducing the DnaK3 Content Affects Cell Growth under Heat Stress Conditions

Next, we tested whether reducing the DnaK3 content affects growth of the mutant strain under low (LL) or high light (HL) growth conditions, respectively (Figure 3A). The *dnaK3KD* and the wt cells had comparable doubling times of 11.2 h ± 0.1 (wt) and 11.1 h ± 0.2 (*dnaK3KD*), and of 8.1 h ± 0.5 (wt) and 9.4 ± 1.2 (*dnaK3KD*) under LL and HL growth conditions, respectively. Thus, reducing the cellular DnaK3 content does not severely affect the growth of *Synechocystis* cells, at least not under standard laboratory growth conditions. Subsequently, growth of the *dnaK3KD* strain was tested under various stress conditions (involving low pH, low temperature, oxidative and osmotic stress; data not shown), but solely increasing the temperature to 42 °C resulted in an obvious growth defect of the mutant strain, with doubling times of 26.4 ± 3.5 h (*dnaK3KD*) and 17.9 ± 0.4 h (wt) (Figure 3B). This observation classifies DnaK3 as a traditional Hsp70 involved in heat-stress responses.

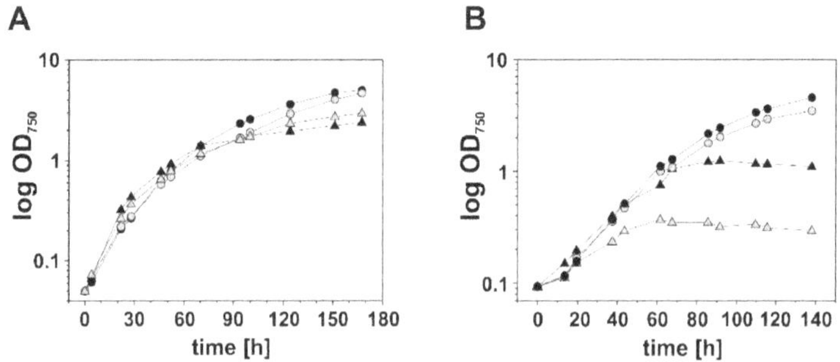

Figure 3. Growth of *Synechocystis* wt and the *dnaK3KD* mutant strain at different growth conditions. *Synechocystis* wt (black) and *dnaK3KD* mutant (gray) cells were grown at (**A**) moderate temperature (30 °C) or (**B**) elevated temperature (42 °C) under low light (circle) or high light (triangle) conditions. Cells were adjusted to $OD_{750} = 0.05$ in BG11 medium containing 5 mM glucose and cell growth was followed over time by measuring the OD_{750}.

3.3. The dnaK3KD Strain Has a Reduced Pigment Content

Photosynthesis is one of the most temperature-sensitive processes in phototrophic organisms and the photosynthetic activity is further impaired when heat-stress is combined with HL [42,43]. Thus, it was well possible that reducing the DnaK3 content affects photosynthetic processes in *Synechocystis*.

As expression of *dnaK3* is light-controlled (Figure 1), we subsequently analyzed the pigment content of the wt and *dnaK3KD* strains after cultivation under LL and HL conditions, respectively. Adaptation of *Synechocystis* cells to HL conditions is typically accompanied by a reduction in the cellular amount of the two PSs, a decreased PS I-to-PS II ratio and a reduced chlorophyll (Chl) content per cell [44,45].

An overall reduction of the pigment content was observed under HL growth conditions when equal amounts of cells were analyzed (Figure 4A). The Chl content was reduced to about half (Figure 4B) and the relative content of plastocyanine (PC) (Figure 4C) and carotenoids (Car) (Figure 4D) were both increased. It has to be noted that while the contents of Chl, PC and Car were decreased under HL growth conditions (Figure 4A), the PC/Chl as well as the Car/Chl ratios were increased in the wt strain, due to the more severely decreased Chl content (Figure 4A,B). Even though light scattering could have contributed to some extent to the determined (absolute) absorbance values used in these analyses, these data clearly show the ability of the wt to reduce the overall pigment content and to adapt it to HL conditions.

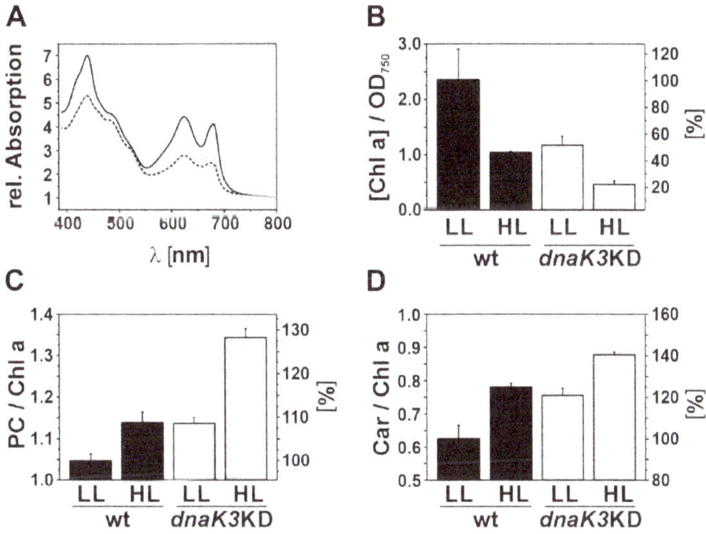

Figure 4. Pigment content and pigment ratios. (**A**) Absorbance spectra of *Synechocystis* wt (black) and *dnaK3KD* (gray) cells (300,000 cells) grown under LL (solid line) or HL (dashed line) conditions. (**B**) The chlorophyll content per OD_{750} was determined as described in "Material and Methods". (**C**) The ratio of PC to Chl was determined as the ratio of the absorptions at 625 and 680 nm. (**D**) The ratio of Car to Chl was determined as the ratio of the absorption at 490 and 440 nm. Error bars represent standard deviation from three independent experiments.

Similarly, the *dnaK3KD* strain adapted to changing light conditions and reduced its pigment content as expected when grown under HL conditions. However, the *dnaK3KD* strain exhibited a severely reduced pigment content already when grown under LL conditions, and the Chl content as well as the pigment ratios were very similar to the ones observed when the wt was grown under HL conditions (Figure 4).

Besides the obvious differences in pigmentation, the TM structure was mostly unaffected, and we only observed a slightly reduced number of TM pairs in the mutant strain when the ultrastructure of *Synechocystis* grown under LL growth conditions was analyzed via electron microscopy (Figure 5).

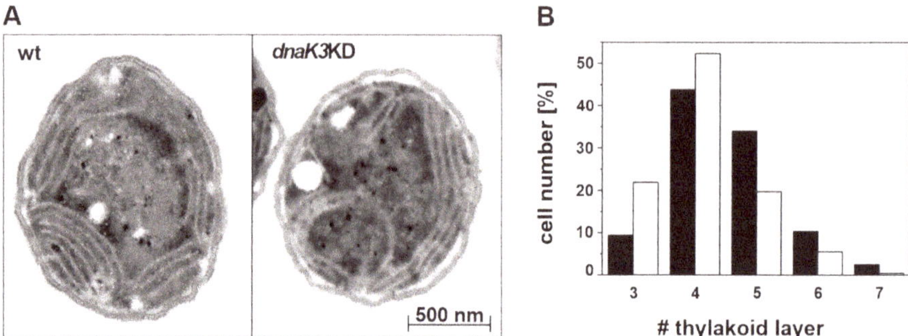

Figure 5. Reducing the cellular DnaK3 content results in fewer thylakoid layers. (**A**) Representative electron micrographs of *Synechocystis* wt and *dnaK3KD* mutant cells cultivated under LL (low light) conditions. (**B**) Cells of the *dnaK3KD* mutant strain (gray) had four thylakoid layers on average, whereas wt cells (black) showed four to five layers and a higher appearance of six and seven layers of TM pairs. Per strain, at least 200 individual cells were counted.

3.4. Reducing the DnaK3 Content Results in an Altered PS I-to-PS II Ratio

Next, the relative amounts of PS II and PS I in the DnaK3 reduced strain were determined via 77 K fluorescence spectroscopy (Figure 6A). Upon chlorophyll excitation at 435 nm, characteristic fluorescence emission maxima were detected at 721 nm (PS I), at 684 nm (CP43, PS II) and 693 nm (CP47, PS II).

Synechocystis wt cells grown under HL conditions showed a decreased PS I-to-PS II ratio compared to LL-adapted cells (Figure 6A), which is a well-documented long-term adaptation to HL [44,46,47]. In contrast, *dnaK3KD* cells had a considerably decreased PS I-to-PS II ratio already under LL growth conditions. This finding is also supported by a Western blot analysis. When an identical quantity of protein was loaded, the Western blot shows that PS core subunits PsaA/B (PS I) and PsbA (PS II) are less abundant in the *dnak3KD* strain (Figure 6C). When the cell extracts were normalized based on the Chl concentration (Figure 6D), no difference in the band intensity was observed in case of PsaA/B, since in *Synechocystis* about 85% of the Chl is bound to PS I (assuming a PS I/PS II ratio of 2.5 [48], 96 chlorophylls per PS I [49], and 35 chl per PS II [50]). However, the PsbA band was more pronounced in the mutant strain when compared to the wt, which further supports the decreased PS I-to-PS II ratio in this strain. The decreased PS I-to-PS II ratio decreases even further when cells were shifted into HL (Figure 6A).

However, at LL as well as at HL conditions, an increased relative fluorescence emission was observed at 684 nm (λ_{ex} = 435nm) in the mutant strain (Figure 6A). This fluorescence emission maximum originates from PS II as well as from the PBSs terminal emitter LCM [51], and thus indicates an increased relative phycobiliprotein content, as already observed in the absorbance measurements (Figure 4A,C). Yet, the increased PBSs fluorescence emission at 684 nm is solely observed when PBSs are uncoupled and do not transfer the harvested light energy to the PSs [51,52]. Thus, to next assess energy transfer from PBSs to PS II, phycobiliproteins were excited at 580 nm and energy transfer to PS II was followed. When PBSs are coupled to PS II, light energy harvested by the PBS is transferred to PS II, resulting in quenching of the PBS fluorescence [53,54]. As can be seen in Figure 6B, upon PBS excitation an increased fluorescence emission at 684 nm (PBSs plus PS II) was observed but not at 693 nm (PS II) (Figure 6B). Thus, the *dnaK3KD* strain indeed contains an increased amount of uncoupled PBSs compared to the wt.

The decreased fluorescence emission at 721 nm results from the decreased PS I content (Figure 6A) and most likely not from light-dependent energy distribution via state transitions, which is supposed to be physiologically important solely under LL growth conditions [55,56].

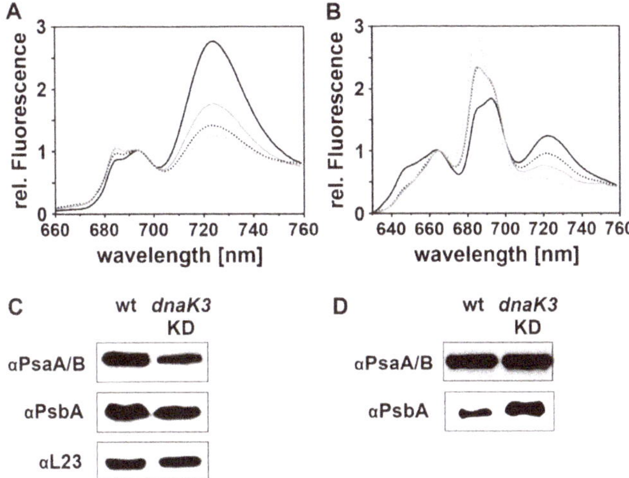

Figure 6. Reduction of the cellular DnaK3 content results in a decreased PS content and a lowered PS I-to-PS II ratio. (**A**) 77 K fluorescence emission spectra of wt (black) and *dnaK3KD* (gray) cultures grown under LL (solid line) and HL (high light) (dashed line) conditions. The spectra were normalized at 695 nm. λ_{Ex} = 435 nm (**B**) 77 K fluorescence emission spectra of wt (black) and *dnaK3KD* (gray) cultures cultivated under LL (solid line) and HL (dashed line) conditions. The spectra were normalized at 665 nm. λ_{Ex} = 580 nm. (**C,D**) Immunoblot analysis of the content of PS I and PS II core subunits (PS I: PsaA/B; PS II: PsbA) in wt and *dnaK3KD* cells grown under LL conditions. Samples were normalized to (C) protein (25 µg) or (D) chlorophyll (0.6 µg). L23 is the loading control.

Taken together, the fluorescence spectra and the Western blot analyses reveal that the mutant has a generally decreased PS content, with a decreased PS I-to-PS II ratio and an increased amount of uncoupled phycobiliproteins. However, the mutant strain still adjusts the PS I-to-PS II ratio to changing light conditions, as observed for the wt strain.

3.5. The Photosynthetic Activity is Impaired in the DnaK3 Depleted Strain

Next, the photosynthetic activity of the mutant strain with a reduced DnaK3 content was studied in greater detail. By measuring oxygen evolution rates in presence of PPBQ, the activity of PS II can be specifically determined (Figure 7A,B). When adapted to HL growth conditions, the O_2 evolution rate per cell (OD_{750}) was reduced in the wt strain compared to LL growth conditions (Figure 7A), in line with the observation that the light-harvesting capacity is generally reduced in cyanobacterial cells under HL conditions [57]. In contrast to wt cells, the *dnaK3KD* strain showed a dramatically decreased O_2 evolution rate already under LL conditions, when compared to the wt, and the activity decreased even further under HL conditions (Figure 7A). However, when the O_2 evolution rates were normalized to the Chl content, the O_2 evolution rate remained essentially stable in the wt, regardless of the light conditions (Figure 7B). In contrast, the O_2 evolution rate was only marginally lower for the mutant strain under LL growth conditions than for the wt, but dramatically decreased under HL growth conditions. Thus, in contrast to the wt, the decreased O_2 evolution in the mutant strain is not only a consequence of the decreased cellular Chl content (Figure 4B), since the O_2 evolution rate was also drastically decreased when the measurements were normalized to the Chl content (Figure 7B).

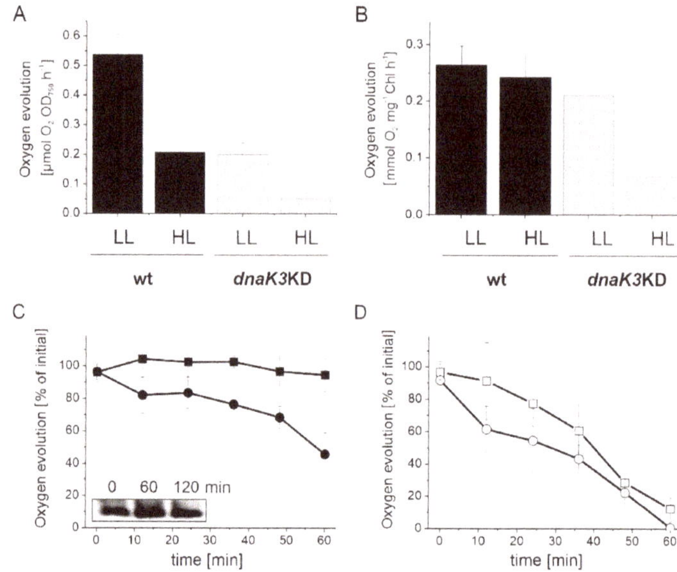

Figure 7. Oxygen evolution rates and relative PS II content of wt and *dnaK3KD Synechocystis* cells. (**A,B**) wt and *dnaK3KD* cells were grown under LL and HL conditions, respectively, and oxygen evolution rates were determined per OD_{750} (**A**) or Chl (**B**). (**C,D**) wt (square) and *dnaK3KD* (circle) cells were exposed to extreme high light (eHL) conditions (1500 μmol photons $m^{-2} \cdot s^{-1}$) either in absence (**C**) or presence (**D**) of lincomycin (100 $\mu g \cdot mL^{-1}$) and thereafter cultured under LL conditions for recovery. Oxygen evolution was measured using 500 μM phenyl-p-benzoquinone (PPBQ) as an electron acceptor at PS II. The recovery rate is given by the slope of a linear regression under LL conditions. Noteworthy, no other antibiotics were present in these experiments. Inlet in (**C**): Immunoblot analysis of the D1 content in the *dnaK3KD* strain after photoinhibition (time 0). Cell extracts with identical Chl contents (0.4 μg) were analyzed. (Error bars represent standard deviation from three independent experiments).

To test whether reducing the DnaK3 content somehow impairs PS II repair, we next determined O_2 evolution rates under extreme HL conditions (1500 μmol photons $m^{-2} \cdot s^{-1}$) in presence or absence of lincomycin, a protein synthesis inhibitor that has already been successfully used to block the PS II repair cycle in *Synechocystis* [58,59]. In absence of lincomycin, the wt strain did not show any changes in the PS II activity under constant extreme HL illumination, i.e., the wt cells harbor an effective PS II repair cycle (Figure 7C). However, in presence of lincomycin, the PS II activity constantly decreased when cells were illuminated with extreme HL (Figure 7D). The decreasing PS II activity, i.e., an impaired PS II protection and/or repair, can be observed for both the wt and the *dnaK3KD* strain in presence of lincomycin (Figure 7D). However, in absence of lincomycin, the PS II activity was already lower in the *dnaK3KD* than in the wt strain after 10 min of illumination and constantly decreased further to about 50% after 1 h of illumination, whereas the wt activity remained about constant (Figure 7C). Thus, PS II repair clearly is severely impaired in *dnaK3KD* cells, albeit the amount of expressed D1 protein did not alter (inlet in Figure 7C). Thus, PS II protection and/or repair is affected especially under light-stress conditions.

The photochemical efficiency of PS II can be specifically assessed using pulse amplitude modulated (PAM) fluorescence measurements. Therefore, dark/light induction curves were recorded (Figure 8A). A minimal fluorescence (F_0) is visible due to the measuring light, which, however, is not strong enough to stimulate photosynthetic electron transfer. Subsequently, a pulse of intense white light is given to reduce all PS II reaction centers, resulting in maximal fluorescence (F_m). The parameter F_v/F_m ($F_v = F_m - F_o$) is used to describe the maximal photochemical efficiency (Figure 8B) [60].

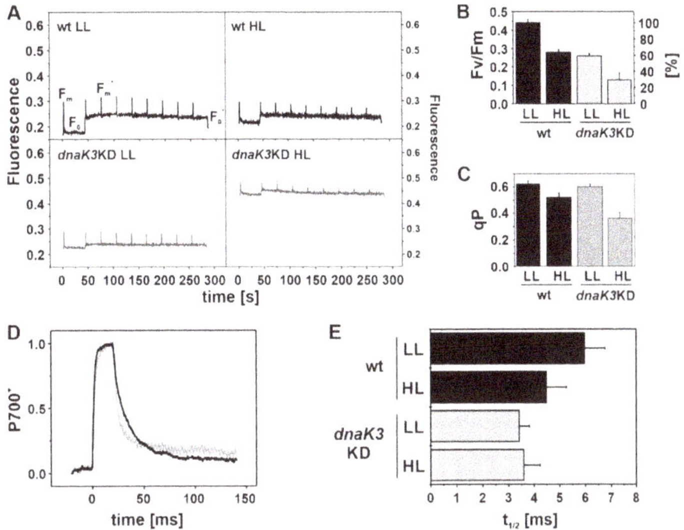

Figure 8. PS II activity and P$_{700}$ re-reduction kinetics. (**A**) Light/dark induction curves were recorded by measuring the pulse amplitude modulated (PAM) fluorescence of wt (black) and *dnaK3KD* cells (gray) grown under LL or HL conditions, respectively. After 40 s of measuring light, the actinic red light was switched on to determine minimal fluorescence values (F$_0$, F$_0{}'$). Pulses of saturating light were applied once during the dark phase and in 30 s intervals during the light phase, to obtain maximal (F$_m$ and F$_m{}'$) fluorescence values. (**B**) Maximal PS II photosynthetic activity of the wt (black) and the *dnaK3KD* strain (gray). (**C**) The coefficient of photochemical quenching of PS II Chl fluorescence (qP) in wt (black) and *dnaK3KD* cells (gray). (B, C) Error bars represent standard deviation from at least four independent experiments. (**D**) P$_{700}{}^+$ re-reduction kinetics of wt (black) and *dnaK3KD* (gray) cells grown under LL conditions. A saturation pulse of 10,000 μmol photons was given for 20 ms to completely oxidize P$_{700}$. The following fluorescence decrease illustrates re-reduction of P$_{700}{}^+$ in the dark. At least ten traces were averaged and normalized. 1 represents completely oxidized and 0 completely reduced P$_{700}$. (**E**) Re-reduction halftimes were determined via fitting the decay curves of the wt (black) and the *dnaK3KD* mutant strain (gray) with single exponential functions. Error bars represent standard deviation from at least three independent experiments.

After switching on actinic light, an increased background fluorescence was detected, and PS II centers became photosynthetically active (Figure 8A). Saturating light pulses resulted in a lowered F$_m{}'$ compared to the maximal fluorescence F$_m$ measured in the dark, due to non-photochemical quenching processes [61]. An apparent increase of the absolute F$_0$ background fluorescence was measured for wt cells grown under HL conditions compared to LL and for the *dnaK3KD* cells (grown under either condition), indicating a more reduced plastoquinone (PQ) pool (Figure 8A). However, determining F$_0$ and F$_0{}'$ is somewhat problematic in cyanobacteria, as the PBS fluorescence can in part also contribute to the determined F$_0$ fluorescence value [62], and thus discussion of solely F$_0$ values is difficult. Hence, we also present the normalized coefficient of photochemical quenching of PS II Chl fluorescence (qP), which is not significantly biased by PBSs fluorescence [63]. qP is defined as 1 in the dark-adapted state and may decrease to 0 when all PS II centers are closed. In line with Figure 7B, in the wt strain slightly less PS II centers are open under HL conditions compared to LL. In the mutant strain, qP is similar to the wt under LL conditions, yet the value was dramatically decreased when *dnaK3KD* cells were grown under HL conditions, indicating an increased amount of closed PS II centers. Thus, the *dnaK3KD* strain can hardly cope with high light treatment. This observation is in excellent agreement with the determined O$_2$ evolution rates (Figure 7), showing an impaired PS II protection and/or repair cycle.

P_{700}^+ re-reduction measurements allow determining the time needed to re-reduce the oxidized PS I reaction center P_{700}^+, which is not only affected by the PS I activity but also by the redox state of the electron transport chain (Figure 8D,E). The P_{700}^+ absorbance signal increased when a saturating light pulse, which completely oxidized P_{700}, was given and subsequently decreased due to P_{700}^+ reduction by PC (Figure 8D). The halftime of the re-reduction kinetic was determined by fitting the changes in the absorbance signal with a single exponential function (Figure 8E). A faster re-reduction rate was observed for wt cells when cells were grown under HL compared to LL conditions, which likely originates from the reduced PS I content and the decreased PS I-to-PS II ratio (Figures 6 and 8). In contrast, the *dnaK3KD* mutant strain had reduced re-reduction halftimes under both tested light conditions, and the halftimes were identical, regardless of the light condition (Figure 8E). The reduced re-reduction halftimes can be explained by a more reduced PQ-pool and thus nicely support the conclusions drawn from the results shown in Figures 6 and 8. Together, these results demonstrate that the activity of PS II, but not of PS I, is impaired in the *dnaK3KD* mutant strain.

4. Discussion

Three different DnaK proteins are expressed in the cyanobacterium *Synechocystis* sp. PCC 6803. While two of the cyanobacterial DnaK proteins, DnaK2 and DnaK3, are essential, solely DnaK2 can be classified as a canonical Hsp70 protein, expression of which can largely alter under various stress conditions [24,64,65]. In contrast, the DnaK3 chaperone of *Synechocystis* has been suggested to be specifically involved in biogenesis and/or maintenance of TMs [16,25]. However, thus far this assumption was essentially exclusively based on the observations that (i) DnaK3 is attached to TMs and (ii) DnaK3 is encoded in all cyanobacterial genomes, except in *Gloeobacter violaceus*, the only cyanobacterium that does not contain an internal TM system [15,29].

Albeit the cellular DnaK3 content clearly is light-regulated (Figure 1), a basal DnaK3 level appears to be required for survival of *Synechocystis* cells not only in the light but also under LAHG conditions (Figure 2), where cells still have rudimentary TMs. Based on the CyanoExpress database the *dnaK3* transcript level does not appear to adjust to changing light conditions in *Synechocystis* [66] or in *Synechococcus* sp. PCC 7942 [16]. Thus, (light-dependent) DnaK3 synthesis likely is post-transcriptionally regulated, as common in cyanobacteria [67,68].

A general decrease in the PS and PBS content per cell as well as a selective down-regulation of PS I is crucially involved in the adaptation of *Synechocystis* cells to HL conditions [46,67]. All these (expected) adjustments were observed when the *Synechocystis* wt strain was shifted from LL to HL growth conditions (Figures 6 and 7). Also, in case of the mutant strain, typical HL-adaptation processes were observed, although the light-induced changes were far less pronounced, since the mutant strain already exhibited characteristics of an HL-stressed strain under LL growth conditions. While in the *dnaK3KD* strain the relative PS I content was reduced and the PS I-to-PS II ratio per cell was lower than in the wt (Figure 6), PS I appears to function normally when DnaK3 was depleted, because re-reduction of PS I was even faster in the mutant strain (Figure 8E), most likely due to the more reduced PQ-pool, which also results in a significant amount of the PBSs being detached from PS II (Figure 6B).

Yet, the activity of PS II was clearly reduced in the mutant strain (Figures 7 and 8). A significant amount of PS II was inactive, potentially due to increased photodamage and/or impaired repair (Figures 7 and 8). Thus, DnaK3 likely is involved in PS II biogenesis and/or repair. In line with this assumption, expression of the *Synechocystis dnaK3* gene was found being enhanced under UV-B stress [68]. Furthermore, the PS II core subunit D1 was proposed to be a substrate for DnaK3, which potentially guides the nascent polypeptide at the ribosome to the TM, where translation is completed [16]. The D1 protein is known to be especially susceptible to photodamage, and photodamaged D1 is rapidly degraded and replaced by newly synthesized protein to maintain a certain level of active PS II centers in cyanobacteria [33]. Thus, the here presented results clearly indicate that biogenesis and/or repair of PS II is impaired when the cellular DnaK3 content is reduced.

However, the observation that DnaK3 appears to be vital also in the dark (Figure 2), where PS II is inactive [7] and the finding that D1 is not essential for survival of *Synechocystis* under photoheterotrophic conditions [69] indicates that DnaK3 likely has additional physiological functions beyond PS II protection and/or repair. In cyanobacteria, TMs also contain the complexes of the respiratory e^--transfer chain [70], and the indications of an over-reduced PQ pool (Figures 6 and 8) suggests that other proteins and protein complexes are also affected when the DnaK3 content is reduced. A broader implication of DnaK3 in TM biogenesis and maintenance, involving biogenesis and/or repair of multiple TM complexes would be a convincing explanation.

Author Contributions: A.T., D.S., R.J. and T.S. designed the research, A.T., R.J., T.S., and U.K. performed the experiments, A.T., D.S., and T.S. wrote the manuscript. All authors have read and agreed to the published version of the manuscript.

Funding: This research was funded by the Deutsche Forschungsgemeinschaft and a grant from the Ernst & Margarete Wagemann Foundation.

Acknowledgments: We thank R. Genswein for excellent technical assistance and Hildegard Pearson as well as Rebecca Keller for critically reading the manuscript. We also thank M. Rögner (Ruhr-University Bochum) for the kind gift of antibodies.

Conflicts of Interest: The authors declare no conflict of interest.

References

1. Rast, A.; Heinz, S.; Nickelsen, J. Biogenesis of thylakoid membranes. *Biochim. Biophys. Acta Bioenerg.* **2015**, *1847*, 821–830. [CrossRef] [PubMed]
2. Vothknecht, U.C.; Westhoff, P. Biogenesis and origin of thylakoid membranes. *Biochim. Biophys. Acta* **2001**, *1541*, 91–101. [CrossRef]
3. Jarvis, P.; López-Juez, E. Biogenesis and homeostasis of chloroplasts and other plastids. *Nat. Rev. Mol. Cell Biol.* **2013**, *14*, 787–802. [CrossRef] [PubMed]
4. Dubreuil, C.; Jin, X.; Barajas-López, J.d.D.; Hewitt, T.C.; Tanz, S.K.; Dobrenel, T.; Schröder, W.P.; Hanson, J.; Pesquet, E.; Grönlund, A.; et al. Establishment of Photosynthesis through Chloroplast Development Is Controlled by Two Distinct Regulatory Phases. *Plant Physiol.* **2018**, *176*, 1199–1214. [CrossRef]
5. Vernotte, C.; Picaud, M.; Kirilovsky, D.; Olive, J.; Ajlani, G.; Astier, C. Changes in the Photosynthetic Apparatus in the Cyanobacterium *Synechocystis* sp Pcc-6714 Following Light-to-Dark and Dark-to-Light Transitions. *Photosyn. Res.* **1992**, *32*, 45–57. [CrossRef]
6. Anderson, S.L.; Mcintosh, L. Light-activated heterotrophic growth of the cyanobacterium *Synechocystis* sp. strain PCC 6803: A blue-light-requiring process. *J. Bacteriol.* **1991**, *173*, 2761–2767. [CrossRef]
7. Barthel, S.; Bernat, G.; Seidel, T.; Rupprecht, E.; Kahmann, U.; Schneider, D. Thylakoid Membrane Maturation and PSII Activation Are Linked in Greening *Synechocystis* sp PCC 6803 Cells. *Plant Physiol* **2013**, *163*, 1037–1046. [CrossRef]
8. Mechela, A.; Schwenkert, S.; Soll, J. A brief history of thylakoid biogenesis. *Open Biol.* **2019**, *9*, 180237. [CrossRef]
9. Spence, E.; Bailey, S.; Nenninger, A.; Moller, S.G.; Robinson, C. A homolog of Albino3/OxaI is essential for thylakoid biogenesis in the cyanobacterium *Synechocystis* sp PCC6803. *J. Biol. Chem.* **2004**, *279*, 55792–55800. [CrossRef]
10. Thurotte, A.; Brüser, T.; Mascher, T.; Schneider, D.; Brueser, T. Membrane chaperoning by members of the PspA/IM30 protein family. *Commun. Integr. Biol.* **2017**, *10*, e1264546. [CrossRef]
11. Heidrich, J.; Thurotte, A.; Schneider, D. Specific interaction of IM30/Vipp1 with cyanobacterial and chloroplast membranes results in membrane remodeling and eventually membrane fusion. *Biochim. Biophys. Acta Biomembr.* **2017**, *1859*, 537–549. [CrossRef] [PubMed]
12. Siebenaller, C.; Junglas, B.; Schneider, D. Functional Implications of Multiple IM30 Oligomeric States. *Front. Plant Sci.* **2019**, *10*, 1500. [CrossRef] [PubMed]
13. Jilly, R.; Khan, N.Z.; Aronsson, H.; Schneider, D. Dynamin-Like Proteins Are Potentially Involved in Membrane Dynamics within Chloroplasts and Cyanobacteria. *Front. Plant Sci.* **2018**, *9*, 206. [CrossRef] [PubMed]

14. Bryan, S.J.; Burroughs, N.J.; Evered, C.; Sacharz, J.; Nenninger, A.; Mullineaux, C.W.; Spence, E.M. Loss of the SPHF Homologue Slr1768 Leads to a Catastrophic Failure in the Maintenance of Thylakoid Membranes in *Synechocystis* sp PCC 6803. *PLoS ONE* **2011**, *6*, e19625. [CrossRef]

15. Rupprecht, E.; Gathmann, S.; Fuhrmann, E.; Schneider, D. Three different DnaK proteins are functionally expressed in the cyanobacterium *Synechocystis* sp. PCC 6803. *Microbiology* **2007**, *153*, 1828–1841. [CrossRef]

16. Katano, Y.; Nimura-Matsune, K.; Yoshikawa, H. Involvement of DnaK3, one of the three DnaK proteins of cyanobacterium *Synechococcus* sp PCC7942, in translational process on the surface of the thylakoid membrane. *Biosci. Biotechnol. Biochem.* **2006**, *70*, 1592–1598. [CrossRef]

17. Frick, G.; Su, Q.X.; Apel, K.; Armstrong, G.A. An Arabidopsis *porB porC* double mutant lacking light-dependent NADPH: protochlorophyllide oxidoreductases B and C is highly chlorophyll-deficient and developmentally arrested. *Plant J.* **2003**, *35*, 141–153. [CrossRef]

18. Hagio, M.; Sakurai, I.; Sato, S.; Kato, T.; Tabata, S.; Wada, H. Phosphatidylglycerol is essential for the development of thylakoid membranes in *Arabidopsis thaliana*. *Plant Cell Physiol.* **2002**, *43*, 1456–1464. [CrossRef]

19. Paddock, T.N.; Mason, M.E.; Lima, D.F.; Armstrong, G.A. Arabidopsis protochlorophyllide oxidoreductase A (PORA) restores bulk chlorophyll synthesis and normal development to a *porB porC* double mutant. *Plant Mol. Biol.* **2010**, *72*, 445–457. [CrossRef]

20. Sato, N.; Hagio, M.; Wada, H.; Tsuzuki, M. Requirement of phosphatidylglycerol for photosynthetic function in thylakoid membranes. *Proc. Natl. Acad. Sci. USA* **2000**, *97*, 10655–10660. [CrossRef]

21. Mayer, M.P.; Bukau, B. Hsp70 chaperones: Cellular functions and molecular mechanism. *Cell. Mol. Life Sci.* **2005**, *62*, 670–684. [CrossRef] [PubMed]

22. Genevaux, P.; Georgopoulos, C.; Kelley, W.L. The Hsp70 chaperone machines of *Escherichia coli*: A paradigm for the repartition of chaperone functions. *Mol. Microbiol.* **2007**, *66*, 840–857. [CrossRef] [PubMed]

23. Rajaram, H.; Chaurasia, A.K.; Apte, S.K. Cyanobacterial heat-shock response: Role and regulation of molecular chaperones. *Microbiology* **2014**, *160*, 647–658. [CrossRef] [PubMed]

24. Rupprecht, E.; Düppre, E.; Schneider, D. Similarities and Singularities of Three DnaK Proteins from the Cyanobacterium *Synechocystis* sp. PCC 6803. *Plant Cell Physiol.* **2010**, *51*, 1210–1218. [CrossRef]

25. Düppre, E.; Rupprecht, E.; Schneider, D. Specific and promiscuous functions of multiple DnaJ proteins in *Synechocystis* sp PCC 6803. *Microbiology* **2011**, *157*, 1269–1278. [CrossRef]

26. Nimura, K.; Takahashi, H.; Yoshikawa, H. Characterization of the *dnaK* multigene family in the cyanobacterium *Synechococcus* sp strain PCC7942. *J. Bacteriol.* **2001**, *183*, 1320–1328. [CrossRef]

27. Nakamoto, H.; Fujita, K.; Ohtaki, A.; Watanabe, S.; Narumi, S.; Maruyama, T.; Suenaga, E.; Misono, T.S.; Kumar, P.K.R.; Goloubinoff, P.; et al. Physical interaction between bacterial heat shock protein (Hsp) 90 and Hsp70 chaperones mediates their cooperative action to refold denatured proteins. *J. Biol. Chem.* **2014**, *289*, 6110–6119. [CrossRef]

28. Oguchi, K.; Nimura, K.; Yoshikawa, H.; Takahashi, H. Sequence and analysis of a *dnaJ* homologue gene in cyanobacterium *Synechococcus* sp. PCC7942. *Biochem. Biophys. Res. Commun.* **1997**, *236*, 461–466. [CrossRef]

29. Nakamura, Y.; Kaneko, T.; Sato, S.; Mimuro, M.; Miyashita, H.; Tsuchiya, T.; Sasamoto, S.; Watanabe, A.; Kawashima, K.; Kishida, Y.; et al. Complete genome structure of *Gloeobacter violaceus* PCC 7421, a cyanobacterium that lacks thylakoids. *DNA Research* **2003**, *10*, 137–145. [CrossRef]

30. Rippka, R.; Waterbury, J.; Cohenbazire, G. Cyanobacterium Which Lacks Thylakoids. *Arch. Microbiol.* **1974**, *100*, 419–436. [CrossRef]

31. Nimura, K.; Yoshikawa, H.; Takahashi, H. DnaK3, one of the three DnaK proteins of cyanobacterium *Synechococcus* sp. PCC7942, is quantitatively detected in the thylakoid membrane. *Biochem. Biophys. Res. Commun.* **1996**, *229*, 334–340. [CrossRef] [PubMed]

32. Aro, E.M.; Virgin, I.; Andersson, B. Photoinhibition of Photosystem II. Inactivation, protein damage and turnover. *Biochim. Biophys. Acta* **1993**, *1143*, 113–134. [CrossRef]

33. Mulo, P.; Sakurai, I.; Aro, E.M. Strategies for *psbA* gene expression in cyanobacteria, green algae and higher plants: From transcription to PSII repair. *Biochim. Biophys. Acta Bioenerg.* **2012**, *1817*, 247–257. [CrossRef] [PubMed]

34. Schroda, M.; Vallon, O.; Wollman, F.A.; Beck, C.F. A chloroplast-targeted heat shock protein 70 (HSP70) contributes to the photoprotection and repair of photosystem II during and after photoinhibition. *Plant Cell* **1999**, *11*, 1165–1178. [CrossRef] [PubMed]

35. Yokthongwattana, K.; Chrost, B.; Behrman, S.; Casper-Lindley, C.; Melis, A. Photosystem II damage and repair cycle in the green alga *Dunaliella salina*: Involvement of a chloroplast-localized HSP70. *Plant Cell Physiol.* **2001**, *42*, 1389–1397. [CrossRef] [PubMed]

36. Rippka, R.; Deruelles, J.; Waterbury, J.B.; Herdman, M.; Stanier, R.Y. Generic Assignments, Strain Histories and Properties of Pure Cultures of Cyanobacteria. *Microbiology* **1979**, *111*, 1–61. [CrossRef]

37. Fuhrmann, E.; Gathmann, S.; Rupprecht, E.; Golecki, J.; Schneider, D. Thylakoid Membrane Reduction Affects the Photosystem Stoichiometry in the Cyanobacterium *Synechocystis* sp PCC 6803. *Plant Physiol.* **2009**, *149*, 735–744. [CrossRef]

38. Fuhrmann, E.; Bultema, J.B.; Kahmann, U.; Rupprecht, E.; Boekema, E.J.; Schneider, D. The Vesicle-inducing Protein 1 from *Synechocystis* sp PCC 6803 Organizes into Diverse Higher-Ordered Ring Structures. *Mol. Biol. Cell* **2009**, *20*, 4620–4628. [CrossRef]

39. Porra, R.J.; Thompson, W.A.; Kriedemann, P.E. Determination of Accurate Extinction Coefficients and Simultaneous-Equations for Assaying Chlorophyll-a and Chlorophyll-B Extracted with 4 Different Solvents—Verification of the Concentration of Chlorophyll Standards by Atomic-Absorption Spectroscopy. *Biochim. Biophys. Acta* **1989**, *975*, 384–394. [CrossRef]

40. Murchie, E.H.; Lawson, T. Chlorophyll fluorescence analysis: A guide to good practice and understanding some new applications. *J. Exp. Bot.* **2013**, *64*, 3983–3998. [CrossRef]

41. van Kooten, O.; Snel, J.F. The use of chlorophyll fluorescence nomenclature in plant stress physiology. *Photosyn. Res.* **1990**, *25*, 147–150. [CrossRef] [PubMed]

42. Salvucci, M.E.; Crafts-Brandner, S.J. Relationship between the heat tolerance of photosynthesis and the thermal stability of rubisco activase in plants from contrasting thermal environments. *Plant Physiol.* **2004**, *134*, 1460–1470. [CrossRef] [PubMed]

43. Asadulghani, M.; Suzuki, Y.; Nakamoto, H. Light plays a key role in the modulation of heat shock response in the cyanobacterium *Synechocystis* sp PCC 6803. *Biochem. Biophys. Res. Commun.* **2003**, *306*, 872–879. [CrossRef]

44. Murakami, A.; Fujita, Y. Regulation of Stoichiometry between PsI and PsII in Response to Light Regime for Photosynthesis Observed with *Synechocystis* Pcc-6714-Relationship between Redox State of Cyt B6-F Complex and Regulation of Psi Formation. *Plant Cell Physiol.* **1993**, *34*, 1175–1180.

45. Muramatsu, M.; Hihara, Y. Transcriptional regulation of genes encoding subunits of photosystem I during acclimation to high-light conditions in *Synechocystis* sp. PCC 6803. *Planta* **2003**, *216*, 446–453. [CrossRef]

46. Hihara, Y.; Sonoike, K.; Ikeuchi, M. A novel gene, *pmgA*, specifically regulates photosystem stoichiometry in the cyanobacterium *Synechocystis* species PCC 6803 in response to high light. *Plant Physiol* **1998**, *117*, 1205–1216. [CrossRef]

47. Muramatsu, M.; Hihara, Y. Acclimation to high-light conditions in cyanobacteria: From gene expression to physiological responses. *J. Plant Res.* **2012**, *125*, 11–39. [CrossRef]

48. Fraser, J.M.; Tulk, S.E.; Jeans, J.A.; Campbell, D.A.; Bibby, T.S.; Cockshutt, A.M. Photophysiological and photosynthetic complex changes during iron starvation in *Synechocystis* sp. PCC 6803 and *Synechococcus elongatus* PCC 7942. *PLoS ONE* **2013**, *8*, e59861. [CrossRef]

49. Jordan, P.; Fromme, P.; Witt, H.T.; Klukas, O.; Saenger, W.; Krauss, N. Three-dimensional structure of cyanobacterial photosystem I at 2.5 angstrom resolution. *Nature* **2001**, *411*, 909–917. [CrossRef]

50. Guskov, A.; Kern, J.; Gabdulkhakov, A.; Broser, M.; Zouni, A.; Saenger, W. Cyanobacterial photosystem II at 2.9-A resolution and the role of quinones, lipids, channels and chloride. *Nat. Struct. Mol. Biol.* **2009**, *16*, 334–342. [CrossRef]

51. Rakhimberdieva, M.G.; Vavilin, D.V.; Vermaas, W.F.J.; Elanskaya, I.V.; Karapetyan, N.V. Phycobilin/ chlorophyll excitation equilibration upon carotenoid-induced non-photochemical fluorescence quenching in phycobilisomes of the *Synechocystis* sp PCC 6803. *Biochim. Biophys. Acta* **2007**, *1767*, 757–765. [CrossRef] [PubMed]

52. Tamary, E.; Kiss, V.; Nevo, R.; Adam, Z.; Bernat, G.; Rexroth, S.; Rogner, M.; Reich, Z. Structural and functional alterations of cyanobacterial phycobilisomes induced by high-light stress. *Biochim. Biophys. Acta Bioenerg.* **2012**, *1817*, 319–327. [CrossRef] [PubMed]

53. Biggins, J.; Bruce, D. Regulation of Excitation-Energy Transfer in Organisms Containing Phycobilins. *Photosyn. Res.* **1989**, *20*, 1–34. [CrossRef] [PubMed]

54. Chang, L.; Liu, X.; Li, Y.; Liu, C.-C.; Yang, F.; Zhao, J.; Sui, S.-F. Structural organization of an intact phycobilisome and its association with photosystem II. *Cell Research* **2015**, *25*, 726–737. [CrossRef]

55. Emlyn-Jones, D.; Ashby, M.K.; Mullineaux, C.W. A gene required for the regulation of photosynthetic light harvesting in the cyanobacterium *Synechocystis* 6803. *Mol. Microbiol.* **1999**, *33*, 1050–1058. [CrossRef]

56. Mullineaux, C.W.; Emlyn-Jones, D. State transitions: An example of acclimation to low-light stress. *J. Exp. Bot.* **2005**, *56*, 389–393. [CrossRef]

57. Kopecna, J.; Komenda, J.; Bucinska, L.; Sobotka, R. Long-term acclimation of the cyanobacterium *Synechocystis* sp. PCC 6803 to high light is accompanied by an enhanced production of chlorophyll that is preferentially channeled to trimeric photosystem I. *Plant Physiol.* **2012**, *160*, 2239–2250. [CrossRef]

58. Komenda, J.; Tichy, M.; Prasil, O.; Knoppova, J.; Kuvikova, S.; de Vries, R.; Nixon, P.J. The exposed N-terminal tail of the D1 subunit is required for rapid D1 degradation during photosystem II repair in *Synechocystis* sp PCC 6803. *Plant Cell* **2007**, *19*, 2839–2854. [CrossRef]

59. Silva, P.; Thompson, E.; Bailey, S.; Kruse, O.; Mullineaux, C.W.; Robinson, C.; Mann, N.H.; Nixon, P.J. FtsH is involved in the early stages of repair of photosystem II in *Synechocystis* sp PCC 6803. *Plant Cell* **2003**, *15*, 2152–2164. [CrossRef]

60. Baker, N.R. Chlorophyll fluorescence: A probe of photosynthesis *in vivo*. *Annu. Rev. Plant Biol* **2008**, *59*, 89–113. [CrossRef]

61. Niyogi, K.K.; Truong, T.B. Evolution of flexible non-photochemical quenching mechanisms that regulate light harvesting in oxygenic photosynthesis. *Curr. Opin. Plant Biol.* **2013**, *16*, 307–314. [CrossRef]

62. Campbell, D.; Bruce, D.; Carpenter, C.; Gustafsson, P.; Oquist, G. Two forms of the photosystem II D1 protein alter energy dissipation and state transitions in the cyanobacterium *Synechococcus* sp PCC 7942. *Photosyn. Res.* **1996**, *47*, 131–144. [CrossRef] [PubMed]

63. Santabarbara, S.; Villafiorita Monteleone, F.; Remelli, W.; Rizzo, F.; Menin, B.; Casazza, A.P. Comparative excitation-emission dependence of the F_V/F_M ratio in model green algae and cyanobacterial strains. *Physiol. Plant* **2019**, *166*, 351–364. [CrossRef] [PubMed]

64. Fulda, S.; Mikkat, S.; Huang, F.; Huckauf, J.; Marin, K.; Norling, B.; Hagemann, M. Proteome analysis of salt stress response in the cyanobacterium *Synechocystis* sp. strain PCC 6803. *Proteomics* **2006**, *6*, 2733–2745. [CrossRef] [PubMed]

65. Mary, I.; Tu, C.J.; Grossman, A.; Vaulot, D. Effects of high light on transcripts of stress-associated genes for the cyanobacteria *Synechocystis* sp PCC 6803 and *Prochlorococcus* MED4 and MIT9313. *Microbiology* **2004**, *150*, 1271–1281. [CrossRef]

66. Hernandez-Prieto, M.A.; Futschik, M.E. CyanoEXpress: A web database for exploration and visualisation of the integrated transcriptome of cyanobacterium *Synechocystis* sp. PCC6803. *Bioinformation* **2012**, *8*, 634–638. [CrossRef]

67. Sonoike, K.; Hihara, Y.; Ikeuchi, M. Physiological significance of the regulation of photosystem stoichiometry upon high light acclimation of *Synechocystis* sp PCC 6803. *Plant Cell Physiol.* **2001**, *42*, 379–384. [CrossRef]

68. Huang, F.; Parmryd, I.; Nilsson, F.; Persson, A.L.; Pakrasi, H.B.; Andersson, B.; Norling, B. Proteomics of *Synechocystis* sp. strain PCC 6803: Identification of plasma membrane proteins. *Mol. Cell Proteomics* **2002**, *1*, 956–966. [CrossRef]

69. Jansson, C.; Debus, R.J.; Osiewacz, H.D.; Gurevitz, M.; Mcintosh, L. Construction of an Obligate Photoheterotrophic Mutant of the Cyanobacterium *Synechocystis* 6803: Inactivation of the *psbA* Gene Family. *Plant Physiol* **1987**, *85*, 1021–1025. [CrossRef]

70. Mullineaux, C.W. Electron transport and light-harvesting switches in cyanobacteria. *Front. Plant Sci.* **2014**, *5*, 7. [CrossRef]

Article

Over Expression of the Cyanobacterial Pgr5-Homologue Leads to Pseudoreversion in a Gene Coding for a Putative Esterase in Synechocystis 6803

Ketty Margulis [1], **Hagit Zer** [1], **Hagar Lis** [1], **Hanan Schoffman** [1], **Omer Murik** [2], **Ginga Shimakawa** [3], **Anja Krieger-Liszkay** [3] and **Nir Keren** [1,*]

[1]   Department of Plant and Environmental Sciences, Edmond J. Safra Campus, The Alexander Silberman Institute of Life Sciences, Hebrew University of Jerusalem, Givat Ram, 9190402 Jerusalem, Israel; ketty.margulis@mail.huji.ac.il (K.M.); hagit.zer@mail.huji.ac.il (H.Z.); hagarlis@gmail.com (H.L.); hanan.schoffman@mail.huji.ac.il (H.S.)

[2]   Medical Genetics Institute, Shaare Zedek Medical Center, 9103102 Jerusalem, Israel; omer.murik@mail.huji.ac.il

[3]   Institute for Integrative Biology of the Cell (I2BC), Commissariat à l'Energie Atomique et aux Energies Alternatives (CEA) Saclay, Centre National de la Recherche Scientifique (CNRS), Université Paris-Saclay, 91198 Gif-sur-Yvette CEDEX, France; ginshimakawa@gmail.com (G.S.); Anja.LISZKAY@i2bc.paris-saclay.fr (A.K.-L.)

*   Correspondence: nir.ke@mail.huji.ac.il; Tel.: +972-2-6585233

Received: 21 July 2020; Accepted: 1 September 2020; Published: 3 September 2020

Abstract: Pgr5 proteins play a major direct role in cyclic electron flow paths in plants and eukaryotic phytoplankton. The genomes of many cyanobacterial species code for Pgr5-like proteins but their function is still uncertain. Here, we present evidence that supports a link between the Synechocystis sp. PCC6803 Pgr5-like protein and the regulation of intracellular redox balance. The knockout strain, *pgr5KO*, did not display substantial phenotypic response under our experimental conditions, confirming results obtained in earlier studies. However, the overexpression strain, *pgr5OE*, accumulated 2.5-fold more chlorophyll than the wild type and displayed increased content of photosystems matching the chlorophyll increase. As a result, electron transfer rates through the photosynthetic apparatus of *pgr5OE* increased, as did the amount of energy stored as glycogen. While, under photoautotrophic conditions, this metabolic difference had only minor effects, under mixotrophic conditions, *pgr5OE* cultures collapsed. Interestingly, this specific phenotype of *pgr5OE* mutants displayed a tendency for reverting, and cultures which previously collapsed in the presence of glucose were now able to survive. DNA sequencing of a *pgr5OE* strain revealed a second site suppression mutation in *slr1916*, a putative esterase associated with redox regulation. The phenotype of the *slr1916* knockout is very similar to that of the strain reported here and to that of the pmgA regulator knockout. These data demonstrate that, in Synechocystis 6803, there is strong selection against overexpression of the Pgr5-like protein. The pseudoreversion event in a gene involved in redox regulation suggests a connection of the Pgr5-like protein to this network.

Keywords: cyanobacteria; electron transport; photosynthesis; carbon metabolism; redox

---

1. Introduction

The major pathway of energy flow through the photosynthetic apparatus begins with light absorption by antenna pigment–protein complexes, leading to charge separation in photosystem reaction centers, and extends to the electron and proton transport process, finally generating NADPH and ATP [1]. This linear pathway is tightly regulated to avoid over-reduction of intermediate electron carriers and to balance ATP/NADPH ratios. Regulatory mechanisms identified at the level of the antenna systems include non-photochemical quenching processes [2–5]. Ion transporters regulate the proton gradient across the thylakoid membrane [6].

In addition to these, a number of cyclic, pseudo-cyclic and alternative electron flow pathways regulate electron transport processes [7,8]. Cyclic electron transport pathways are defined as reactions that involve only photosystem I (PSI) photochemistry. They support the production of a ΔpH without producing NADPH. Two such major pathways which funnel electrons from reduced ferredoxins (Fd) back to the plastoquinone (PQ) pool [9] have been identified in chloroplasts: one through the action of the chloroplast NDH (NADH dehydrogenase-like) complex and the other via PGR5/PGRL1 systems.

Chloroplasts contain a homologue of the mitochondrial NADH dehydrogenase complex. The chloroplast complex (the NADH dehydrogenase-like complex) was shown to be involved in cyclic electron flow. Current data suggest that, unlike its mitochondrial counterpart, the chloroplast NDH complex is able to interact directly with Fd [10–12]. In angiosperms, this route is considered to be minor. The major cyclic route in angiosperms is suggested to take place via PGR5/PGRL1 proteins [13]. The PGR5 mutant (proton gradient regulator) was identified in a screen for chlorophyll fluorescence phenotypes in Arabidopsis [14]. The picture was completed with the identification of its membrane embedded counterpart PGRL1 [13]. A putative ferredoxin:plastoquinone reductase (FQR) was also suggested as a path for PQ reduction in chloroplasts [15]. Together, they were mapped to the antimycin A sensitive electron transport pathway that was originally discovered by Arnon and coworkers in the 1950s [16].

The phenotypes arising from disruptions of these genes are varied and depend on the species of photosynthetic organisms and their growth conditions, but they are all related to sensitivity under excess or fluctuating light conditions [17,18]. The study of the function of PGR5/PGRL1 is still ongoing and hypotheses for their function range from direct and substantial involvement in cyclic electron transfer to regulatory functions [19–21].

The photosynthetic pathway in chloroplasts is relatively simple as compared to that of cyanobacteria, where respiratory and photosynthetic electron transport pathways occur in the same compartment and intersect at the PQ pool [22,23]. Apart from photosystem II (PSII), the PQ pool can be reduced by respiratory succinate dehydrogenase (SDH) and NAD(P)H dehydrogenases (both NDH-1 and NDH-2) [24]. Furthermore, several cyanobacterial species, including *Synechocystis* sp. PCC6803, have sulfide quinone reductase (SQR) genes [25,26]. The PQ pool can be oxidized via the linear cytochrome b_6f-plastocyanin route, which, in cyanobacteria, is shared between photosynthetic and respiratory electron transport chains. It can also be oxidized by respiratory terminal oxidases such as quinol oxidase (Cyd), alternate respiratory terminal oxidases (ARTO) or plastoquinol terminal oxidase (PTOX) that were shown to be able to accept electrons directly from PQ [27–31]. This maze of pathways opens up a range of possible routes for cyclic electron flow intertwined amongst the photosynthetic and respiratory processes. To simplify the discussion in this manuscript, we will refer to the electron transport pathway starting at PSII and ending at NADPH as linear and to other pathways connecting to the PQ pool as alternatives to the linear path.

The discovery of the plant PGR5/PGRL1 pathway sparked interest in homologous proteins in cyanobacteria. The genomes of some cyanobacteria code for proteins that show a certain degree of similarity to plant PGR5 (*ssr2016* in *Synechocystis sp.* PCC6803; [32]). A number of studies indicated that the pgr5-like gene is expressed under oxidative stress conditions [33,34]. It was also suggested that it is part of the regulon controlled by Hik33 and PerR response regulators [35]. However, deletion mutants in the Pgr5-like protein resulted in very minor phenotypic responses, as compared to the M55 mutant of the NDH-1 complex [36,37]. The most detailed analysis was performed by Yeremenko and coworkers [32], who were able to detect an antimycin A dependent effect on electron transport in a Δpgr5/M55 double knockout strain. In addition, the double mutant exhibited light sensitivity. The identity of the PGRL1 counterpart of the cyanobacterial Pgr5-like protein is still unresolved. A recent study suggested that, in Synechocystis 6803, ORF Sll1217 may play this role [38].

In our studies, we identified a locus controlling the expression of the Pgr5-like gene. Disruption of a previously uncharacterized putative two-component system gene, *slr1658*, reduced the ability to recover from iron limitation [39]. Transcriptomic analysis of *slr1658* mutants placed Pgr5 as the topmost overexpressed protein. Interestingly, Pgr5 overexpression in these mutants was constitutive regardless of growth or external stress conditions. Here, we describe a study of both knockout and overexpression strains of the Pgr5 homologue in *Synechocystis* 6803 that provides insight into its functional importance in redox regulation.

2. Materials and Methods

2.1. Growth Conditions

Stock cultures of the glucose tolerant *Synechocystis* sp. strain PCC 6803 [40] (wild type (WT)) and mutants were grown in YBG11 (an EDTA (ethylenediaminetetraacetic acid) amended BG11 medium [41]), containing 6 μM iron. Stock cultures of *pgr*5KO and *pgr*5OE strains were supplemented with 50 μg/mL kanamycin. In several experiments, glucose was added to a final concentration of 5 mM, as indicated. Cultures were grown at 30 °C with constant shaking. Light intensity was set at 40 μmol photons m^{-2} s^{-1}.

2.2. Strain Construction

Strains in this study were generated using the restriction–ligation method with Takara's DNA Ligation Kit (Cat.# 6023) [42]. Restriction enzymes used in this work were SacI, SalI, SpeI and SacII by NEB for *pgr*5KO and NdeI and HpaI by NEB for *pgr*5OE. Kanamycin resistance caste was added for selection in both mutant strains. A map of the insertion sites is shown in Figure 1, and primer sequences are listed in Table 1. Transformation was performed as described in [43] for both strains. The vectors used for the construction were pGEM Teasy (Promega) for *pgr*5KO and pTKP2031v vector [44] for *pgr*5OE. The pTKP2031v vector carrying both upstream and coding region of the *slr2031* gene (Figure 1). The *slr2031* gene is not expressed and is often used as an insertion site [45]. The construct includes the strong constitutive promotor of *psbAII* upstream of the *slr2031* start codon, where a NdeI site was introduced. The *slr2031* gene contains a HpaI site, allowing the introduction of pgr5 gene (excised with NdeI and HpaI) within the open reading frame *slr2031*, as described by Satoh and coworkers [44]. It has been shown that no significant differences in chlorophyll content were observed in *Synechocystis sp.* strain PCC 6803 with *psbAII* promotor and a kanamycin resistance cascade compared to the wild type [44]. The *pgr*5OE vector was sequenced prior to introducing it to the genome in order to ensure its integrity (not shown).

Table 1. Primers.

Mutant Name	Genomic Location		Primer
*pgr*5KO	890467–890486	F	5′-CACCATTGGCCTGGTATTGG-3′
	890848–890867	R	5′-TTGGTTCGTCAACAGTTAGG-3′
	890915–890938	F	5′-GCCAGACCATCACCAACTTTTGTA-3′
	891404–891424	R	5′-AAATGCCAGGTAACTAATTTG-3′
*pgr*5KO segregation check	890268–890287	F	5′-ACGTCACGTCCTTTGAGGTC-3′
	891290–891309	R	5′-GGATGACCAGGAAGCCAACC-3′
*pgr*5OE	890816–890835	F	5′-GAGTCACTGCCATGTTCGCC-3′
	891025–891049	R	5′-CTCTTCGTTTTCAATAATTCTTGCC-3′
slr2030–slr2031 segregation check	781363–781383	F	5′-TGGGCACAACCATTTACCCTG-3′
	782328–782348	R	5′-AACTATGACCAACTGCGCCAG-3′
*pgr*5OE RT-PCR	890827–890852	F	5′-ATGTTCGCCCCCATCGTTATCTTGG-3′
	891289–891314	R	5′-GAGGGTTTTGCCGTTGGACTTAGCT-3′
slr1916 mutation verification	619827–619847	F	5′-CCCGTTCAGAATATGACCTGG-3′
	620881–620902	R	5′-GCCGTACTTATTGGCAATTCC-3′

Primers used for this project.

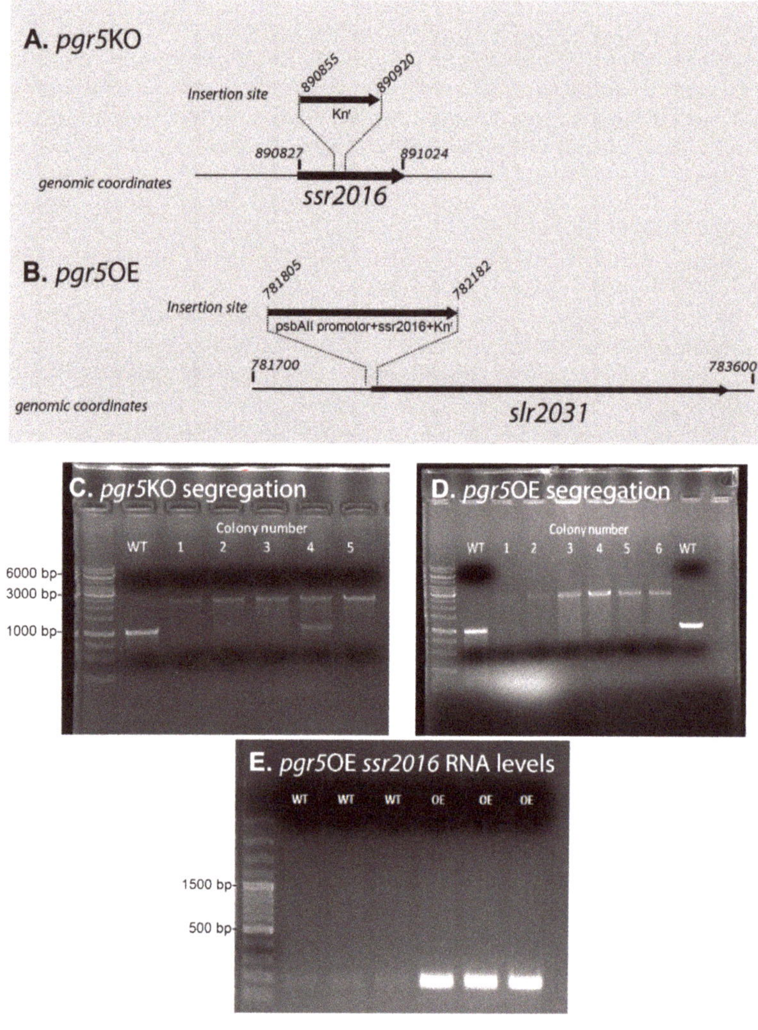

Figure 1. Genomic structure of mutants in the *ssr2016* gene. Panels (**A**) and (**B**) describe the genomic coordinates and insertion sites. The *slr2031* gene is not functional due to a deletion in its coding region [44]. The *pgr*5OE vector was sequenced prior to introducing it to the genome in order to ensure its integrity (not shown). Segregation of *pgr*5KO (wild type—~1000bp, *pgr*5KO—~2000bp) (**C**) and *pgr*5OE (wild type—~1000bp, *pgr*5OE—~3000bp) (**D**) was verified by PCR performed on genomic DNA (primer list in Table 1); GeneRuler 1 kb DNA Ladder—ThermoFisher (Cat. # SM0311) was used. Colony #5 of the *pgr*5KO and colony #4 of the *pgr*5OE were used for further work. (**E**) Amplified RNA expression of ssr2016 (~200bp) in wild type and *pgr*5OE (three repeats) by RT-PCR using the gene specific primers *pgr*5OE RT-PCR shown in Table 1. GeneRuler 100 bp plus DNA Ladder—ThermoFisher was used (Cat. # SM0321). Total RNA was extracted from cultures grown for three days in YBG11.

2.3. Spectroscopy and Microscopy

Cell density was measured as OD_{730} [46] using a Carry 300Bio spectrophotometer (Varian, CA, USA). Additional measurements of cell density and size were performed by direct cell counting using a hemocytometer. PSI activity was measured as P700 photo-oxidation using a JTS-10 spectrophotometer [46], using 10 μM 3-(3,4-dichlorophenyl)-1,1- dimethylurea (DCMU) to block PSII electron transport and 10 μM 2,5-dibromo-3-methyl-6-isopropylbenzoquinone (DBMIB) to block cytochrome b6f electron transport. Photochemical efficiency (Fv/Fm) of PSII was measured using the Satlantic FIRe (Fluorescence Induction and Relaxation) System [47]. Chlorophyll fluorescence spectra at 77K was measured using a Quantamaster 8075 Spectrofluorometer (HORIBA Jobin Yvon PTI, NJ, USA) and NADPH oxidoreduction using the NADPH/9-AA module of a DUAL-PAM (Walz, Effeltrich, Germany).

2.4. Chlorophyll Extraction

Samples were centrifuged at 16,000× *g* for 10 min; 100% methanol was added to the pellet, and samples were incubated in the dark for 30 min and then spun down. Absorbance was measured using a Carry 300Bio spectrophotometer (Varian, CA, USA) at 665 nm. Chlorophyll concentrations were calculated according to Porra et al. [48].

2.5. Biochemical Assays

Protein quantification was conducted according to the Bradford method [49]. Samples for determining glycogen content were adjusted to OD_{730} 0.1 and then were collected after 3 days of growth with or without glucose. Cultures were washed two times with YBGll media before breakage. Glycogen concentrations were determined as described in [50] using the Glucose (GO) Assay Kit (Sigma Aldrich) (Cat. # G3660).

2.6. DNA Extraction, Library Preparation and Sequencing

DNA was extracted as described before [39]. Sequencing libraries were prepared using Celero DNA-seq kit (Tecan) and then sequenced on an Illumina HiSeq × 300 (2 × 150), generating 10 million paired-end reads per sample, giving an estimated average coverage of × 400.

The mutation was then verified by PCR (primers in Table 1), followed by Sanger sequencing.

2.7. Bioinformatics Analysis

Raw sequencing reads were filtered and adaptors trimmed using Trimmomatic with default parameters [51]. Quality of filtered reads was assessed using fastQC (http://www.bioinformatics.babraham.ac.uk/projects/fastqc/) [52]. The filtered reads were mapped to the reference genome (accession CP012832.1) using bwa [53]. Variations from the reference sequence were calculated using VarScan (Version 2.3.9) [54]. The effect of the called variants on the amino acid sequences was evaluated with bcftools [55].

2.8. RNA Extraction

Cultures of WT and *pgr5*OE were collected from the logarithmic growth phase and extracted as described before [39]. To avoid DNA contamination, RNA was treated with DNase using "TURBO DNA-free kit" ThermoFisher (Cat. # AM1907) and converted to cDNA using "RevertAID first strand cDNA synthesis kit" ThermoFisher (Cat. # K1622). RT was then conducted using Hy-taq polymerase by Hylabs (Cat. # EZ1012) (primers used for RT in primer list—Table 1).

3. Results

Following up on the results from our previous work [39], we studied here the function of the *Synechocystis* 6803 PGR5-like protein. We constructed two strains: a deletion strain, *pgr5*KO, and an overexpression strain, *pgr5*OE (Figure 1A–D). The *pgr5*OE strain constitutively overexpressed the *pgr5* transcript (Figure 1E) at levels similar to those previously observed in the Δ*slr1658* strain [39]. Initially, we measured growth during the transition into and out of iron limitation. In these experiments, both mutants exhibited growth rates similar to those of the wild type (not shown). This indicated that the growth phenotype observed in the Δ*slr1658* cannot be recreated by overexpression of *pgr5* alone.

However, while the mechanism(s) responsible for the Δ*slr1658* phenotype remain to be identified, we did observe significant changes in the function of the photosynthetic apparatus in the *pgr5*OE strain. The cellular chlorophyll content was 2.5 times higher on average in the *pgr5*OE strain as compared to the wild type (Figure 2A). At the same time, the content of active PSI units increased. The maximal P700 oxidation value, ΔAmax [46], exhibited a similar 2.5 ratio to wild type values (Figure 2B). The fraction of electrons reducing PSI that arrive from alternative, non-PSII sources remained similar in standard YBG11 media (Figure 2C). PSII parameters, measured by fluorescence induction techniques, were also unchanged in both *pgr5*KO and *pgr5*OE strains as compared to wild type (Figure 2D, Supplementary Table S1).

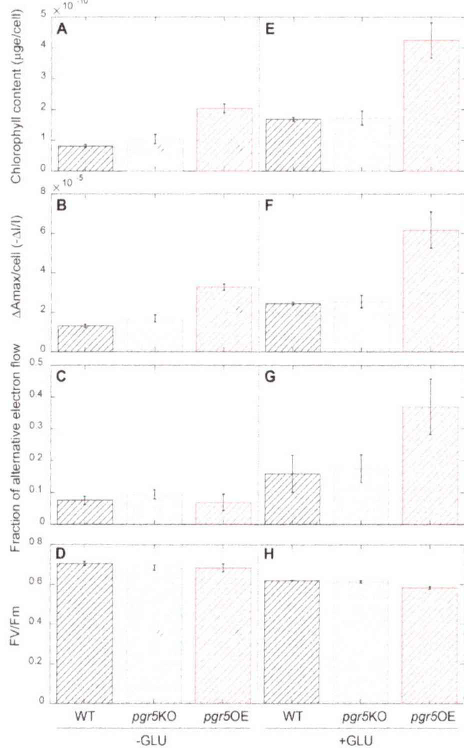

Figure 2. Photosynthetic parameters. Photosynthetic parameters were measured after 3 days of growth in YBG11 media (**A–D**) and in YBG11 media with 5 mM glucose (**E–H**). Chlorophyll extraction was performed and normalized to number of cells (**A,E**). The activity of PSI was measured as the maximal change in P700 absorbance—ΔAmax [46] (sample of raw data in Supplementary Figure S3) normalized to number of cells (**B,F**). Alternative electron flow was calculated as the area trapped between the DCMU and DCMU and DBMIB measurements, normalized to the DCMU and DBMIB measurement (**C,G**). Fv/Fm of photosystem II was measured in 3 min dark adapted samples (**D,H**). Additional fluorescence parameters from the FIRe measurements are included in Supplementary Table S1. Error bars represent standard deviation with *n* = 3. Cell numbers for panels A-B, E-F: wild type $2.2 \times 10^8 \pm 1.6 \times 10^7$, *pgr5*KO $2.1 \times 10^8 \pm 2.9 \times 10^7$ and *pgr5*OE $6.9 \times 10^7 \pm 8.8 \times 10^6$. The average cell size of *pgr5*OE was larger than that of wild type (Supplementary Figure S4).

Based on the role of PGR5 in plant systems and the observed effect on PSI in our experiments, we suspected a change in cellular redox regulation in *pgr5*OE mutants. To challenge the electron transport network, we measured the same parameters in mixotrophic cultures grown for three days in YBGII + 5 mM glucose. In the wild type, both chlorophyll and active PSI content increased under these conditions. *pgr5*KO strain values remained indistinguishable from wild type values. The *pgr5*OE strain retained 2.5 higher chlorophyll and active PSI parameters (Figure 2E,F). PSII fluorescence parameters were slightly lower than wild type and *pgr5*KO strains (Figure 2H). While, for these parameters, the effect of *pgr5*OE was not modified by the addition of glucose, the fraction of electrons reducing PSI that arrive from alternative, non-PSII sources increased dramatically in this strain and reached values of close to 40% of the electrons passing through PSI (Figure 2G).

We also examined the strains under high light conditions: WT and *pgr5*OE were very similar in their responses (Figure S1).

To show whether the ratio between PSI and PSII was altered in *pgr5*OE compared to wild type, we performed Western blot analysis, using antibodies directed against PsaA and PsbA as representatives of PSI and PSII. These blots indicated that, qualitatively, the levels of the two proteins changed in accordance with changes in the cellular chlorophyll content, indicating no alteration in PSI/PSII ratio in *pgr5*OE and wild type (Figure 3A,B). This observation was further supported by 77K chlorophyll fluorescence emission spectroscopy indicating constant PSI/PSII fluorescence intensity ratios regardless of the treatment (Figure 3C). In the absence of glucose, no significant change was observed between the curves. In the *pgr5*OE + Glu trace, there is a 2 nm blue-shift in the position of the PSI peak and an additional fluorescence band at ~655 nm not observed in the wild type. The PSI shift is consistent with partially assembled photosystems [56]. The 655 nm band excitation spectra were measured and a peak at 620 nm, corresponding to phycocyanin (PC) absorption, was observed (not shown). While the absorption cross-section of PC at 430 nm is low, the fluorescence intensity of uncoupled phycobilisome is high. Both changes are consistent with a deteriorating state of *pgr5*OE cells on glucose, indicating breakdown of photosystems and uncoupling of antenna complexes.

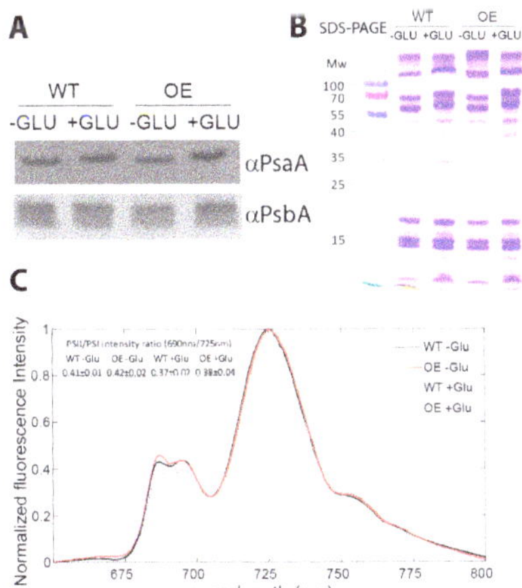

Figure 3. Photosystem ratios in the WT and in the *Pgr5*OE mutant. Western blot analysis using PsaA and PsbA antibodies (quantification of the blot is given in Supplementary Table S2) (**A**). SDS-PAGE analysis (**B**) of proteins in the WT and *pgr5*OE mutant. The gels were loaded on an equal chlorophyll basis (1.5 µg per lane). The experiment was repeated three times with comparable results (Table S2). (**C**) 77 K chlorophyll emission fluorescence spectra. WT and *pgr5*OE cultures were grown on YBG11 medium with 5mM glucose for 3 days. Excitation at 430 ± 5 nm, mainly at the Soret band of chlorophyll but also exciting the far blue tail of phycocyanin absorption. Graphs are an average of three independent repeats. Statistical values for the ratio of PSII to PSI fluorescence intensity are presented in the inserted table.

Downstream of PSI, we detected effects on the redox state of the NAD(P)H pool in the absence and in the presence of glucose (Figure 4A). In dark adapted glucose free (–Glu) cultures, the NAD(P)H pool was in a more oxidized state in wild type than in *pgr5*OE, as seen by the large increase in the fluorescence level upon the onset of light. In the presence of glucose, the NAD(P)H pool was almost fully reduced in the dark in the wild type, while *pgr5*OE exhibited a more oxidized pool (Figure 4A). In combination, these results suggest significant electron flux through the photosynthetic apparatus, from both linear and alternative sources in *pgr5*OE cells. A major sink for excess energy sources in cyanobacterial cells is glycogen [57–59] and, indeed, *pgr5*OE contained more glycogen. This effect was amplified considerably by the addition of glucose to the media (Figure 4B).

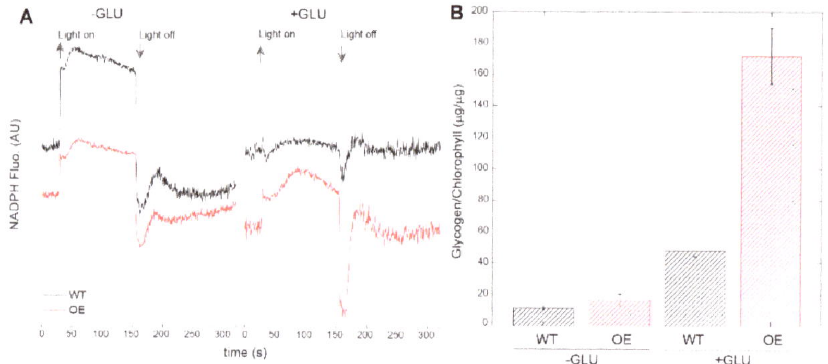

Figure 4. Energy flow downstream of PSI. (**A**) NAD(P)H fluorescence measurements. Black: control; red: *pgr5*OE. NAD(P)H fluorescence was measured in the dark and during exposure to actinic light (200 µmol photons m^{-2} s^{-1}) of cultures with equal 4–5 µg chl ml^{-1} concentrations. Arrows indicate the illumination period. (**B**) Glycogen content measured in cells grown with and without glucose (5 mM) after 3 days of growth. Cellular glycogen concentrations were normalized to the chlorophyll concentration.

The *pgr5*KO strain growth was identical to that of the wild type in glucose supplemented media (Figure 5). However, in the presence of glucose, *pgr5*OE cultures collapsed (Figure 5). This phenotype was observed both by optical density and microscopy cell counts. However, a turn of events occurred when we repeated the growth experiments—*pgr5*OE cultures stopped collapsing when exposed to 5 mM glucose (Figure S2). Interestingly, other aspects of the phenotype were retained: chlorophyll content and PSI activity per cell remained higher in the *pgr5*OE mutant ($3 \times 10^{-11} \pm 4 \times 10^{-12}$ and $1.8 \times 10^{-5} \pm 2 \times 10^{-6}$ respectively). Resequencing of the *pgr5*OE strain revealed that it had adopted a second site suppression mutation. The mutation is a single codon substitution resulting in a phenylalanine to a serine mutation in *slr1916* (Figure 6), a protein that was previously identified as part of the redox regulatory pathway, with a similar phenotype to that of the *pmgA* mutant [60]. Under high light conditions, it is glucose sensitive and has high PSI and chlorophyll content [60]. Going back to our original glycerol *pgr5*OE stocks resulted in the same outcome—initial glucose sensitivity that was lost over time.

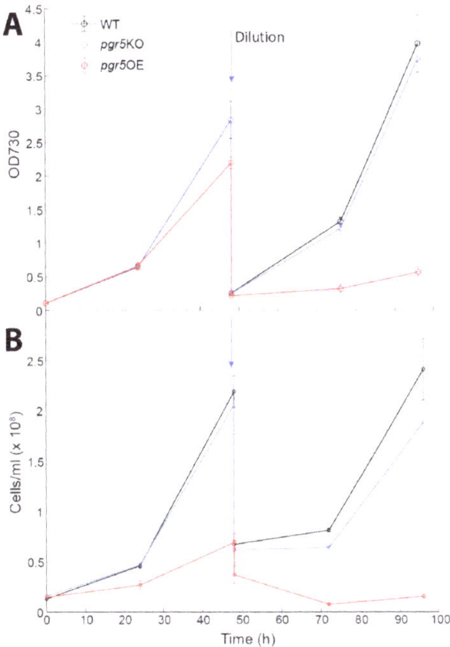

Figure 5. Effects of glucose on biomass accumulation. Cultures were grown on YBG11 medium with 5 mM glucose for five days. Biomass was monitored as optical density at 730 nm (**A**) and cells/mL (**B**). Error bars represent standard deviation derived from three repeats. Cultures were diluted to ensure that they remained in logarithmic phase and to ensure that the cultures did not exhaust media nutrients. The arrow marks the point at which the cultures were diluted.

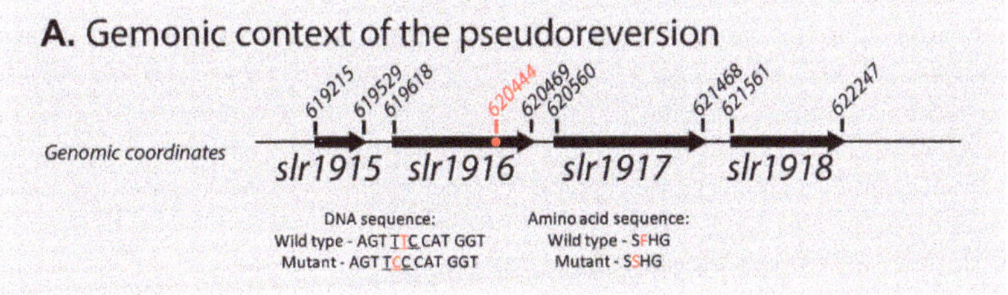

B. Putative structure of the slr1916 protein

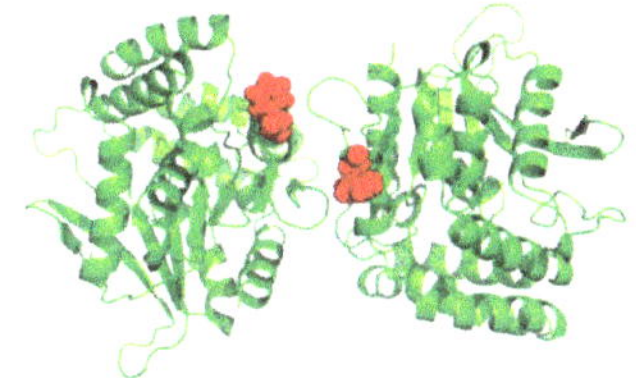

Figure 6. The genomic context and the putative structure of *slr1916*. (**A**) Genomic map—the start and end points of each gene are marked by their genomic coordinates. The *slr1916* missense mutation position is marked by a red dot. (**B**) The top scoring model from the Robetta analysis suggests that Slr1619 is a dimer. The mutated amino acid in the *pgr5*OE strain is shown in red.

4. Discussion

Our study of the *Synechocystis* 6803 Pgr5-like protein resulted in a number of observations. (a) Its deletion did not result in any major observable effect. These results are in line with previously published data [32]. (b) Overexpression resulted in glucose sensitivity but this phenotype was not stable and reverted on two occasions. Since we observed recurring events of loss of glucose sensitivity, we suggest that there is a strong selection against the overexpression of the Pgr5-like protein. The polyploidy of the *Synechocystis* 6803 genome allows these events to progress slowly to the point of complete loss of the glucose sensitivity phenotype. (c) Genomic analysis of one of the reversion events led to the identification of a second site suppressor mutation in *slr1916*, coding for a protein involved in redox regulation. The similarity in the phenotype of the *pgr5OE* strain (harboring the *slr1916* point mutation; Figure 6) and the *slr1916* knockout strain [60] raises the distinct possibility that the additional aspect of the phenotype is a result of the point mutation rather than the over expression.

slr1916 is a very interesting gene with respect to redox regulation. It was annotated as an esterase, identified in a transposon library screen [60]. Disruption of *slr1916* resulted in an altered chlorophyll fluorescence kinetic profile, high chlorophyll and PSI content under high light and glucose sensitivity [60], similar to that observed in *pmgA* [61–63]. It was shown that *slr1916* is essential for growth under photomixotrophic conditions [64]. *slr1916* is strongly induced under different types of environmental stress (acid and heat shock stresses) [65,66].

To better understand these observations, we used Robetta to predict the structure of WT Slr1916 protein [67]. In all top predictions, F242, the amino acid that corresponds to the point mutation discussed above, is located on the surface of the Slr1619 protein (details in Figure 6). According to mCSM, a sever that evaluates the potential effect of mutations on protein stability [68], a mutation of F242 to serine would significantly destabilize Slr1619's structure. The phenotype observed on the *pgr5OE* background could, therefore, be the result of an inability of Slr1916 to interact with binding partners to perform its function in redox regulation.

Pseudoreversions are not unique to this study as they were observed in numerous mutants involved in photosynthetic redox regulation as well as mutants downstream of the photosynthetic pathway [69,70]. For example, growth of the knockout strain of the *pgmA* redox regulator on glucose results in numerous pseudorevertant colonies, mostly mutated in NDH-1 complex components [71]. This is not surprising, as the tendency of the original glucose-sensitive *Synechocystis* 6803 strain to revert when exposed to glucose is what made it an appealing model system in the early days of cyanobacterial genetic research [72]. While this may be frustrating, we nevertheless argue that it is not arbitrary. The tendency of mutant strains in *pmgA* [71], *Slr1658* [39], *pgr5OE* and of other redox regulation related mutants to succumb to pseudoreversions testifies to their pivotal role. This tendency must be taken into account when considering reports on their physiological importance in this organism.

Failing to regulate energy flow can have detrimental effects, leading to cell death. However, the main cause for this cytotoxicity in glucose sensitive mutants is still being elucidated [73]. The discovery of a relation of Pgr5 to these processes adds another connection to this regulatory network controlling energy flow in cyanobacterial cells. This study also provides another example of the risk involved in the mutational analysis of major genes coding for proteins involved in redox regulation.

Supplementary Materials: Include additional experimental data on the effect of high light exposure, examples of raw data for the physiological measurements, quantification of cell sizes and examples for the tendency of the mutant to revert. The following are available online at http://www.mdpi.com/2075-1729/10/9/174/s1.

Author Contributions: K.M. designed and performed experiments and wrote the manuscript; H.Z., H.S., G.S., O.M. and A.K.-L. performed experiments; H.L. designed experiments; N.K. designed experiments and wrote the manuscript. All authors have read and agreed to the published version of the manuscript.

Funding: This work was supported by the Israeli Science Foundation 1182/19 and NSFC-ISF (2466/18) grants. The I2BC benefits from the support of Saclay Plant Sciences-SPS (ANR-17-EUR-0007).

Acknowledgments: We would like to thank Noa Hazoni and Emma Joy Dodson for their assistance in this project.

Conflicts of Interest: The authors declare that there is no conflict of interest.

References

1. Nugent, J.H.A. Photosynthetic electron transport in plants and bacteria. *Trends Biochem. Sci.* **1984**, *9*, 354–357. [CrossRef]
2. Sonoike, K.; Hihara, Y.; Ikeuchi, M. Physiological Significance of the Regulation of Photosystem Stoichiometry upon High Light Acclimation of Synechocystis sp. PCC 6803. *Plant Cell Physiol.* **2001**, *42*, 379–384. [CrossRef] [PubMed]
3. Fujita, Y.; Murakami, A.; Aizawa, K.; Ohki, K. Short-term and Long-term Adaptation of the Photosynthetic Apparatus: Homeostatic Properties of Thylakoids. In *The Molecular Biology of Cyanobacteria*; Springer: Dordrecht, The Netherlands, 1994; pp. 677–692.
4. El Bissati, K.; Delphin, E.; Murata, N.; Etienne, A.-L.; Kirilovsky, D. Photosystem II fluorescence quenching in the cyanobacterium Synechocystis PCC 6803: Involvement of two different mechanisms. *Biochim. Biophys. Acta Bioenerg.* **2000**, *1457*, 229–242. [CrossRef]
5. Horton, P.; Ruban, A.V. Regulation of Photosystem II. *Photosynth. Res.* **1992**, *34*, 375–385. [CrossRef]
6. Höhner, R.; Aboukila, A.; Kunz, H.-H.; Venema, K. Proton Gradients and Proton-Dependent Transport Processes in the Chloroplast. *Front. Plant Sci.* **2016**, *7*, 218. [CrossRef]
7. Kramer, D.M.; Evans, J.R. The Importance of Energy Balance in Improving Photosynthetic Productivity. *Plant Physiol.* **2011**, *155*, 70–78. [CrossRef]
8. Mullineaux, C.W. Electron transport and light-harvesting switches in cyanobacteria. *Front. Plant Sci.* **2014**, *5*, 7. [CrossRef]
9. Johnson, G.N. Physiology of PSI cyclic electron transport in higher plants. *Biochim. Biophys. Acta-Bioenerg.* **2011**, *1807*, 384–389. [CrossRef]
10. Guedeney, G.; Corneille, S.; Cuiné, S.; Peltier, G. Evidence for an association of ndh B, ndh J gene products and ferredoxin-NADP-reductase as components of a chloroplastic NAD(P)H dehydrogenase complex. *FEBS Lett.* **1996**, *378*, 277–280. [CrossRef]
11. Yamamoto, H.; Peng, L.; Fukao, Y.; Shikanai, T. An Src homology 3 domain-like fold protein forms a ferredoxin binding site for the chloroplast NADH dehydrogenase-like complex in Arabidopsis. *Plant Cell* **2011**, *23*, 1480–1493. [CrossRef]
12. Schuller, J.M.; Birrell, J.A.; Tanaka, H.; Konuma, T.; Wulfhorst, H.; Cox, N.; Schuller, S.K.; Thiemann, J.; Lubitz, W.; Sétif, P.; et al. Structural adaptations of photosynthetic complex I enable ferredoxin-dependent electron transfer. *Science* **2019**, *363*, 257–260. [CrossRef] [PubMed]
13. DalCorso, G.; Pesaresi, P.; Masiero, S.; Aseeva, E.; Schünemann, D.; Finazzi, G.; Joliot, P.; Barbato, R.; Leister, D. A Complex Containing PGRL1 and PGR5 Is Involved in the Switch between Linear and Cyclic Electron Flow in Arabidopsis. *Cell* **2008**, *132*, 273–285. [CrossRef] [PubMed]
14. Munekage, Y.; Hojo, M.; Meurer, J.; Endo, T.; Tasaka, M.; Shikanai, T. PGR5 is involved in cyclic electron flow around photosystem I and is essential for photoprotection in Arabidopsis. *Cell* **2002**, *110*, 361–371. [CrossRef]
15. Cleland, R.E.; Bendall, D.S. Photosystem I cyclic electron transport: Measurement of ferredoxin-plastoquinone reductase activity. *Photosynth. Res.* **1992**, *34*, 409–418. [CrossRef]
16. Arnon, D.I.; Allen, M.B.; Whatley, F.R. Photosynthesis by isolated chloroplasts. *Nature* **1954**, *174*, 394–396. [CrossRef]
17. Suorsa, M.; Järvi, S.; Grieco, M.; Nurmi, M.; Pietrzykowska, M.; Rantala, M.; Kangasjärvi, S.; Paakkarinen, V.; Tikkanen, M.; Jansson, S.; et al. PROTON GRADIENT REGULATION5 is essential for proper acclimation of Arabidopsis photosystem I to naturally and artificially fluctuating light conditions. *Plant Cell* **2012**, *24*, 2934–2948. [CrossRef]
18. Johnson, X.; Steinbeck, J.; Dent, R.M.; Takahashi, H.; Richaud, P.; Ozawa, S.I.; Houille-Vernes, L.; Petroutsos, D.; Rappaport, F.; Grossman, A.R.; et al. Proton gradient regulation 5-mediated cyclic electron flow under ATP- or redox-limited conditions: A study of ΔATPase pgr5 and ΔrbcL pgr5 mutants in the green alga Chlamydomonas reinhardtii. *Plant Physiol.* **2014**, *165*, 438–452. [CrossRef]
19. Nandha, B.; Finazzi, G.; Joliot, P.; Hald, S.; Johnson, G.N. The role of PGR5 in the redox poising of photosynthetic electron transport. *Biochim. Biophys. Acta-Bioenerg.* **2007**, *1767*, 1252–1259. [CrossRef]
20. Labs, M.; Rühle, T.; Leister, D. The antimycin A-sensitive pathway of cyclic electron flow: From 1963 to 2015. *Photosynth. Res.* **2016**, *129*, 231–238. [CrossRef]

21. Shikanai, T. Central role of cyclic electron transport around photosystem I in the regulation of photosynthesis. *Curr. Opin. Biotechnol.* **2014**, *26*, 25–30. [CrossRef]
22. Vermaas, W.F. Photosynthesis and Respiration in Cyanobacteria. *Encycl. Life Sci.* **2001**. [CrossRef]
23. Lea-Smith, D.J.; Bombelli, P.; Vasudevan, R.; Howe, C.J. Photosynthetic, respiratory and extracellular electron transport pathways in cyanobacteria. *Biochim. Biophys. Acta-Bioenerg.* **2016**, *1857*, 247–255. [CrossRef] [PubMed]
24. Cooley, J.W.; Vermaas, W.F.J. Succinate Dehydrogenase and Other Respiratory Pathways in Thylakoid Membranes of Synechocystis sp. Strain PCC 6803: Capacity Comparisons and Physiological Function. *J. Bacteriol.* **2001**, *183*, 4251–4258. [CrossRef] [PubMed]
25. Bronstein, M.; Schütz, M.; Hauska, G.; Padan, E.; Shahak, Y. Cyanobacterial sulfide-quinone reductase: Cloning and heterologous expression. *J. Bacteriol.* **2000**, *182*, 3336–3344. [CrossRef]
26. Theissen, U.; Hoffmeister, M.; Grieshaber, M.; Martin, W. Single Eubacterial Origin of Eukaryotic Sulfide:Quinone Oxidoreductase, a Mitochondrial Enzyme Conserved from the Early Evolution of Eukaryotes During Anoxic and Sulfidic Times. *Mol. Biol. Evol.* **2003**, *20*, 1564–1574. [CrossRef]
27. Berry, S.; Schneider, D.; Vermaas, W.F.J.; Rögner, M. Electron transport routes in whole cells of Synechocystis sp. Strain PCC 6803: The role of the cytochrome bd-type oxidase. *Biochemistry* **2002**, *41*, 3422–3429. [CrossRef]
28. McDonald, A.E.; Ivanov, A.G.; Bode, R.; Maxwell, D.P.; Rodermel, S.R.; Hüner, N.P.A. Flexibility in photosynthetic electron transport: The physiological role of plastoquinol terminal oxidase (PTOX). *Biochim. Biophys. Acta-Bioenerg.* **2011**, *1807*, 954–967. [CrossRef]
29. Pisareva, T.; Shumskaya, M.; Maddalo, G.; Ilag, L.; Norling, B. Proteomics of Synechocystis sp. PCC 6803. *FEBS J.* **2007**, *274*, 791–804. [CrossRef] [PubMed]
30. Huang, F.; Parmryd, I.; Nilsson, F.; Persson, A.L.; Pakrasi, H.B.; Andersson, B.; Norling, B. Proteomics of Synechocystis sp. Strain PCC 6803. *Mol. Cell. Proteomics* **2002**, *1*, 956–966. [CrossRef] [PubMed]
31. Feilke, K.; Ajlani, G.; Krieger-Liszkay, A. Overexpression of plastid terminal oxidase in *Synechocystis* sp. PCC 6803 alters cellular redox state. *Philos. Trans. R. Soc. B Biol. Sci.* **2017**, *372*, 20160379. [CrossRef] [PubMed]
32. Yeremenko, N.; Jeanjean, R.; Prommeenate, P.; Krasikov, V.; Nixon, P.J.; Vermaas, W.F.J.; Havaux, M.; Matthijs, H.C.P. Open Reading Frame ssr2016 is Required for Antimycin A-sensitive Photosystem I-driven Cyclic Electron Flow in the Cyanobacterium Synechocystis sp. PCC 6803. *Plant Cell Physiol.* **2005**, *46*, 1433–1436. [CrossRef] [PubMed]
33. Los, D.A.; Zorina, A.; Sinetova, M.; Kryazhov, S.; Mironov, K.; Zinchenko, V.V.; Los, D.A.; Zorina, A.; Sinetova, M.; Kryazhov, S.; et al. Stress Sensors and Signal Transducers in Cyanobacteria. *Sensors* **2010**, *10*, 2386–2415. [CrossRef] [PubMed]
34. Kanesaki, Y.; Yamamoto, H.; Paithoonrangsarid, K.; Shumskaya, M.; Suzuki, I.; Hayashi, H.; Murata, N. Histidine kinases play important roles in the perception and signal transduction of hydrogen peroxide in the cyanobacterium, Synechocystis sp. PCC 6803. *Plant J.* **2007**, *49*, 313–324. [CrossRef] [PubMed]
35. Murata, N.; Los, D.A. Histidine kinase Hik33 is an important participant in cold-signal transduction in cyanobacteria. *Physiol. Plant.* **2006**, *126*, 17–27. [CrossRef]
36. Ogawa, T. A gene homologous to the subunit-2 gene of NADH dehydrogenase is essential to inorganic carbon transport of Synechocystis PCC6803. *Proc. Natl. Acad. Sci. USA* **1991**, *88*, 4275–4279. [CrossRef]
37. Allahverdiyeva, Y.; Mustila, H.; Ermakova, M.; Bersanini, L.; Richaud, P.; Ajlani, G.; Battchikova, N.; Cournac, L.; Aro, E.-M. Flavodiiron proteins Flv1 and Flv3 enable cyanobacterial growth and photosynthesis under fluctuating light. *Proc. Natl. Acad. Sci. USA* **2013**, *110*, 4111–4116. [CrossRef]
38. Dann, M.; Leister, D. Evidence that cyanobacterial Sll1217 functions analogously to PGRL1 in enhancing PGR5-dependent cyclic electron flow. *Nat. Commun.* **2019**, *10*, 1–7. [CrossRef]
39. Zer, H.; Margulis, K.; Georg, J.; Shotland, Y.; Kostova, G.; Sultan, L.D.; Hess, W.R.; Keren, N. Resequencing of a mutant bearing an iron starvation recovery phenotype defines Slr1658 as a new player in the regulatory network of a model cyanobacterium. *Plant J.* **2018**, *93*, 235–245. [CrossRef]
40. Williams, J.G.K. Construction of Specific Mutations in Photosystem II Photosynthetic Reaction Center by Genetic Engineering Methods in Synechocystis 6803. *Methods Enzymol.* **1988**, *167*, 766–778. [CrossRef]
41. Shcolnick, S.; Shaked, Y.; Keren, N. A role for mrgA, a DPS family protein, in the internal transport of Fe in the cyanobacterium Synechocystis sp. PCC6803. *Biochim. Biophys. Acta Bioenerg.* **2007**, *1767*, 814–819. [CrossRef]

42. Hayashi, K.; Nakazawa, M.; Ishizaki, Y.; Hiraoka, N.; Obayashi, A. Regulation of inter- and intramolecular ligation with T4 DNA ligase in the presence of polyethylene glycol. *Nucleic Acids Res.* **1986**, *14*, 7617–7631. [CrossRef] [PubMed]

43. Eaton-Rye, J.J. Construction of Gene Interruptions and Gene Deletions in the Cyanobacterium Synechocystis sp. Strain PCC 6803. In *Methods in Molecular Biology (Clifton, N.J.)*; Humana Press: Totowa, NJ, USA, 2011; Volume 684, pp. 295–312.

44. Satoh, S.; Ikeuchi, M.; Mimuro, M.; Tanaka, A. Chlorophyll b expressed in Cyanobacteria functions as a light-harvesting antenna in photosystem I through flexibility of the proteins. *J. Biol. Chem.* **2001**, *276*, 4293–4297. [CrossRef] [PubMed]

45. Kamei, A.; Ogawa, T.; Ikeuchi, M. *Photosynthesis: Mechanism and Effects*; Garab, G., Ed.; Kluwer Academic Publishers: Dordrecht, The Netherlands, 1998.

46. Salomon, E.; Keren, N. Manganese limitation induces changes in the activity and in the organization of photosynthetic complexes in the cyanobacterium Synechocystis sp. strain PCC 6803. *Plant Physiol.* **2011**, *155*, 571–579. [CrossRef] [PubMed]

47. Gorbunov, M.Y.; Kolber, Z.S.; Falkowski, P.G. Measuring photosynthetic parameters in individual algal cells by Fast Repetition Rate fluorometry. *Photosynth. Res.* **1999**, *62*, 141–153. [CrossRef]

48. Porra, R.J.; Thompson, W.A.; Kriedemann, P.E. Determination of accurate extinction coefficients and simultaneous equations for assaying chlorophylls a and b extracted with four different solvents: Verification of the concentration of chlorophyll standards by atomic absorption spectroscopy. *Biochim. Biophys. Acta Bioenerg.* **1989**, *975*, 384–394. [CrossRef]

49. Zor, T.; Selinger, Z. Linearization of the Bradford protein assay increases its sensitivity: Theoretical and experimental studies. *Anal. Biochem.* **1996**, *236*, 302–308. [CrossRef]

50. De Porcellinis, A.; Frigaard, N.-U.; Sakuragi, Y. Determination of the Glycogen Content in Cyanobacteria. *J. Vis. Exp.* **2017**, e56068. [CrossRef]

51. Bolger, A.M.; Lohse, M.; Usadel, B. Trimmomatic: A flexible trimmer for Illumina sequence data. *Bioinformatics* **2014**, *30*, 2114–2120. [CrossRef]

52. Babraham Bioinformatics—FastQC A Quality Control tool for High Throughput Sequence Data. Available online: http://www.bioinformatics.babraham.ac.uk/projects/fastqc/ (accessed on 24 June 2020).

53. Li, H. Aligning sequence reads, clone sequences and assembly contigs with BWA-MEM. *arXiv* **2013**, arXiv:1303.3997v2.

54. Koboldt, D.C.; Zhang, Q.; Larson, D.E.; Shen, D.; McLellan, M.D.; Lin, L.; Miller, C.A.; Mardis, E.R.; Ding, L.; Wilson, R.K. VarScan 2: Somatic mutation and copy number alteration discovery in cancer by exome sequencing. *Genome Res.* **2012**, *22*, 568–576. [CrossRef]

55. Li, H. A statistical framework for SNP calling, mutation discovery, association mapping and population genetical parameter estimation from sequencing data. *Bioinformatics* **2011**, *27*, 2987–2993. [CrossRef] [PubMed]

56. Zak, E.; Norling, B.; Maitra, R.; Huang, F.; Andersson, B.; Pakrasi, H.B. The initial steps of biogenesis of cyanobacterial photosystems occur in plasma membranes. *Proc. Natl. Acad. Sci. USA* **2001**, *98*, 13443–13448. [CrossRef] [PubMed]

57. Ball, S.G.; Morell, M.K. From Bacterial Glycogen to Starch: Understanding the Biogenesis of the Plant Starch Granule. *Annu. Rev. Plant Biol.* **2003**, *54*, 207–233. [CrossRef] [PubMed]

58. Klotz, A.; Forchhammer, K. Glycogen, a major player for bacterial survival and awakening from dormancy. *Future Microbiol.* **2017**, *12*, 101–104. [CrossRef] [PubMed]

59. Damrow, R.; Maldener, I.; Zilliges, Y. The Multiple Functions of Common Microbial Carbon Polymers, Glycogen and PHB, during Stress Responses in the Non-Diazotrophic Cyanobacterium Synechocystis sp. PCC 6803. *Front. Microbiol.* **2016**, *7*, 966. [CrossRef] [PubMed]

60. Ozaki, H.; Ikeuchi, M.; Ogawa, T.; Fukuzawa, H.; Sonoike, K. Large-Scale Analysis of Chlorophyll Fluorescence Kinetics in Synechocystis sp. PCC 6803: Identification of the Factors Involved in the Modulation of Photosystem Stoichiometry. *Plant Cell Physiol.* **2007**, *48*, 451–458. [CrossRef] [PubMed]

61. Hihara, Y.; Ikeuchi, M. Mutation in a novel gene required for photomixotrophic growth leads to enhanced photoautotrophic growth of Synechocystis sp. PCC 6803. *Photosynth. Res.* **1997**, *53*, 243–252. [CrossRef]

62. Hihara, Y.; Sonoike, K.; Ikeuchi, M.; Bryant, D.A. A Novel Gene, *pmgA*, Specifically Regulates Photosystem Stoichiometry in the Cyanobacterium *Synechocystis* Species PCC 6803 in Response to High Light. *Plant Physiol.* **1998**, *117*, 1205–1216. [CrossRef] [PubMed]

63. Yao, L.; Shabestary, K.; Björk, S.M.; Asplund-Samuelsson, J.; Joensson, H.N.; Jahn, M.; Hudson, E.P. Pooled CRISPRi screening of the cyanobacterium Synechocystis sp PCC 6803 for enhanced industrial phenotypes. *Nat. Commun.* **2020**, *11*, 1–13. [CrossRef] [PubMed]

64. de Porcellinis, A.J.; Klähn, S.; Rosgaard, L.; Kirsch, R.; Gutekunst, K.; Georg, J.; Hess, W.R.; Sakuragi, Y. The Non-Coding RNA Ncr0700/PmgR1 is Required for Photomixotrophic Growth and the Regulation of Glycogen Accumulation in the Cyanobacterium *Synechocystis* sp. PCC 6803. *Plant Cell Physiol.* **2016**, *57*, 2091–2103. [CrossRef]

65. Ohta, H.; Shibata, Y.; Haseyama, Y.; Yoshino, Y.; Suzuki, T.; Kagasawa, T.; Kamei, A.; Ikeuchi, M.; Enami, I. Identification of genes expressed in response to acid stress in Synechocystis sp. PCC 6803 using DNA microarrays. *Photosynth. Res.* **2005**, *84*, 225–230. [CrossRef] [PubMed]

66. Singh, A.K.; Summerfield, T.C.; Li, H.; Sherman, L.A. The heat shock response in the cyanobacterium Synechocystis sp. Strain PCC 6803 and regulation of gene expression by HrcA and SigB. *Arch. Microbiol.* **2006**, *186*, 273–286. [CrossRef] [PubMed]

67. Kim, D.E.; Chivian, D.; Baker, D. Protein structure prediction and analysis using the Robetta server. *Nucleic Acids Res.* **2004**, *32*, W526–W531. [CrossRef] [PubMed]

68. Pires, D.E.V.; Ascher, D.B.; Blundell, T.L. mCSM: Predicting the effects of mutations in proteins using graph-based signatures. *Bioinformatics* **2014**, *30*, 335–342. [CrossRef]

69. Yu, J.; Mcintosh, L. Isolation and genetic characterization of pseudorevertants from site- directed PSI mutants in Synechocystis 6803. *Methods Enzymol.* **1998**, *297*, 18–26. [CrossRef]

70. Vermaas, W. Molecular genetics of the cyanobacterium Synechocystis sp. PCC 6803: Principles and possible biotechnology applications. *J. Appl. Phycol.* **1996**, *8*, 263–273. [CrossRef]

71. Nishijima, Y.; Kanesaki, Y.; Yoshikawa, H.; Ogawa, T.; Sonoike, K.; Nishiyama, Y.; Hihara, Y. Analysis of spontaneous suppressor mutants from the photomixotrophically grown pmgA-disrupted mutant in the cyanobacterium Synechocystis sp. PCC 6803. *Photosynth. Res.* **2015**, *126*, 465–475. [CrossRef]

72. Ermakova, S.Y.; Elanskaya, I.V.; Kallies, K.U.; Weihe, A.; Börner, T.; Shestakov, S.V. Cloning and sequencing of mutant psbB genes of the cyanobacterium Synechocystis PCC 6803. *Photosynth. Res.* **1993**, *37*, 139–146. [CrossRef] [PubMed]

73. Cano, M.; Holland, S.C.; Artier, J.; Burnap, R.L.; Ghirardi, M.; Morgan, J.A.; Yu, J. Glycogen Synthesis and Metabolite Overflow Contribute to Energy Balancing in Cyanobacteria. *Cell Rep.* **2018**, *23*, 667–672. [CrossRef] [PubMed]

Article

Diurnal Regulation of In Vivo Localization and CO$_2$-Fixing Activity of Carboxysomes in *Synechococcus elongatus* PCC 7942

Yaqi Sun, Fang Huang, Gregory F. Dykes and Lu-Ning Liu

Institute of Systems, Molecular and Integrative Biology, University of Liverpool, Liverpool L69 7ZB, UK;
Yaqi.Sun@liverpool.ac.uk (Y.S.); fang.huang@liverpool.ac.uk (F.H.); Gregory.Dykes@liverpool.ac.uk (G.F.D.)
* Correspondence: luning.liu@liverpool.ac.uk

Received: 20 July 2020; Accepted: 27 August 2020; Published: 29 August 2020

Abstract: Carboxysomes are the specific CO$_2$-fixing microcompartments in all cyanobacteria. Although it is known that the organization and subcellular localization of carboxysomes are dependent on external light conditions and are highly relevant to their functions, how carboxysome organization and function are actively orchestrated in natural diurnal cycles has remained elusive. Here, we explore the dynamic regulation of carboxysome positioning and carbon fixation in the model cyanobacterium *Synechococcus elongatus* PCC 7942 in response to diurnal light-dark cycles, using live-cell confocal imaging and Rubisco assays. We found that carboxysomes are prone to locate close to the central line along the short axis of the cell and exhibit a greater preference of polar distribution in the dark phase, coupled with a reduction in carbon fixation. Moreover, we show that deleting the gene encoding the circadian clock protein KaiA could lead to an increase in carboxysome numbers per cell and reduced portions of pole-located carboxysomes. Our study provides insight into the diurnal regulation of carbon fixation in cyanobacteria and the general cellular strategies of cyanobacteria living in natural habitat for environmental acclimation.

Keywords: carboxysome; carbon fixation; circadian clock; cyanobacteria; confocal microscopy; diurnal regulation; environmental acclimation; *Synechococcus elongatus*

---

1. Introduction

The extraordinary ability of cyanobacteria to survive in diverse ecosystems and adapt to extremes of environmental stress is ascribed to their metabolic robustness and tunability [1]. As cyanobacterial cells rely directly on light for photosynthesis, their abilities to respond to changes in the environmental light conditions are indispensable [2–10]. A typical example is the natural diurnal cycles that synchronize to the rotation of the Earth. It has been shown that the expression of many genes and metabolic activities in cyanobacteria are subject to the circadian rhythm that are regulated by an intrinsic circadian clock [11,12]. This regulation is of physiological importance to improve fitness and facilitate adaptation to diurnal light-dark cycles [13–17]. However, most of laboratory studies on cyanobacterial physiology are still performed under constant light, given the limitations of practical operations and considerations [18].

Carboxysomes are the essential CO$_2$-fixing microcompartments present in all cyanobacteria [19–23]. The carboxysome is composed of a polyhedral protein shell that encapsulates the CO$_2$-fixing enzymes ribulose-1,5-bisphosphate carboxylase/oxygenase (Rubisco) and carbonic anhydrases, as well as structural proteins and chaperones [24–29]. Impairment of carboxysome formation led to complete loss of CO$_2$ fixation ability in cyanobacterial cells grown in ambient air conditions [30]. The cyanobacterial CO$_2$-concentrating mechanisms (CCM) also comprise bicarbonate transporters in the cytoplasmic membrane and thylakoid-integrated CO$_2$-converting complexes that function in accumulation of bicarbonate in the cytoplasm and preventing CO_2 leakage from the cell [31,32]. Elevated bicarbonate then diffuses passively into the carboxysome through the shell and is dehydrated by carbonic anhydrases to CO_2 near Rubisco enzymes [33]. Overall, this CCM system concentrates CO_2 around Rubisco up to 1000-fold, facilitating Rubisco carboxylation and inhibiting oxygenation that leads to "wasteful" photorespiration [23,34].

Spatial distribution of carboxysomes within the cyanobacterial cell is pivotal for carboxysome biogenesis, functionality, and inheritance. It has been shown that multiple carboxysomes are equally distributed along the longitudinal axis of the rod-shaped cells of cyanobacterium *Synechococcus elongatus* PCC 7942 (Syn7942) [35]. Recent work has further revealed the spatial dynamics of carboxysomes in *Synechococcus* sp. PCC 7002 and the role of cell poles in in carboxysome activity modification [36]. This equidistant carboxysome partitioning was initially found to be determined by cytoskeleton protein ParA [35] (also termed McdA [37]) and was recently revealed to be mediated by McdB that can interact with McdA and carboxysome shells [37]. The McdAB system is widespread among β-cyanobacteria [38]. Our recent studies revealed that the biosynthesis, organization, and regulation of carboxysomes in Syn7942 cells are highly sensitive to changes in light [28,39]. Increase in light intensity could accelerate carboxysome biosynthesis, resulting in a higher abundance of carboxysomes and enhanced carbon-fixation activities of Syn7942 cells [39]. It could also alter the protein stoichiometry, diameter, and mobility of carboxysomes in cells [28]. Moreover, we showed that the spatial organization of carboxysomes in Syn7942 is correlated with the redox state of photosynthetic electron transport chain [39], a key controller for circadian rhythm in light-dark cycles [40]. Based on these results and the findings revealing that the expression of carboxysome genes and their encoded proteins is diurnally oscillated [41–43], we question whether in vivo organization and function of carboxysomes in Syn7942 are regulated under diurnal light-dark cycles.

Here, we use live-cell confocal fluorescence imaging to probe the localization of fluorescently labeled carboxysomes in the model cyanobacterium Syn7942 that grow in diurnal cycles. Additionally, we assess the impact of the circadian clock on the carboxysome biosynthesis and distribution in Syn7942 without *kaiA* that encoded the essential protein to sustain oscillation of the circadian clock [17]. We also determine the real-time carbon fixation activities and capacities of Syn7942 cells under diurnal conditions using in vivo Rubisco assays. Our results shed light on the physiological regulation of carboxysome organization and functionality in cyanobacteria that grow in natural environment with regular light-dark cycles.

2. Materials and Methods

2.1. Strains, Generation of Constructs, and Culture Conditions

Escherichia coli (*E. coli*) DH5α and BW25113 strains were used to generate plasmids for homologous recombination in cyanobacteria through λ-red recombination system [44], as described in previous work [3,28,32,39]. In brief, the plasmid containing the target gene coding sequence, *eyfp* coding sequence and apramycin resistant operon as well as 1500 bp upstream/downstream sequences of the target gene amplified from WT Syn7942 genome, was transformed into *E. coli* to replace the endogenous gene via homologous recombination for fluorescence labeling. The plasmid containing the spectinomycin resistant operon and 1500 bp upstream/downstream sequences of the target gene was transformed into *E. coli* to replace the endogenous gene for gene deletion. The diagram of recombination is shown in Figure S1. The generated plasmids were extracted from *E. coli* and were then transformed into Syn7942. The successful modification and segregation status of Syn7942 strains were confirmed by polymerase chain reaction (PCR) with primers designed up/downstream of the modified region. Agarose gel electrophoresis was performed with the standardized amounts of PCR products. The RbcL-eYFP Syn7942 mutant was generated in previous studies [28], with the C-terminus of the Rubisco large subunit RbcL fused with enhanced yellow fluorescent protein (eYFP) after a 11 amino acid linker (LPGPELPGPGP), at the native chromosomal locus and under the control of endogenous promoter.

All strains used were listed in Table S1. Primers used for construct generation and screening were listed in Table S2.

Syn7942 cells were cultivated in BG-11 medium [45] or on BG-11 agar plates with TES buffer pH 8.2 (10 mM $C_6H_{15}NO_6S$) and sodium thiosulphate (20 mM $Na_2S_2O_3$), solidified by 1.5% Agar-agar (*w/v*). For constant moderate light treatment (CL), 30 mL of cultures were added in filter capped culture flasks (Nunc^{TM} Cell Culture Treated EasYFlasks^{TM}, Thermo-Fisher Scientific, Waltham, MA, USA) and kept in 30 °C culture room with constant shaking at 120 rpm under warm white growth light (3200 K) at an intensity of 50 μmole photons·m^{-2}·s^{-1}. Diurnal light treatments (DL) with 12-h dark and 12-h illumination were performed according to method previously described [46]. Light intensity and other parameters were kept the same with CL treatment. Antibiotics were supplied at following concentrations: apramycin at 50 μg·mL^{-1}, spectinomycin at 50 μg·mL^{-1}, and kanamycin at 50 μg·mL^{-1} and chloramphenicol at 10 μg·mL^{-1} in ethanol.

2.2. Carbon-Fixation Assays

The maximum CO_2 fixation capacity measurement was carried out as described previously [28,39]. The in vivo carbon fixation rate measurements were carried out with BG-11 growth medium (nitrogen saturated via bubbling) containing 2 mM of radioactive sodium bicarbonate ($NaH^{14}CO_3$). Gas exchanges with the atmosphere were minimized in air-tight containers during assays. The cultures in sealed flasks were placed into light treatments for 30 min of growth. Cultures were sampled at a volume of 500 μL and mixed with 200 μL 10% formic acid. The mixture was then placed on heat blocks at 95 °C to remove unfixed $NaH^{14}CO_3$. The pellets were resuspended in distilled water then mixed with scintillation cocktail (Ultima Gold XR; PerkinElmer, Waltham, MA, USA). Radioactivity measurements were carried out using a scintillation counter (Tri-Carb; Perkin-Elmer, Waltham, MA, USA). Raw readings were processed to determine the amount of fixed ^{14}C, calibrated by pre-permeabilized cell samples treated with mixed alkyltrimethylammonium bromide (MTA), and then converted as total carbon fixation rates. Carbon fixation rates were normalized by cell density, indicated by measured OD_{750} readings.

2.3. Circadian Bioluminescence Monitoring

Detection of bioluminescence from the luciferase reporters in Syn7942 was performed using a protocol adapted from previous work [47]. pAM2195 introduces the bioluminescence-generating genes *luxAB* and *luxCDE* with circadian-controlled *psbAI* promoter into the Neutral Insertion site II (NSII) of Syn7942 genome as described [47]. Successful pAM2195 transformant was inoculated in BG-11 medium for 2 days of initial growth with DL treatments. Cultures were then pipetted on solid medium (BG-11 1.5% agar) to form a droplet and then placed back to CL treatment for further growth before imaging. For signal quantification, the petri dish containing cell droplets was placed in a light-tight imaging box for bioluminescence capturing for 1 min manually by ImageQuant LAS 4000 (GE Healthcare Life Sciences, Waltham, MA, USA) with a 2-h imaging interval and placed back to CL before next imaging over a tracking period of 22 h.

2.4. Fluorescence Microscopy and Data Analysis

Sample preparation was done as described earlier [3]. For quantitative imaging, laser scanning confocal microscopy used a Zeiss LSM780 with alpha Plan-Fluor 100 × 1.45 Oil objective and excitation at 514 nm from Argon laser. Emissions of YFP signal were captured at 520–550 nm. Chlorophyll auto-fluorescence signals were captured at 660–700 nm. Images were recorded as 512 × 512 pixels images in 16 bits. KaiA-eYFP/RbcL-CFP dual fluorescence imaging was performed as described in [27]. The sample platform was pre-incubated and thermo-controlled at 30 °C before and during imaging. The laser power and imaging settings were maintained the same for all samples for quantitative comparison of fluorescence signals. Images were captured with all pixels below saturation.

Intensity profiling, carboxysome recognition was carried out using Fiji (ImageJ 1.52p, National Institute of Mental Health, Bethesda, MD, USA) [48]. Raw data were processed by Origin 2018 (OriginLab, Northampton, MA, USA) and MATLAB R2018a (Mathworks, Natick, MA, USA) for profile extraction and statistical analysis and the goodness-of-fit parameter for Violin plot visualization. Image SXM [49] was used for statistical analysis of carboxysome numbers per cell, carboxysome distribution within cells, as well as dimensions of cell length/width measurements, as performed previously [39]. Carboxysome distribution profiles along the longitudinal axis and short axis of the cells were analyzed following the method described previously [39]. Analysis of standard deviation of the distribution profiles along the longitudinal axis was performed according to [2,3]. To evaluate the effectiveness of sampling, sampling errors were calculated from three randomized sub-dataset at each timepoint. For each timepoint, a minimum of 300 cells was analyzed. Differences were analyzed with two-sided student *t*-test for significance in pairs or one-way ANOVA and Tukey test for multiple-group comparison. Polar distribution frequency was analyzed based on [36] using ImageSXM. Ten per cent of cell length at both cell ends was considered as the polar region. Polar distribution frequencies were calculated as the percentage of carboxysomes that located in the polar region of all the cells analyzed at each timepoint during diurnal cycles.

3. Results

3.1. Carboxysome Biosynthesis Is Regulated during Diurnal Cycles in Syn7942

To determine whether carboxysome abundance and subcellular organization are regulated during diurnal cycles, we first made a Syn7942 mutant by transforming a luciferase reporter plasmid pAM2195 [47] into wild-type (WT) Syn7942 cells. The intensity profiles of luciferase bioluminescence exhibit a peak at 10–14 h during the 22 h period (Figure S2), consistent with previous findings [47]. This confirmed the proper DL treatments and the circadian regulation in Syn7942 under our established growth conditions. The cell dimensions are relatively constant within experimental error during DL (Figure S3).

We then grew the RbcL-eYFP Syn7942 cells under DL (Figure 1A). The *eyfp* gene was fused to the 3′-end of *rbcL* at the native chromosomal locus and under control of the endogenous promoter (Figure S1). This ensures that the proteins are expressed in context and at physiological levels [27,39]. We performed live-cell confocal imaging on the RbcL-eYFP Syn7942 strain at selected timepoints that covered 1 h before/after light transition as well as quarter marks in a cycle at 1H, 4H, 8H, and 11H from −24 h to 0 h, and then counted the carboxysome number per cell [39] (Figure 1B). A higher carboxysome number per cell was detected in the Syn7942 cells during the light period of diurnal cycles (4.1 ± 1.9 for L1H, 4.1 ± 2.2 for L4H, 3.9 ± 2.0 for L8H, $n = 200$ as cell counts for each timepoint) than those in the dark period (3.3 ± 1.5 for D1H, 3.2 ± 1.4 for D4H, 3.5 ± 1.4 for D8H, 3.3 ± 1.4 for D11H, $n = 200$ as cell counts for each timepoint) (Figure 1B, $p < 0.05$), except for L11H (3.4 ± 1.5 carboxysomes per cell, $n = 200$).

We also determined the contents of Rubisco in the RbcL-eYFP mutant during DL cycles, by quantifying the YFP signal per cell [27]. The cellular levels of Rubisco remain relatively constant (Figure 1C, $p = 0.86$, $n = 200$ as cell counts for each timepoint). The average Rubisco content per carboxysome in cells, as indicated by the peak value ± half-width at half-maximum (HWHM) [28], was relatively lower at L1H, L4H, and L8H than that determined at D1H, D4H, D8H, D11H, and L11H (0.77 ± 0.47, 0.83 ± 0.42, and 0.81 ± 0.51 compared with 0.94 ± 0.56, 0.94 ± 0.51, 0.96 ± 0.62, 1.05 ± 0.59, and 0.93 ± 0.47) (Figure 1D, $n = 200$ as cell counts for each timepoint).

3.2. Subcellular Localization of Carboxysomes Is Diurnally Regulated in Syn7942

We evaluated the spatial localization of carboxysome within the DL-adapted cells (Figure 2A). Carboxysome distribution profiles along the short axis of the cell [28] and analysis of the relative areas under distribution curves indicated that at the later stages of the dark period (D8H and D11H), carboxysomes exhibit a more central positioning along the short axis of the cell, in contrast to the carboxysome distribution at other timepoints of DL cycles (Figure 2B,C).

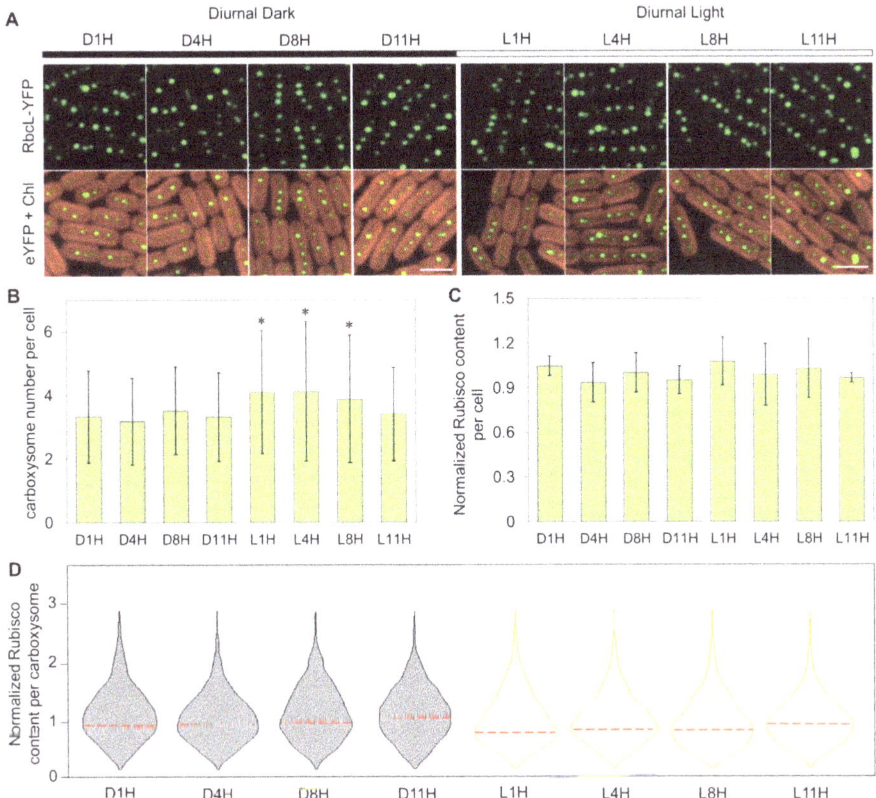

Figure 1. Carboxysome abundance and Rubisco content in Syn7942 cells grown during diurnal light-dark cycles. (**A**) Representative confocal images taken at respective time points during DL. Merged images show carboxysomes in green and Chl fluorescence in red. Scale bar = 2 μm. (**B–D**) Analysis of carboxysome number per cell, total Rubisco content per cell (estimated by RbcL-eYFP content from fluorescence microscopy), and Rubisco content per carboxysome during the dark-light cycle in A. Violin plots were generated by R illustrate the fluorescence intensity distribution of RbcL-eYFP during selected time points. The representative values and deviations were represented by Peak value from kernel density fitting and half-width at half maximum (HWHM). Error bars represent standard deviations. A total of 200 cells were analyzed for each timepoint. * $p < 0.05$.

Analysis of the distribution profiles of carboxysomes along the longitudinal axis of the cell (Figure 2D) and standard deviations of the distribution profiles showed that segregation of carboxysomes was reinforced during the dark period from D1H to D11H (Figure 2E). It appears that more random distribution occurred during the light-dark transition; after light adaptation, the carboxysomes are prone to be segregated to specific cellular positions along the longitudinal axis of the cell (Figure 2E). In addition, carboxysomes exhibit a greater preference of polar distribution during the dark period than during the light period (Figure 2F). At D11H, carboxysomes have the highest tendency to be positioned at the cell poles (polar distribution frequency = 22.7 ± 2.2%, as average ± SE), which have been suggested to be the biogenic sites of carboxysomes and which accommodate inactive carboxysomes [36,50,51].

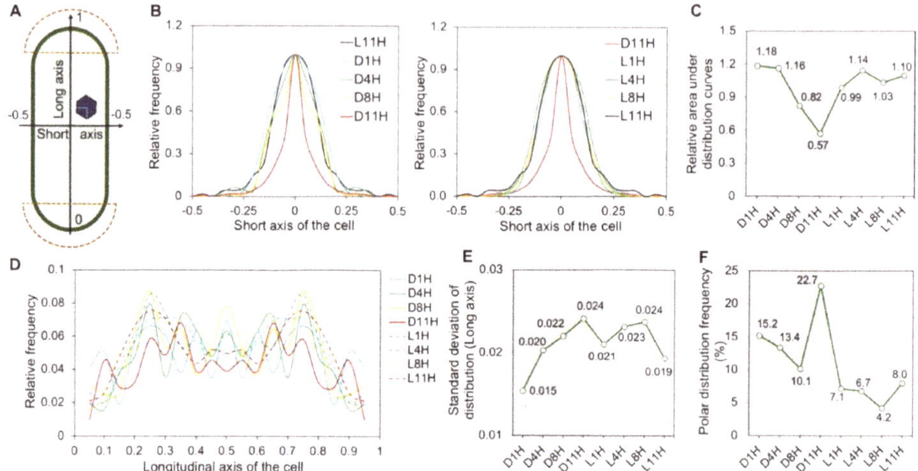

Figure 2. Carboxysome localization under diurnal light-dark conditions. (**A**) Diagram of the carboxysome localization analysis within the rod-shaped Syn7942 cells. Cell pole regions were marked in dash line, covering the 10% of cell length along the longitudinal axis from each end of the cell. (**B**) Distribution profiles of carboxysomes along the short axis of the cell. (**C**) Quantitative analysis of the area under the distribution profile curves in (**B**). (**D**) Distribution profiles of carboxysomes along the longitudinal axis of the cell. (**E**) Standard deviation (SD) analysis of the distribution profiles along the longitudinal axis in (**D**). (**F**) Polar distribution frequency of carboxysomes (located within the polar region marked in (**A**). For each timepoint, 200 cells were analyzed.

3.3. KaiA Deletion Alters the Carboxysome Localization and Abundance in Syn7942

To evaluate the regulation of circadian clock on carboxysome biogenesis in Syn7942 cells, we generated the circadian null strains, Δ*kaiA* and Δ*kaiA*/RbcL-eYFP, by deleting the core oscillator gene *kaiA* [52]. Successful deletion of *kaiA* was confirmed by PCR (Figure S4). We first characterized the carboxysome localization in the Δ*kaiA* mutant in CL conditions (Figure 3). Confocal images were taken using the Syn7942 cells that have been fully adapted to CL for two days. It showed that both Δ*kaiA*/RbcL-eYFP and RbcL-eYFP strains possess canonical carboxysome distributions (Figure 3A). However, the in-depth analysis revealed that carboxysomes in the Δ*kaiA*/RbcL-eYFP mutant exhibited a relatively more centralized distribution along the short axis of the cell compared to the RbcL-eYFP strain (Figure 3B,C, *n* = 500 as cell counts for each strain). Carboxysomes possess more defined localization at specific regions along the longitudinal axis of the Δ*kaiA*/RbcL-eYFP cell, compared with those in the RbcL-eYFP cell (Figure 3D,E, *n* = 500 as cell counts for each strain). Moreover, carboxysomes in the Δ*kaiA*/RbcL-eYFP strain exhibited a lower tendency of the polar localization than in the RbcL-eYFP strain (Figure 3F).

Confocal image analysis also revealed that deletion of *kaiA* induced an increase in the copy number of carboxysomes (Figure 3G, 4.6 ± 1.1 in Δ*kaiA*/RbcL-eYFP, 3.1 ± 0.8 in RbcL-eYFP, *n* = 500 as cell counts, *p* < 0.05). A ~1-fold decrease in the YFP fluorescence intensity per carboxysome was observed in Δ*kaiA*/RbcL-eYFP (*n* = 1000 as carboxysome counts for each strain, *p* < 0.05) (Figure 3H), indicative of the reduced Rubisco content per carboxysome in the Δ*kaiA*/RbcL-eYFP cell.

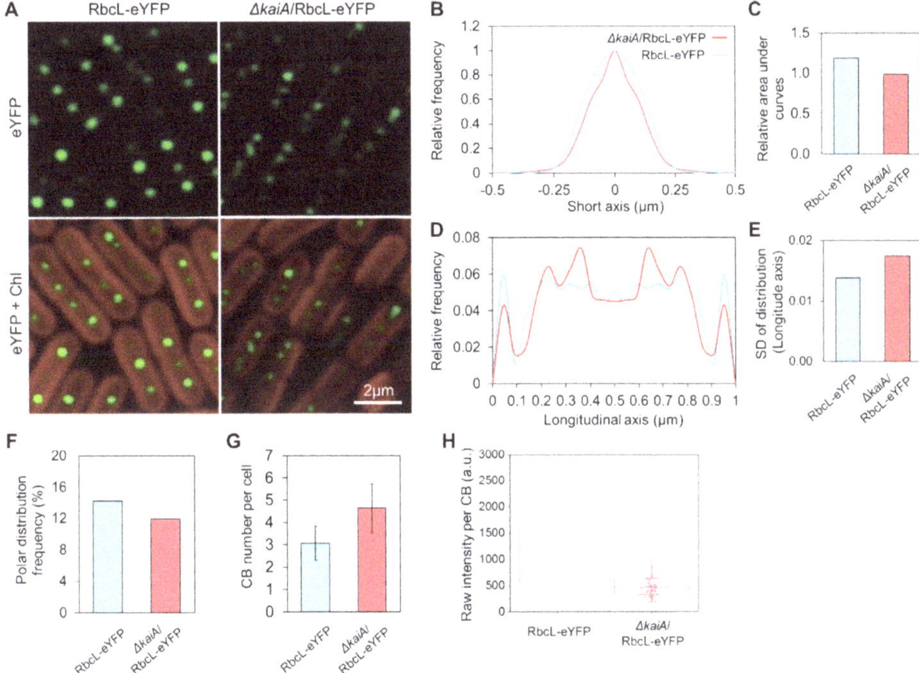

Figure 3. Carboxysome localization and abundance in the Δ*kaiA* Syn7942 mutant. (**A**) Representative confocal images for Syn7942 RbcL-eYFP and Δ*kaiA*/RbcL-eYFP cells grown in CL. Scale bar = 2 μm. (**B**,**C**) Carboxysome distribution profiles along the short axis of the cell and quantitative comparisons of the relative area under distribution profile curves. (**D**,**E**) Carboxysome distribution profiles along the longitudinal axis of the cell and quantitative comparisons of SD of distribution profiles. (**F**) Polar distribution frequency of carboxysomes. (**G**) Carboxysome (CB) number per cell measured from confocal images in A. Data are shown as mean ± SD. A total of 500 cells were analyzed for each strain in (**B**–**G**). (**H**) YFP Signal quantifications for on each carboxysome in two strains. $n = 1000$ as carboxysome number, $p < 0.05$. Data are shown in an arbitrary unit (a.u.). The averaging standard errors for Figure 3C,E,F are 0.13, 0.002, and 2.2, respectively.

3.4. Carbon Fixation of Syn7942 Cells Is Rhythmically Alternated during Diurnal Cycles

To study the regulation of carbon fixation of Syn7942 cells under DL cycles, we measured the whole-cell maximum CO_2-fixation capacities for four days (two days in DL and CL, respectively) using radioactive CO_2-fixation assays (Figure 4A). Rhythmic changes in the CO_2-fixation capacities were observed in DL: The cellular CO_2-fixation capacities were gradually reduced during the dark periods from D1H to D11H and were then rescued suddenly after entering the light period (L1H) and sequentially reached the highest at L4H; the rise in CO_2-fixation capacities was then followed by a decrease from the 2nd half of light period at L8H and L11H. Similar changes were also recorded in the 2nd DL cycle. On average, the whole-cell CO_2-fixation capacities during the light period (5.0 ± 0.5 nmol·mL^{-1}·min^{-1}, $n = 15$) were higher ($p < 0.05$) than those during the dark period (4.1 ± 0.6 nmol·mL^{-1}·min^{-1}, $n = 15$). We further recorded the CO_2-fixation capacities of Syn7942 cells when cells were transferred to CL (Figure 4A). The periodic variations of the cellular carbon fixation were retained during CL, including the increase at the initial timepoints of the subjective light periods during 12–24 and 37–40 h followed by a gradual daily decrease in CO_2-fixation activities (daily averages as 4.65 ± 0.50 to 4.45 ± 0.94, 4.18 ± 0.36, and 3.18 ± 0.28, respectively). However, the average CO_2-fixation activity during each subjective light period at CL was not elevated compared to that measured during the previous subjective dark period. These results indicate that both the circadian clock and light-dark transition play roles in the carbon-fixation regulation of Syn7942 cells.

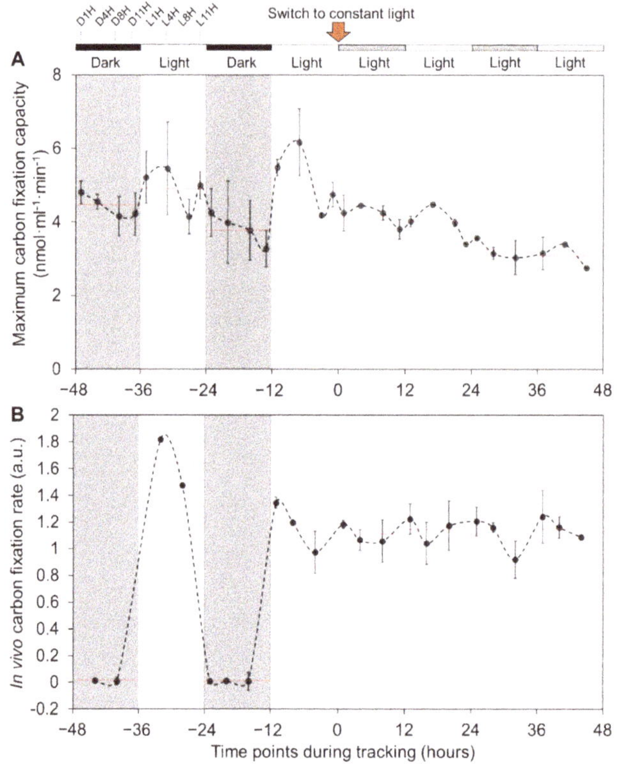

Figure 4. Maximum CO_2-fixation capacities and in vivo CO_2-fixation rates in WT Syn7942 under DL/CL conditions. (**A**) CO_2-fixation capacities of cells grown in DL from −48 to 0 h and additional two days from 0 h to 48 h in CL. The black-white bars above and grey-white background indicate the dark and light cycles with corresponding time-point marks, respectively. Red arrow indicates the time-point of the switch from DL to CL. (**B**) In vivo ^{14}C CO_2-fixation rates measured during growth in $NaH^{14}CO_3$ containing BG-11 medium under DL and CL. Relative CO_2-fixation rates are displayed in an arbitrary unit (a.u.). The red dashed lines indicate time point averages within the 12-h phase for fixation capacities and rates, respectively. Cell contents are normalized by cell density inferred through OD_{750} readings. Data are shown as mean ± SD. $n = 3$ (three independent biological replicates). Cell density OD_{750} was used for normalization.

We also performed real-time in vivo CO_2 fixation assays of Syn7942 cells in both DL and CL conditions (Figure 4B). Unlike the maximum CO_2-fixation assays that were performed by adding exogenous ribulose 1,5-bisphosphate (RuBP) and bicarbonate at saturated concentrations to the permeabilized cells, in vivo CO_2 fixation assays were conducted with endogenous RuBP and bicarbonate in Syn7942 cells. A notable decrease in the CO_2-fixation rate (at a magnitude of ~100-fold) to almost zero was measured during diurnal dark periods (Figure 4B). After switching to CL, the average CO_2-fixation rates of the cells became relatively constant regardless of the subjective light and dark periods ($p = 0.19$). Given that the maximum CO_2-fixation assays indicated the functionality of these carboxysomes in Syn7942 cells (Figure 4A), in vivo CO_2-fixation assays revealed that Syn7942 cells in the dark have a largely restricted Rubisco activity (Figure 4B), probably due to the limited levels of intracellular RuBP and bicarbonate in the dark-adapted Syn7942 cells. To address whether circadian clock is involved in the CO_2-fixation regulation of Syn7942 during DL, we compared the Rubisco activities of $\Delta kaiA$ and WT cells (Figure 5). The Rubisco activities of $\Delta kaiA$ cells were significantly decreased during the light period of DL, in contrast to WT ($p < 0.05$, $n = 4$). No significant difference was detected between $\Delta kaiA$ and WT during the dark period of DL and under CL (Table S3). The results implicated that circadian regulation on the CO_2-fixation activities of carboxysomes specifically occurs during the light phase of DL.

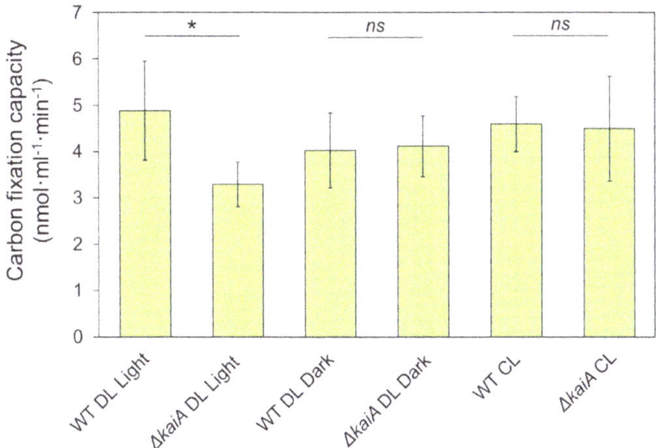

Figure 5. CO_2-fixation assays of WT and $\Delta kaiA$ Syn7942 cells during DL and CL conditions. Reduced CO_2 fixation was detected in $\Delta kaiA$ cells compared to WT during the light period of DL ($p < 0.05$). Error bar represents SD from a minimum of 4 independent biological replicates. Statistics for pair-wise comparison by Tukey test were shown in Table S3. (* as significant difference, $p < 0.05$; *ns* as not significant, $p \geq 0.05$)

4. Discussion

In this work, we characterized the effects of diurnal light-dark cycles on carboxysome biosynthesis, subcellular localization and function in Syn7942. We showed that Syn7942 cells adapt to diurnal cycles by orchestrating carboxysome abundance and spatial localization and CO_2-fixation activities. Moreover, we evaluated the role of the circadian clock in regulating carboxysome biosynthesis and positioning using a $\Delta kaiA$ Syn7942 mutant. Our results provide insight into the natural regulatory strategies evolved in cyanobacteria to control the assembly and functionality of carboxysomes, a key "biofactory" in global carbon fixation. A deeper understanding of the diurnal regulation of cyanobacterial metabolisms may also inform industrial applications to grow cyanobacteria that are facing the natural light-dark cycles in the outdoors [53].

Previous studies indicated that there were no significant changes in the percentage of tagged/non-tagged RbcL in the RbcL-eYFP strains grown under different light conditions [39]. The cellular levels of Rubisco detected in this study remain relatively constant under DL conditions (Figure 1C), in agreement with published proteomic data [54,55]. In contrast, transcriptional assays showed that the levels of cyanobacterial Rubisco genes *rbcL* and *rbcS*, together with other genes located in a *ccm* operon (*ccmK2*, *ccmL*, *ccmM*, *ccmN*, and *ccmO*), were rhythmically alternated under DL conditions [42,56]. This discrepancy may suggest possible post-transcriptional regulation of Rubisco [57]. Due to the imaging limit, it remains unclear whether free Rubisco proteins in the cytoplasm, as reported in marine cyanobacteria [58], were omitted in the cellular Rubisco quantification (Figure 1B,D), which merits future investigation.

Studies on the spatial localization of carboxysomes within the rod-shaped Syn7942 cells have suggested its significance in the biogenesis, function, and inheritance of carboxysomes [35–38]. Carboxysomes possess equal distribution along the longitudinal axis of the Syn7942 cell, which was indicated to be mediated by the McdAB system that is widespread among β-cyanobacteria [37,38]. Beyond these findings, here we showed the tunable subcellular positioning of carboxysomes during diurnal cycles, confirming the importance of light in determining the in vivo localization of carboxysomes. Under diurnal light-dark conditions, cell elongation and division of Syn7942 took place during the mid-phase of light period [59]. The activity of McdAB system might be determined by the cellular levels of ATP, which was known to accumulate throughout the light period [60]. Indeed, we observed gradually strengthened localization control from L1H to L8H, quantified as standard deviation of the distribution along the longitudinal axis of the cell (Figure 2E), which appear synchronously with the cell elongation and division events and rising levels of ATP. Carboxysomes and chromosomes are mutually exclusive in the cytoplasm [61]. Chromosome compaction mainly happened during the light period in the diurnal cycle, whereas during the dark period chromosomes were evenly distributed in the cytoplasm [62]. The restrained localization of carboxysomes in dark was therefore unlikely to be a result of space exclusion from chromosome positioning. The mechanisms that define the dynamic carboxysome distribution during diurnal cycles remain to be answered. In addition, we showed that carboxysomes have a high preference to locate at the cell poles during the dark phase in the RbcL-eYFP mutant; in the dark period, the WT Syn7942 cells show a reduced CO_2-fixation activity (Figures 2F and 4). These observations are consistent with the previous studies suggesting that the cell poles may serve as the sites for accommodating carboxysome precursors [50] or inactive or degrading carboxysomes [36].

In contrast to the better-understood distribution along the longitudinal axis of Syn7942 cells, the positioning of carboxysome along the short axis of the cell has remained poorly characterized. Our previous study has indicated that the central localization of carboxysomes along the short axis of the Syn7942 cell was ascribed to the reduced plastoquinone pool in photosynthetic electron transport chain [39]. The redox status is also a key signal in circadian regulation of cyanobacteria and can be modulated in light-dark transitions [63]. The plastoquinone pool is prone to be oxidized in the light phase [64] and cumulatively reduced throughout the major dark phase [65]. In agreement with this, we observed that a more central distribution of carboxysomes along the short axis of the cell appears in the dark period (Figure 2B,C), confirming the role of the redox state of the plastoquinone pool in mediating carboxysome positioning in Syn7942. Whether the circadian clock was involved in the redox-coupled positioning modulation of carboxysomes remains to be explored.

Interestingly, confocal images of the dual-labeled Syn7942 mutant KaiA-eYFP/RbcL-CFP (Figure S5) displayed that in addition to the polar localization of KaiA in Syn7942 as reported earlier [46], several KaiA fluorescent puncta are spatially close to carboxysomes in the cytoplasm (Figure S5A). Time-lapse imaging showed the dynamic formation process of KaiA assemblies in the dark period of DL (Figure S5B). The functional relevance of KaiA foci close to carboxysome in Syn7942 merits further investigations.

Our data suggested the regulation of circadian clock in carboxysome biogenesis and function in Syn7942 (Figures 3 and 5). The circadian control of carbon assimilation has also been reported in higher plants [66–68]. In the chloroplast of dinoflagellates, Rubisco carboxylation is regulated by circadian clock through rearrangement of Rubiscos localization inside chloroplasts while maintaining constant levels of Rubisco proteins [57]. In C3 plants, the enzymatic activity of Rubisco is regulated by a series of Rubisco activases, of which the oscillated expression is circadian controlled [69,70]. In Crassulacean acid metabolism (CAM) plants, the CAM genes possess daily regulation by the circadian clock [71] and phosphoenolpyruvate carboxylase kinase represents the well-defined circadian control in primary CO_2 fixation [72]. Meanwhile, the alternated distribution patterns of Rubisco in Syn7942 cells (higher numbers of carboxysomes that each contained fewer Rubisco), represented by RbcL-eYFP in Δ*kaiA*/RbcL-eYFP, together with unchanged cellular levels of Rubiscos (data not shown) and carbon-fixation capacity compared to WT Syn7942 in CL (Figures 3 and 5) might be an outcome of carboxysome compensation without KaiA to achieve similar levels of cellular carbon-fixation capacities, likely indicating the KaiA-involved modulation of carboxysomes [28]. It would be interesting in future research to survey the protein content of other carboxysome components to gain a complete picture of circadian clock-based structural plasticity. In conclusion, these studies highlighted the general regulation of carbon assimilation in cells in response to the natural diurnal cycles.

Supplementary Materials: The following are available online at http://www.mdpi.com/2075-1729/10/9/169/s1, Figure S1: The strategy of fluorescent proteins (FPs) fusion and knock-out (KO) using REDIRECT protocol, Figure S2: Bioluminescence assays confirm the diurnal treatment and circadian control in the Syn7942 strain containing pAM2195, Figure S3: Cell dimensions of Syn7942 during DL, Figure S4: PCR screening of the segregation of *kaiA* mutants, Figure S5: Fluorescence images of the KaiA-eYFP/RbcL-CFP mutant show the distribution of carboxysomes and KaiA assemblies in Syn7942, Table S1: Strains used in this work, Table S2: Primers used in this work, Table S3: *p*-values of Tukey test on differences of maximum carbon fixation capacities listed in Figure 5.

Author Contributions: Conceptualization, Y.S. and L.-N.L.; formal analysis, Y.S., F.H., and L.-N.L.; investigation, Y.S., F.H., and G.F.D.; writing—original draft preparation, Y.S. and L.-N.L.; writing—review and editing, Y.S., F.H., G.F.D., and L.-N.L.; supervision, L.-N.L. All authors have read and agreed to the published version of the manuscript.

Funding: This research was funded by the Royal Society, grant numbers URF\R\180030, RGF\EA\181061, RGF\EA\180233, and the Biotechnology and Biological Sciences Research Council, grant numbers BB/M024202/1 and BB/R003890/1.

Acknowledgments: The authors would like to thank the Liverpool Centre for Cell Imaging for technical assistance and provision of confocal microscopy.

Conflicts of Interest: The authors declare no conflict of interest. The funders had no role in the design of the study; in the collection, analyses, or interpretation of data; in the writing of the manuscript; or in the decision to publish the results.

References

1. Dvornyk, V.; Vinogradova, O.; Nevo, E. Origin and evolution of circadian clock genes in prokaryotes. *Proc. Natl. Acad. Sci. USA* **2003**, *100*, 2495–2500. [CrossRef] [PubMed]

2. Casella, S.; Huang, F.; Mason, D.; Zhao, G. Y.; Johnson, G.N.; Mullineaux, C.W.; Liu, L.-N. Dissecting the Native Architecture and Dynamics of Cyanobacterial Photosynthetic Machinery. *Mol. Plant* **2017**, *10*, 1434–1448. [CrossRef] [PubMed]

3. Liu, L.N.; Bryan, S.J.; Huang, F.; Yu, J.; Nixon, P.J.; Rich, P.R.; Mullineaux, C.W. Control of electron transport routes through redox-regulated redistribution of respiratory complexes. *Proc. Natl. Acad. Sci. USA* **2012**, *109*, 11431–11436. [CrossRef] [PubMed]

4. Wiltbank, L.B.; Kehoe, D.M. Diverse light responses of cyanobacteria mediated by phytochrome superfamily photoreceptors. *Nat. Rev. Microbiol.* **2019**, *17*, 37–50. [CrossRef] [PubMed]

5. Schuergers, N.; Lenn, T.; Kampmann, R.; Meissner, M.V.; Esteves, T.; Temerinac-Ott, M.; Korvink, J.G.; Lowe, A.R.; Mullineaux, C.W.; Wilde, A. Cyanobacteria use micro-optics to sense light direction. *Elife* **2016**, *5*, e12620. [CrossRef]

6. Sanfilippo, J.E.; Garczarek, L.; Partensky, F.; Kehoe, D.M. Chromatic Acclimation in Cyanobacteria: A Diverse and Widespread Process for Optimizing Photosynthesis. *Annu. Rev. Microbiol.* **2019**, *73*, 407–433. [CrossRef]

7. Liu, L.N. Distribution and dynamics of electron transport complexes in cyanobacterial thylakoid membranes. *Biochim. Biophys. Acta* **2016**, *1857*, 256–265. [CrossRef]

8. Mullineaux, C.W.; Liu, L.N. Membrane dynamics in phototrophic bacteria. *Annu. Rev. Microbiol.* **2020**, *74*, 633–654. [CrossRef]

9. Zhao, L.S.; Huokko, T.; Wilson, S.; Simpson, D.M.; Wang, Q.; Ruban, A.V.; Mullineaux, C.W.; Zhang, Y.Z.; Liu, L.N. Structural variability, coordination, and adaptation of a native photosynthetic machinery. *Nat. Plants* **2020**, *6*, 869–882. [CrossRef]

10. Mahbub, M.; Hemm, L.; Yang, Y.; Kaur, R.; Carmen, H.; Engl, C.; Huokko, T.; Riediger, M.; Watanabe, S.; Liu, L.-N.; et al. mRNA localisation, reaction centre biogenesis and thylakoid membrane targeting in cyanobacteria. *Nat. Plants* **2020**, in press. [CrossRef]

11. Vijayan, V.; Zuzow, R.; O'Shea, E.K. Oscillations in supercoiling drive circadian gene expression in cyanobacteria. *Proc. Natl. Acad. Sci. USA* **2009**, *106*, 22564–22568. [CrossRef] [PubMed]

12. Yang, Q.; Pando, B.F.; Dong, G.; Golden, S.S.; van Oudenaarden, A. Circadian gating of the cell cycle revealed in single cyanobacterial cells. *Science* **2010**, *327*, 1522–1526. [CrossRef] [PubMed]

13. Johnson, C.H.; Zhao, C.; Xu, Y.; Mori, T. Timing the day: What makes bacterial clocks tick? *Nat. Rev. Microbiol.* **2017**, *15*, 232–242. [CrossRef]

14. Rust, M.J.; Golden, S.S.; O'Shea, E.K. Light-driven changes in energy metabolism directly entrain the cyanobacterial circadian oscillator. *Science* **2011**, *331*, 220–223. [CrossRef] [PubMed]

15. Golden, S.S.; Canales, S.R. Cyanobacterial circadian clocks—Timing is everything. *Nat. Rev. Microbiol.* **2003**, *1*, 191–199. [CrossRef]

16. Woelfle, M.A.; Ouyang, Y.; Phanvijhitsiri, K.; Johnson, C.H. The adaptive value of circadian clocks: An experimental assessment in cyanobacteria. *Curr. Biol. CB* **2004**, *14*, 1481–1486. [CrossRef] [PubMed]

17. Welkie, D.G.; Rubin, B.E.; Chang, Y.G.; Diamond, S.; Rifkin, S.A.; LiWang, A.; Golden, S.S. Genome-wide fitness assessment during diurnal growth reveals an expanded role of the cyanobacterial circadian clock protein KaiA. *Proc. Natl. Acad. Sci. USA* **2018**, *115*, E7174–E7183. [CrossRef]

18. Welkie, D.G.; Rubin, B.E.; Diamond, S.; Hood, R.D.; Savage, D.F.; Golden, S.S. A Hard Day's Night: Cyanobacteria in Diel Cycles. *Trends Microbiol.* **2019**, *27*, 231–242. [CrossRef]

19. Turmo, A.; Gonzalez-Esquer, C.R.; Kerfeld, C.A. Carboxysomes: Metabolic modules for CO_2 fixation. *FEMS Microbiol. Lett.* **2017**, *364*, fnx176. [CrossRef]

20. Yeates, T.O.; Tsai, Y.; Tanaka, S.; Sawaya, M.R.; Kerfeld, C.A. Self-assembly in the carboxysome: A viral capsid-like protein shell in bacterial cells. *Biochem. Soc. Trans.* **2007**, *35*, 508–511. [CrossRef]

21. Kerfeld, C.A.; Erbilgin, O. Bacterial microcompartments and the modular construction of microbial metabolism. *Trends Microbiol.* **2015**, *23*, 22–34. [CrossRef] [PubMed]

22. Sui, N.; Huang, F.; Liu, L.N. Photosynthesis in Phytoplankton: Insights from the Newly Discovered Biological Inorganic Carbon Pumps. *Mol. Plant* **2020**, *13*, 949–951. [CrossRef]

23. Rae, B.D.; Long, B.M.; Forster, B.; Nguyen, N.D.; Velanis, C.N.; Atkinson, N.; Hee, W.Y.; Mukherjee, B.; Price, G.D.; McCormick, A.J. Progress and challenges of engineering a biophysical CO_2-concentrating mechanism into higher plants. *J. Exp. Bot.* **2017**, *68*, 3717–3737. [CrossRef] [PubMed]

24. Faulkner, M.; Rodriguez-Ramos, J.; Dykes, G.F.; Owen, S.V.; Casella, S.; Simpson, D.M.; Beynon, R.J.; Liu, L.-N. Direct characterization of the native structure and mechanics of cyanobacterial carboxysomes. *Nanoscale* **2017**, *9*, 10662–10673. [CrossRef]

25. Long, B.M.; Badger, M.R.; Whitney, S.M.; Price, G.D. Analysis of carboxysomes from Synechococcus PCC7942 reveals multiple Rubisco complexes with carboxysomal proteins CcmM and CcaA. *J. Biol. Chem.* **2007**, *282*, 29323–29335. [CrossRef] [PubMed]

26. Fang, Y.; Huang, F.; Faulkner, M.; Jiang, Q.; Dykes, G.F.; Yang, M.; Liu, L.N. Engineering and Modulating Functional Cyanobacterial CO2-Fixing Organelles. *Front. Plant Sci.* **2018**, *9*, 739. [CrossRef]

27. Huang, F.; Vasieva, O.; Sun, Y.; Faulkner, M.; Dykes, G.F.; Zhao, Z.; Liu, L.N. Roles of RbcX in Carboxysome Biosynthesis in the Cyanobacterium Synechococcus elongatus PCC7942. *Plant Physiol.* **2019**, *179*, 184–194. [CrossRef]

28. Sun, Y.; Wollman, A.J.M.; Huang, F.; Leake, M.C.; Liu, L.N. Single-organelle quantification reveals the stoichiometric and structural variability of carboxysomes dependent on the environment. *Plant Cell* **2019**, *31*, 1648–1664. [CrossRef]

29. Huang, F.; Kong, W.; Sun, Y.; Chen, T.; Dykes, G.F.; Jiang, Y.L.; Liu, L.N. Rubisco accumulation factor 1 (Raf1) plays essential roles in mediating Rubisco assembly and carboxysome biogenesis. *Proc. Natl. Acad. Sci. USA* **2020**, *117*, 17418–17428. [CrossRef]

30. Rae, B.D.; Long, B.M.; Badger, M.R.; Price, G.D. Structural determinants of the outer shell of beta-carboxysomes in *Synechococcus elongatus* PCC 7942: Roles for CcmK2, K3-K4, CcmO, and CcmL. *PLoS ONE* **2012**, *7*, e43871. [CrossRef]

31. Wang, C.; Sun, B.; Zhang, X.; Huang, X.; Zhang, M.; Guo, H.; Chen, X.; Huang, F.; Chen, T.; Mi, H.; et al. Structural mechanism of the active bicarbonate transporter from cyanobacteria. *Nat. Plants* **2019**, *5*, 1184–1193. [CrossRef] [PubMed]

32. Hennacy, J.H.; Jonikas, M.C. Prospects for Engineering Biophysical CO_2 Concentrating Mechanisms into Land Plants to Enhance Yields. *Annu. Rev. Plant Biol.* **2020**, *71*, 461–485. [CrossRef] [PubMed]

33. Faulkner, M.; Szabó, I.; Weetman, S.L.; Sicard, F.; Huber, R.G.; Bond, P.J.; Rosta, E.; Liu, L.-N. Molecular simulations unravel the molecular principles that mediate selective permeability of carboxysome shell protein. *bioRxiv* **2020**. 2020.06.14.151241. Available online: https://www.biorxiv.org/content/10.1101/2020.06.14.151241v1.abstract (accessed on 14 June 2020).

34. Burnap, R.L.; Hagemann, M.; Kaplan, A. Regulation of CO_2 Concentrating Mechanism in Cyanobacteria. *Life* **2015**, *5*, 348–371. [CrossRef] [PubMed]

35. Savage, D.F.; Afonso, B.; Chen, A.H.; Silver, P.A. Spatially ordered dynamics of the bacterial carbon fixation machinery. *Science* **2010**, *327*, 1258–1261. [CrossRef] [PubMed]

36. Hill, N.C.; Tay, J.W.; Altus, S.; Bortz, D.M.; Cameron, J.C. Life cycle of a cyanobacterial carboxysome. *Sci. Adv.* **2020**, *6*, eaba1269. [CrossRef] [PubMed]

37. MacCready, J.S.; Hakim, P.; Young, E.J.; Hu, L.; Liu, J.; Osteryoung, K.W.; Vecchiarelli, A.G.; Ducat, D.C. Protein gradients on the nucleoid position the carbon-fixing organelles of cyanobacteria. *Elife* **2018**, *7*, e39723. [CrossRef]

38. MacCready, J.S.; Basalla, J.L.; Vecchiarelli, A.G. Origin and Evolution of Carboxysome Positioning Systems in Cyanobacteria. *Mol. Biol. Evol.* **2020**, *37*, 1434–1451. [CrossRef]

39. Sun, Y.; Casella, S.; Fang, Y.; Huang, F.; Faulkner, M.; Barrett, S.; Liu, L.N. Light modulates the biosynthesis and organization of cyanobacterial carbon fixation machinery through photosynthetic electron flow. *Plant Physiol.* **2016**, *171*, 530–541. [CrossRef]

40. Cohen, S.E.; Golden, S.S. Circadian Rhythms in Cyanobacteria. *Microbiol. Mol. Biol. Rev.* **2015**, *79*, 373–385. [CrossRef]

41. Watson, G.M.; Tabita, F.R. Regulation, unique gene organization, and unusual primary structure of carbon fixation genes from a marine phycoerythrin-containing cyanobacterium. *Plant Mol. Biol.* **1996**, *32*, 1103–1115. [CrossRef]

42. Ito, H.; Mutsuda, M.; Murayama, Y.; Tomita, J.; Hosokawa, N.; Terauchi, K.; Sugita, C.; Sugita, M.; Kondo, T.; Iwasaki, H. Cyanobacterial daily life with Kai-based circadian and diurnal genome-wide transcriptional control in *Synechococcus elongatus*. *Proc. Natl. Acad. Sci. USA* **2009**, *106*, 14168–14173. [CrossRef] [PubMed]

43. Aryal, U.K.; Stockel, J.; Krovvidi, R.K.; Gritsenko, M.A.; Monroe, M.E.; Moore, R.J.; Koppenaal, D.W.; Smith, R.D.; Pakrasi, H.B.; Jacobs, J.M. Dynamic proteomic profiling of a unicellular cyanobacterium Cyanothece ATCC51142 across light-dark diurnal cycles. *BMC Syst. Biol.* **2011**, *5*, 194. [CrossRef] [PubMed]

44. Datsenko, K.A.; Wanner, B.L. One-step inactivation of chromosomal genes in Escherichia coli K-12 using PCR products. *Proc. Natl. Acad. Sci. USA* **2000**, *97*, 6640–6645. [CrossRef]

45. Rippka, R.; Deruelles, J.; Waterbury, J.B.; Herdman, M.; Stanier, R.Y. Generic Assignments, Strain Histories and Properties of Pure Cultures of Cyanobacteria. *Microbiology* **1979**, *111*, 1–61. [CrossRef]

46. Cohen, S.E.; Erb, M.L.; Selimkhanov, J.; Dong, G.; Hasty, J.; Pogliano, J.; Golden, S.S. Dynamic localization of the cyanobacterial circadian clock proteins. *Curr. Biol.* **2014**, *24*, 1836–1844. [CrossRef] [PubMed]

47. Mackey, S.R.; Ditty, J.L.; Clerico, E.M.; Golden, S.S. Detection of rhythmic bioluminescence from luciferase reporters in cyanobacteria. *Methods Mol. Biol.* **2007**, *362*, 115–129. [CrossRef] [PubMed]

48. Schindelin, J.; Arganda-Carreras, I.; Frise, E.; Kaynig, V.; Longair, M.; Pietzsch, T.; Preibisch, S.; Rueden, C.; Saalfeld, S.; Schmid, B.; et al. Fiji: An open-source platform for biological-image analysis. *Nat. Methods* **2012**, *9*, 676–682. [CrossRef]

49. Barrett, S.D. Image SXM. 2015. Available online: http://www.ImageSXM.org.uk (accessed on 29 August 2020).

50. Cameron, J.C.; Wilson, S.C.; Bernstein, S.L.; Kerfeld, C.A. Biogenesis of a bacterial organelle: The carboxysome assembly pathway. *Cell* **2013**, *155*, 1131–1140. [CrossRef]

51. Chen, A.H.; Robinson-Mosher, A.; Savage, D.F.; Silver, P.A.; Polka, J.K. The bacterial carbon-fixing organelle is formed by shell envelopment of preassembled cargo. *PLoS ONE* **2013**, *8*, e76127. [CrossRef]

52. Clerico, E.M.; Cassone, V.M.; Golden, S.S. Stability and lability of circadian period of gene expression in the cyanobacterium Synechococcus elongatus. *Microbiology* **2009**, *155*, 635–641. [CrossRef]

53. Diamond, S.; Jun, D.; Rubin, B.E.; Golden, S.S. The circadian oscillator in Synechococcus elongatus controls metabolite partitioning during diurnal growth. *Proc. Natl. Acad. Sci. USA* **2015**, *112*, E1916–E1925. [CrossRef] [PubMed]

54. Waldbauer, J.R.; Rodrigue, S.; Coleman, M.L.; Chisholm, S.W. Transcriptome and Proteome Dynamics of a Light-Dark Synchronized Bacterial Cell Cycle. *PLoS ONE* **2012**, *7*, e43432. [CrossRef] [PubMed]

55. Guerreiro, A.C.L.; Benevento, M.; Lehmann, R.; van Breukelen, B.; Post, H.; Giansanti, P.; Maarten Altelaar, A.F.; Axmann, I.M.; Heck, A.J.R. Daily Rhythms in the Cyanobacterium Synechococcus elongatus Probed by High-resolution Mass Spectrometry–based Proteomics Reveals a Small Defined Set of Cyclic Proteins. *Mol. Cell. Proteom.* **2014**, *13*, 2042–2055. [CrossRef] [PubMed]

56. Vijayan, V.; Jain, I.H.; O'Shea, E.K. A high resolution map of a cyanobacterial transcriptome. *Genome Biol.* **2011**, *12*, R47. [CrossRef] [PubMed]

57. Nassoury, N.; Fritz, L.; Morse, D. Circadian Changes in Ribulose-1,5-Bisphosphate Carboxylase/Oxygenase Distribution Inside Individual Chloroplasts Can Account for the Rhythm in Dinoflagellate Carbon Fixation. *Plant Cell* **2001**, *13*, 923. [CrossRef]

58. Dai, W.; Chen, M.; Myers, C.; Ludtke, S.J.; Pettitt, B.M.; King, J.A.; Schmid, M.F.; Chiu, W. Visualizing Individual RuBisCO and Its Assembly into Carboxysomes in Marine Cyanobacteria by Cryo-Electron Tomography. *J. Mol. Biol.* **2018**, *430*, 4156–4167. [CrossRef] [PubMed]

59. Martins, B.M.C.; Tooke, A.K.; Thomas, P.; Locke, J.C.W. Cell size control driven by the circadian clock and environment in cyanobacteria. *Proc. Natl. Acad. Sci. USA* **2018**, *115*, E11415–E11424. [CrossRef]

60. Saha, R.; Liu, D.; Hoynes-O'Connor, A.; Liberton, M.; Yu, J.; Bhattacharyya-Pakrasi, M.; Balassy, A.; Zhang, F.; Moon, T.S.; Maranas, C.D.; et al. Diurnal Regulation of Cellular Processes in the Cyanobacterium Synechocystis sp. Strain PCC 6803: Insights from Transcriptomic, Fluxomic, and Physiological Analyses. *mBio* **2016**, *7*, e00464-16. [CrossRef]

61. Jain, I.H.; Vijayan, V.; O'Shea, E.K. Spatial ordering of chromosomes enhances the fidelity of chromosome partitioning in cyanobacteria. *Proc. Natl. Acad. Sci. USA* **2012**, *109*, 13638–13643. [CrossRef]

62. Murata, K.; Hagiwara, S.; Kimori, Y.; Kaneko, Y. Ultrastructure of compacted DNA in cyanobacteria by high-voltage cryo-electron tomography. *Sci. Rep.* **2016**, *6*, 34934. [CrossRef]

63. Tamoi, M.; Miyazaki, T.; Fukamizo, T.; Shigeoka, S. The Calvin cycle in cyanobacteria is regulated by CP12 via the NAD(H)/NADP(H) ratio under light/dark conditions. *Plant J.* **2005**, *42*, 504–513. [CrossRef] [PubMed]

64. Ivanov, B.; Mubarakshina, M.; Khorobrykh, S. Kinetics of the plastoquinone pool oxidation following illumination Oxygen incorporation into photosynthetic electron transport chain. *FEBS Lett.* **2007**, *581*, 1342–1346. [CrossRef] [PubMed]

65. Kim, Y.I.; Vinyard, D.J.; Ananyev, G.M.; Dismukes, G.C.; Golden, S.S. Oxidized quinones signal onset of darkness directly to the cyanobacterial circadian oscillator. *Proc. Natl. Acad. Sci. USA* **2012**, *109*, 17765–17769. [CrossRef]

66. Carmo-Silva, A.E.; Keys, A.J.; Andralojc, P.J.; Powers, S.J.; Arrabaca, M.C.; Parry, M.A. Rubisco activities, properties, and regulation in three different C4 grasses under drought. *J. Exp. Bot.* **2010**, *61*, 2355–2366. [CrossRef]

67. Lan, Y.; Woodrow, I.E.; Mott, K.A. Light-dependent changes in ribulose bisphosphate carboxylase activase activity in leaves. *Plant Physiol.* **1992**, *99*, 304–309. [CrossRef] [PubMed]

68. Li, J.; Yokosho, K.; Liu, S.; Cao, H.R.; Yamaji, N.; Zhu, X.G.; Liao, H.; Ma, J.F.; Chen, Z.C. Diel magnesium fluctuations in chloroplasts contribute to photosynthesis in rice. *Nat. Plants* **2020**, *6*, 848–859. [CrossRef]

69. To, K.Y.; Suen, D.F.; Chen, S.C. Molecular characterization of ribulose-1,5-bisphosphate carboxylase/oxygenase activase in rice leaves. *Planta* **1999**, *209*, 66–76. [CrossRef]

70. Liu, Z.; Taub, C.C.; McClung, C.R. Identification of an Arabidopsis thaliana ribulose-1,5-bisphosphate carboxylase/oxygenase activase (RCA) minimal promoter regulated by light and the circadian clock. *Plant Physiol.* **1996**, *112*, 43–51. [CrossRef]

71. Hartwell, J.; Dever, L.V.; Boxall, S.F. Emerging model systems for functional genomics analysis of Crassulacean acid metabolism. *Curr. Opin. Plant Biol.* **2016**, *31*, 100–108. [CrossRef]

72. Boxall, S.F.; Dever, L.V.; Knerova, J.; Gould, P.D.; Hartwell, J. Phosphorylation of Phosphoenolpyruvate Carboxylase Is Essential for Maximal and Sustained Dark CO_2 Fixation and Core Circadian Clock Operation in the Obligate Crassulacean Acid Metabolism Species Kalanchoe fedtschenkoi. *Plant Cell* **2017**, *29*, 2519–2536. [CrossRef]

Article

On the Role and Production of Polyhydroxybutyrate (PHB) in the Cyanobacterium *Synechocystis* sp. PCC 6803

Moritz Koch [1], Kenneth W. Berendzen [2] and Karl Forchhammer [1,*]

[1] Interfaculty Institute of Microbiology and Infection Medicine Tübingen, Eberhard-Karls-Universität Tübingen, 72076 Tübingen, Germany; moritz.koch@uni-tuebingen.de

[2] Center for Plant Molecular Biology, Eberhard-Karls-Universität Tübingen, 72076 Tübingen, Germany; kenneth.berendzen@zmbp.uni-tuebingen.de

* Correspondence: karl.forchhammer@uni-tuebingen.de; Tel.: +49-7071-29-72096

Received: 17 March 2020; Accepted: 21 April 2020; Published: 22 April 2020

Abstract: The cyanobacterium *Synechocystis* sp. PCC 6803 is known for producing polyhydroxybutyrate (PHB) under unbalanced nutrient conditions. Although many cyanobacteria produce PHB, its physiological relevance remains unknown, since previous studies concluded that PHB is redundant. In this work, we try to better understand the physiological conditions that are important for PHB synthesis. The accumulation of intracellular PHB was higher when the cyanobacterial cells were grown under an alternating day–night rhythm as compared to continuous light. In contrast to previous reports, a reduction of PHB was observed when the cells were grown under conditions of limited gas exchange. Since previous data showed that PHB is not required for the resuscitation from nitrogen starvation, a series of different abiotic stresses were applied to test if PHB is beneficial for its fitness. However, under none of the tested conditions did cells containing PHB show a fitness advantage compared to a PHB-free-mutant ($\Delta phaEC$). Additionally, the distribution of PHB in single cells of a population *Synechocystis* cells was analyzed via fluorescence-activated cell sorting (FACS). The results showed a considerable degree of phenotypic heterogeneity at the single cell level concerning the content of PHB, which was consistent over several generations. These results improve our understanding about how and why *Synechocystis* synthesizes PHB and gives suggestions how to further increase its production for a biotechnological process.

Keywords: cyanobacteria; bioplastic; PHB; sustainable; resuscitation; chlorosis; bacterial survival; *Synechocystis*; biopolymers

---

1. Introduction

Cyanobacteria have colonized our planet for more than two billion years and are widespread within the light-exposed biosphere [1]. Their ability to conduct oxygenic photosynthesis enables them to survive under extreme environmental conditions, even in the absence of organic carbon sources. In adaption to these diverse environments, many cyanobacteria have evolved the ability to produce a variety of biopolymers [2]. Most of the mentioned polymers serve to store macro-nutrients, like carbon (in the form of glycogen), phosphate (polyphosphate), or nitrogen (cyanophycin). Although it has been known since 1966 that cyanobacteria also possess polyhydroxybutyrate (PHB), its function remains puzzling [3–5].

Around 30 years later, Hein et al. found that the required biosynthetic genes for the PHB synthesis are also present in the model organism *Synechocystis* sp. PCC 6803 (hereafter "*Synechocystis*" or "WT" for wild-type) [6]. Soon after, it has been shown that this strain is indeed capable of producing PHB under nutrient limited conditions [7]. It has been hypothesized that the polymer serves as an additional carbon and energy storage, similarly to glycogen, which could help to survive environmental stress conditions. However, until today, the true physiological function remains unknown [5].

In other organisms, PHB can fulfill manifold functions. The polymer often accumulates under nutrient limitation or unbalanced conditions (e.g., an excess of carbon) [8]. In certain organisms like *Ralstonia eutropha*, it is also accumulated during normal growth phase [9]. In the strain *Azospirillum brasilense*, for example, heat, UV irradiation, desiccation, osmotic shock, and osmotic pressure affect the growth of a PHB deficient strain [10]. In the strain *Herbaspirillum seropedicae*, PHB is able to reduce redoxstress. Hence, PHB could serve as an electron sink to eliminate a surplus of reducing equivalents [11]. A similar behavior has been shown in the anoxygenic phototrophic bacterium *Chromatium vinosum*: it converts glycogen to PHB (under anerobic, dark metabolism) [12]. Thereby, the strain does not lose carbon, compared to other bacteria, which secrete their fermentation products. In agreement, it has been shown that excess NADPH sustains PHB accumulation [13].

A better understanding about the production of PHB in cyanobacteria would be beneficial for the biotechnological production of the biopolymer. PHB from cyanobacteria is suggested as a sustainable source for biodegradable plastics [14]. However, the current production rates are rather low, making it difficult to commercially compete with the PHB production in heterotrophic organisms [15]. So far, most attempts to increase the PHB yield focused on either medium optimization or metabolic engineering approaches. At the same time, fundamental questions about the PHB metabolism and PHB-forming conditions have been neglected. For instance, it has just recently been discovered that PHB is mostly formed from intracellular glycogen [16,17]. This work therefore aims to better understand the conditions, under which PHB is produced in *Synechocystis*, as well as gaining further insights in the physiological function of PHB within the cyanobacterial metabolism.

This knowledge will be helpful for both basic and applied research, since *Synechocystis* is considered a promising host for the industrial production of bioplastic from PHB [18].

2. Materials and Methods

2.1. Cyanobacterial Cultivation Conditions

Synechocystis sp. PCC 6803 cells were grown in standard BG_{11} medium as described before [19]. Additionally, 5 mM $NaHCO_3$ were added. All used strains are listed in Table A1. To ensure the preservation of the genetic modifications, appropriate antibiotics were added to the different mutant strains. All cells were pre-adapted to their growth conditions by growing a pre-culture for 3 days at the same condition. For normal growth, cells were grown under constant illumination of ~50 µE m^{-2} s^{-1} and at 28 °C. Aeration was provided by continuous shaking at 120 rpm. Either 50 or 200 mL bacterial culture were grown in baffle free Erlenmayer flasks. Whenever nitrogen starvation was required, cells were shifted to nitrogen depleted medium as previously described [20]. In short, 200 mL exponentially growing cells at an OD_{750} of ~0.8 were centrifuged at 4000 g for 10 min. The pellet was resuspended in 100 mL BG_0 (BG11 without any sodium-nitrate) and centrifuged again. The pellet was then resuspended in BG_0 once more and the OD_{750} was adjusted to 0.4. For resuscitation experiments, a chlorotic culture was spun down and resuspended in BG_{11} medium.

2.2. Physical Stress Conditions

To test whether the formation of PHB is advantageous under conditions of physical stress, chlorotic WT and Δ*phaEC* cells, which were starved from nitrogen for ~2 weeks, were used. The cells were treated with the conditions described in category 1 of Table 1 and subsequently recovered on BG_{11} plates. Afterwards, a serial dilution of cell suspension, from OD_{750} ~1 (= 100) until 10-4, was prepared. From each dilution, 5 µL were dropped on an BG11 agar plates containing 1.5 % agar and incubated for 1–2 weeks under continuous light, until visible colonies formed. Alternatively, chlorotic cells were recovered on BG_{11} plates with additional ingredients (category 2 of Table 1) which can cause stress, e.g., high salt concentrations. The number of formed colonies was compared between the WT and ΔphaEC cells.

Table 1. List of abiotic stresses applied to WT and $\Delta phaEC$ cells, which were starved for two weeks of combined nitrogen sources. A detailed description is listed under "Treatment". For the first category of experiments, cells either treated for a specific time with the conditions listed under (1). After the treatment, cells were grown on BG_{11} agar plates according to the drop plate method as depicted in Figure 4. Alternatively, chlorotic cells were transferred to BG_{11} plates containing the ingredients listed under category (2) to apply the abiotic stress during the recovery process. Observed differences are listed under "Effect". When "no difference" was observed, both strains (WT and $\Delta phaEC$) showed similar amounts of colonies. "All cells were dead" indicates experiments, where no colonies appeared. All treatments were tested in three different biological replicates.

Abiotic Stress	Treatment	Effect
	(1)	
Cold	4 °C over night	No difference
	1 h at −20°C; 14 h at −20 °C	
Heat	45 min at 40 °C;	No difference
	45 min at 40°C and 20 min at 50°C	All cells were dead
		No difference
Physical Force	Centrifugation for 30 min at 20,000 g; 1 h at 25,000 g	
	3 × 5 min at 4 m/s glass-bead milling in Ribolyser;	No difference
	3 × 5 min at 7 m/s Ribolyser	
		All cells were dead
Darkness	1 day; 2 days, 5 days; 8 days; 10 days; 15 days	No difference
Low Light	14 days at 5 µE	No difference
Alternating Light	12/12 light/dark	No difference
Drought	30 min at 30 °C SpeedVac	All cells were dead
		No difference
Nitrogen Starvation	3 weeks	$\Delta phaEC$ showed weak
	10 weeks	growth advantage
High Light	1 day incubation at 500 µE	No difference
	(2)	
Buffered Medium	BG_{11} agar plates containing 300 µL, 1 mL, 3 mL TES buffer	No difference
Salt	Recovery at BG_{11} agar plates with: 100, 150, 300 mM NaCl	No difference
	2 × BG_{11} salts	No difference
Carbon Availability	0, 10, 50, 150 mM bicarbonate (added to BG_{11} agar plates)	No difference

2.3. Oxygen Measurements

To measure oxygen levels in a liquid culture, an oxygen detecting sensor was placed at the bottom of a standing culture. The readout was performed using an OXY-1 SMA device (PreSens, Regensburg, Germany). At the beginning of the measurements, *Synechocystis* cells were shifted to nitrogen free BG0 medium and the oxygen levels were monitored for three constitutive days. The equilibrium of dissolved oxygen within the cultures was measured at 360 µM/L.

2.4. Microscopy and Staining Procedures

To visualize cell morphology and PHB granules within the cells, 100 µL of cyanobacterial culture were centrifuged. The resulting pellet was resuspended in a mixture of 10 µL Nile red and 20 µL water. From the resuspended mixture, 10 µL were used and dropped on an agarose-coated microscopy slide. A Leica DM5500B microscope (Leica, Wetzlar, Germany) with a 100 × /1.3 oil objective was used for fluorescence microscopy. For the detection of Nile red stained PHB granules, a suppression filter BP 610/75 was used, together with an excitation filter BP 535/50. The pictures were taken by a Leica DFC360FX.

2.5. Electron Microscopy

For detailed pictures of the intracellular PHB granules, electron microscopy was used. For this, glutaraldehyde and potassium permanganate were used to fix and postfix *Synechocystis* cells. Citrate and uranyl acetate were used to stain ultrathin sections, as described before [21]. Finally, a Philips Tecnai 10 electron microscope (Philips, Amsterdam, Netherlands) was used to examine the samples at 80 kHz.

2.6. Spectral Analysis

To measure the whole-cell absorption spectrum, a Specord 50 with the software WinAspect (Analytik Jena, Jena, Germany) was used. The absorption was measured between 350 and 750 nm. The spectra were normalized to the OD_{750}. To determine the recovery of the photopigments, the change in absorption between the induction of the recovery (t0) and after three days (t72) was determined (normalized to OD_{750}). For this, the specific wavelengths for phycobilisomes and chlorophyll (absorption at 630 nm and 680, respectively) were compared.

2.7. PAM

To detect the photosynthetic activity, pulse–amplitude–modulation fluorometry (PAM) was used. This measures the relative quantum yield of the photosystem II, Y(II). A Heinz Walz GmbH (Effeltrich, Germany) WATER-PAM Chlorophyll Fluorometer with WinControl Software was used. For the measurements, a cell suspension with an OD_{750} between 0.4 and 1 was used and diluted 20-fold. After 5 min incubation in the dark, the maximum PSII quantum yield (Fv/Fm) was determined applying the saturation pulse method [22]. For each time point, three measurements with a time constant of 30 s were taken.

2.8. PHB Quantification

To determine the intracellular PHB content, ~15 mL of cells were centrifuged at 4000 g for 10 min. The pellet was dried for at least 2 h at 60 °C in a speed vac until all pellets were dry. Next, 1 mL of concentrated sulfuric acid (18 M H_2SO_4) were added and the mixture was boiled for 1 h at 100 °C. This process releases PHB from the cells and converts it to crotonic acid. From this, 100 µL were taken and diluted with 900 µL of 14 mM H_2SO_4. The sample was centrifuged for 5 min at 20,000 g. From the supernatant, 500 µL were transferred into a new tube and combined with 500 µL H_2SO_4. The samples were centrifuged once more for 5 min at max speed and 400 µL of the resulting supernatant was used for further HPLC analysis. For this, a Nucleosil 100 C 18 column was used (125 by 3 mm). For a liquid phase, 20 mM phosphate buffer with pH 2.5 was used. Crotonic acid was detected at 250 nm. As a standard, commercially available crotonic acid was used, with a conversion rate to PHB of 0.893.

2.9. FACS and Flow Cytometry

Synechocystis cells that were starved from nitrogen for ~2 weeks before FACS experiments. Cells were sorted with a MoFlo XDP (Beckman Coulter, Munich, Germany) into 500 µL PBS buffer using a 70 uM CytoNozzle at 30 p.s.i. and PBS [pH 7.0] as sheath. Before FACS or analysis, 1 µL of BODIPY (5 mg/mL) was added to 500 µL of cells and incubated for 10 min. Cells were identified based on their scatter (SSC-LA vs. FSC-LA), chlorophyll-fluorescence (670/30) captured from a 488 nm (70 mW) laser. Cells were divided into low and high producers based on their emission profile and at 534/30 (BODIPY) when compared to unstained and PHB deficient cells. For analysis, the software Summit FACS and FCS Express was used. BODIPY staining was also scored using a Cytoflex analyzer (Beckmann Coulter, Munich, Germany) equipped with a single 488 nm laser. Principle BODIPY emission was captured with a 525/40 bandpass and plotted against scatter to remove clumplets and distinguish BODIPY. PHB content was inferred when compared to unstained and PHB deficient cells.

For analysis and illustration of the date, the software programs CytoExpert (Beckman Coulter, Munich, Germany) and FlowJo (FlowJo LLC, Oregon, USA) were used.

3. Results

3.1. The Influence of Environmental Conditions and Central Pathways on PHB Production

In PHB producing cyanobacteria, PHB synthesis is efficiently induced by depleting the nitrogen source from the medium. This is also the case for the model cyanobacterium *Synechocystis* sp. PCC 6803 used in this study. To further understand the conditions, under which PHB is produced, cells, which were transferred to nitrogen-depleted medium, were grown under different conditions of aeration and illumination (Figure 1).

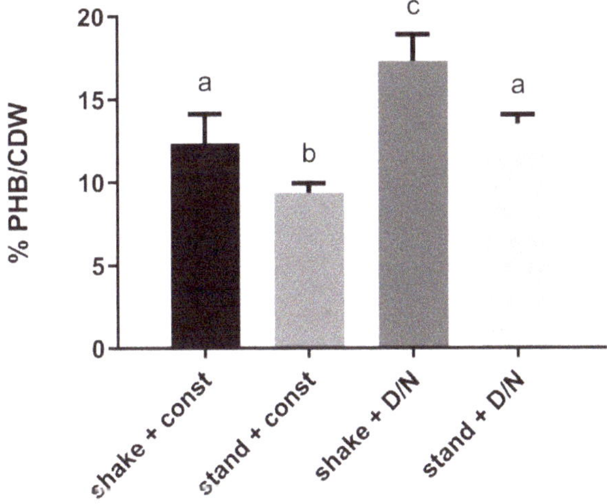

Figure 1. Quantification of polyhydroxybutyrate (PHB) content after 13 days of nitrogen starvation. WT cells were grown under different conditions of aeration and illumination. All cells were pre-adapted to these conditions three days before the shift to nitrogen starvation. Shake = continuous shaking at 120 rpm. Stand = cultures were standing without any shaking. Const = constant illumination with ~50 µE. D/N = altering illumination with 12 h of light (50 µE) and 12 h dark. Data shown as mean ± SD of three biological experiments; levels not connected by the same letter are significantly different ($p \leq 0.05$). CDW = cell dry weight.

It was shown in a previous study that the limitation of gas exchange can boost the PHB production [23]. To verify this observation, cells were grown without any shaking to create a situation of reduced aeration. Under these standing conditions, the cells produced oxygen and were oxygen-oversaturated during the day, whereas they consumed it during the night, resulting in transient periods of limited oxygen availability (Figure A2). Furthermore, cyanobacteria are naturally adapted to day-night rhythms, while they are commonly grown under continuous light in the laboratory. To test whether this affects the PHB production, PHB was quantified from cells grown under both light regimes. It turned out that, compared to standard laboratory conditions of continuous light and shaking, a limitation in gas exchange always led to a reduced PHB content. In contrast, the growth under day-night rhythm resulted, both in standing and shaken cultures, and in an increased content of PHB per cell-dry-weight (CDW). Therefore, the maximal PHB production was achieved in shaken cultures in a 12 h light/dark regime.

To test whether these effects are linked to a specific carbon pathway, knockout mutants of two central pathways were used: The mutant Δ*pfk* is unable to metabolize carbon via the the EMP (Embden–Meyerhof–Parnas) pathway and the mutant Δ*zwf* cannot use the OPP (oxidative pentose phosphate) and ED (Entner–Doudoroff) pathway. Both strains and a WT control were grown under the same conditions as in the previous experiment and their PHB content was then compared (Figure 2).

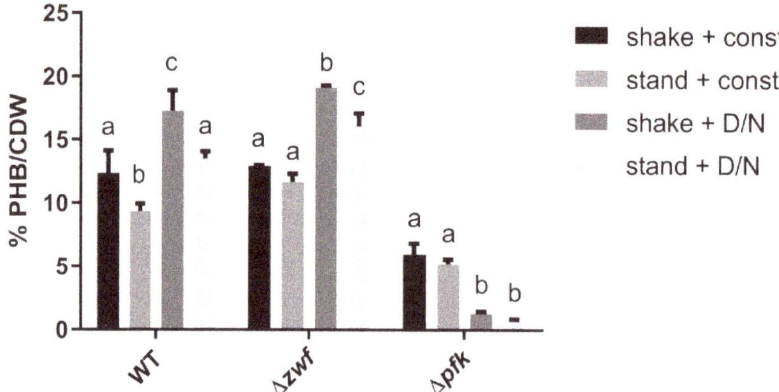

Figure 2. Quantification of PHB content after 13 days of nitrogen starvation. Mutant strains lacking either the EMP (Δ*pfk*) or the OPP and ED pathway (Δ*zwf*) were compared to the WT strain to test the influence of these carbon pathways on the PHB production. The cells were grown under different conditions of aeration and illumination. Shake = continuous shaking at 120 rpm. Stand = cultures were standing without any shaking. Const = constant illumination with ~50 µE. D/N = altering illumination with 12 h of light (50 µE) and 12 h dark. Data shown as mean ± SD of three biological experiments; levels not connected by the same letter are significantly different ($p \leq 0.05$; only within the same genetic background).

The Δ*zwf* mutant lacks glucose-6P dehydrogenase, which feeds sugar catabolites into the two central pathways ED and OPP, which are crucial for vegetative growth. These pathways are neglectable for the production of PHB under nitrogen starvation, since the Δ*zwf* mutant showed the same production pattern as the WT under all tested conditions (see Figure 2). In contrast, the Δ*pfk* mutant, which is unable to use the EMP pathway, showed a strongly impaired PHB production under all conditions. Interestingly, growth with a 12 h light/dark regime caused increased PHB production in the WT and the Δ*zwf* mutant, whereas it resulted in a severe reduction of PHB in the Δ*pfk* strain.

3.2. Recovery from Nitrogen Starvation

To test whether the formation of PHB plays a role for the recovery from nitrogen chlorosis under specific conditions, WT and a PHB-free mutant (Δ*phaEC*) were nitrogen-starved for 18 days and were transferred to standard BG_{11} medium to induce resuscitation. Thereafter, parameters, which indicate the process of the resuscitation, such as the yield of the photosynthetic activtity (Y(II)) and reconstruction of light absorbing pigments, were analyzed over three days (Figure 3).

During the course of resuscitation, the photosynthetic activity of the WT recovered at the same pace as the Δ*phaEC* mutant (Figure 3A). Likewise, there was no difference in the re-appearance of photosynthetic pigments, indicated by the absorption at 630 nm and 680 nm (Figure 3B and 3C, respectively). The full spectra between 600 and 750 nm are shown in Figure A1.

To test if the formation of PHB is beneficial for chlorotic cells under conditions of certain abiotic stresses, WT and Δ*phaEC* cells were nitrogen starved for two weeks to induce PHB production. Subsequently, one of the following experimental setups was performed: (1) One specific abiotic stress was applied to the chlorotic culture for a specific time, before the culture was plated on BG_{11} agar plates and the formation of CFU was determined. (2) Chlorotic cells were plated on BG_{11} plates, which contained additional components causing stress (for example, higher salt concentrations). The results are summarized in Table 1.

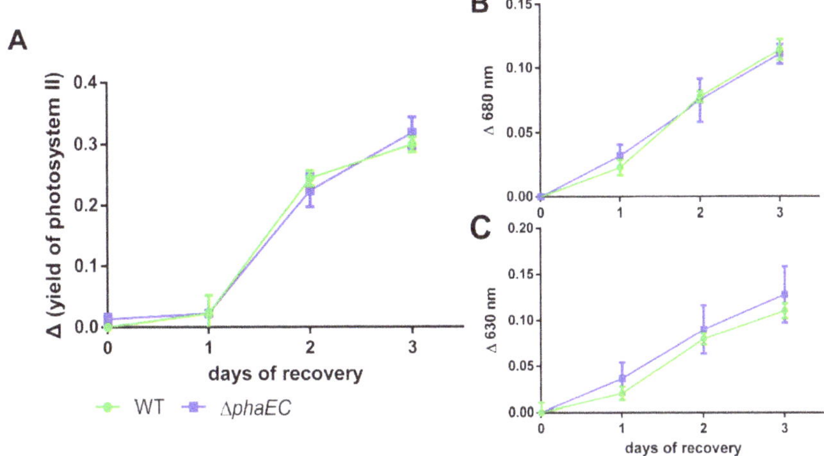

Figure 3. Physiological parameters during the resuscitation after 18 days of nitrogen starvation. Cells were grown under alternating day-night rhythm and continuous shaking; (**A**) PAM measurements to determine the maximum PSII quantum yield (Fv/Fm). The measurements were normalized to the yield at timepoint 0; (**B,C**): difference in absorption of normalized spectra during resuscitation. Spectra, which were normalized to OD_{750} (Figure A1), were used to calculate the difference between the initial absorption values (at wavelength of 630 and 680 nm) at time point 0 (day 0) and the various time points during resuscitation. All samples represent three individual biological replicates.

Under all tested conditions, no growth advantage was observed for the WT compared to the Δ*phaEC* mutant strain. Some conditions were too harsh for any cells to survive—for example, the simulation of heat at 50 °C. To illustrate what a typical result looked like, a representative drop plate assay is depicted in Figure 4. In this example, no viability difference between WT and Δ*phaEC* cells was observed after resuscitation from prolonged chlorosis and growth on 100 mM NaCl. In summary, there was no condition found where the possession of PHB was advantageous for the WT.

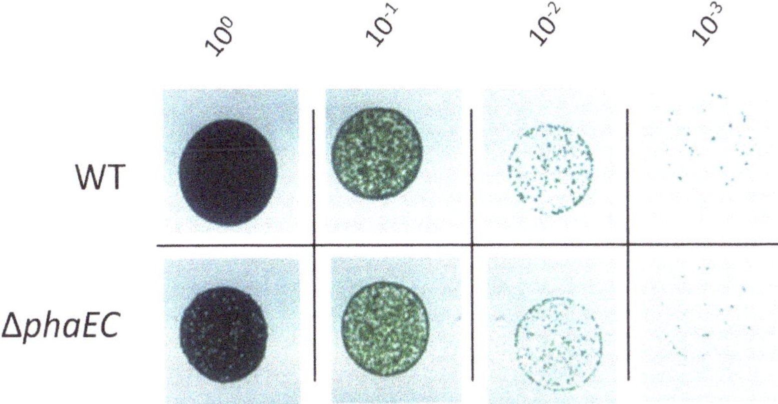

Figure 4. Viability assay of *Synechocystis* WT and Δ*phaEC* mutant cells using the drop plate method after two weeks of nitrogen starvation and subsequent growth in BG_{11} plates with 100 mM NaCl. Several dilution steps were dropped on BG_{11} agar plates, ranging from OD_{750} = 1 (represents dilution 10^0) to OD_{750} = 0,001 (represents dilution 10^{-3}).

3.3. Heterogeneity of PHB Production

In microscopic studies, we noticed that the number of PHB granules varied strongly between individual WT cells during nitrogen starvation. While most cells did contain PHB, the amount varied greatly, both in the size as well as in the number of PHB granules (Figure 5).

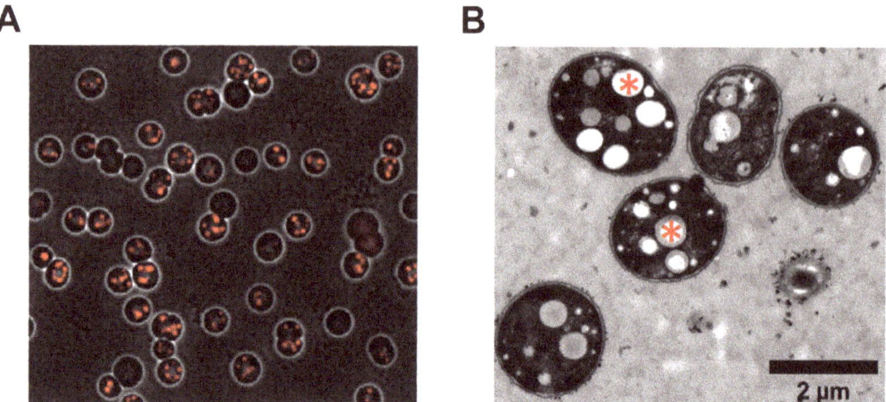

Figure 5. Microscopic analysis of varying PHB contents in WT cells; (**A**) fluorescence microscopy of WT cells after 14 days of nitrogen starvation. PHB granules are visualized by staining with Nile red; (**B**) TEM picture of WT cells after 17 days of nitrogen starvation. Representative PHB granules are indicated in two different cells by red asterisks.

Since this observation has not been systematically addressed before, we further investigated this phenomenon via flow cytometry (FC). Therefore, *Synechocystis* cells were starved for two weeks from nitrogen and stained with Bodipy. To distinguish the fluorescence signal of Bodipy stained PHB from background signals, two controls were performed: (1) determination of the unspecific background fluorescence from unstained WT cells as well as (2) the fluorescence of Bodipy stained PHB free Δ*phaEC* cells (Figure 6).

Compared to the unstained WT cells, the Δ*phaEC* cells showed a higher fluorescence signal, which is caused by unspecific staining of hydrophobic structures within the cells by Bodipy. Compared to the Bodipy stained Δ*phaEC* cells, a large portion of the Bodipy stained WT cell showed a fluorescence signal that was partially overlapping with that of Δ*phaEC* cells, but, on average, shifted to higher intensities (Figure 6, blue circle). This corresponds to a major population of cells that contained only low to medium amounts of PHB. From these, a second part of the WT population could clearly be distinguished, which showed much higher fluorescence signals (Figure 6, brown circle). This corresponds to a subpopulation of high PHB producing cells. Since the cells used for PHB production are derived from a single clone, they are assumed to be genetically identical. To definitively clarify that the different PHB synthesis phenotypes are not caused by (epi)genetic differences but are more likely, and are based on stochastic regulation of PHB synthesis, we isolated low and high PHB producing cells by FACS sorting. Single cells were recovered and grown on BG_{11} agar plates until colonies appeared. Several colonies derived from high- or low-producing cells were separately pooled and used to inoculate a fresh BG_{11} culture. Afterwards, the cultures were shifted again to nitrogen free BG_0 medium to trigger PHB synthesis and the cells were again investigated using FC as described above. The results are shown in Figure 7.

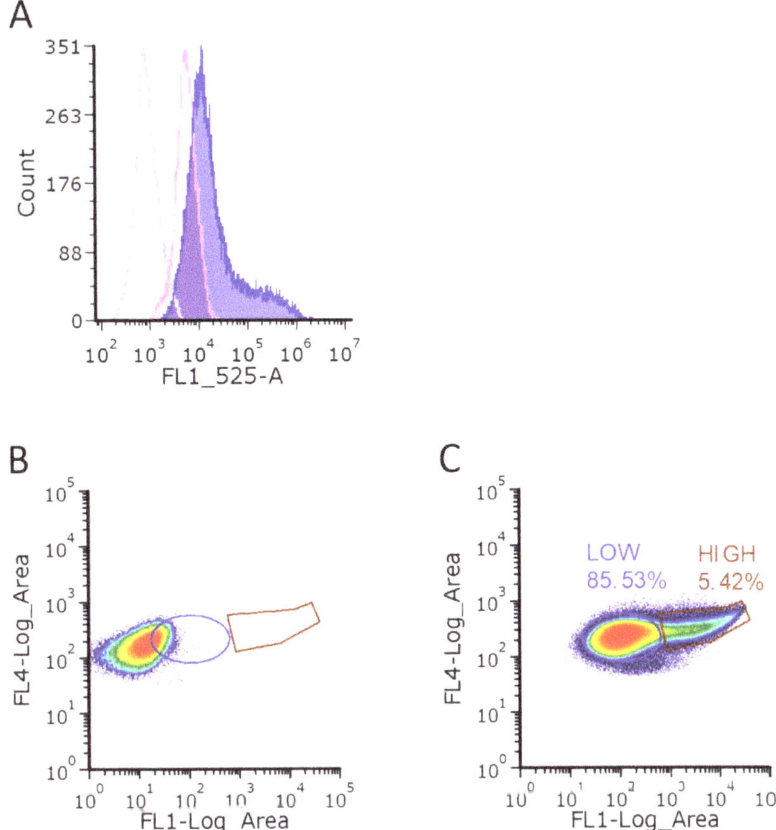

Figure 6. Analysis of intracellular PHB content using FC to detect *Synechocystis* cells stained with Bodipy. (**A**) overlay of the measurements for unstained WT (grey), Δ*phaEC* mutant strain stained with Bodipy (red) and WT stained with Bodipy (blue). To illustrate how the cells were separated with FACS, the sort regions for Δ*phaEC* mutant strain and WT are shown in (**B,C**), respectively. The red peak of Δ*phaEC* cells in (A) corresponds to the cell population in (B), while the blue peak of WT cells in (A) corresponds to the cell population in (C). Sort regions in WT for low- and high producers are indicated as a blue or brown circle, respectively, in (B) and (C). FL4 = Chlorophyll emission and FL1 = Bodipy emission.

Undoubtedly, the cells in these new cultures established again the same PHB heterogeneity as in the previous experiment, regardless if they were derived from previously low- or high-producing cells. As previously described, all Bodipy stained WT cells showed a higher overall fluorescence compared to the unstrained control and the Bodipy stained Δ*phaEC* mutant strain, indicating that most cells did contain PHB. Furthermore, all replicates showed the characteristic shoulder in the FC analysis, representing cells with high PHB contents. This result demonstrates that the regulation of PHB synthesis in nitrogen-starved cells follows a stochastic program, resulting in a mixed population with a majority of low- to medium PHB producers and a minority of high PHB producing cells.

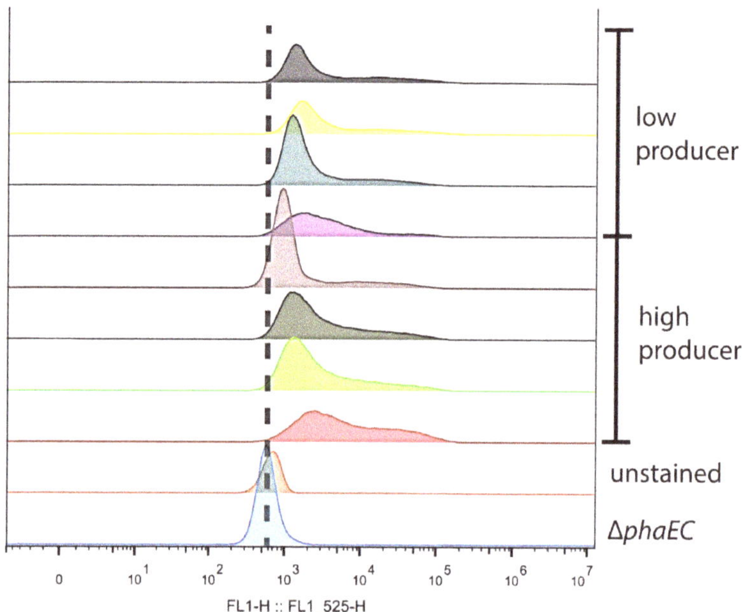

Figure 7. FC analysis of intracellular PHB content. "Low" and "high" represent four individual biological replicates from FACS, which were based on cells that contained low or high amounts of PHB, respectively. Cells were starved from nitrogen for two weeks and stained with Bodipy (except for the unstained control).

4. Discussion

4.1. Heterogeneity of PHB Content

When living in a changing environment, being prepared for different potential outcomes can be beneficial. In order to be prepared for unpredictable future scenarios, many bacteria have evolved a strategy of phenotypical heterogeneity for bed hedging [24]. This allows them to have a certain part of their population being well prepared for either outcome. In this work, we could show that the formation of PHB in *Synechocystis* also represents this form of phenotypical heterogeneity. We observed this via fluorescence and electron microscopy (Figure 5) and further confirmed it by flow-cytometry (Figure 6). As already hypothesized, this heterogeneity has no inheritable reason, since cells that were separated into low- and high producers produced progeny with the same heterogeneity in PHB content as the ancestral generation (Figure 7). It is likely that the heterogeneity results from a probabilistic genetic program (bet-hedging) that results in the observed distribution of PHB production. The expression of key enzymes (such as PhaEC) might be quite variable, resulting in some cells that do produce more PHB than others. It was shown before that transcriptional infidelity promotes heritable phenotypic changes, which could also explain the PHB heterogeneity [25]. The phenomenon of a bet-hedging strategy in PHB contents has also been described in a different manner for some heterotrophic bacteria, such as *Sinorhizobium meliloti*. Here, when dividing *S. meliloti* cells face starvation, they form two daughter cells with different phenotypes, one with low and one with high PHB content. These daughter cells are adapted to either short- or long-term starvation, respectively [26].

4.2. Physiological Function of PHB

Since the majority of cells contained only minor or medium amounts of PHB, it has to be assumed that the conditions, under which it is beneficial to contain large PHB quantities, are rather seldom.

To better understand PHB metabolism, this work aimed at providing additional information about when and why PHB is formed. We show clear evidence that the PHB content within the cells is increased when the bacteria are grown under alternating day–night rhythm (Figurs 1 and 2). This is in coherence with a recently published transcriptomic data set [27]. Here, the authors have analyzed the transcription of genes under two consecutive dark- and light phases. We extracted all known PHB related genes from this data set and found a clear correlation between the genes of PHB metabolism and the diurnal rhythm (Figure A3). The assumption that PHB might play a role during dark, anaerobic conditions was already hypothesized before [28].

Since PHB is formed from intracellular glycogen pools [17] and cyanobacteria catabolize glycogen during the night [29], this phenotype might be explained by an increased carbon availability from glycogen during the night. Furthermore, *Synechocystis* accumulated less PHB when grown under standing conditions (Figure 1). These findings are in line with previous reports, which showed a generally decreased metabolic activity under conditions of limited gas-exchange and CO_2 availability [30]. In addition, photosynthetically active *Synechocystis* cells produce oxygen, which accumulates in the medium when grown under standing conditions (Figure A2). Excess of oxygen is known to impair the efficiency of the RuBisCO enzyme by causing the oxygenase reaction, which likely resulted in a reduced PHB reduction due to the carbon loss caused by photorespiration.

The results fit to the observation that the EMP pathway proved to be the most important carbon pathway for PHB production (Figure 2). Studies have shown that cyanobacteria employ the EMP pathway for degradation of glucose residues to pyruvate [30]. This is also the case for growth in dark phases [31]. Furthermore, the EMP produces less NADPH and more ATP, compared to other glycolytic routes, such as the OPP (oxidative-pentose-phosphate pathway) or the ED (Entner-Doudoroff) pathway. Although the cultivation under standing conditions is not a strictly anaerobic condition (Figure A2), the O_2 limitation during the dark phase could be sufficient to induce fermentation-like processes. It is known that cyanobacteria carry out fermentation under dark/anoxic conditions and produce a variety of different fermentation products [32]. Although PHB is so far not considered a cyanobacterial fermentation product, these conditions of limited oxygen availability and absence of light could explain the observed increase of intracellular PHB (Figure 1).

In contrast to the common belief, the ability to produce PHB under conditions of nitrogen starvation was not shown to be beneficial since we could not detect any physiological differences between WT and Δ*phaEC* cells during the resuscitation process (Figure 3). Even when additional abiotic stresses were added, no growth advantage was observed (Table 1). It might be that growth under the controlled environment of laboratory conditions is not limiting the cells, neither in carbon nor energy, and hence no phenotypes were visible.

It was previously shown that excess NADPH under nitrogen-starved conditions sustains PHB accumulation [13]. Hence, PHB might serve as an intracellular pool for electrons under conditions of excess reduction equivalents. The fact that we did not observe any phenotypical difference under conditions of electron excess (for example under high amounts of light).

Table 1 might be explained by other regulatory mechanisms compensating for the lack of PHB. Since the correct regulation of intracellular redox-state is crucial for the cell physiology, *Synechocystis* has evolved various strategies to cope with high levels of reduction equivalents. One example for such a mechanism is the flavodiiron protein Flv3, which serves as a sink for excess electrons from the photosynthetic light reaction, by converting O_2 to H_2O. A recent publication showed indeed increased PHB synthesis in a Flv3 deficient strain [33]. If PHB is indeed serving as an intracellular electron sink, its absence might be compensated by a higher activity of Flv3. Alternative ways of getting rid of reduction equivalents could be the secretion of reduced organic molecules, such as acetate. It can be assumed that the correct regulation of the ATP/NADPH ratio is crucial for cyanobacterial cells. Since the production of PHB from glycogen provides ATP but consumes NADPH, the biopolymer could help the cells to regulate this ratio under conditions of electron excess [33].

5. Conclusions

This work describes different factors that influence the formation of PHB. However, the conditions where PHB is advantageous during nitrogen starvation have yet to be discovered. The results of this work can help to create strains with enhanced PHB contents. Besides applying dark phases, finding regulators of the EMP pathway to unlock the carbon flow from glycogen to PHB could further boost the production [17]. One alternative strategy could be the deletion of other NAPDH consuming pathways, as already shown in a Δflv3 strain. Finally, a deeper understanding of the PHB-heterogeneity might result in a more homogenous culture of high-producing cells, which is beneficial for the overall yield.

Author Contributions: Conceptualization, M.K. and K.F.; Methodology, M.K., K.W.B., and K.F.; Investigation, M.K. and K.W.B.; Writing—Original Draft Preparation, M.K., K.W.B. and K.F.; Writing—Review and Editing, M.K. and K.F.; Supervision, K.F.; Project Administration, M.K. and K.F. All authors have read and agreed to the published version of the manuscript.

Funding: This research was funded by the Studienstiftung des Deutschen Volkes and the RTG 1708 "Molecular principles of bacterial survival strategies". We acknowledge support by the Open Access Publishing Fund of University of Tübingen.

Acknowledgments: We would like to thank Eva Nussbaum for maintaining the strain collection and Andreas Kulick for assistance with HPLC analysis. We also thank Markus Maisch and Andreas Kappler for instructions for the oxygen measurement device. Furthermore, we thank Claudia Menzel for technical assistance with the electron microscope.

Conflicts of Interest: The authors declare no conflict of interest.

Appendix A

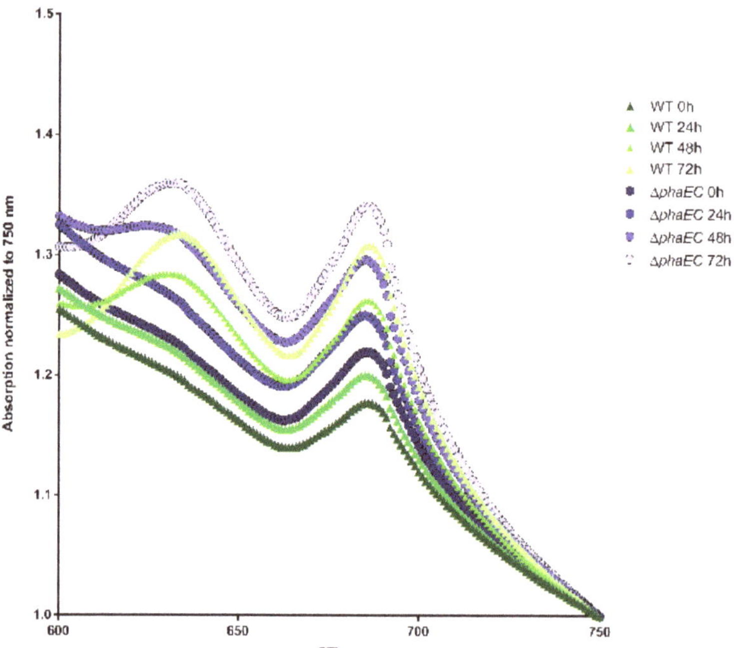

Figure A1. Cell spectra from 600 nm to 750 nm of WT and Δ*phaEC* after 0–4 days of resuscitation from nitrogen starvation. Each line represents the mean of the spectra from three independent biological triplicates. The spectra were normalized to their corresponding value at 750 nm.

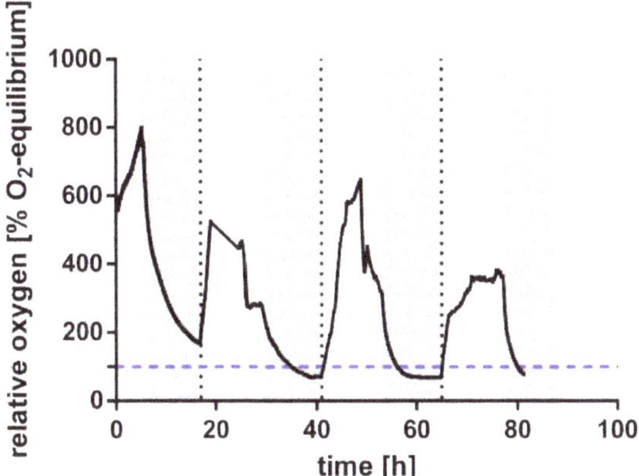

Figure A2. Oxygen measurements of a standing WT culture at light/dark regime. At timepoint 0, nitrogen starvation was induced. The amount of oxygen (green) within the culture was measured during three adjacent days. Grey bars in the back represent the dark phases. The equilibrated oxygen levels are shown in a blue dashed line.

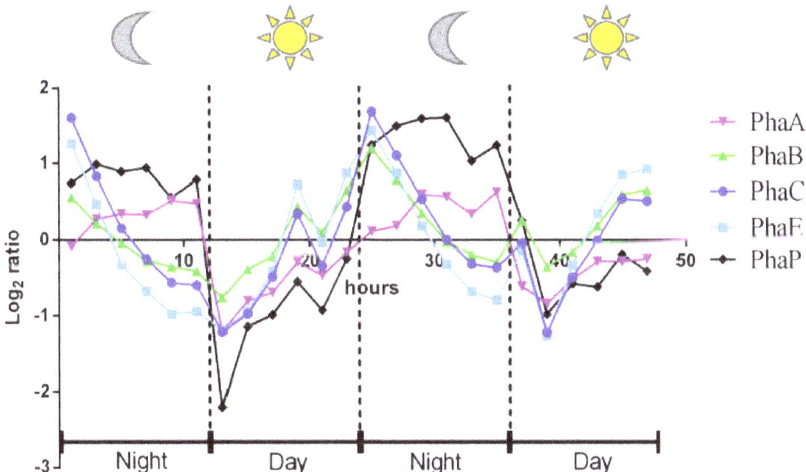

Figure A3. Transcriptomic data of PHB related genes [28]. The data are based on two consecutive days, including two 12 h phases with and without light.

Table A1. List of used strains.

Strain	Relevant Marker of Genotype	Reference
Synechocystis sp. PCC 6803	-	Pasteur culture collection
Δ*zwf*	*Slr1843::cmR*	Chen et al. 2016
Δ*pfkB1/2*	*sll1196::kmR, sll0745::spR*	Chen et al. 2016
Δ*phaEC*	*slr1829,slr1830::kmR*	Klotz et al. 2016

References

1. Herrero, A.; Flores, E. *The Cyanobacteria: Molecular Biology, Genomics and Evolution*; Caister Academic Press: Norfolk, UK, 2008.
2. Allen, M.M. Cyanobacterial cell inclusions. *Annu. Rev. Microbiol.* **1984**, *38*. [CrossRef] [PubMed]
3. Carr, N.G. The occurrence of poly-beta-hydroxybutyrate in the blue-green alga, Chlorogloea fritschii. *Biochim. Biophys. Acta* **1966**, *120*, 308–310. [CrossRef]
4. Damrow, R.; Maldener, I.; Zilliges, Y. The Multiple Functions of Common Microbial Carbon Polymers, Glycogen and PHB, during Stress Responses in the Non-Diazotrophic Cyanobacterium *Synechocystis* sp. PCC 6803. *Front. Microbiol.* **2016**, *7*, 966. [CrossRef] [PubMed]
5. Klotz, A.; Georg, J.; Bučinská, L.; Watanabe, S.; Reimann, V.; Januszewski, W.; Sobotka, R.; Jendrossek, D.; Hess, W.R.; Forchhammer, K. Awakening of a Dormant Cyanobacterium from Nitrogen Chlorosis Reveals a Genetically Determined Program. *Curr. Biol.* **2016**, *26*, 2862–2872. [CrossRef]
6. Hein, S.; Tran, H.; Steinbuchel, A. *Synechocystis* sp. PCC6803 possesses a two-component polyhydroxyalkanoic acid synthase similar to that of anoxygenic purple sulfur bacteria. *Arch. Microbiol.* **1998**, *170*, 162–170. [CrossRef]
7. Wu, G.F.; Wu, Q.Y.; Shen, Z.Y. Accumulation of poly-beta-hydroxybutyrate in cyanobacterium *Synechocystis* sp. PCC6803. *Bioresour. Technol.* **2001**, *76*, 85–90. [CrossRef]
8. Anderson, A.J.; Dawes, E.A. Occurrence, Metabolism, Metabolic Role, and Industrial Uses of Bacterial Polyhydroxyalkanoates. *Microbiol. Rev.* **1990**, *54*, 450–472. [CrossRef]
9. Jendrossek, D.; Pfeiffer, D. New insights in the formation of polyhydroxyalkanoate granules (carbonosomes) and novel functions of poly(3-hydroxybutyrate). *Environ. Microbiol.* **2014**, *16*, 2357–2373. [CrossRef]
10. Kadouri, D.; Jurkevitch, E.; Okon, Y. Involvement of the reserve material poly-beta-hydroxybutyrate in *Azospirillum brasilense* stress endurance and root colonization. *Appl. Environ. Microbiol.* **2003**, *69*, 3244–3250. [CrossRef]
11. Batista, M.B.; Teixeira, C.S.; Sfeir, M.Z.T.; Alves, L.P.S.; Valdameri, G.; de Oliveira Pedrosa, F.; Sassaki, G.L.; Steffens, M.B.R.; de Souza, E.M.; Dixon, R.; et al. PHB Biosynthesis Counteracts Redox Stress in *Herbaspirillum seropedicae*. *Front. Microbiol.* **2018**, *9*, 472. [CrossRef]
12. Van Gemerden, H. On the ATP generation by Chromatium in darkness. *Arch. Mikrobiol.* **1968**, *64*, 118–124. [CrossRef] [PubMed]
13. Hauf, W.; Schlebusch, M.; Hüge, J.; Kopka, J.; Hagemann, M.; Forchhammer, K. Metabolic Changes in *Synechocystis* PCC6803 upon Nitrogen-Starvation: Excess NADPH Sustains Polyhydroxybutyrate Accumulation. *Metabolites* **2013**, *3*, 101–118. [CrossRef] [PubMed]
14. Balaji, S.; Gopi, K.; Muthuvelan, B. A review on production of poly β hydroxybutyrates from cyanobacteria for the production of bio plastics. *Algal Res.* **2013**, *2*, 278–285. [CrossRef]
15. Knöttner, S.; Drosg, B.; Ellersdorfer, M.; Meixner, K.; Fritz, I. Photoautotrophic production of poly-hydroxybutyrate—First detailed cost estimations. *Algal Res.* **2019**, *41*, 101558.
16. Dutt, V.; Srivastava, S. Novel quantitative insights into carbon sources for synthesis of poly hydroxybutyrate in *Synechocystis* PCC 6803. *Photosynth. Res.* **2018**, *136*, 303–314. [CrossRef] [PubMed]
17. Koch, M.; Doello, S.; Gutekunst, K.; Forchhammer, K. PHB is Produced from Glycogen Turn-over during Nitrogen Starvation in *Synechocystis* sp. PCC 6803. *Int. J. Mol. Sci.* **2019**, *20*, 1942. [CrossRef] [PubMed]
18. Martin, K.; Lukas, M. Cyanobacterial Polyhydroxyalkanoate Production: Status Quo and Quo Vadis? *Curr. Biotechnol.* **2015**, *4*, 464–480.
19. Rippka, R.; Deruelles, J.; aterbury, J.B.; Herdman, M.; Stanier, R.Y. Generic Assignments, Strain Histories and Properties of Pure Cultures of Cyanobacteria. *Microbiol. Soc.* **1979**, *111*. [CrossRef]
20. Schlebusch, M.; Forchhammer, K. Requirement of the nitrogen starvation-induced protein Sll0783 for polyhydroxybutyrate accumulation in *Synechocystis* sp. strain PCC 6803. *Appl. Environ. Microbiol.* **2010**, *76*, 6101–6107. [CrossRef]
21. Fiedler, G.; Arnold, M.; Hannus, S.; Maldener, I. The DevBCA exporter is essential for envelope formation in heterocysts of the cyanobacterium *Anabaena* sp. strain PCC 7120. *Mol. Microbiol.* **1998**, *27*, 1193–1202. [CrossRef]

22. Schreiber, U.; Endo, T.; Mi, H.; Asada, K. Quenching Analysis of Chlorophyll Fluorescence by the Saturation Pulse Method: Particular Aspects Relating to the Study of Eukaryotic Algae and Cyanobacteria. *Plant Cell Physiol.* **1995**, *36*, 873–882. [CrossRef]

23. Panda, B.; Mallick, N. Enhanced poly-beta-hydroxybutyrate accumulation in a unicellular cyanobacterium, *Synechocystis* sp. PCC 6803. *Lett. Appl. Microbiol.* **2007**, *44*, 194–198. [CrossRef] [PubMed]

24. Veening, J.W.; Smits, W.K.; Kuipers, O.P. Bistability, epigenetics, and bet-hedging in bacteria. *Annu. Rev. Microbiol.* **2008**, *62*, 193–210. [CrossRef] [PubMed]

25. Gordon, A.J.E.; Halliday, J.A.; Blankschien, M.D.; Burns, P.A.; Yatagai, F.; Herman, C. Transcriptional Infidelity Promotes Heritable Phenotypic Change in a Bistable Gene Network. *PLoS Biol.* **2009**, *7*, e1000044. [CrossRef]

26. Ratcliff, W.C.; Denison, R.F. Individual-level bet hedging in the bacterium *Sinorhizobium meliloti*. *Curr. Biol.* **2010**, *20*, 1740–1744. [CrossRef]

27. Saha, R.; Liu, D.; Hoynes-O'Connor, A.; Liberton, M.; Yu, J.; Bhattacharyya-Pakrasi, M.; Balassy, A.; Zhang, F.; Moon, T.S.; Maranas, C.D.; et al. Diurnal Regulation of Cellular Processes in the Cyanobacterium *Synechocystis* sp. Strain PCC 6803: Insights from Transcriptomic, Fluxomic, and Physiological Analyses. *mBio* **2016**, *7*, e00464-16. [CrossRef]

28. Ueda, S.; Kawamura, Y.; Iijima, H.; Nakajima, M.; Shirai, T.; Okamoto, M.; Kondo, A.; Hirai, M.Y.; Osanai, T. Anionic metabolite biosynthesis enhanced by potassium under dark, anaerobic conditions in cyanobacteria. *Sci. Rep.* **2016**, *6*, 32354. [CrossRef]

29. Smith, A.J. Modes of cyanobacterial carbon metabolism. *Ann. Inst. Pasteur/Microbiol.* **1983**, *134B*, 93–113. [CrossRef]

30. Stal, L.J.; Moezelaar, R. Fermentation in cyanobacteria1. *FEMS Microbiol. Rev.* **1997**, *21*, 179–211. [CrossRef]

31. Makowka, A.; Nichelmann, L.; Schulze, D.; Spengler, K.; Wittmann, C.; Forchhammer, K.; Gutekunst, K. Glycolytic Shunts Replenish the Calvin-Benson-Bassham Cycle as Anaplerotic Reactions in Cyanobacteria. *Mol. Plant* **2020**, *13*, 471–482. [CrossRef]

32. Heyer, H.; Krumbein, W.E. Excretion of Fermentation Products in Dark and Anaerobically Incubated Cyanobacteria. *Arch. Microbiol.* **1991**, *155*, 284–287. [CrossRef]

33. Thiel, K.; Patrikainen, P.; Nagy, C.; Fitzpatrick, D.; Pope, N.; Aro, E.-M.; Kallio, P. Redirecting photosynthetic electron flux in the cyanobacterium *Synechocystis* sp. PCC 6803 by the deletion of flavodiiron protein Flv3. *Microb. Cell Fact.* **2019**, *18*, 189. [CrossRef] [PubMed]

Review

Structural Determinants and Their Role in Cyanobacterial Morphogenesis

Benjamin L. Springstein [1,*], **Dennis J. Nürnberg** [2], **Gregor L. Weiss** [3], **Martin Pilhofer** [3] and **Karina Stucken** [4]

1   Department of Microbiology, Blavatnik Institute, Harvard Medical School, Boston, MA 02115, USA
2   Department of Physics, Biophysics and Biochemistry of Photosynthetic Organisms, Freie Universität Berlin, 14195 Berlin, Germany; dennis.nuernberg@fu-berlin.de
3   Department of Biology, Institute of Molecular Biology & Biophysics, ETH Zürich, 8092 Zürich, Switzerland; gregor.weiss@mol.biol.ethz.ch (G.L.W.); pilhofer@mol.biol.ethz.ch (M.P.)
4   Department of Food Engineering, Universidad de La Serena, La Serena 1720010, Chile; kstucken@userena.cl
*   Correspondence: benjamin_springstein@hms.harvard.edu

Received: 2 November 2020; Accepted: 9 December 2020; Published: 17 December 2020

Abstract: Cells have to erect and sustain an organized and dynamically adaptable structure for an efficient mode of operation that allows drastic morphological changes during cell growth and cell division. These manifold tasks are complied by the so-called cytoskeleton and its associated proteins. In bacteria, FtsZ and MreB, the bacterial homologs to tubulin and actin, respectively, as well as coiled-coil-rich proteins of intermediate filament (IF)-like function to fulfil these tasks. Despite generally being characterized as Gram-negative, cyanobacteria have a remarkably thick peptidoglycan layer and possess Gram-positive-specific cell division proteins such as SepF and DivIVA-like proteins, besides Gram-negative and cyanobacterial-specific cell division proteins like MinE, SepI, ZipN (Ftn2) and ZipS (Ftn6). The diversity of cellular morphologies and cell growth strategies in cyanobacteria could therefore be the result of additional unidentified structural determinants such as cytoskeletal proteins. In this article, we review the current advances in the understanding of the cyanobacterial cell shape, cell division and cell growth.

Keywords: cyanobacteria; morphology; cell division; cell shape; cytoskeleton; FtsZ; MreB; IF proteins

---

1. Introduction

One of the major perquisites of cellular functions is a structured and coordinated internal organization. Cells have to build and sustain or, if appropriate, modify their shape, which allows them to rapidly change their behavior in response to external factors. During different life cycle stages, such as cell growth, cell division or cell differentiation, internal structures must dynamically adapt to the current requirements. In eukaryotes, these manifold tasks are fulfilled by the cytoskeleton: proteinaceous polymers that assemble into stable or dynamic filaments or tubules in vivo and in vitro. The eukaryotic cytoskeleton is historically divided into three classes: the actin filaments (consisting of actin monomers), the microtubules (consisting of tubulin subunits) and the intermediate filaments (IFs), although other cytoskeletal classes have been identified in recent years [1,2]. Only the collaborative work of all three cytoskeletal systems enables proper cell mechanics [3]. The long-lasting dogma that prokaryotes, based on their simple cell shapes, do not require cytoskeletal elements was finally abolished by the discovery of FtsZ, a prokaryotic tubulin homolog [4–6] and MreB, a bacterial actin homolog [7,8]. These discoveries started an intense search for other cytoskeletal proteins in bacteria and archaea which finally led to the identification of bacterial IF-like proteins such as Crescentin from *Caulobacter crescentus* [9] and even bacterial-specific cytoskeletal protein classes, including bactofilins [10]. Constant influx of new findings finally established that numerous prokaryotic cellular functions, including cell division, cell elongation or bacterial microcompartment segregation are governed by the prokaryotic cytoskeleton (reviewed by [11,12]).

Cyanobacteria are today's only known prokaryotes capable of performing oxygenic photosynthesis. Based on the presence of an outer membrane, cyanobacteria are generally considered Gram-negative bacteria. However, unlike other Gram-negative bacteria, cyanobacteria contain an unusually thick peptidoglycan (PG) layer between the inner and outer membrane, thus containing features of both Gram phenotypes [13–15]. Additionally, the degree of PG crosslinking is much higher in cyanobacteria than in other Gram-negative bacteria, although teichoic acids, typically present in Gram-positive bacteria, are absent [16]. The processes of PG biosynthesis and the proteinaceous components involved in the composition of the cyanobacterial PG were previously reviewed [17] and will not be part of this review.

While Cyanobacteria are monophyletic [18], their cellular morphologies are extremely diverse and range from unicellular species to complex cell-differentiating, multicellular species. Based on this observation, cyanobacteria have been classically divided into five subsections [19]. Subsection I cyanobacteria (*Chroococcales*) are unicellular and divide by binary fission or budding, whereas subsection II cyanobacteria (*Pleurocapsales*) are also unicellular but can undergo multiple fission events, giving rise to many small daughter cells termed baeocytes. Subsection III comprises multicellular, non-cell differentiating cyanobacteria (*Oscillatoriales*) and subsection IV and V cyanobacteria (*Nostocales* and *Stigonematales*) are multicellular, cell differentiating cyanobacteria that form specialized cell types in the absence of combined nitrogen (heterocysts), during unfavorable conditions (akinetes) or to spread and initiate symbiosis (hormogonia). Whereas subsections III and IV form linear cell filaments (termed trichomes) that are surrounded by a common sheath, subsection V can produce lateral branches and/or divide in multiple planes, establishing multiseriate trichomes [19]. Considering this complex morphology, it was postulated that certain subsection V-specific (cytoskeletal) proteins could be responsible for this phenotype. However, no specific gene was identified whose distribution was specifically correlated with the cell morphology among different cyanobacterial subsections [20,21]. Therefore, it seems more likely that differential expression of cell growth and division genes rather than the presence or absence of a single gene is responsible for the cyanobacterial morphological diversity [20,22]. In the heterocystous cyanobacterium *Anabaena* sp. PCC 7120 (hereafter *Anabaena*), the multicellular shape is strictly dependent on cell-cell communication through gated proteinaceous complexes, termed septal junctions, which resemble eukaryotic gap junctions [23] and allow the diffusion of small regulators and metabolites such as sucrose [24]. Septal junctions pierce through the nanopores in the septal PG mesh, which are drilled by AmiC amidases [25]. Additionally, isolated PG sacculi suggest that the PG mesh from neighboring cells is connected, allowing the isolation of seemingly multicellular PG sacculi [25,26]. Several components are putatively involved in the function, formation and integrity of septal junctions, including septal-localized proteins such as SepJ (formerly also known as FraG [27]), SepI, FraC and FraD as well as AmiC1/2/3 and the PG-binding protein SjcF1, but only FraD was unambiguously shown to be a direct component of the septal junctions [23]. For reviews on cell–cell communication and the general multicellular nature of *Anabaena* see [28–30]. Here we will review the structural and environmental determinants of cyanobacterial shape, division and growth, focusing on the role of cytoskeletal proteins.

2. How Do Cyanobacteria Modify Their Cell Shape?

2.1. Morphology and Environmental Cues

Cyanobacteria show a high degree of morphological diversity and can undergo a variety of cellular differentiation processes in order to adapt to certain environmental conditions. This helps them thrive in almost every habitat on Earth, ranging from freshwater to marine and terrestrial habitats, including even symbiotic interactions [31]. One factor which can drive morphological changes in cyanobacteria is light.

As cyanobacteria are bacteria that use light to fuel their energy-producing photosynthetic machinery they depend on perceiving light in order to optimize their response and to avoid harmful light that could result in the formation of reactive oxygen species (ROS) and subsequently in their death (reviewed by [32]). Optimal light conditions may be defined by quantity (irradiance), duration (day-night cycle) and wavelength (i.e., the color of light). The photosynthetically useable light range of the solar spectrum is generally referred to as PAR (photosynthetically active radiation), but some cyanobacteria may expand on PAR by not only absorbing in the visible spectrum, but also the near-infrared light spectrum. This employs a variety of chlorophylls and allows phototrophic growth up to a wavelength of 750 nm (reviewed by [33]). To sense the light across this range of wavelengths, cyanobacteria possess various photoreceptors of the phytochrome superfamily (reviewed by [34]).

Some filamentous cyanobacteria such as *Fremyella diplosiphon* (also called *Calothrix sp. PCC 7601* or *Tolypothrix* sp. *PCC 7610*) are predicted to encode up to 27 unique phytochrome superfamily photoreceptors, mostly of unknown function [35]. *F. diplosiphon* has been, however, well studied regarding a process called complementary chromatic acclimation (CCA) [36,37]. When the organism is grown under green light, it changes the composition of its light harvesting antennas, the phycobilisomes, by synthesizing phycoerythrin. Under red light, the pigment phycocyanin is introduced instead. This allows the organism to fine-tune its photosynthetic machinery and to generate sufficient energy for growth under changing light conditions. The photoreceptor that is linked with the pigment change was identified as the sensor kinase RcaE [38,39]. *rcaE* deletion mutants *of F. diplosiphon* furthermore revealed the importance of RcaE for cell morphology and shape [40]. When grown under red light, cells show generally a more coccoid cell shape and trichomes are short, whereas under green light the filaments are long, and cells become rod-shaped [36]. This change is regulated by RcaE [40] through its regulatory effect on the expression of the transcriptional regulator BolA. Under red light, the auto-kinase activity of RcaE activity promotes the upregulation of *bolA* expression. BolA in turn binds to the promoter region of *mreB*, inhibiting transcription while under green light, lower levels of BolA result in an accumulation of MreB leading to rod-shaped cells [41]. *rcaE* deletion mutants were unable to change their morphology and remained coccoid under the different light conditions [40].

Phytochromes are also involved in phototaxis and motility, as it has been shown in *Synechocystis* sp. PCC 6803 (hereafter *Synechocystis*) (reviewed by [42]). In some filamentous cyanobacteria of the order *Nostocales* phototaxis is facilitated by the formation of motile hormogonia in response to the colour of light and its irradiance [43–46]. In *F. diploshiphon* for example, the formation of hormogonia is induced by red light while green light suppresses their induction [45,47]. However, hormogonia formation differs from other cyanobacterial differentiation processes such as heterocyst formation by the fact that it can be initiated by various other environmental cues, including nutrient concentrations and signals from symbiotic partners [48–51]. Once hormogonia formation has been initiated, a cascade of cellular development follows. This includes the synchronous division of cells, the fragmentation of trichomes at heterocyst-vegetative cell junctions and necridia (apoptotic cells), a reduction in cell volume and sometimes the formation of tapered filament termini [44]. We have shown in hormogonia from the branching cyanobacterium *Mastigocladus laminosus* SAG 4.84 that molecular exchange between neighboring cells is fast and that the reduction in cell volume could even further accelerate signal transduction to coordinate movement [52]. During gliding motility cell growth and division are arrested [53] but once the hormogonium reaches its destination, e.g., the host, cells elongate, divide and potentially differentiate into heterocysts. Various hosts have been identified for cyanobacteria, including bryophytes (hornworts, liverworts), the angiosperm *Gunnera*, the aquatic fern *Azolla*, fungi (forming lichens), the fungus *Geosiphon*, cycads and diatoms [54,55]. The cyanobiont (cyanobacterial symbiont) can either grow intracellularly or extracellularly in specialized compartments. For example in the case of the *Anabaena-Azolla* symbiosis the cyanobacterium resides in the leaf pockets of the fern [56]. Internal symbionts have additional challenges. Once inside the host, cell division needs to be regulated to avoid bursting of the cells. Indeed, it has been observed that the cyanobacterium

Calothrix rhizosoleniae grows in shorter trichomes when inside the host cell than in its free-living state [57]. The signals and mechanisms behind the restricted growth remain basically unknown.

Although not further discussed in this review, cellular morphology can be changed by nutrient composition [58] as well as stresses such as temperature, salt, ultraviolet (UV) light and drought and are connected with the formation of ROS (reviewed by [59]). The aforementioned examples show the complexity and diversity of the environmental cues that influence morphology and cell shape in cyanobacteria. In the following paragraphs we will address the underlying genetic and structural components that are associated with the different morphologies.

2.2. Morphological Plasticity in Cyanobacteria

Morphological plasticity, or the ability of one cell to alternate between different shapes, is a common strategy of many bacteria in response to environmental changes or as part of their normal life cycle (reviewed by [60–62]). Bacteria may alter their shape by simpler transitions from rod to coccoid (and vice versa) as in *Escherichia coli* [63], by more complex transitions while establishing multicellularity (reviewed by [60]) or by the development of specialized cells, structures or appendages where the population presents a pleomorphic lifestyle [64]. The precise molecular circuits that govern those morphological changes are yet to be identified, however, a so-far constant factor is that the cell shape is determined by the rigid PG sacculus which consists of glycan strands crosslinked by peptides. To grow, cells must synthesize new PG while breaking down the existent polymer to insert the newly synthesized material. How cells grow and elongate has been extensively reviewed in model organisms of both, rod-shaped [65,66] and coccoid bacteria [67]. The molecular basis for morphological plasticity and pleomorphism in more complex bacteria, however, is slowly being elucidated as well (see a recent review by [62]). The protein complex responsible for cell wall elongation in rod-shaped bacteria is referred to as the elongasome and is composed of, among others, MreB, MreC, MreD, PBP2, PBP1A, RodA and RodZ [65,68,69]. MreB polymerizes into dynamic filaments that act as a scaffold ion which the PG synthesis machinery assembles [70,71]. MreB orchestrates elongasome assembly through interaction with transmembrane proteins, such as RodZ and MreC/D [65] and the direct involvement of MreB in cell wall morphogenesis was described upon the correlation between MreB polymers and PG deposition along the lateral cell wall [72–74].

The phylogenetic distribution of MreB seems to be ubiquitous in rod-shaped bacteria, which encode for at least one *mreB* homolog whereas, with few exceptions, coccoid bacteria lack *mreB*, supporting the theory that coccoid bacteria evolved from rods [75]. In fact, out of 253 sequenced bacterial genomes representing all possible shapes, 63% of the transitions from rod to coccoid were related to the loss of *mreB* [76]. In accordance with this hypothesis, an analysis of 141 fully sequenced cyanobacterial genomes found that only four species lack *mreB*. The four species included *Synechocystis*, *Crocosphaera watsonii* WH 8501, *Atelocyanobacterium thalassa* ALOHA and *Gloeocapsa* sp. PCC 73106, all of which are unicellular and coccoid. In all these cyanobacteria, the loss of *mreB* was accompanied by the loss of the complete *mreBCD* operon [77]. Among the myriad cyanobacterial shapes, all multicellular or baeocyte-forming cyanobacteria, independent of their cell shape, had a complete *mreBCD* operon, suggesting that the coccoid shape in some of those multicellular cyanobacteria is achieved by alternative mechanisms than simply a lack of MreB [77]. Few other cyanobacterial taxa have retained only a copy of *mreB* on their genomes and lack *mreC* and *mreD*. For example, the multicellular cyanobacterium *Trichodesmium erythraeum* contains only a partial *mreBCD* operon that lacks *mreD*, while the filamentous helical shaped *Arthrospira maxima* CS-328 only encodes for *mreB* [77]. Likewise, the helical shaped *Helicobacter pylori* contains both MreB and MreC but lacks MreD. However, *mreB* is not essential in this bacterium and is not involved in cell shape-determination [78]. Instead, a family of endopeptidases actively remodels and flexibilizes the PG crosslinks that enable the helical cell curvature needed for the successful colonization of the human stomach [79]. Likewise, the transition from helical to straight trichomes [80], could be governed by other proteins than MreB instead.

There are few studies that have tried to elucidate the role of MreB in cyanobacteria and even less that succeeded in obtaining *mreB* deletion mutants either partially or completely segregated. However, the common denominator of these studies is a function of MreB in cell shape maintenance, independent of the cyanobacterial morphology [81–84]. In the rod-shaped *Synechococcus elongatus* PCC 7942 (hereafter *Synechococcus*) and *Synechococcus* sp. PCC 7002, MreB appears to be essential [81,82] as only partially segregated mutants could be obtained. In both *Synechococcus* species partial loss of *mreB* resulted in cell shape defects where cells became more coccoid (Figure 1). As a result of the polyploid nature of cyanobacteria and their asynchronous DNA replication (reviewed by [85]), most cyanobacterial studies on MreB have focused on elucidating the role of this protein in chromosome partitioning. Indeed, *mreB* knockdown mutants in *Synechococcus* show disarranged chromosomal replication origin (ori)-foci, suggesting that MreB is involved in chromosomal positioning [83]. However, the role of MreB in chromosomal positioning seems to be species-specific as chromosome partitioning was not affected in a *mreB Synechococcus* sp. PCC 7002 depletion mutant [82]. Furthermore, MreB seems to be involved in cellular compartmentalization in *Synechococcus* as *mreB* mutants show altered carboxysome placements. However, this effect is likely indirect in which the function of MreB in cell shape determination provides the necessary structural framework to organize carboxysomes [81]. A similar pleiotropic and indirect effect might also explain the alterations in chromosome positioning in the *Synechococcus mreB* knockdown mutant [83]. A notable exception of the essential nature of *mreB* is *Anabaena* where *mreB* affects cell shape but was found to be dispensable for cell viability with combined nitrogen and did not affect chromosome segregation or placement [84]. *Anabaena mreB*, *mreC* and *mreD* deletion mutants were all characterized by an alteration of cell size, regardless of the growth conditions [58]. In wild type *Anabaena*, single cells are longer than they are wide (in respect to the trichome growth axis). In the *mreB*, *mreC* and *mreD* mutants, however, cells became more coccoid and seemingly inverted their orientation within the trichome, being wider than long (Figure 1). MreB, MreC and MreD additionally affected the *Anabaena* trichome length, possibly through a strengthening of the septal cell wall, which was found to be increased in diameter in the three mutants [58].

Cells within the trichomes of the multiseriate and branching cyanobacteria *Fischerella muscicola* PCC 7414 may display rod, coccoid or tapered shapes while also differing in cell size [19,93]. *F. muscicola* also shows alternative growth modes that include apical, septal, and lateral trichome growth, although it is still not known how MreB contributes to cell shape or PG synthesis in this cyanobacterium. Deletions of *mreB* could not be obtained in *F. muscicola*, but overexpression of GFP-MreB from the copper inducible *petE* promoter showed alternative MreB localization in the different cell morphotypes (hormogonia, young and mature trichomes) from *F. muscicola* [77]. Further assessment of *mreB* regulation and localization dynamics in the different morphotypes is necessary to elucidate the role of this protein in the morphogenesis of complex multicellular and branching cyanobacteria.

The above-described observations attribute a largely structural function to MreB in cyanobacteria, however, MreB has also been indicated to be involved in other cellular processes. *Spiroplasma eriocheiris*, a cell wall-less helical bacterium with swimming motility encodes for five MreB variants [94]. Together with the fibril protein, MreB was proposed to contribute to the propelling mechanism of *S. eriocheiris* by coordinating the length changes of their cytoskeletal ribbons [95]. Unlike any other cyanobacterium, some marine *Synechococcus* move by swimming using a still unidentified propulsion mechanism while surprisingly lacking apparent flagella systems [42]. Mechanisms such as the expulsion of a Newtonian fluid were excluded early on and instead a swimming mechanism was proposed to resemble the helical rotor mechanism propelling myxobacteria [96]. However, the involvement of MreB in cyanobacterial motility has so far not been demonstrated.

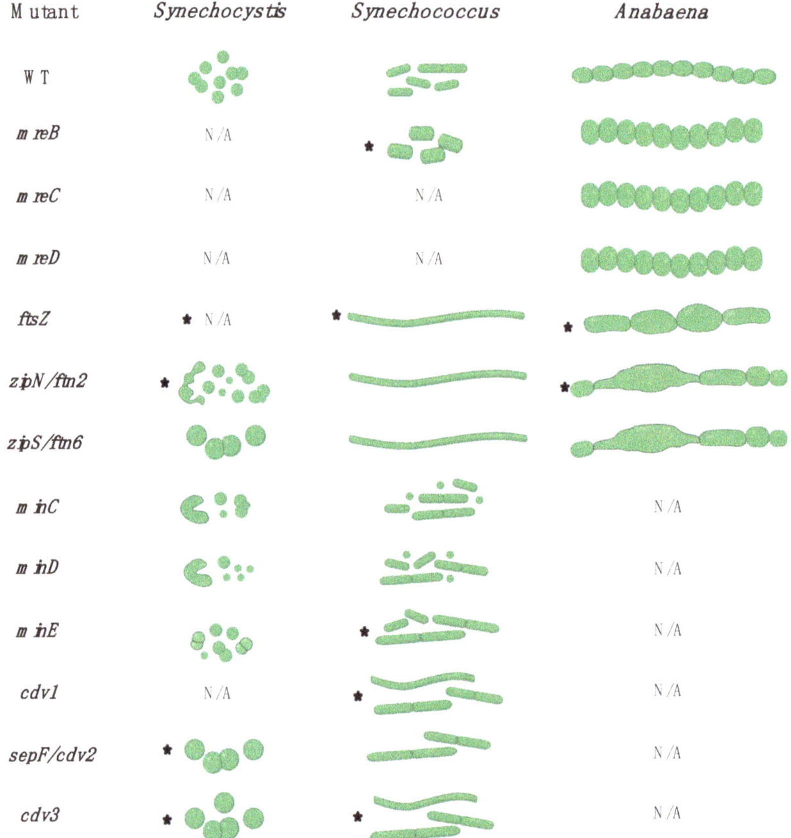

Figure 1. Cyanobacterial cell division and cell growth mutant phenotypes in *Synechocystis*, *Synechococcus*, and *Anabaena*. Stars indicate gene essentiality in the respective organism. Of note: while one gene can be essential in one cyanobacterial organism/morphotype, it does not necessarily mean it is essential in all other cyanobacteria. N/A indicates that no mutant phenotypes have been described. WT: wild type. Image created with BioRender.com. WT [19]; *mreB* [58,81,83,84]; *mreC* [58]; *mreD* [58]; *ftsZ* [86,87]; *zipN/ftn2* [86,88,89]; *zipS/ftn6* [86,89,90]; *minC* [88,91]; *minD* [88,91]; *minE* [88,91]; *cdv1* [86]; *sepF/cdv2* [86,90]; *cdv3* [86,92].

2.3. Different Modes of Cell Shape Regulation in Cyanobacteria

Despite their morphological complexity, cyanobacteria contain all conserved and so far known bacterial morphogens (Table 1). Understanding cyanobacterial morphogenesis is challenging, as there are numerous morphotypes among cyanobacterial taxa, which can also vary within a given strain during its life cycle [19]. Changes in cellular or even trichome morphologies are tasks that would require active cell wall remodeling and thus far no genes attributed to the different morphotypes have been identified in cyanobacteria [20]. Therefore, the most likely scenario is that genes or their products are differentially regulated during these cell morphology transitions [22], as it has been hypothesized for most bacteria [62]. In multicellular cyanobacteria, division of labor between cells within a trichome is achieved by different cell programing strategies. Thus, gene regulation occurs differentially in these specific cell types [30,97,98].

Table 1. Proteins involved in cyanobacterial cell division and cell morphology.

| Proteins | Cyanobase Locus Tags and NCBI Accession Numbers | | | Function |
	Synechocystis	*Synechococcus*	*Anabaena*	
FtsZ	Sll1633 (WP_010872126.1)	Synpcc7942_2378 (WP_011244037.1)	Alr3858 (WP_010997999.1)	Cell division
ZipN (Ftn2)	Sll0169 (WP_010873289.1)	Synpcc7942_1943 (WP_011244461.1)	All2707 (WP_010996860.1)	Cell division
ZipS (Ftn6)	Sll1939 (WP_010871735.1)	Synpcc7942_1707 (WP_011244694.1)	All1616 (BAB77982.1)	Cell division
Cdv1	Sll0227 (WP_010871341.1)	Synpcc7942_0653 (WP_011243187.1)	All4287 (BAB75986.1)	Cell division
SepF (Cdv2)	Slr2073 (WP_010872037.1)	Synpcc7942_2059 (WP_011378295.1)	Alr0487 (WP_010994663.1)	Cell division
Cdv3	Slr0848 (WP_010873766.1)	Synpcc7942_2006 (WP_011244399.1)	Alr4701 (WP_010998832.1)	Cell division
YlmD	Slr1573 (WP_010874196.1) [a]	Synpcc7942_0346 (WP_011243479.1)	All5255 (WP_010999379.1)	Cell wall synthesis
YlmE	Slr0556 (WP_010874100.1)	Synpcc7942_2060 (WP_011244343.1)	Alr0486 (WP_010994662.1)	Unknown
YlmG	Ssr2142 (WP_010871471.1)	Synpcc7942_0477 (WP_011243354.1) [b]	Asl2061 (WP_010996222.1)	Cell division
	Ssl0353 (WP_010873648.1)	Synpcc7942_2017 (WP_011244388.1)	Asl0940 (WP_010995114.1)	
YlmH	Sll1252 (WP_010872783.1) [c]	Synpcc7942_1503 (WP_011378057.1) [c]	Alr2890 (WP_010997041.1) [c]	Unknown
MinC	Sll0288 (WP_010873891.1)	Synpcc7942_2001 (ABB58031.1)	Alr3455 (BAB75154.1)	Cell division
MinD	Sll0289 (WP_010873890.1)	Synpcc7942_0896 (WP_011242956.1) [d]	Alr3456 (WP_010997606.1) [e]	Cell division
	—	Synpcc7942_0220 (WP_011243604.1)	All2033 (WP_010996194.1)	
	—	—	All2797 (WP_010996948.1)	
MinE	Ssl0546 (WP_010873889.1)	Synpcc7942_0897 (WP_011242955.1)	Asr3457 (WP_010997607.1)	Cell division
SulA	Slr1223 (WP_014407090.1)	Synpcc7942_2477 (WP_011243937.1)	All2390 (WP_010996546.1)	Cell division
FtsE	Slr0544 (WP_010874063.1) [f]	Synpcc7942_1414 (WP_011242455.1) [f]	Alr1706 (BAB78072.1)	Cell division
FtsI	Sll1833 (WP_010871772.1)	Synpcc7942_0482 (WP_011243349.1)	Alr0718 (WP_010994893.1) [g]	Cell division
FtsK/SpoIIIE	Sll0284 (WP_010873902.1) [h]	Synpcc7942_0981 (WP_011242875.1) [h]	Alr3799 (WP_010997940.1) [h]	Cell division
	—	—	All7666 (WP_010993994.1) [i]	
FtsN	Slr0702 (WP_010873961.1)	N/A	N/A	Cell division
FtsQ	Sll1632 (WP_010872127.1)	Synpcc7942_2377 (WP_011378434.1)	Alr3857 (WP_010997998.1)	Cell division
FtsW	Slr1267 (WP_010872891.1)	Synpcc7942_0324 (WP_011377535.1)	All0154 (WP_010994331.1)	Cell division
FtsX	N/A	N/A	All1757 (WP_010995925.1)	Cell division
CyDiv	N/A	N/A	All2320 (WP_010996476.1)	Cell division
SepI	N/A	N/A	Alr3364 (BAB75063.1) [j]	Cell–cell contact
RodA	N/A	Synpcc7942_1104 (WP_011377865.1)	Alr0653 (WP_010994829.1)	Cell elongation
MreB	N/A	Synpcc7942_0300 (WP_011243524.1)	All0087 (BAB77611.1)	Cell elongation
MreC	N/A	Synpcc7942_0299 (WP_011243525.1)	All0086 (WP_010994263.1)	Cell elongation
MreD	N/A	Synpcc7942_0298 (ABB56330.1)	All0085 (BAB77609.1)	Cell elongation
BolA	Ssr3122 (WP_010871705.1)	Synpcc7942_1146 (ABB57176.1)	Asr0798 (WP_010994972.1)	Cell elongation

Table 1. *Cont.*

Proteins	Synechocystis	Synechococcus	Anabaena	Function
CikA	Slr1969 (WP_010872820.1)	Synpcc7942_0644 (WP_011243194.1)	All1688 (WP_010995857.1)	Circadian rhythm
PBP1	Sll0002 (WP_010873436.1)	Synpcc7942_2000 (WP_011378270.1)	Alr5101 (WP_010999227.1)	Cell wall synthesis
PBP2	Slr1710 (WP_010871874.1)	Synpcc7942_0785 (ABB56817.1)	Alr4579 (WP_010998711.1)	Cell wall synthesis
PBP3	Sll1434 (WP_010872930.1)	Synpcc7942_2571 (WP_011243849.1)	All2981 (WP_010997132.1)	Cell wall synthesis
PBP4	Sll1833 (WP_010871772.1)	Synpcc7942_0580 (WP_011377631.1)	Alr5326 (BAB77025.1)	Cell wall synthesis
PBP5	Slr0646 (WP_010873596.1)	Synpcc7942_1934 (ABB57964.1)	Alr5324 (WP_010999448.1)	Cell wall synthesis
PBP6	Sll1167 (WP_010872913.1)	Synpcc7942_0482 (WP_011243349.1)	All2981 (WP_010997132.1)	Cell wall synthesis
PBP7	Slr1924 (WP_010873199.1)	N/A	Alr5045 (WP_010999171.1)	Cell wall synthesis
PBP8	Slr0804 (WP_010872730.1)	N/A	Alr0718 (WP_010994893.1) [g]	Cell wall synthesis
PBP9	N/A	N/A	Alr0153 (WP_010994330.1)	Cell wall synthesis
PBP10	N/A	N/A	Alr1666 (WP_010995835.1)	Cell wall synthesis
PBP11	N/A	N/A	Alr0054 (WP_010994231.1)	Cell wall synthesis
PBP12	N/A	N/A	All2656 (WP_010996812.1)	Cell wall synthesis

Absent in cyanobacteria according to [28,86,88]: FtsA, FtsB, ZapA, ZapB, ZapC, ZipA, EzrA, FtsB, FtsL, FtsN (although FtsN was reported in *Synechocystis* by [92]). a: Note that [92] identified Slr1593 as YlmD homolog, while we found this protein to be not the closest relative to YlmD from *Bacillus subtilis* or *Staphylococcus aureus* YlmD. b: YlmG as identified by [99]. c: predicted as photosystem II S4 domain protein. d: MinD identified by [100]; 27.5% sequence identity to Synpcc7942_0220. e: *Anabaena* MinD sequence identities: Alr3456+All2033: 23.1%; Alr3456+All2797: 25.9%; All2033+All2797: 59.4%. f: No FtsE was predicted in *Synechocystis* and *Synechococcus elongatus* according to [86]. g: Identified as FtsI by [101]. h: Predicted as YjgR family proteins of the HerA clade, relatives of FtsK [102]. i: Present on the *Anabaena* plasmid pCC7120beta. j: CyDiv is proposed to be part of an essential late divisome protein complex [103]. N/A: not available. **Note**: Differences in identified penicillin-binding proteins (PBPs) were found between [92,101,104,105]. Here, we present the data from [104] as it presents the most comprehensive analysis of cyanobacterial PBPs. "—" indicates absence of additional homologs.

The multiplicity of mechanisms and life strategies displayed by cyanobacteria such as photosynthetic lifestyle, the presence of thylakoid membranes (with the exception of *Gloeobacter* [106]), carboxysome assembly, motility, nitrogen fixation and cell differentiation (i.e., hormogonia, akinetes, heterocysts and necridia) are associated with specific regulatory mechanisms that coordinate the different processes [30,97,98,107–111]. These regulatory mechanisms are complex and often intricate. Given this vast regulatory network, mutations that affect processes such as cell wall synthesis [26,101,105,112,113], intercellular transport [114,115] and cell division [88,90,116,117] may alter cell shape. Consequently, the function of the *mreB* gene or the entire *mreBCD* operon [58,84] was analyzed in gene overexpression or gene deletion mutants of these aforementioned processes (reviewed by [59]). In all studied cases, upregulation of *mreB* is associated with the transition from coccoid to rod shaped cells. MreB is also involved in the morphological transition during *N. punctiforme* hormogonia differentiation. Transcriptomic studies revealed that *mreB* and *rodA* were both upregulated in hormogonia from *N. punctiforme*. Similar to *M. laminosus* hormogonia [118], *N. punctiforme* hormogonia are characterized for having a smaller cell size and rod-shape in comparison with the larger and more coccoid cells of the mature trichome. Upregulation of *mreB* was also observed in a "branchless" morphotype of *F. muscicola* induced under sucrose supplementation [22], indicating that environmental growth conditions play a crucial role in cell shape regulation. Branch-less cultures were characterized by long trichomes that appear as nascent hormogonia previous to the detachment from the parent trichome [19,22]. Cells in the branchless cultures are longer, narrower and display a rod shape with tapered cells at the tip of the trichome compared to the more diverse cell morphologies (e.g., elliptical, rod-shaped, coccoid-shaped) in the parent trichome [22].

As photosynthetic microorganisms, iron has a pivotal role in cyanobacterial photosynthesis and defense against oxidative stress [119]. The transcriptional regulator FurA has been demonstrated as the master regulator of iron homeostasis in *Anabaena* [120] and was also shown to be involved in several other processes such as heterocyst differentiation and programmed cell death [119]. Overexpression of FurA in *Anabaena* lead to alterations in the cell shape, possibly through its positive regulatory function of the *mreBCD* operon [121]. Another prominent environmental factor affecting cyanobacterial cell shape is the availability of fixed nitrogen sources. Similar to the essential cell division gene *ftsZ* (discussed further below), *mreB* is differentially regulated during heterocyst formation [84]. Unlike *ftsZ*, *mreB* is upregulated during heterocyst formation in *Anabaena* pro-heterocysts [84,97] and an N-terminally GFP-tagged MreB localized to the cell poles in both vegetative cells and heterocysts [84]. The increase in MreB levels during heterocyst formation possibly provides the framework for the increase in cell size, which requires de-novo synthesis and integration of PG into the cell wall. In agreement with this, a recent study found that the incorporation of fluorescently labelled amino acids [122] into the *Anabaena* cell wall was elevated during heterocyst maturation [123]. PG biosynthesis enzymes, which are associated with the MreB-driven elongasome [68,124], were furthermore identified by several different groups to be essential for heterocyst formation [13,125,126], strengthening the importance of MreB function for heterocyst development. Additionally, *mreB* and *mreC* but not *mreD* are essential for diazotrophic growth of *Anabaena*, with a supposable function subsequent to heterocyst formation as *Anabaena mreB*, *mreC* and *mreD* mutants still differentiated heterocysts [58]. In the *Anabaena* wild type, cells are shorter during diazotrophic growth and longer in the presence of combined nitrogen [58]. This phenomenon can be explained by an increase in the levels of the global transcriptional regulator NtcA during diazotrophic growth. NtcA negatively regulates the *mreBCD* operon, leading to a reduced cell length. Consistently, an *ntcA* deletion mutant was characterized by an increased cell length [58].

Other factors that might regulate cell shape in cyanobacteria could be the interplay between the FtsZ and MreB cytoskeleton. In *E. coli* FtsZ and MreB can physically interact and this interaction is important for the progression from cell growth to cell division [127], whereas no direct effect of MreB on Z-ring placement and septum formation was observed in *Anabaena* [84]. This finding is in concert with the lack of interaction between MreB and FtsZ in *Anabaena* [77]. Notably, we recently showed that in the complex multicellular cyanobacteria *F. muscicola* and *Chlorogloeopsis fritschii* PCC 6912, MreB physically interacted with FtsZ [77], suggesting that their complex trichome and cell phenotypes could, in part, rely on the crosstalk between the elongasome and the FtsZ-driven divisome.

3. The Cyanobacterial Cell Division Complex—Function and Regulation

Numerous studies over the past years have conclusively shown that cyanobacteria not only possess a hybrid Gram phenotype in terms of their cell envelope but also possess proteinaceous structural determinants otherwise restricted to a single Gram type. The processes of PG and cell wall remodeling as well as cell septation rely on other divisome components that are recruited to the Z-ring. The Z-ring functions as a scaffolding structure for other divisome components but also potentially exerts constrictive force as indicated by FtsZ's ability to bend liposomes [128,129]. In *E. coli*, more than 30 proteins have been identified as divisome or divisome-associated components, among those, 12 are essential and commonly associated with the divisome in the order: FtsZ → FtsA/ZipA → FtsE/FtsX → FtsK → FtsQ/FtsL/FtsB → FtsW/FtsI → FtsN (for reviews on bacterial cell division processes, please see [68,130,131]). The arrival of FtsN primes the divisome for septal PG synthesis and cell division. Homologs to some of those divisome proteins have been identified in cyanobacteria, including FtsE, FtsQ, FtsW and FtsI, while FtsA, ZipA, FtsL and FtsB are absent in cyanobacteria [28,86,88]. Other divisome-associated proteins from *E. coli* or *Bacillus subtilis* are likewise absent in cyanobacteria, including ZapA, ZapB, ZapC and EzrA. With one exception identified in *Synechocystis* [92], FtsN cannot be found in cyanobacteria. Similarly, a FtsX and FtsK homolog was so far only identified in *Anabaena* [86,132]. In agreement with their enormous morphological diversity, several morphological determinants specific to the Cyanobacteria phylum were also described, among those Ftn2 (ZipN) and Ftn6 (ZipS) [88–90,92,132–134]. Given the nonuniform nomenclature of cyanobacterial protein identifiers and to ease future research on cyanobacterial morphologies, we have collected a comprehensive list of important cyanobacterial structural determinants from the three widely used cyanobacteria *Anabaena*, *Synechocystis* and *Synechococcus* (Table 1). In the following sections, we will further elucidate the currently available information on some of those proteins, including their cellular context and functional properties.

3.1. Polymerization Properties of Cyanobacterial FtsZ

Cell division in bacteria is, with a few exceptions, strictly dependent on the function of the tubulin homolog FtsZ and its associated multiprotein complex, termed the divisome. FtsZ is an essential and highly conserved GTPase in almost all bacteria, Euryarchaeota, photosynthetic eukaryotes (i.e., in their chloroplasts) and even in some mitochondria [4,5,12,135–137]. Upon completion of chromosome segregation, FtsZ is the first protein to assemble at the future division site, forming a ring-like structure (the Z-ring) through GTP-dependent polymerization of FtsZ monomers into short protofilaments. Both, *Anabaena* and *Synechocystis* FtsZ contain the conserved glycine-rich GTP-binding domain, which is crucial for in vivo Z-ring formation and in vitro polymerization [88,138–140]. Unlike other bacterial FtsZ proteins, purified *Synechocystis* FtsZ assembles into a mixture of straight bundles, similar to chloroplast FtsZ, and toroidal filaments, indicating that the curved cyanobacterial FtsZ polymers could bend the cytoplasmic membrane [139,140]. Many cyanobacteria contain a highly variable N-terminal sequence extension of the FtsZ protein (between 20 and 80 amino acids long) that is absent in other bacteria but strikingly conserved among heterocystous cyanobacteria [88,140]. The N-terminal sequence is essential for *Anabaena* viability and, although FtsZ and an N-terminally truncated FtsZ (ΔN-FtsZ) interact with each other, a ΔN-FtsZ-GFP fusion protein could not integrate into native Z-rings, possibly a result of its inability to interact with the FtsZ membrane anchor SepF [140,141]. While native *Anabaena* FtsZ forms toroids in vitro, ΔN-FtsZ only associates into filament bundles. As a consequence, the N-terminal peptide of *Anabaena* FtsZ, and possibly that of other cyanobacteria, likely promotes filament curling and decreases lateral filament bundling [140]. FtsZ filament curling or toroid-formation in *Anabaena* and *Synechocystis* FtsZ supports a constriction force of the cyanobacterial Z-ring itself, which is also described for *E. coli* FtsZ [139,142]. However, studies on *Prochlorococcus* Z-ring assembly suggest that it is likely not contractile in this species and possibly merely functions as a scaffold in oval-shaped cyanobacteria [143]. Straight filament bundles and toroids were previously reported for *E. coli* and *M. tuberculosis* FtsZ but only in the presence of crowding agents such as methylcellulose or polyvinyl alcohol. Given the larger cell diameter of *Synechocystis* and *Anabaena* (2–3 μm), filament bundling could be beneficial for their increased cell size compared to, for example, rod-shaped *E. coli* or *Synechococcus*, which are considerably smaller (1 μm cell diameter). Whether filament bundling exists in filamentous cyanobacteria with diameters less than 1 μm such as species of the genus *Halomicronema* [144] remains yet to be investigated. Many small rod-shaped bacteria also lack a signature motif in the H8 helix, which is likely responsible for filament bundling [139,140]. Consequently, cell shape possibly poses an evolutionary constraint on the functional diversification of proteins important for cell division. Considering the different cyanobacterial morphotypes, it will be interesting to test whether a similar observation also exists for processes that regulate cell growth in cyanobacteria, i.e., for the cell shape determining protein MreB.

3.2. FtsZ is Essential in Cyanobacteria

In cyanobacteria, *ftsZ* homologs were detected in every sequenced species and *ftsZ* was found to be essential in *Anabaena*, *Synechocystis* and *Synechococcus* [86,88,138,145]. Partial inactivation of *ftsZ* or addition of a tubulin assembly inhibitor (thiabendazole) causes cell filamentation (elongation) in the rod-shaped *E. coli* and *Synechococcus* and cell swelling in the coccoid *Synechocystis* [145]. Contrasting this, partial depletion of *ftsZ* or overexpression of the FtsZ assembly inhibitor *minC* results in a mixed filamentous/elongated and swollen cell shape in *Anabaena* [87,116] (Figure 1). Given that the ellipsoid cell shape of *Anabaena* can be considered a hybrid phenotype between the coccoid and rod-shape phenotype of *Synechocystis* and *Synechococcus*, formation of a hybrid cell shape defect upon impairment of cell division seems consistent. In the baeocytes-forming subsection II cyanobacterium *Chroococcidiopsis* sp. CCMEE 029, *ftsZ* is also essential and partial deletion disrupts the regularity in daughter cell arrangements, leading to cell aggregates. These aggregates, however, did not enlarge in cell volume compared to the wild type [146]. Therefore, the impact of impaired cell division appears to be highly dependent on the respective cyanobacterial morphotype and could result in different shapes in other so far understudied cyanobacterial subsections.

3.3. Cellular Localization of FtsZ in Cyanobacteria

As in other bacteria, FtsZ localizes to the middle of the cell in cyanobacteria, forming the typical Z-ring structure [86,88,116]. Consecutive Z-rings from neighbouring cells in the *Anabaena* trichome align parallel to each other. Z-rings from deeply-constricted *Synechocystis* daughter cells, however, form Z-rings that are perpendicular to each other [116], reminiscent of what we observed for true-branching subsection V cyanobacteria [77]. Using photobleaching of cyanobacterial autofluorescence coupled to super-resolution microscopy (STORM; stochastic optical reconstruction microscopy) of the unicellular, coccoid-shaped *Prochlorococcus* sp. MED4, a lateral resolution of 10 nm of Z-ring assembly was achieved [143]. Liu and colleagues found that FtsZ rings all contained small gaps, being non-continuous assemblies, and that FtsZ first polymerizes into incomplete and then complete rings, resembling the observations from *E. coli*. Consequently, their study thus supports the so-called patchy band model, where FtsZ assembles into discontinuous strings during cell division in contrast to the lateral association model that states that FtsZ polymers interact laterally to assemble into a complete Z-ring [143]. Studying other cyanobacterial morphotypes could consequently shed more light onto the debate about the FtsZ polymerization mechanisms in bacteria and might even indicate different assembly properties based on different cell shapes.

3.4. Transcriptional and Posttranslational Control of Cell Division in Cyanobacteria

E. coli ftsZ is transcribed in an operon together with *ftsA* (absent in cyanobacteria) and *ftsQ* (i.e., the *ftsQAZ* operon), whereas no *ftsQZ* operon was observed in *Anabaena*, *Synechocystis* nor *Synechococcus*, where *ftsZ* is independently transcribed from *ftsQ* instead [109,147], contrasting the identified *ftsQZ* operon structure in *M. aeruginosa* [148]. Given that cyanobacteria are photosynthetic organisms, it is not surprising that cell division and consequently *ftsZ* expression patterns are dependent on the circadian clock with an expression peak near dusk in *Synechococcus* and *Synechocystis* [147,149]. This circadian rhythmicity in *Synechococcus* is governed by the essential circadian clock protein kinase KaiC, through inhibition of Z-ring formation without impacting the cellular FtsZ protein levels [150]. In contrast, *ftsZ* transcription and cell division occur during the light cycle in the diazotrophic (nitrogen-fixing), unicellular *Cyanothece* sp. ATCC 51142 [151]. Diurnal control of *ftsZ* transcription also occurs in the marine, filamentous and nitrogen-fixing *T. erythraeum* IMS101. There, cell division and cell differentiation (into nitrogen-fixing diazocytes) occurs early during the dark period, which is preceded by an upregulation of *ftsZ* expression (and FtsZ protein level) [151]. Reminiscent of the dependency of cell division for heterocyst-development in *Anabaena* [152], diazocyte-development could be dependent on cell division in *T. erythraeum* [151]. Consequently, cell differentiation in cyanobacteria appears to be strongly connected to cell division and is halted in differentiated cells through a downregulation of *ftsZ* transcription and/or a decrease in FtsZ protein levels [116,153–155]. This notion is supported by the absence of Z-rings in terminally differentiated mature heterocysts in *Anabaena* [19,116,133]. Notably, loss of Z-rings precedes loss of detectable *ftsZ* transcripts, with the former taking place in immature heterocysts and the latter in mature heterocysts [133]. Thus, FtsZ is a cell division factor specific to vegetative cells in *Anabaena* [155]. The arrested cell division after heterocyst development is also apparent in true-branching filamentous cyanobacteria such as *M. laminosus* and other *Fischerella* species. During the life cycle the initially narrow trichomes with cylindrically shaped cells mature to wide trichomes with coccoid cells that give rise to true branches of cylindrical cells [52,156]. Once a heterocyst has formed within a certain type of trichome its morphology remains unchanged even when neighboring vegetative cells undergo the maturing process. An intriguing hypothesis is that in heterocysts, proteolytic FtsZ degradation is specifically increased, thus abolishing Z-ring formations (i.e., cell division). Although not shown to be specific to or increased in heterocysts, FtsZ-specific proteases were discovered in *Anabaena*, *F. muscicola* and *C. fritschii* cell extracts but not in the extracts of the non-cell differentiating *Synechocystis* or *E. coli* [77,157]. The precise nature of these proteases remains unknown but it was found to only cleave natively folded FtsZ in *Anabaena*, thus being structure and not sequence-specific [157]. Recently, PatA, a protein involved in the differentiation of intercalary heterocysts under nitrogen-deprived growth conditions and that localizes to the Z-ring and the cell poles [133], was found to function in destabilizing the Z-ring in *Anabaena* [133]. PatA interacts with ZipN and SepF, two crucial cyanobacterial cell division factors (discussed in more detail below) and it was hypothesized that this interaction ultimately promotes the loss of Z-ring structures during heterocysts development (Figure 1). Thus, PatA could be one component responsible for the loss of Z-ring structures in immature heterocysts, ultimately promoting cell differentiation progression [133]. In *T. erythraeum*, *ftsZ* transcription occurs only after DNA replication (extrapolated from *dnaA* gene expression) [151] as it was also shown in the unicellular, bloom-forming cyanobacterium *Microcystis aeruginosa* NIES298. There, *ftsZ* expression is repressed upon halt of DNA replication, suggesting that there are factors in the cell that sense the DNA content and regulate *ftsZ* transcription in response [148]. *Synechococcus* cells in stationary phase cultures rarely divide and elongate instead, largely due to the inhibition of DNA replication and consistent with a requirement of DNA biosynthesis for cell division [158]. Similar observations were also made for *Synechocystis* cultures in the stationary phase that revealed 4 to 10 times lower *ftsZ* transcript levels compared to log phase cultures; a mechanism that has been linked to cell density sensing [149]. In *Synechocystis*, two transcription factors (Sll0822 and Sll0359), which regulate *ftsZ* and *ftsQ* transcription, belong to the cyAbrB clade B of transcriptional regulators. Deletion of *sll0822* results in a cell division defect with swollen cells [159], similar to the *Synechocystis* *ftsZ* depletion strain [145]. In *Anabaena*, the cyAbrB transcriptional regulator CalA specifically regulates *ftsZ* expression in vegetative cells [160]. Whether the regulation of *ftsZ* expression by those transcription factors results in response to DNA content or other factors remains to be elucidated. Studies in *Anabaena* and *Synechococcus* have independently highlighted a positive correlation of DNA content with cell size and not with cell division, indicating that some cyanobacteria can sense their cell volume and adapt their chromosome content accordingly [116,161,162]. Consequently, it is conceivable that not FtsZ but MreB is indirectly involved in chromosome copy number determination, possibly through its regulatory function on cell shape and size. It is worth noting that the overall protein concentration within *Synechococcus* cells remained constant, regardless of the growth rate and is positively correlated with cell volume and DNA content [100].

3.5. FtsZ-Associated Regulators Control Cell Division in Cyanobacteria

In *E. coli* and *B. subtilis*, a number of factors, including the Zap proteins (ZapA/B/C/D) crosslink FtsZ polymers with each other or to the chromosome ends. The actin homolog FtsA, which, like FtsZ, is capable of filamentous assembly [163,164], and ZipA are major contributors of the first stage of Z-ring assembly. They anchor FtsZ to the cytoplasmic membrane through their interactions with FtsZ's C-terminal peptide (CCTP) (reviewed by [165]). Both, ZipA and FtsA regulate divisome dynamics and recruit downstream divisome components to the Z-ring [166]. While FtsA is essential in *E. coli*, it can be deleted in *B. subtilis*, which results in filamentous cells that reveal a disturbed Z-ring formation [167]. Cyanobacteria lack ZipA, FtsA, ZapA and EzrA (a presumed Gram-positive-specific FtsZ membrane anchor [68]) homologs [86,148] and instead contain the cyanobacterial-specific protein ZipS (also known as Ftn6; [89]) and the cyanobacterial and plant-specific protein ZipN (also known as Ftn2) [92,132] (for a current model of the *Anabaena* divisome see Figure 2). Additionally, cyanobacteria possess SepF (also termed YlmF or Cdv2 [86]), a protein otherwise restricted to Gram-positive bacteria [168]. ZipN, ZipS and SepF all localize to the midcell in *Synechocystis* [88,90] suggesting that all three are important factors of the cell division machinery. They are, however, characterized by different levels of essentiality, depending on the respective cyanobacterial morphotype. *sepF* is essential in *Synechococcus* and *Synechocystis* [86,90], *zipN* is dispensable for *Synechococcus* but is essential in *Synechocystis* and *Anabaena* [88,89,132], whereas *zipS* can be deleted in *Synechococcus* and *Anabaena* but not in *Synechocystis* [89]. This inconsistency could reflect adaptations of the specific proteins to the respective morphology of its host.

SepF from *B. subtilis* is recruited early to the Z-ring and functions as a specific FtsZ membrane anchor, regulating the late septum formation [141,168]. In *E. coli*, FtsA or ZipA alone are sufficient to establish Z-ring anchorage to the cytoplasmic membrane, which is only lost upon simultaneous deletion of both [166]. The situation seems to be quite a bit more complex in cyanobacteria, possibly a result of the hybrid Gram phenotype and the morphological diversity. ZipN and its plant homolog ARC6 [169] contain a C-terminal transmembrane domain potentially suitable for membrane attachment [92,132], reminiscent of the amphipathic helix that mediates membrane localization of FtsA [165]. Furthermore, ZipN homologs contain a chaperone-like N-terminal DnaJ domain and a tetratricopeptide repeat (TPR) domain, suggesting that ZipN could function in mediating protein-protein interactions and/or affect protein folding [89]. Indeed, ZipN interacts with FtsZ in vitro and localizes to the Z-ring in *Synechocystis* and *Anabaena* (see Figure 2) [112,133], which is likely mediated by the DnaJ domain of ZipN, as removal of this domain results in diffuse cytoplasmic GFP-ZipN signals [112]. Reminiscent of FtsAs' function in *E. coli*, ZipN is able to self-interact and functions as a *de novo* anchor of FtsZ to the cytoplasmic membrane in cyanobacteria [92,132]. Similar to *E. coli* FtsA, *B. subtilis* SepF assembles into round protein filaments and associates and bundles with FtsZ filaments in vitro [170]. *Synechocystis* SepF and ZipS directly interact with FtsZ filaments in vitro but only SepF is able to stimulate the assembly of FtsZ filaments [90]. Based on these observations, it was suggested that ZipS functions downstream of SepF, i.e., after the Z-ring is functionally assembled [90].

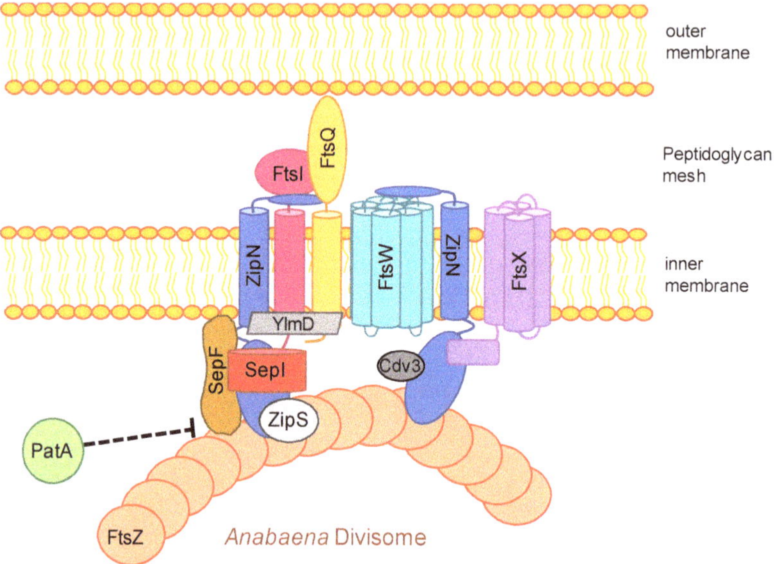

Figure 2. Proposed model of the *Anabaena* divisome. Proteins in grey shades are inferred from a previous model described for *Synechocystis* [92]. PatA is assumed to negatively interfere with the linkage of FtsZ to the cytoplasmic membrane through the loss of interaction with its presumed cytoplasmic membrane anchors SepF and ZipN.

A direct function of SepF or ZipS in FtsZ membrane anchoring has not yet been described and depletion of *sepF* and *zipS* did not affect Z-ring formation but considerably altered the Z-ring structure and delayed cytokinesis, leading to swollen *Synechocystis* cells [90] (Figure 1). In the coccoid *Synechocystis*, *zipN* is required for normal cytokinesis and a *zipN*-depleted strain formed minicells or spiral-shaped cells [62] (see also Figure 1). Deletion of *zipN* and *zipS* but not depletion of *sepF* abrogates Z-ring formation in *Synechococcus*, leading to a patchy and diffuse pattern of FtsZ at the septum site [86], suggesting functional differences of *zipS* and *zipN* between the rod-shaped *Synechococcus* and the coccoid *Synechocystis*. Similar to the filamentous *B. subtilis sepF* mutant [168], rod-shaped *Synechococcus* cells deleted of *zipN* or *zipS* or depleted of *sepF* became filamentous or elongated [86,89]. Filamentous *Synechococcus zipN* or *zipS* mutants divided irregularly and can be up to 100 or 20 times longer than wild type *Synechococcus*, respectively [89]. Using light microscopy, both mutants appeared normal, however, ultrastructural analysis using scanning and transmission electron microscopy discovered that they are characterized by irregular cell bending and spiralization and have a decreased cell wall rigidity that is not a result of a PG layer defect [134]. None of this, however, affected the growth rate of the *zipN* and *zipS Synechococcus* mutants [89]. In the multicellular, ellipsoid-shaped *Anabaena*, *zipN* is essential while *zipS* can be deleted and both depletion of *zipN* or deletion of *zipS* lead to aberrant elongated and swollen *Anabaena* cells [89,132]. Thus, it seems that in those strains cell division but not cell growth is impaired. Confirming this, FtsZ was found to localize in a patchy and delocalized pattern around the cell division septa in the *zipN* and *zipS Synechococcus* mutants [86]. Likewise, depletion of *zipN* leads to a delocalization of FtsZ and a loss of Z-ring formation in *Anabaena*, implying a dysfunctional Z-ring assembly in strains lacking *zipN* [132].

The observation of swollen (*Synechocystis*, *Anabaena*) or filamentous (*Synechococcus*, *Anabaena* and *E. coli*) cells for the respective cell division mutants (for a depiction of several cyanobacterial cell division and cell shape mutant phenotypes see Figure 1) is considered to occur when cell septation is slowed down (or impaired) in relation to cell growth (i.e., lateral PG insertion) [4,17]. This idea is supported by a study that analyzed the proteome of filamentous *Synechococcus* cells deleted of *zipN* or *zipS*, which detected an upregulation of proteins involved in nucleotide biosynthesis like *dnaN* (DNA polymerase III beta subunit) and cell growth, including *mreB* [117]. As a result of the upregulation of cell-cycle-specific genes, the authors conclude that ZipN and ZipS likely act in a stage prior to cell division [117], which correlates with their occurrence early at the Z-ring during cell division in *Synechocystis* [88,90]. Notably, carboxysome-associated genes were also differentially expressed in the *Synechococcus zipN* and *zipS* mutants [117], being in concert with the observed decreased carboxysome count and the appearance of abnormal carboxysome-like structures not present in the wild type [134]. Consequently, both ZipN and ZipS have pleiotropic functions besides cell division and could be involved in carbon fixation. Given that carboxysome-segregation is dependent on McdA/B [171] and carboxysome subunit expression is affected by the deletion of *zipS* and *zipN* [117], a functional relationship between the cell division apparatus (i.e., divisome) and the carboxysome-segregation mechanism is possible and worth future investigation. More recently, ZipN was also attributed a function in cell differentiation in *Anabaena*. It was shown that ZipN protein levels, similar to FtsZ levels, are downregulated during heterocyst development, albeit at an earlier stage of heterocyst development (i.e., pre-heterocysts) [133]. Based on the interaction of PatA with ZipN (and SepF), the authors hypothesize that the initial binding of PatA to ZipN leads to a destabilization and loss of ZipN early during heterocyst-formation, followed by a destabilization and loss of FtsZ due to the lack of its membrane anchor (i.e., ZipN) in mature heterocysts. The subsequent downregulation of *ftsZ* transcription then seals the fate for the irreversible cell differentiation into heterocysts [133].

Highlighting its essential role for cyanobacterial viability and morphology, ZipN was found to interact with ZipS, SepF, FtsI and FtsQ in *Synechocystis* and *Anabaena* [92,112,132,172] with FtsK, FtsW, SepJ and SepI specifically in *Anabaena* [87,132,172] and with Cdv3 specifically in *Synechocystis* [92] (Figure 2). A more condensed summary of this and other known interaction networks of morphological determinants in *Synechocystis* and *Anabaena* is also given in Figure 3. Not much is currently known about the interaction profile of SepF and ZipS in cyanobacteria and unlike the FtsZ membrane-tethering function of ZipN, the precise function of SepF and ZipS remains to be elucidated. However, given that ZipS contains an N-terminal DnaD-like domain, which is involved in DNA binding, it could putatively act to bridge DNA replication with cell division in cyanobacteria [173]. It will be interesting to see whether this assumed function might provide the functional basis explaining the lack of a nucleoid occlusion system (explained in more detail below) in cyanobacteria.

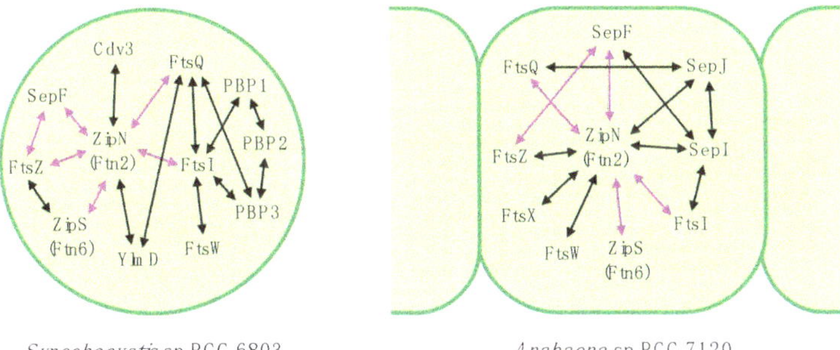

Figure 3. Interaction network of cell division/growth proteins in *Synechocystis* and *Anabaena*. Depiction of protein-protein interactions as identified by bacterial adenylate cyclase two hybrid assays and co-immunoprecipitation experiments. Interactions were identified in [87,88,90,92,132,172]. Black arrows indicate interactions solely found in one species so far, while purple arrows mark interactions found in both *Synechocystis* and *Anabaena*. Interactions only attributed to one species do not necessarily imply these interactions do not exist in the other but rather that these interactions were not yet tested for. Image created with BioRender.com.

3.6. The Divisome is Linked to the Sites of Cell-cell Connections in ANABAENA

Connection of multicellular cyanobacteria, including *Anabaena*, is, in part, mediated by an incomplete cell division and likely also through cell-cell-joining structures between neighboring cells (reviewed in [28,29,60]). The latter is indicated by a trichome fragmentation phenotype of mutants of the septal-localized proteins SepJ and SepI or septal junction proteins FraC and FraD [114,115,172]. Besides septal localization, SepJ, SepI and FraC additionally localize to the midcell in rings, reminiscent of the Z-ring [105,114,115,172,174,175], giving rise to the compelling connection between the cell division apparatus and the cell septa in *Anabaena*. This idea was followed up by numerous protein-protein interaction and fluorescent protein localization studies that revealed that, although SepJ does not interact with FtsZ, it does bind the cell division protein FtsQ [87], a bitopic membrane protein that putatively links the periplasmic to the cytoplasmic divisome proteins [176]. SepI also interacts with SepJ and with FtsI, SepF but not with FtsZ, ZipS, FraC, FraD, MreB, MinC and other Fts proteins (Figure 1) [172]. Unlike in unicellular bacteria, multicellularity in *Anabaena* is achieved through incomplete cleavage of the septal PG [177], hinting for a modified function and composition of the *Anabaena* divisome [87]. Consistent with a function of the divisome in septal junction integrity and thus *Anabaena* multicellularity, SepJ interacts with FtsQ and ZipN and its septal localization was largely lost in *ftsZ* and *zipN* depleted *Anabaena* strains, being mostly dispersed in patches [87,132]. Considering that FtsQ recruits numerous proteins to the *E. coli* divisome [178] and that ZipN interacts with an extensive set of divisome proteins [132], it is conceivable that both FtsQ and ZipN recruit SepJ to the *Anabaena* divisome, which then remains in the septa upon completed cell division [28,87]. Deletion of *sepI*, which, like *sepJ*, also functions in *Anabaena* multicellularity, nanopore formation and cell-cell communication did not alter Z-ring placement and only mildly affected cell shape and growth. These results suggest that SepI is a late divisome protein in *Anabaena* and its function is rather associated with septum integrity than divisome function [172]. Notably, and in contrast to the other characterized *Anabaena* divisome proteins, SepI was found to affect the colony morphology on agar plates. The implications of this for the growth of *Anabaena* in its natural habitat remain elusive.

3.7. Z-Ring—All in One Place

There are several main factors that function as negative regulators of and restrict Z-ring formation to the correct midcell placement [179]. They include the nucleoid occlusion (NO) system, consisting of SlmA in *E. coli* and Noc in *B. subtilis*, two chromosome-associated proteins that prevent Z-ring formation at sites occupied by DNA. Cyanobacteria lack SlmA and Noc homologs [17] and in *Synechococcus*, Z-rings were identified at sites occupied by chromosomal DNA, suggesting that cyanobacteria lack NO systems [86]. MipZ from *C. crescentus* belongs to the ParA/MinD family of ATPases and inhibits FtsZ assembly near the cell poles and the nucleoid through its association with the DNA-binding protein ParB [180]. Using bioinformatic searches, we could not identify MipZ homologs in *Anabaena*, *Synechococcus* and *Synechocystis*, consistent with the lack of a classical ParA/B/S-based DNA-segregation mechanism in cyanobacteria [83,171]. SulA, a FtsZ antagonist that sequesters FtsZ monomers and prevents Z-ring formation [181,182], is part of the SOS response as a reaction to DNA damage [179]. Homologs to SulA were identified in cyanobacteria ([86]. see also Table 1) and SulA was found to be essential in *Synechocystis*. Partial deletion of *sulA* resulted in cell division defects, prevented proper daughter cell segregation and led to cloverleaf-like cell aggregates [183]. *Anabaena* cells overexpressing *E. coli* but not *Anabaena* SulA became elongated, showed diminished Z-ring formations (like in *E. coli*), did not divide and, in accordance with an essential role of cell division for heterocyst-formation, did not differentiate heterocysts under nitrogen-deprived growth conditions [152]. The inability to differentiate heterocysts likely stems from the observation that *ftsZ* is downregulated 24 h after nitrogen-deprivation while *sulA* is upregulated. This argues for an additive effect to limit cell division through less FtsZ protein, which is additionally sequestered by SulA to prevent Z-ring formation [154]. An inhibition of Z-ring formation is likely also promoted through inhibition of FtsZ's GTPase activity by SulA [152], which is essential for FtsZ filament-formation [184]. The Min system is the fourth known system that restricts the Z-ring to the midcell and works to prevent aberrant cell division planes in *E. coli* and *B. subtilis*. In *E. coli*, the Min system consists of three major proteins, MinC, MinD and MinE. MinC is the mechanistic antagonist to FtsZ polymerization through its interaction with the GTPase domain in FtsZ and concurrently with FtsZ's CCTP domain, thus competing with FtsA to prevent membrane anchorage of FtsZ. MinC is recruited to the plasma membrane by its interaction with MinD, a Walker A-type ATPase and member of the ParA/MinD family. MinE, in the form of a MinE ring, associates to the membrane bound MinCD, causing its detachment from the membrane and giving rise to a spatiotemporal dynamic pole-to-pole oscillation of MinCDE, which is highest at the cell poles and lowest at the midcell. This gradient of the FtsZ inhibitor MinC ultimately restricts Z-ring formation at the correct midcell location (reviewed in [185]). *B. subtilis* does not contain a MinE homolog and instead possesses a coiled-coil protein called DivIVA, which localizes to areas of negative curvature—the cell poles or the division site of constricting cells—and recruits MinCD through a linker protein called MinJ. Consequently, no Min oscillation exists in the Gram-positive *B. subtilis* but Z-ring formation is statically inhibited by MinCD and DivIVA (reviewed in [68]). So far, the functional properties of the Min system were only elucidated in *Synechococcus* that, besides MinCDE, also contains a DivIVA-like protein called Cdv3. While one report suggested that Cdv3 is essential for *Synechococcus* [86], another report more recently reported the complete deletion of *cdv3* [91]. Although, the lack of functional domains essential for DivIVA function (e.g., membrane curvature sensing) in Cdv3 suggests that it is likely not a direct homolog of DivIVA [91]. Nonetheless, Cdv3 homologs, like DivIVA, are absent from other Gram-negative bacteria [86]. Given that *Synechococcus* contains MinCDE, essential for pole-to-pole oscillation, but also contains a sophisticated thylakoid membrane system that could potentially inhibit oscillation, it was unknown whether pole-to-pole oscillation can be recapitulated in cyanobacteria. MacCready and colleagues, however, elegantly modelled the existence of Min system oscillation in *Synechococcus* under the prerequisite that thylakoid membranes have a minimal permeability. They further showed robust Min system oscillation that spatiotemporally restricts Z-ring placement to the midcell in *Synechococcus*, demonstrating that the Min system can differentiate between the cytoplasmic membrane and the thylakoid systems [91]. Notably, they found two different modes that utilizes MinC's ability to inhibit FtsZ polymerization: one dynamic, *E. coli*-like mode that employs dynamic MinC distribution through MinDE oscillation, although with a longer periodicity, and another, *B. subtilis*-like static mode in which Cdv3 and MinD recruit MinC rings adjacent to the Z-ring at the midcell position [91]. *Synechococcus* MinD is highly conserved and its C-terminal amphipathic helix but not the N-terminal ATPase domain is involved in membrane-targeting, while both are essential for MinD function [88]. MinE only shows low sequence similarity to *E. coli* MinE, and unlike in the *E. coli minB* operon, *minC* is not encoded together with *minDE* in *Synechococcus* [91]. Searching for MinD homologs, we identified one homolog in *Synechocystis* but two and three MinD homologs in *Synechococcus* and *Anabaena*, respectively (Table 1), raising the question of the function of the other MinD homologs. As in *E. coli*, *minE* is essential in *Synechococcus* [91], but non-essential in *Synechocystis* where deletion of *minE* has only a mild phenotype with rare minicell formations [88]. Although one study reported a fully segregated *Synechococcus minE* mutant with a 5′ inserted transposon, thus it is debatable whether *minE* is essential in *Synechococcus* [86]. Among the Min proteins, MinE likely functions as the essential regulator of Z-ring formation as *minE* depleted *Synechococcus* cells were filamentous and lacked clear Z-ring formation, which could still be observed in *minC* and *minD* knockout strains [91]. MinC overexpression induced cell enlargement (elongation in *Synechococcus* [91] and swelling in *Anabaena* [116]), and similar to SulA overexpression also halted cell division and cell differentiation in *Anabaena* [116], thus, attributing an important role of MinC in proper cell division and cell differentiation. Analogous to the situation in *E. coli*, *minC* and *minD* deletion causes defects in FtsZ placement resulting in a mixed population of minicells and elongated cells in *Synechococcus*. Analogous to this effect, a fully segregated *Synechococcus cdv3* mutant was also filamentous but did not form minicells [91]. In the coccoid *Synechocystis*, *minC*, *minD* and *minE* deletion strains did not enlarge in cell volume and instead became spiral-shaped (Δ*minC* Δ*minD*) or formed minicells (Δ*minC*,

Δ*minD* and Δ*minE*) [88], whereas depletion of *cdv3* produced giant cells [92] (Figure 1). This suggests that in *Synechocystis*, Cdv3 could be of more importance for the control of cell division (i.e., inhibition of FtsZ polymerization, indicated by swollen cells) than the MinCDE system. Hence, while the deletion of the MinCDE pathway is possible, the Cdv3/MinD pathway is essential for *Synechocystis*; the exact opposite of the situation in *Synechococcus*. The most apparent difference between those two species is cell shape (coccoid vs. rod-shaped), consequently the different MinC-driven modes of FtsZ antagonism could be of different importance for different cyanobacterial cell morphotypes. It will be interesting to analyze the effect of the Min-system in *Anabaena*, which somewhat displays a hybrid morphotype between coccoid and rod-shaped.

Similar to MinC, Cdv3 in *Synechococcus* localizes to the midcell in rings [91], likely through an interaction with ZipN [92]. Overexpression of MinC and Cdv3 leads to the formation of remarkably long cell filaments, which can reach near millimeter-length for Cdv3-overexpressing strains [91,186]. As overexpression of MinC and Cdv3 does not inhibit cell growth, the increased sedimentation rates of those strains are now being exploited to optimize biomass harvesting procedures in cyanobacterial biotechnology [186]. Interestingly, low-light conditions or extended culture period (i.e., stationary phase cultures) are also associated with cell elongation in *Synechococcus*, leading to Min system-controlled asymmetric cell divisions [158]. The Min system enforces asymmetric division in elongated cells but ensures symmetric division in short daughter cells [187]. Notably, elongated cells produce more progeny cells than shorter ones and could act as storage units to overcome unfavorable conditions [158]. Considering all that information, it becomes apparent that cyanobacteria employ numerous mechanisms to regulate faithful cell division and utilize systems previously described to be restricted to either Gram-positive or Gram-negative bacteria. Finally, bearing in mind that many more cyanobacterial morphotypes have not yet been studied, it is conceivable that other, cyanobacterial-specific mechanisms to control cell division are yet to be discovered. Moreover, although a multitude of interactions have been identified among cell division/growth-related proteins, many other interactions are likely yet to be identified, making room to further explore the cyanobacterial cell division processes.

4. Coiled-Coil-Rich Proteins in Cyanobacteria

Despite relatively poor sequence conservation [188], eukaryotic intermediate filament (IF) proteins, the third major class of eukaryotic cytoskeletal proteins [2], reveal a robust tripartite building plan. IF proteins consist of highly variable N- and C-termini that flank a central α-helical rod-domain of conserved size (Figure 4). The rod domain consists of different coil segments that mediate the assembly into the characteristic coiled-coil (CC) structures with other IF proteins through lateral and longitudinal association, ultimately forming long IFs with a diameter of 11 nm (reviewed by [189]). About two decades ago, a functional involvement of an IF-like bacterial coiled-coil-rich protein (CCRP) in cell shape was described in the curved Gram-negative bacterium *C. crescentus* [9]. Although Crescentin is generally considered to be the first discovered bacterial IF-like cytoskeletal protein [190,191], the TlpA protein from *Salmonella enterica* was previously described as a bacterial CCRP with IF-like functions [192]. Nonetheless, Crescentin remains the best studied bacterial IF-like protein and has been shown to be essential for the typical crescent-like shape of *C. crescentus*. It aligns at the inner cell curvature [9], possibly mechanically controlling PG biosynthesis through a functional and potentially direct association with the MreB cytoskeleton [190]. Crescentin, reminiscent of eukaryotic IF proteins, forms filaments in vitro with a width of approximately 10 nm [9]. Although revealing compelling structural and domain similarities to eukaryotic IF proteins (Figure 4), given its restricted distribution to only one identified organism, Crescentin is considered to be likely no direct homologue of eukaryotic IF proteins but could rather be acquired by *C. crescentus* by horizontal gene transfer or as a result of convergent evolution [193–195]. The convergent evolutionary theory is supported by the ongoing discoveries of unrelated but structurally similar bacterial CCRPs that reveal IF-like characteristics. These proteins were shown to be involved in numerous different cellular functions, including cell shape (RsmP; [196]), cellular rigidity and polar PG biosynthesis (FilP, Scy and DivIVA; [197–199]), chemotaxis (Scc; [200]), gliding motility (AglZ; [201]), swimming motility and cell shape (*Helicobacter pylori* Ccrps; [78,202]), reminiscent of their eukaryotic counterparts (reviewed for example by [3,203]).

Given their seemingly ubiquitous involvement in cell shape, we recently searched for cyanobacterial CCRPs [204] that could be functionally involved in the manifestation of the enormous morphological diversity in the Cyanobacteria phylum [19]. In this study, we found that CCRPs are more prevalent in multicellular filamentous cyanobacteria compared to unicellular species. A specific reduction in CCRP proportion was identified in the genomes of the marine Picocyanobacteria, which could coincide with their reduced genome sizes [205]. The intriguing observation of higher CCRP counts in more complex cyanobacteria could indicate that CCRPs, at least in part, are important for the establishment of sophisticated morphological features in cyanobacteria. In fact, several septal junction-associated proteins, which are essential for the multicellular phenotype in *Anabaena*, contain CC domains [28,172]. Using a streamlined approach to readily test several candidate CCRPs with a newly developed in vitro polymerization assay allowed us to detect four novel filament-forming CCRPs in cyanobacteria. In *Synechocystis*, Slr1301 (termed HmpF$_{Syn}$) is a homologous protein to HmpF from *Nostoc punctiforme* [206], which, similar to its homolog, was found to be involved in *Synechocystis* twitching motility (as also previously identified by [207]), possibly through its interaction with the pilus ATPase PilB [204]. Despite its high CC content, HmpF$_{Syn}$ did not assemble into IF-like polymers in vitro and in vivo, highlighting that the pure presence of many CC domains is not sufficient to predict IF-like properties. Another *Synechocystis* CCRP, Slr7083 is encoded on a plasmid (the large toxin-antitoxin plasmid pSYSA) similar to TlpA from *Salmonella enterica* [192]. In contrast to HmpF$_{Syn}$, Slr7083 assembles into a honeycomb-like web of protein filaments in vitro and localizes circumferentially to the cell envelope. Slr7083 also affected the cellular motility of *Synechocystis* (although to a lesser extent) and as it directly interacted with HmpF$_{Syn}$, both CCRPs could be involved in cellular motility [204], reminiscent of the *H. pylori* CCRPs that regulate swimming motility [202] and AglZ, which is involved in gliding motility [201]. We also showed that a protein specific to multicellular cyanobacteria, Fm7001, polymerizes into extremely stable filamentous sheets at 4.5 M urea, a concentration where the eukaryotic IF protein vimentin only exist as tetramers [208]. This incredibly strong self-association capacity could function in the manifestation and stabilization of the *F. muscicula* trichome phenotype [204]. Interestingly, we also showed for the first time that a bacterial tetratricopeptide repeat (TPR) protein, All4981 from *Anabaena*, assembles into filamentous structures in vivo and in vitro, while interacting with a number of S-layer proteins. Notably, no S-layer has been detected in *Anabaena* [209], and a deletion of *all4981* could not be obtained, hampering a functional dissection of All4981. In our study [204], we further found two *Synechococcus* CCRPs of which Synpcc7942_1139 (HmpF$_{Syc}$), a *Synechococcus* homolog to HmpF, is essential and has a severe impact on colony morphology, a novel trait of prokaryotic CCRPs. This essential property is in contrast to the non-essentiality of HmpF and HmpF$_{Syn}$, suggesting specific functional adaptations to *Synechococcus*. Both mutants of *hmpF$_{Syc}$* and *synpcc7942_2039* (hereafter *syc2039*) resulted in an elongated cell morphotype, reminiscent of other cell division genes previously identified in *Synechococcus*, including *ftsZ*, *ftn2* (*zipN*), *ftn6* (*zipS*), *cikA*, *cdv1*, *cdv2*, *cdv3*, *clpX*, and *minE* [86,91,210], indicating an impact of HmpF$_{Syc}$ and Syc2039 on cell division. However, the cell elongation effect was not as severe as in the *zipN* and *zipS* mutants [86]. While the localization of HmpF$_{Syc}$ was largely inconclusive, Syc2039-GFP formed spindle-like filamentous structures within several cyanobacterial strains and in *E. coli*, suggesting a strong self-sufficient assembly property. Nonetheless, in vitro Syc2039 filaments were not observed. Instead, Syc2039 seemed to rather be involved in DNA segregation as cells lacking syc2039 revealed an altered DNA distribution within the cell, reminiscent of *Synechococcus* cells treated with thiabendazole, a tubulin assembly inhibitor [145]. Although membrane association of bacterial CCRPs was described before (reviewed by [190]), no bacterial CCRP containing a transmembrane domain has been identified so far (reviewed by [12]). Consequently, the existing N-terminal TDM in Syc2039 further suggests that Syc2039 does not itself form filaments but rather associates with another filamentous system in bacteria [204]. Collectively, the myriad of different functional properties of cyanobacterial CCRPs, including cell and colony shape, cell division, motility, DNA segregation, and trichome integrity provide an initial foundation for future studies on the impact of CCRPs on the morphological and functional diversification in cyanobacteria. The employed in vitro polymerization assay using an unspecific NHS-fluorescein dye proved to be a valuable tool to conveniently detect polymerizing proteins in vitro (for a list of polymerizing cyanobacterial proteins see Figure 4) and could facilitate the identification of other filamentous proteins. We are currently working on the additional characterization of several *Anabaena* CCRPs and initial results indicate that some of those CCRPs, including ZicK and ZacK, could be involved in the stabilization of the linear trichome phenotype in *Anabaena*, extending the known impact of CCRPs from cell shape to trichome shape [211]. Noteworthy, ZicK and ZacK were also observed to be strictly interdependent to form heteropolymers in vitro and in vivo, describing a novel trait of bacterial CCRPs [211].

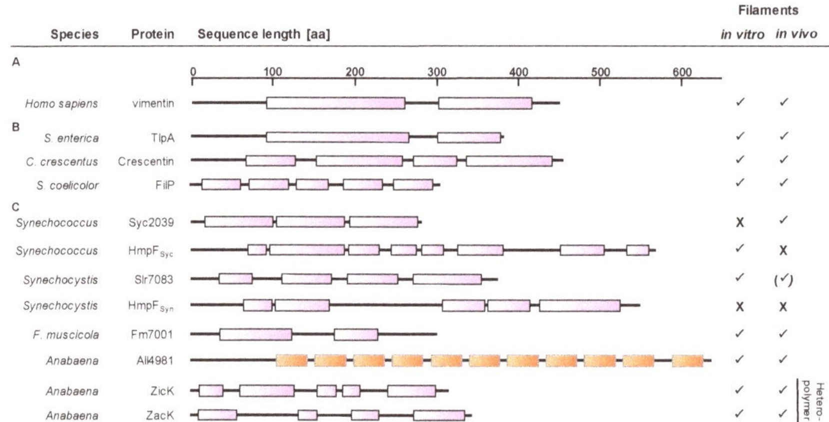

Figure 4. Bacterial coiled-coil-rich proteins. Depiction of coiled-coil-rich-regions (purple rectangles) in (A) the human vimentin, (B) previously described bacterial CCRPs, and (C) recently identified cyanobacterial CCRPs. Coiled-coil-rich regions were predicted using the COILS algorithm [212] or were obtained from [9,197,204,211]. Orange rectangles indicate TPR repeats that are also identified as coiled-coils by the COILS algorithm. Vimentin [188]; TlpA [192]; Crescentin [9]; FilP [197], Syc2039, HmpF$_{Syc}$, Slr7083, HmpF$_{Syn}$, Fm7001, All4981 [204]; ZicK, ZacK [211].

5. Undescribed Filamentous Systems in Cyanobacteria

Despite our analysis at the gene level, several conventional electron microscopy studies reported tubular, possibly cytoskeletal features in diverse cyanobacteria; however, they always lacked a clear identification of their protein composition. These observations can be divided into two subclasses: Microtubule-like structures (10–22 nm in diameter, length up to more than 1 µm) or thinner (3–8 nm in diameter) and less rigid microfilaments. Already in the late 1960s, a study observed microtubule-resembling, ~300 nm long structures with a diameter of 15 nm in an uncharacterized *Synechococcus* strain [213]. Thin sections of *Nostoc* strains revealed an even more intricate arrangement of tubules, consisting of an amorphous, ~1 µm long base plate parallel to the septum and numerous microtubule-like filaments perpendicular to it, protruding towards the cell center [214]. A similar complex could be visualized in *Anabaena* ([215] and reviewed in [216]); however, the described architecture shows strong similarities to phycobilisome arrays bound to thylakoid membranes, which were recently resolved in a native state in *Synechocystis* with cryo-electron tomography (cryoET) [217]. In *Anabaena*, it was further speculated that microtubular filaments could be important for the positioning of carboxysomes [215], whereas, the reported striated microtubules and sleeve bodies (Figure 5A) more resemble a membranous compartment or vesicles if re-analyzed with today's knowledge and to a lesser extent cytoskeletal features [215]. In contrast, the finding of tubular structures bound to the cytoplasmic membrane in two *Nostoc* strains is even more remarkable nowadays [218], as they show striking similarities to the phage tail-like apparatus of the bacterial type VI secretion system, which was structurally discovered more than 30 years later and has not yet been identified in cyanobacteria [219]. The second prominent observations, which were termed microfilaments, could be visualized only in *Anabaena* [215] and in *Cyanothece* [220]. These finer filaments were observed in all areas of the cytoplasm and similar findings were recently made in our lab (unpublished), after artefact free-thinning of frozen-hydrated *Anabaena* cells with cryo-focused ion beam (FIB) milling [221] and subsequent cryoET. Multiple 5 nm wide and >500 nm long filaments bundled up in the cytoplasm and a repetitive subunit every 5.5 nm was discernible (Figure 5B). No discrete anchoring towards a membrane was detectable, although one end often co-localized with a thylakoid membrane stack. In cross-sections, the filaments revealed a tight packing with a center to center spacing of 11 nm (Figure 5C). We could observe these filaments in ~2% of our tomograms ($n > 500$ tomograms), which does not reasonably allow the suggestion of a function. Nevertheless, these data show that the cyanobacterial cytoskeleton is not yet fully understood and an integrative, multi-scale approach, from molecular biology to near-native imaging techniques like cryoET, is crucial to elucidate its diverse functions.

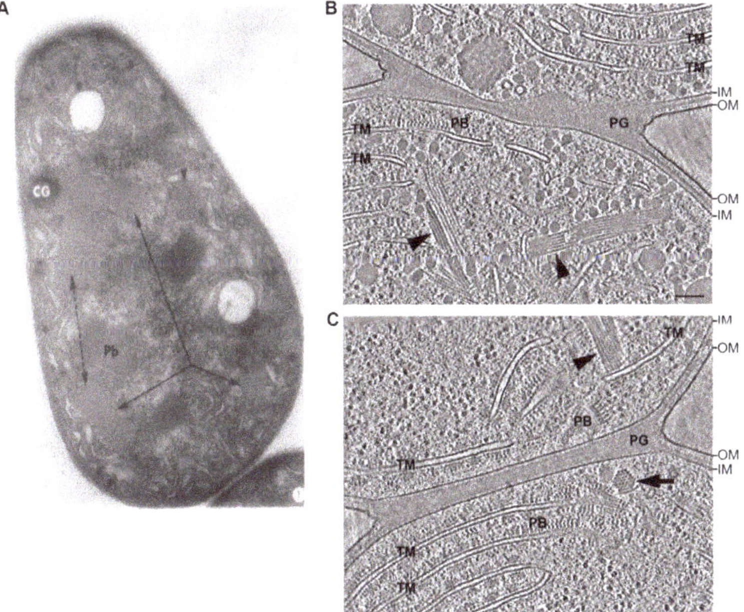

Figure 5. Electron micrographs of *Anabaena*. (**A**) Thin section of *Anabaena minutissima* showing striated microtubules (indicated by arrows). Pb, polyhedral bodies; CG, cyanophycin granule. (**B,C**) Cryo-electron tomograms of *Anabaena* showing uncharacterized intracellular filaments. (**B**) 5 nm wide filaments, with repetitive units every 5.5 nm, bundled together in the cytoplasm (arrowheads) and spanning parts of the cell. (**C**) Cross-section of these bundles revealed a tight packing with 11 nm spacing between filament centres (arrow). IM, inner membrane; OM, outer membrane; PB, phycobilisomes; PG, peptidoglycan; TM, thylakoid membrane. Bars, 100 nm. Shown are 13.5 nm thick slices. Figure 5A is reprinted from "The fine structure of striated microtubules and sleeve bodies in several species of *Anabaena*" [215], vol. 57, 1976, by Thomas E. Jensen and Robert P. Ayala with permission from Elsevier.

6. Conclusions and Future Perspectives

Recent scientific advances in the field of cyanobacterial research have started to unravel the mysteries and the evolution of cyanobacterial multicellularity, the cell–cell communication in multicellular cyanobacteria as well as cyanobacterial motility and provided growing insight into the molecular mechanisms that govern cell division and growth. This led to the insight that cyanobacteria are not just a mixture of Gram-positive and Gram-negative bacteria based on their cell wall characteristics but also because they harbor cell division genes specific to both Gram types and additionally possess cyanobacteria-specific cell division genes. Especially in multicellular cyanobacteria, a direct interplay between cell division processes and the establishment of cell–cell communication and ultimately multicellularity seems to exist. In the future, it will be of interest to analyze whether the cell shape-determining and MreB-driven elongasome is also linked to the incomplete cell separation process and intercellular communication in multicellular cyanobacteria. However, even apart from multicellular cyanobacteria and as a result of their mixed Gram phenotypes, it is intriguing to speculate that unicellular cyanobacteria have also evolved cell elongation processes and thus, elongasome functions in an alternative way to the well described model systems such as *E. coli* and *B. subtilis*. Nonetheless, given that several cell division processes appear to be cell shape-specific rather than phylum-specific, general statements about the functional properties of single proteins will likely remain restricted to the species or subsection level. The recent identification of cyanobacterial CCRPs with the property to form filament-like structures and their seemingly diverse cellular roles suggests that, at least in part, these proteins, like in other bacteria, could contribute to the special phenotype of cyanobacteria. The characterization of cell growth, cell division, and cytoskeletal processes in cyanobacteria has just begun and will likely provide us with unique insights into those fascinating bacteria. Technical advances like artefact-free sample thinning with cryo-FIB milling combined with correlative cryo-light microscopy/electron tomography will further allow in vivo visualization of these cytoskeletal features in a native state. Given their enormous ecological importance as primary inventors of oxygenic photosynthesis and their increasing importance due to the emerging climate change, cyanobacteria will likely receive more and more attention in the future that will also allow us to better understand their molecular circuits and consequently their unique adaptation strategies to the vast habitats that cyanobacteria populate.

Author Contributions: Conceptualization, B.L.S. and K.S.; Writing-Original Draft Preparation, B.L.S., D.J.N., G.L.W., M.P. and K.S.; Writing-Review & Editing, B.L.S., D.J.N., G.L.W., M.P. and K.S. All authors have read and agreed to the published version of the manuscript.

Funding: This research was funded by B.L.S. was supported by a "Maximizing Investigators' Research Award" (MIRA; No. R35GM136247) awarded to Ann Hochschild (Harvard Medical School, Boston, USA). D.J.N. is supported by the German Research Foundation (DFG; Deutsche Forschungsgemeinschaf) for an Emmy Noether project. K.S. was supported by the DFG (No. STU513/2–1) and ANID, ANID Fondecyt (No. 11170842). The APC was funded by the funding agencies of K.S., G.L.W. was supported by a Boehringer Ingelheim Fonds PhD fellowship. M.P. was supported by the Swiss National Science Foundation (#31003A_179255), the European Research Council (#679209) and the NOMIS foundation.

Conflicts of Interest: The authors declare no conflict of interest. The funders had no role in the design of the study; in the collection, analyses, or interpretation of data; in the writing of the manuscript, or in the decision to publish the results.

Data Availability: No new data were created or analyzed in this study. Data sharing is not applicable to this article.

References

1.   Moseley, J.B. An expanded view of the eukaryotic cytoskeleton. *Mol. Biol. Cell* **2013**, *24*, 1615–1618. [CrossRef] [PubMed]

2.   Alberts, B.; Johnson, A.; Lewis, J.; Morgan, D.; Raff, M.; Roberts, K.; Walter, P. *Molecular Biology of the Cell*, 6th ed.; Garland Science: New York, NY, USA, 2014.

3.   Huber, F.; Boire, A.; López, M.P.; Koenderink, G.H. Cytoskeletal crosstalk: When three different personalities team up. *Curr. Opin. Cell Biol.* **2015**, *32*, 39–47. [CrossRef]

4.   Bi, E.; Lutkenhaus, J. FtsZ ring structure associated with division in Escherichia coli. *Nature* **1991**, *354*, 161–164. [CrossRef] [PubMed]

5.   De Boer, P.; Crossley, R.; Rothfield, L. The essential bacterial cell-division protein FtsZ is a GTPase. *Nature* **1992**, *359*, 254–256. [CrossRef] [PubMed]

6.   Löwe, J.; Amos, L.A. Crystal structure of the bacterial cell division protein FtsZ. *Nature* **1998**, *391*, 203–206. [CrossRef] [PubMed]

7.   Bork, P.; Sander, C.; Valencia, A. An ATPase domain common to prokaryotic cell cycle proteins, sugar kinases, actin, and hsp70 heat shock proteins. *Proc. Natl. Acad. Sci. USA* **1992**, *89*, 7290–7294. [CrossRef] [PubMed]

8.   Van den Ent, F.; Amos, L.A.; Löwe, J. Prokrayotic origin of the actin cytoskeleton. *Nature* **2001**, *413*, 39–44. [CrossRef]

9.   Ausmees, N.; Kuhn, J.R.; Jacobs-Wagner, C. The bacterial cytoskeleton: An intermediate filament-like function in cell shape. *Cell* **2003**, *115*, 705–713. [CrossRef]

10.   Kühn, J.; Briegel, A.; Mörschel, E.; Kahnt, J.; Leser, K.; Wick, S.; Jensen, G.J.; Thanbichler, M. Bactofilins, a ubiquitous class of cytoskeletal proteins mediating polar localization of a cell wall synthase in Caulobacter crescentus. *EMBO J.* **2010**, *29*, 327–339. [CrossRef]

11.   Lin, L.; Thanbichler, M. Nucleotide-independent cytoskeletal scaffolds in bacteria. *Cytoskeleton* **2013**, *70*, 409–423. [CrossRef]

12.   Wagstaff, J.; Löwe, J. Prokaryotic cytoskeletons: Protein filaments organizing small cells. *Nat. Rev. Microbiol.* **2018**, *16*, 187–201. [CrossRef] [PubMed]

13.   Videau, P.; Rivers, O.S.; Ushijima, B.; Oshiro, R.T.; Kim, M.J.; Philmus, B.; Cozy, L.M. Mutation of the murC and murB genes impairs heterocyst differentiation in Anabaena sp. strain PCC 7120. *J. Bacteriol.* **2016**, *198*, 1–12. [CrossRef] [PubMed]

14.   Hoiczyk, E.; Baumeister, W. Envelope structure of four gliding filamentous cyanobacteria. *J. Bacteriol.* **1995**, *177*, 2387–2395. [CrossRef] [PubMed]

15.   Gumbart, J.C.; Beeby, M.; Jensen, G.J.; Roux, B. Escherichia coli peptidoglycan structure and mechanics as predicted by atomic-scale simulations. *PLoS Comput. Biol.* **2014**, *10*, e1003475. [CrossRef]

16.   Hoiczyk, E.; Hansel, A. Cyanobacterial cell walls: News from an unusual prokaryotic envelope. *J. Bacteriol.* **2000**, *182*, 1191–1199. [CrossRef]

17.   Cassier Chauvat, C.; Chauvat, F. Cell division in cyanobacteria. In *The Cell Biology of Cyanobacteria*; Flores, E., Herrero, A., Eds.; Caister Academic Press: Rover, UK, 2014.

18.   Schirrmeister, B.E.; Antonelli, A.; Bagheri, H.C. The origin of multicellularity in cyanobacteria. *BMC Evol. Biol.* **2011**, *11*, 45. [CrossRef]

19.   Rippka, R.; Stanier, R.Y.; Deruelles, J.; Herdman, M.; Waterbury, J.B. Generic assignments, strain histories and properties of pure cultures of cyanobacteria. *Microbiology* **1979**, *111*, 1–61. [CrossRef]

20.   Dagan, T.; Roettger, M.; Stucken, K.; Landan, G.; Koch, R.; Major, P.; Gould, S.B.; Goremykin, V.V.; Rippka, R.; De Marsac, N.T.; et al. Genomes of stigonematalean cyanobacteria (subsection V) and the evolution of oxygenic photosynthesis from prokaryotes to plastids. *Genome Biol. Evol.* **2013**, *5*, 31–44. [CrossRef]

21.   Shih, P.M.; Wu, D.; Latifi, A.; Axen, S.D.; Fewer, D.P.; Talla, E.; Calteau, A.; Cai, F.; Tandeau de Marsac, N.; Rippka, R.; et al. Improving the coverage of the cyanobacterial phylum using diversity-driven genome sequencing. *Proc. Natl. Acad. Sci. USA* **2013**, *110*, 1053–1058. [CrossRef]

22.   Koch, R.; Kupczok, A.; Stucken, K.; Ilhan, J.; Hammerschmidt, K.; Dagan, T. Plasticity first: Molecular signatures of a complex morphological trait in filamentous cyanobacteria. *BMC Evol. Biol.* **2017**, *17*, 1–11. [CrossRef]

23.   Weiss, G.L.; Kieninger, A.-K.; Maldener, I.; Forchhammer, K.; Pilhofer, M. Structure and function of a bacterial gap junction analog. *Cell* **2019**, *178*, 374–384.e15. [CrossRef] [PubMed]

24.   Nürnberg, D.J.; Mariscal, V.; Bornikoel, J.; Nieves-Morión, M.; Krauß, N.; Herrero, A.; Maldener, I.; Flores, E.; Mullineaux, C.W. Intercellular diffusion of a fluorescent sucrose analog via the septal junctions in a filamentous cyanobacterium. *MBio* **2015**, *6*, 1–12. [CrossRef] [PubMed]

25.   Kieninger, A.K.; Forchhammer, K.; Maldener, I. A nanopore array in the septal peptidoglycan hosts gated septal junctions for cell-cell communication in multicellular cyanobacteria. *Int. J. Med. Microbiol.* **2019**, *309*, 151303. [CrossRef] [PubMed]

26. Lehner, J.; Zhang, Y.; Berendt, S.; Rasse, T.M.; Forchhammer, K.; Maldener, I. The morphogene AmiC2 is pivotal for multicellular development in the cyanobacterium Nostoc punctiforme. *Mol. Microbiol.* **2011**, *79*, 1655–1669. [CrossRef] [PubMed]

27. Nayar, A.S.; Yamaura, H.; Rajagopalan, R.; Risser, D.D.; Callahan, S.M. FraG is necessary for filament integrity and heterocyst maturation in the cyanobacterium Anabaena sp. strain PCC 7120. *Microbiology* **2007**, *153*, 601–607. [CrossRef] [PubMed]

28. Herrero, A.; Stavans, J.; Flores, E. The multicellular nature of filamentous heterocyst-forming cyanobacteria. *FEMS Microbiol. Rev.* **2016**, *40*, 831–854. [CrossRef] [PubMed]

29. Flores, E.; Nieves-Morión, M.; Mullineaux, C. Cyanobacterial septal junctions: Properties and regulation. *Life* **2018**, *9*, 1. [CrossRef]

30. Flores, E.; Herrero, A. Compartmentalized function through cell differentiation in filamentous cyanobacteria. *Nat. Rev. Microbiol.* **2010**, *8*, 39–50. [CrossRef]

31. Gaysina, L.A.; Saraf, A.; Singh, P. Cyanobacteria in diverse habitats. In *Cyanobacteria—From Basic Science to Applications*; Mishra, A.K., Tiwari, D.N., Rai, A.N.B.T.-C., Eds.; Academic Press: Cambridge, MA, USA, 2019; pp. 1–28.

32. Pospíšil, P. Production of reactive oxygen species by photosystem II as a response to light and temperature stress. *Front. Plant Sci.* **2016**, *7*, 1950. [CrossRef]

33. Larkum, A.W.D.; Ritchie, R.J.; Raven, J.A. Living off the sun: Chlorophylls, bacteriochlorophylls and rhodopsins. *Photosynthetica* **2018**, *56*, 11–43. [CrossRef]

34. Wiltbank, L.B.; Kehoe, D.M. Diverse light responses of cyanobacteria mediated by phytochrome superfamily photoreceptors. *Nat. Rev. Microbiol.* **2019**, *17*, 37–50. [CrossRef] [PubMed]

35. Yerrapragada, S.; Shukla, A.; Hallsworth-Pepin, K.; Choi, K.; Wollam, A.; Clifton, S.; Qin, X.; Muzny, D.; Raghuraman, S.; Ashki, H.; et al. Extreme sensory complexity encoded in the 10-megabase draft genome Sequence of the chromatically acclimating cyanobacterium tolypothrix sp. PCC 7601. *Genome Announc.* **2015**, *3*, e00355-15. [CrossRef] [PubMed]

36. Bennett, A.; Bogorad, L. Complementary chromatic adaptation in a filamentous blue-green alga. *J. Cell Biol.* **1973**, *58*, 419. [CrossRef] [PubMed]

37. Sanfilippo, J.E.; Garczarek, L.; Partensky, F.; Kehoe, D.M. Chromatic acclimation in cyanobacteria: A diverse and widespread process for optimizing photosynthesis. *Annu. Rev. Microbiol.* **2019**, *73*, 407–433. [CrossRef] [PubMed]

38. Kehoe, D.M.; Grossman, A.R. Similarity of a chromatic adaptation sensor to phytochrome and ethylene receptors. *Science* **1996**, *273*, 1409–1412. [CrossRef]

39. Terauchi, K.; Montgomery, B.L.; Grossman, A.R.; Lagarias, J.C.; Kehoe, D.M. RcaE is a complementary chromatic adaptation photoreceptor required for green and red light responsiveness. *Mol. Microbiol.* **2004**, *51*, 567–577. [CrossRef]

40. Bordowitz, J.R.; Montgomery, B.L. Photoregulation of cellular morphology during complementary chromatic adaptation requires sensor-kinase-class protein RcaE in Fremyella diplosiphon. *J. Bacteriol.* **2008**, *190*, 4069–4074. [CrossRef]

41. Singh, S.P.; Montgomery, B.L. Morphogenes bolA and mreB mediate the photoregulation of cellular morphology during complementary chromatic acclimation in Fremyella diplosiphon. *Mol. Microbiol.* **2014**, *93*, 167–182. [CrossRef]

42. Wilde, A.; Mullineaux, C.W. Light-controlled motility in prokaryotes and the problem of directional light perception. *FEMS Microbiol. Rev.* **2017**, *41*, 900–922. [CrossRef]

43. Robinson, B.L.; Miller, J.H. Photomorphogenesis in the blue-green alga Nostoc commune 584. *Physiol. Plant.* **1970**, *23*, 461–472. [CrossRef]

44. Meeks, J.C.; Campbell, E.L.; Summers, M.L.; Wong, F.C. Cellular differentiation in the cyanobacterium Nostoc punctiforme. *Arch. Microbiol.* **2002**, *178*, 395–403. [CrossRef]

45. Damerval, T.; Guglielmi, G.; Houmard, J.; De Marsac, N.T. Hormogonium differentiation in the cyanobacterium Calothrix: A photoregulated developmental process. *Plant Cell* **1991**, *3*, 191–201. [CrossRef] [PubMed]

46. Zepu, Z.; Yuhuan, W.; Jie, X.; Lijie, F.; Dingji, S. Differentiation of hormogonia and photosynthetic characterization of Nostoc flagelliforme. *Acta Bot. Sin.* **2000**, *42*, 570–575.

47. Campbell, D.; Houmard, J.; De Marsac, N.T. Electron transport regulates cellular differentiation in the filamentous cyanobacterium Calothrix. *Plant Cell* **1993**, *5*, 451–463. [CrossRef] [PubMed]

48. Khamar, H.J.; Breathwaite, E.K.; Prasse, C.E.; Fraley, E.R.; Secor, C.R.; Chibane, F.L.; Elhai, J.; Chiu, W.-L. Multiple roles of soluble sugars in the establishment of Gunnera-Nostoc endosymbiosis. *Plant Physiol.* **2010**, *154*, 1381–1389. [CrossRef] [PubMed]

49. Meeks, J.C.; Elhai, J. Regulation of cellular differentiation in filamentous cyanobacteria in free-living and plant-associated symbiotic growth states. *Microbiol. Mol. Biol. Rev.* **2002**, *66*, 94–121. [CrossRef] [PubMed]

50. Liaimer, A.; Helfrich, E.J.N.; Hinrichs, K.; Guljamow, A.; Ishida, K.; Hertweck, C.; Dittmann, E. Nostopeptolide plays a governing role during cellular differentiation of the symbiotic cyanobacterium Nostoc punctiforme. *Proc. Natl. Acad. Sci. USA* **2015**, *112*, 1862–1867. [CrossRef] [PubMed]

51. Hashidoko, Y.; Nishizuka, H.; Tanaka, M.; Murata, K.; Murai, Y.; Hashimoto, M. Isolation and characterization of 1-palmitoyl-2-linoleoyl-sn-glycerol as a hormogonium-inducing factor (HIF) from the coralloid roots of Cycas revoluta (Cycadaceae). *Sci. Rep.* **2019**, *9*, 1–12. [CrossRef]

52. Nürnberg, D.J.; Mariscal, V.; Parker, J.; Mastroianni, G.; Flores, E.; Mullineaux, C.W. Branching and intercellular communication in the Section V cyanobacterium Mastigocladus laminosus, a complex multicellular prokaryote. *Mol. Microbiol.* **2014**, *91*, 935–949. [CrossRef]

53. Campbell, E.L.; Meeks, J.C. Characteristics of hormogonia formation by symbiotic Nostoc spp. in response to the presence of anthoceros punctatus or its extracellular products. *Appl. Environ. Microbiol.* **1989**, *55*, 125–131. [CrossRef]

54. Adams, D.G.; Bergman, B.; Nierzwicki-Bauer, S.A.; Duggan, P.S.; Rai, A.N.; Schüßler, A. Cyanobacterial-plant symbioses. In *The Prokaryotes: Prokaryotic Biology and Symbiotic Associations*; Rosenberg, E., DeLong, E.F., Lory, S., Stackebrandt, E., Thompson, F., Eds.; Springer: Berlin/Heidelberg, Germany, 2013; pp. 359–400.

55. Decelle, J.; Colin, S.; Foster, R.A. Photosymbiosis in marine planktonic protists. In *Marine Protists: Diversity and Dynamics*; Ohtsuka, S., Suzaki, T., Horiguchi, T., Suzuki, N., Not, F., Eds.; Springer: Tokyo, Japan, 2015; pp. 465–500.

56. Peters, G.A.; Perkins, S.K. The Azolla–Anabaena symbiosis: Endophyte continuity in the Azolla life-cycle is facilitated by epidermal trichomes. *New Phytol.* **1993**, *123*, 65–75. [CrossRef]

57. Caputo, A.; Nylander, J.A.A.; Foster, R.A. The genetic diversity and evolution of diatom-diazotroph associations highlights traits favoring symbiont integration. *FEMS Microbiol. Lett.* **2019**, *366*, fny297. [CrossRef] [PubMed]

58. Velázquez-Suárez, C.; Luque, I.; Herrero, A. The inorganic nutrient regime and the mre genes regulate cell and filament size and morphology in the phototrophic multicellular bacterium Anabaena. *MSphere* **2020**, *5*, e00747-20. [CrossRef] [PubMed]

59. Singh, S.P.; Montgomery, B.L. Determining cell shape: Adaptive regulation of cyanobacterial cellular differentiation and morphology. *Trends Microbiol.* **2011**, *19*, 278–285. [CrossRef] [PubMed]

60. Claessen, D.; Rozen, D.E.; Kuipers, O.P.; Søgaard-Andersen, L.; Van Wezel, G.P. Bacterial solutions to multicellularity: A tale of biofilms, filaments and fruiting bodies. *Nat. Rev. Microbiol.* **2014**, *12*, 115–124. [CrossRef] [PubMed]

61. Zhu, Z.; Piao, S.; Myneni, R.B.; Huang, M.; Zeng, Z.; Canadell, J.G.; Ciais, P.; Sitch, S.; Friedlingstein, P.; Arneth, A.; et al. Greening of the Earth and its drivers. *Nat. Clim. Chang.* **2016**, *6*, 791–795. [CrossRef]

62. Caccamo, P.D.; Brun, Y.V. The molecular basis of noncanonical bacterial morphology. *Trends Microbiol.* **2018**, *26*, 191–208. [CrossRef]

63. Lange, R.; Hengge-Aronis, R. Growth phase-regulated expression of bolA and morphology of stationary-phase Escherichia coli cells are controlled by the novel sigma factor σ(S). *J. Bacteriol.* **1991**, *173*, 4474–4481. [CrossRef]

64. Young, K.D. The selective value of bacterial shape. *Microbiol. Mol. Biol. Rev.* **2006**, *70*, 660–703. [CrossRef]

65. Typas, A.; Banzhaf, M.; Gross, C.A.; Vollmer, W. From the regulation of peptidoglycan synthesis to bacterial growth and morphology. *Nat. Rev. Microbiol.* **2012**, *10*, 123. [CrossRef]

66. Egan, A.J.F.; Errington, J.; Vollmer, W. Regulation of peptidoglycan synthesis and remodelling. *Nat. Rev. Microbiol.* **2020**, *18*, 446–460. [CrossRef] [PubMed]

67. Pinho, M.G.; Kjos, M.; Veening, J.W. How to get (a)round: Mechanisms controlling growth and division of coccoid bacteria. *Nat. Rev. Microbiol.* **2013**, *11*, 601–614. [CrossRef] [PubMed]

68. Errington, J.; Wu, L.J. Cell cycle machinery in bacillus subtilis. In *Prokaryotic Cytoskeletons: Filamentous Protein Polymers Active in the Cytoplasm of Bacterial and Archaeal Cells*; Löwe, J., Amos, L.A., Eds.; Springer: Berlin/Heidelberg, Germany, 2017; pp. 67–101.

69. Den Blaauwen, T.; De Pedro, M.A.; Nguyen-Disteche, M.; Ayala, J.A. Morphogenesis of rod-shaped sacculi. *FEMS Microbiol. Rev.* **2008**, *32*, 321–344. [CrossRef]

70. Jones, L.J.F.; Carballido-López, R.; Errington, J. Control of cell shape in bacteria: Helical, actin-like filaments in Bacillus subtilis. *Cell* **2001**, *104*, 913–922. [CrossRef]

71. Kruse, T.; Bork-Jensen, J.; Gerdes, K. The morphogenetic MreBCD proteins of Escherichia coli form an essential membrane-bound complex. *Mol. Microbiol.* **2005**, *55*, 78–89. [CrossRef]

72. Domínguez-Escobar, J.; Chastanet, A.; Crevenna, A.H.; Fromion, V.; Wedlich-Söldner, R.; Carballido-López, R. Processive movement of MreB-associated cell wall biosynthetic complexes in bacteria. *Science* **2011**, *333*, 225–228. [CrossRef]

73. Garner, E.C.; Bernard, R.; Wang, W.; Zhuang, X.; Rudner, D.Z.; Mitchison, T. Circumferential motions of the cell wall synthesis machinery drive cytoskeletal dynamics in B. subtilis. *Science* **2011**, *333*, 222–225. [CrossRef]

74. Van Teeffelen, S.; Wang, S.; Furchtgott, L.; Huang, K.C.; Wingreen, N.S.; Shaevitz, J.W.; Gitai, Z. The bacterial actin MreB rotates, and rotation depends on cell-wall assembly. *Proc. Natl. Acad. Sci. USA* **2011**, *108*, 15822-7. [CrossRef]

75. Siefert, J.L.; Fox, G.E. Phylogenetic mapping of bacterial morphology. *Microbiology* **1998**, *144*, 2803–2808. [CrossRef]

76. Yulo, P.R.J.; Hendrickson, H.L. The evolution of spherical cell shape; progress and perspective. *Biochem. Soc. Trans.* **2019**, *47*, 1621–1634. [CrossRef]

77. Springstein, B.L.; Weissenbach, J.; Koch, R.; Stücker, F.; Stucken, K. The role of the cytoskeletal proteins MreB and FtsZ in multicellular cyanobacteria. *FEBS Open Biol.* **2020**, *10*, 2510–2531. [CrossRef] [PubMed]

78. Waidner, B.; Specht, M.; Dempwolff, F.; Haeberer, K.; Schaetzle, S.; Speth, V.; Kist, M.; Graumann, P.L. A novel system of cytoskeletal elements in the human pathogen Helicobacter pylori. *PLoS Pathog.* **2009**, *5*, e1000669. [CrossRef] [PubMed]

79. Sycuro, L.K.; Pincus, Z.; Gutierrez, K.D.; Biboy, J.; Stern, C.A.; Vollmer, W.; Salama, N.R. Peptidoglycan crosslinking relaxation promotes helicobacter pylori's helical shape and stomach colonization. *Cell* **2010**, *141*, 822–833. [CrossRef]

80. Zhi, P.W.; Zhao, Y. Morphological reversion of Spirulina (Arthrospira) platensis (Cyanophyta): From linear to helical. *J. Phycol.* **2005**, *41*, 622–628.

81. Savage, D.F.; Afonso, B.; Chen, A.H.; Silver, P.A. Spatially ordered dynamics of the bacterial carbon fixation machinery. *Science* **2010**, *327*, 1258–1261. [CrossRef] [PubMed]

82. Moore, K.A.; Tay, J.W.; Cameron, J.C. Multi-generational analysis and manipulation of chromosomes in a polyploid cyanobacterium. *BioRxiv* **2019**, 661256. [CrossRef]

83. Watanabe, S.; Noda, A.; Ohbayashi, R.; Uchioke, K.; Kurihara, A.; Nakatake, S.; Morioka, S.; Kanesaki, Y.; Chibazakura, T.; Yoshikawa, H. ParA-like protein influences the distribution of multi-copy chromosomes in cyanobacterium Synechococcus elongatus PCC 7942. *Microbiology* **2018**, *164*, 45–56. [CrossRef] [PubMed]

84. Hu, B.; Yang, G.; Zhao, W.; Zhang, Y.; Zhao, J. MreB is important for cell shape but not for chromosome segregation of the filamentous cyanobacterium Anabaena sp. PCC 7120. *Mol. Microbiol.* **2007**, *63*, 1640–1652. [CrossRef]

85. Watanabe, S. Cyanobacterial multi-copy chromosomes and their replication. *Biosci. Biotechnol. Biochem.* **2020**, *84*, 1309–1321. [CrossRef]

86. Miyagishima, S.Y.; Wolk, P.P.; Osteryoung, K.W. Identification of cyanobacterial cell division genes by comparative and mutational analyses. *Mol. Microbiol.* **2005**, *56*, 126–143. [CrossRef]

87. Ramos-León, F.; Mariscal, V.; Frías, J.E.; Flores, E.; Herrero, A. Divisome-dependent subcellular localization of cell-cell joining protein SepJ in the filamentous cyanobacterium Anabaena. *Mol. Microbiol.* **2015**, *96*, 566–580. [CrossRef] [PubMed]

88. Mazouni, K.; Domain, F.; Cassier-Chauvat, C.; Chauvat, F. Molecular analysis of the key cytokinetic components of cyanobacteria: FtsZ, ZipN and MinCDE. *Mol. Microbiol.* **2004**, *52*, 1145–1158. [CrossRef] [PubMed]

89. Koksharova, O.A.; Wolk, C.P. A novel gene that bears a DnaJ motif influences cyanobacterial cell division. *J. Bacteriol.* **2002**, *184*, 5524–5528. [CrossRef] [PubMed]

90. Marbouty, M.; Saguez, C.; Cassier-Chauvat, C.; Chauvat, F. Characterization of the FtsZ-interacting septal proteins SepF and Ftn6 in the spherical-celled cyanobacterium Synechocystis strain PCC 6803. *J. Bacteriol.* **2009**, *191*, 6178–6185. [CrossRef]

91. MacCready, J.S.; Schossau, J.; Osteryoung, K.W.; Ducat, D.C. Robust min-system oscillation in the presence of internal photosynthetic membranes in cyanobacteria. *Mol. Microbiol.* **2017**, *103*, 483–503. [CrossRef]

92. Marbouty, M.; Saguez, C.; Cassier-Chauvat, C.; Chauvat, F. ZipN, an FtsA-like orchestrator of divisome assembly in the model cyanobacterium Synechocystis PCC6803. *Mol. Microbiol.* **2009**, *74*, 409–420. [CrossRef]

93. Stucken, K.; Ilhan, J.; Roettger, M.; Dagan, T.; Martin, W.F. Transformation and conjugal transfer of foreign genes into the filamentous multicellular cyanobacteria (subsection V) Fischerella and Chlorogloeopsis. *Curr. Microbiol.* **2012**, *65*, 552–560. [CrossRef]

94. Liu, P.; Zheng, H.; Meng, Q.; Terahara, N.; Gu, W.; Wang, S.; Zhao, G.; Nakane, D.; Wang, W.; Miyata, M. Chemotaxis without conventional two-component system, based on cell polarity and aerobic conditions in helicity-switching swimming of Spiroplasma eriocheiris. *Front. Microbiol.* **2017**, *8*, 1–13. [CrossRef]

95. Kürner, J.; Frangakis, A.S.; Baumeister, W. Cryo-electron tomography reveals the cytoskeletal structure of Spiroplasma melliferum. *Science* **2005**, *307*, 436–438. [CrossRef]

96. Ehlers, K.; Oster, G. On the mysterious propulsion of synechococcus. *PLoS ONE* **2012**, *7*, e36081. [CrossRef]

97. Campbell, E.L.; Summers, M.L.; Christman, H.; Martin, M.E.; Meeks, J.C. Global gene expression patterns of Nostoc punctiforme in steady-state dinitrogen-grown heterocyst-containing cultures and at single time points during the differentiation of akinetes and hormogonia. *J. Bacteriol.* **2007**, *189*, 5247–5256. [CrossRef] [PubMed]

98. Campbell, E.L.; Christman, H.; Meeks, J.C. DNA microarray comparisons of plant factor- and nitrogen deprivation-induced hormogonia reveal decision-making transcriptional regulation patterns in Nostoc punctiforme. *J. Bacteriol.* **2008**, *190*, 7382LP–7391LP. [CrossRef]

99. Kabeya, Y.; Nakanishi, H.; Suzuki, K.; Ichikawa, T.; Kondou, Y.; Matsui, M.; Miyagishima, S.-Y. The YlmG protein has a conserved function related to the distribution of nucleoids in chloroplasts and cyanobacteria. *BMC Plant Biol.* **2010**, *10*, 57. [CrossRef]

100. Zheng, X.; O'Shea, E.K. Cyanobacteria maintain constant protein concentration despite genome copy-number variation. *Cell Rep.* **2017**, *19*, 497–504. [CrossRef]

101. Burnat, M.; Schleiff, E.; Flores, E. Cell envelope components influencing filament length in the heterocyst-forming cyanobacterium Anabaena sp. strain PCC 7120. *J. Bacteriol.* **2014**, *196*, 4026–4035. [CrossRef] [PubMed]

102. Iyer, L.M.; Makarova, K.S.; Koonin, E.V.; Aravind, L. Comparative genomics of the FtsK-HerA superfamily of pumping ATPases: Implications for the origins of chromosome segregation, cell division and viral capsid packaging. *Nucleic Acids Res.* **2004**, *32*, 5260–5279. [CrossRef] [PubMed]

103. Mandakovic, D.; Trigo, C.; Andrade, D.; Riquelme, B.; Gómez-Lillo, G.; Soto-Liebe, K.; Díez, B.; Vásquez, M. CyDiv, a conserved and novel filamentous cyanobacterial cell division protein involved in septum localization. *Front. Microbiol.* **2016**, *7*, 1–11. [CrossRef] [PubMed]

104. Leganés, F.; Blanco-Rivero, A.; Fernández-Piñas, F.; Redondo, M.; Fernández-Valiente, E.; Fan, Q.; Lechno-Yossef, S.; Wolk, C.P. Wide variation in the cyanobacterial complement of presumptive penicillin-binding proteins. *Arch. Microbiol.* **2005**, *184*, 234–248. [CrossRef]

105. Berendt, S.; Lehner, J.; Zhang, Y.V.; Rasse, T.M.; Forchhammer, K.; Maldener, I. Cell wall amidase amiC1 is required for cellular communication and heterocyst development in the cyanobacterium Anabaena PCC 7120 but not for filament integrity. *J. Bacteriol.* **2012**, *194*, 5218–5227. [CrossRef]

106. Rexroth, S.; Mullineaux, C.W.; Ellinger, D.; Sendtko, E.; Rögner, M.; Koenig, F. The plasma membrane of the cyanobacterium Gloeobacter violaceus contains segregated bioenergetic domains. *Plant Cell* **2011**, *23*, 2379LP–2390. [CrossRef]

107. Mitschke, J.; Georg, J.; Scholz, I.; Sharma, C.M.; Dienst, D.; Bantscheff, J.; Voß, B.; Steglich, C.; Wilde, A.; Vogel, J.; et al. An experimentally anchored map of transcriptional start sites in the model cyanobacterium Synechocystis sp. PCC6803. *Proc. Natl. Acad. Sci. USA* **2011**, *108*, 2124LP–2129LP. [CrossRef] [PubMed]

108. Mitschke, J.; Vioque, A.; Haas, F.; Hess, W.R.; Muro-Pastor, A.M. Dynamics of transcriptional start site selection during nitrogen stress-induced cell differentiation in Anabaena sp. PCC7120. *Proc. Natl. Acad. Sci. USA* **2011**, *108*, 20130–20135. [CrossRef] [PubMed]

109. Flaherty, B.L.; Van Nieuwerburgh, F.; Head, S.R.; Golden, J.W. Directional RNA deep sequencing sheds new light on the transcriptional response of Anabaena sp. strain PCC 7120 to combined-nitrogen deprivation. *BMC Genom.* **2011**, *12*, 332. [CrossRef] [PubMed]

110. Stöckel, J.; Elvitigala, T.R.; Liberton, M.; Pakrasi, H.B. Carbon availability affects diurnally controlled processes and cell morphology of cyanothece 51142. *PLoS ONE* **2013**, *8*, 1–10. [CrossRef]

111. Khayatan, B.; Meeks, J.C.; Risser, D.D. Evidence that a modified type IV pilus-like system powers gliding motility and polysaccharide secretion in filamentous cyanobacteria. *Mol. Microbiol.* **2015**, *98*, 1021–1036. [CrossRef]

112. Marbouty, M.; Mazouni, K.; Saguez, C.; Cassier-Chauvat, C.; Chauvat, F. Characterization of the Synechocystis strain PCC 6803 penicillin-binding proteins and cytokinetic proteins FtsQ and FtsW and their network of interactions with ZipN. *J. Bacteriol.* **2009**, *191*, 5123–5133. [CrossRef]

113. Bornikoel, J.; Staiger, J.; Madlung, J.; Forchhammer, K.; Maldener, I. LytM factor Alr3353 affects filament morphology and cell–cell communication in the multicellular cyanobacterium Anabaena sp. PCC 7120. *Mol. Microbiol.* **2018**, *108*, 187–203. [CrossRef]

114. Flores, E.; Pernil, R.; Muro-Pastor, A.M.; Mariscal, V.; Maldener, I.; Lechno-Yossef, S.; Fan, Q.; Wolk, C.P.; Herrero, A. Septum-localized protein required for filament integrity and diazotrophy in the heterocyst-forming cyanobacterium Anabaena sp. strain PCC 7120. *J. Bacteriol.* **2007**, *189*, 3884–3890. [CrossRef]

115. Merino-Puerto, V.; Mariscal, V.; Mullineaux, C.W.; Herrero, A.; Flores, E. Fra proteins influencing filament integrity, diazotrophy and localization of septal protein SepJ in the heterocyst-forming cyanobacterium Anabaena sp. *Mol. Microbiol.* **2010**, *75*, 1159–1170. [CrossRef]

116. Sakr, S.; Thyssen, M.; Denis, M.; Zhang, C.C. Relationship among several key cell cycle events in the developmental cyanobacterium Anabaena sp. strain PCC 7120. *J. Bacteriol.* **2006**, *188*, 5958–5965. [CrossRef]

117. Koksharova, O.A.; Klint, J.; Rasmussen, U. Comparative proteomics of cell division mutants and wild-type of Synechococcus sp. strain PCC 7942. *Microbiology* **2007**, *153*, 2505–2517. [CrossRef] [PubMed]

118. Hernández-Muñiz, W.; Stevens, S.E. Characterization of the motile hormogonia of Mastigocladus laminosus. *J. Bacteriol.* **1987**, *169*, 218LP–223LP. [CrossRef]

119. González, A.; Fillat, M.F.; Bes, M.-T.; Peleato, M.-L.; Sevilla, E. The challenge of Iron stress in Cyanobacteria. In *Cyanobacteria*; IntechOpen: London, UK, 2018.

120. González, A.; Bes, M.T.; Valladares, A.; Peleato, M.L.; Fillat, M.F. FurA is the master regulator of iron homeostasis and modulates the expression of tetrapyrrole biosynthesis genes in A nabaena sp. PCC 7120. *Environ. Microbiol.* **2012**, *14*, 3175–3187. [CrossRef] [PubMed]

121. González, A.; Bes, M.T.; Barja, F.; Peleato, M.L.; Fillat, M.F. Overexpression of FurA in Anabaena sp. PCC 7120 reveals new targets for this regulator involved in photosynthesis, iron uptake and cellular morphology. *Plant Cell Physiol.* **2010**, *51*, 1900–1914. [CrossRef] [PubMed]

122. Hsu, Y.-P.; Rittichier, J.; Kuru, E.; Yablonowski, J.; Pasciak, E.; Tekkam, S.; Hall, E.; Murphy, B.; Lee, T.K.; Garner, E.C.; et al. Full color palette of fluorescent d-amino acids for in situ labeling of bacterial cell walls. *Chem. Sci.* **2017**, *8*, 6313–6321. [CrossRef] [PubMed]

123. Zhang, J.Y.; Lin, G.M.; Xing, W.Y.; Zhang, C.C. Diversity of growth patterns probed in live cyanobacterial cells using a fluorescent analog of a peptidoglycan precursor. *Front. Microbiol.* **2018**, *9*, 1–10. [CrossRef] [PubMed]

124. Laddomada, F.; Miyachiro, M.M.; Dessen, A. Structural insights into protein-protein interactions involved in bacterial cell wall biogenesis. *Antibiotics* **2016**, *5*, 14. [CrossRef]

125. Lázaro, S.; Fernández-Piñas, F.; Fernández-Valiente, E.; Blanco-Rivero, A.; Leganés, F. PbpB, a gene coding for a putative penicillin-binding protein, is required for aerobic nitrogen fixation in the cyanobacterium Anabaena sp. strain PCC7120. *J. Bacteriol.* **2001**, *183*, 628–636. [CrossRef]

126. Zhu, J.; Jäger, K.; Black, T.; Zarka, K.; Koksharova, O.; Wolk, C.P. HcwA, an autolysin, is required for heterocyst maturation in Anabaena sp. strain PCC 7120. *J. Bacteriol.* **2001**, *183*, 6841LP–6851LP. [CrossRef]

127. Fenton, A.K.; Gerdes, K. Direct interaction of FtsZ and MreB is required for septum synthesis and cell division in Escherichia coli. *EMBO J.* **2013**, *32*, 1953–1965. [CrossRef]

128. Osawa, M.; Anderson, D.E.; Erickson, H.P. Reconstitution of contractile FtsZ rings in liposomes. *Science* **2008**, *320*, 792–794. [CrossRef] [PubMed]

129. Osawa, M.; Anderson, D.E.; Erickson, H.P. Curved FtsZ protofilaments generate bending forces on liposome membranes. *EMBO J.* **2009**, *28*, 3476–3484. [CrossRef] [PubMed]

130. Lutkenhaus, J.; Du, S.E. coli Cell Cycle Machinery. In *Prokaryotic Cytoskeletons: Filamentous Protein Polymers Active in the Cytoplasm of Bacterial and Archaeal Cells*; Löwe, J., Amos, L.A., Eds.; Springer: Berlin/Heidelberg, Germany, 2017; pp. 27–65.

131. Den Blaauwen, T.; Hamoen, L.W.; Levin, P.A. The divisome at 25: The road ahead. *Curr. Opin. Microbiol.* **2017**, *36*, 85–94. [CrossRef] [PubMed]

132. Camargo, S.; Picossi, S.; Corrales-Guerrero, L.; Valladares, A.; Arévalo, S.; Herrero, A. ZipN is an essential FtsZ membrane tether and contributes to the septal localization of SepJ in the filamentous cyanobacterium Anabaena. *Sci. Rep.* **2019**, *9*, 1–15. [CrossRef] [PubMed]

133. Valladares, A.; Velázquez-Suárez, C.; Herrero, A. Interactions of PatA with the divisome during heterocyst differentiation in Anabaena. *mSphere* **2020**, *5*, e00188-20. [CrossRef] [PubMed]

134. Gorelova, O.A.; Baulina, O.I.; Rasmussen, U.; Koksharova, O.A. The pleiotropic effects of ftn2 and ftn6 mutations in cyanobacterium Synechococcus sp. PCC 7942: An ultrastructural study. *Protoplasma* **2013**, *250*, 931–942. [CrossRef]

135. Chen, C.; MacCready, J.S.; Ducat, D.C.; Osteryoung, K.W. The molecular machinery of chloroplast division. *Plant Physiol.* **2018**, *176*, 138–151. [CrossRef]

136. Yoshida, Y.; Mogi, Y.; TerBush, A.D.; Osteryoung, K.W. Chloroplast FtsZ assembles into a contractible ring via tubulin-like heteropolymerization. *Nat. Plants* **2016**, *2*, 16095. [CrossRef]

137. Miyagishima, S.; Nakamura, M.; Uzuka, A.; Era, A. FtsZ-less prokaryotic cell division as well as FtsZ- and dynamin-less chloroplast and non-photosynthetic plastid division. *Front. Plant Sci.* **2014**, *5*, 1–13. [CrossRef]

138. Zhang, C.-C.; Hugenin, S.; Friry, A.; Huguenin, S.; Friry, A. Analysis of genes encoding the cell division protein FtsZ and a glutathione synthetase homologue in the cyanobacterium Anabaena sp. PCC 7120. *Res. Microbiol.* **1995**, *146*, 445–455. [CrossRef]

139. Wang, N.; Bian, L.; Ma, X.; Meng, Y.; Chen, C.S.; Ur Rahman, M.; Zhang, T.; Li, Z.; Wang, P.; Chen, Y. Assembly properties of the bacterial tubulin homolog FtsZ from the cyanobacterium Synechocystis sp. PCC 6803. *J. Biol. Chem.* **2019**, *294*, 16309–16319. [CrossRef]

140. Corrales-Guerrero, L.; Camargo, S.; Valladares, A.; Picossi, S.; Luque, I.; Ochoa De Alda, J.A.G.; Herrero, A. FtsZ of filamentous, heterocyst-forming cyanobacteria has a conserved N-terminal peptide required for normal FtsZ polymerization and cell division. *Front. Microbiol.* **2018**, *9*, 1–20. [CrossRef] [PubMed]

141. Duman, R.; Ishikawa, S.; Celik, I.; Strahl, H.; Ogasawara, N.; Troc, P.; Löwe, J.; Hamoen, L.W. Structural and genetic analyses reveal the protein SepF as a new membrane anchor for the Z ring. *Proc. Natl. Acad. Sci. USA* **2013**, *110*, E4601–E4610. [CrossRef] [PubMed]

142. Erickson, H.P.; Anderson, D.E.; Osawa, M. FtsZ in bacterial cytokinesis: Cytoskeleton and force generator all in one. *Microbiol. Mol. Biol. Rev.* **2010**, *74*, 504–528. [CrossRef] [PubMed]

143. Liu, R.; Liu, Y.; Liu, S.; Wang, Y.; Li, K.; Li, N.; Xu, D.; Zeng, Q. Three-dimensional superresolution imaging of the FtsZ ring during cell division of the cyanobacterium prochlorococcus. *MBio* **2017**, *8*, 1–11. [CrossRef] [PubMed]

144. Abed, R.M.; Garcia-Pichel, F.; Hernández-Mariné, M. Polyphasic characterization of benthic, moderately halophilic, moderately thermophilic cyanobacteria with very thin trichomes and the proposal of Halomicronema excentricum gen. nov., sp. nov. *Arch. Microbiol.* **2002**, *177*, 361–370. [CrossRef]

145. Sarcina, M.; Mullineaux, C.W. Effects of tubulin assembly inhibitors on cell division in prokaryotes in vivo. *FEMS Microbiol. Lett.* **2000**, *191*, 25–29. [CrossRef]

146. Billi, D. Loss of topological relationships in a Pleurocapsalean cyanobacterium (Chroococcidiopsis sp.) with partially inactivatedftsZ. *Ann. Microbiol.* **2009**, *59*, 235. [CrossRef]

147. Mori, T.; Johnson, C.H. Independence of circadian timing from cell division in cyanobacteria. *J. Bacteriol.* **2001**, *183*, 2439–2444. [CrossRef]

148. Yoshida, T.; Maki, M.; Okamoto, H.; Hiroishi, S. Coordination of DNA replication and cell division in Cyanobacteria Microcystis aeruginosa. *FEMS Microbiol. Lett.* **2005**, *251*, 149–154. [CrossRef]

149. Esteves-Ferreira, A.A.; Inaba, M.; Obata, T.; Fort, A.; Fleming, G.T.A.; Araújo, W.L.; Fernie, A.R.; Sulpice, R. A novel mechanism, linked to cell density, largely controls cell division in synechocystis. *Plant Physiol.* **2017**, *174*, 2166–2182. [CrossRef] [PubMed]

150. Dong, G.; Yang, Q.; Wang, Q.; Kim, Y.-I.; Wood, T.L.; Osteryoung, K.W.; van Oudenaarden, A.; Golden, S.S. Elevated ATPase activity of KaiC applies a circadian checkpoint on cell division in Synechococcus elongatus. *Cell* **2010**, *140*, 529–539. [CrossRef] [PubMed]

151. Sandh, G.; El-Shehawy, R.; Díez, B.; Bergman, B. Temporal separation of cell division and diazotrophy in the marine diazotrophic cyanobacterium Trichodesmium erythraeum IMS101. *FEMS Microbiol. Lett.* **2009**, *295*, 281–288. [CrossRef] [PubMed]

152. Sakr, S.; Jeanjean, R.; Zhang, C.-C.; Arcondeguy, T. Inhibition of cell division suppresses heterocyst development in Anabaena sp. strain PCC 7120. *J. Bacteriol.* **2006**, *188*, 1396–1404. [CrossRef]

153. Klint, J.; Rasmussen, U.; Bergman, B. FtsZ may have dual roles in the filamentous cyanobacterium Nostoc/Anabaena sp. strain PCC 7120. *J. Plant Physiol.* **2007**, *164*, 11–18. [CrossRef]

154. Wang, Y.; Xu, X. Regulation by hetC of genes required for heterocyst differentiation and cell division in Anabaena sp. strain PCC 7120. *J. Bacteriol.* **2005**, *187*, 8489–8493. [CrossRef]

155. Kuhn, I.; Peng, L.; Bedu, S. Developmental regulation of the cell division protein FtsZ in Anabaena sp. strain PCC 7120, a cyanobacterium capable of terminal differentiation. *J. Bacteriol.* **2000**, *182*, 4640–4643. [CrossRef]

156. Golubić, S.; Hernández-Mariné, M.; Hoffmann, L. Developmental aspects of branching in filamentous Cyanophyta/cyanobacteria. *Algol. Stud. Für Hydrobiol. Suppl. Vol.* **1996**, *83*, 303–329. [CrossRef]

157. Lopes Pinto, F.; Erasmie, S.; Blikstad, C.; Lindblad, P.; Oliveira, P. FtsZ degradation in the cyanobacterium Anabaena sp. strain PCC 7120. *J. Plant Physiol.* **2011**, *168*, 1934–1942. [CrossRef]

158. Goclaw-Binder, H.; Sendersky, E.; Shimoni, E.; Kiss, V.; Reich, Z.; Perelman, A.; Schwarz, R. Nutrient-associated elongation and asymmetric division of the cyanobacterium Synechococcus PCC 7942. *Environ. Microbiol.* **2012**, *14*, 680–690. [CrossRef]

159. Yamauchi, Y.; Kaniya, Y.; Kaneko, Y.; Hihara, Y. Physiological roles of the cyAbrB transcriptional regulator pair Sll0822 and Sll0359 in synechocystis sp. strain PCC 6803. *J. Bacteriol.* **2011**, *193*, 3702–3709. [CrossRef] [PubMed]

160. He, D.; Xu, X. CalA, a cyAbrB protein, binds to the upstream region of ftsZ and is down-regulated in heterocysts in Anabaena sp. PCC 7120. *Arch. Microbiol.* **2010**, *192*, 461–469. [CrossRef] [PubMed]

161. Jain, I.H.; Vijayan, V.; O'Shea, E.K. Spatial ordering of chromosomes enhances the fidelity of chromosome partitioning in cyanobacteria. *Proc. Natl. Acad. Sci. USA* **2012**, *109*, 13638–13643. [CrossRef] [PubMed]

162. Chen, A.H.; Afonso, B.; Silver, P.A.; Savage, D.F. Spatial and temporal organization of chromosome duplication and segregation in the cyanobacterium synechococcus elongatus PCC 7942. *PLoS ONE* **2012**, *7*, 1–10. [CrossRef] [PubMed]

163. Szwedziak, P.; Wang, Q.; Freund, S.M.; Löwe, J. FtsA forms actin-like protofilaments. *EMBO J.* **2012**, *31*, 2249–2260. [CrossRef]

164. Loose, M.; Mitchison, T.J. The bacterial cell division proteins ftsA and ftsZ self-organize into dynamic cytoskeletal patterns. *Nat. Cell Biol.* **2014**, *16*, 38–46. [CrossRef]

165. Du, S.; Lutkenhaus, J. Assembly and activation of the Escherichia coli divisome. *Mol. Microbiol.* **2017**, *105*, 177–187. [CrossRef]

166. Pichoff, S.; Lutkenhaus, J. Unique and overlapping roles for ZipA and FtsA in septal ring assembly in Escherichia coli. *EMBO J.* **2002**, *21*, 685–693. [CrossRef]

167. Jensen, S.O.; Thompson, L.S.; Harry, E.J. Cell division in Bacillus subtilis: FtsZ and FtsA association is Z-ring independent, and FtsA is required for efficient midcell Z-ring assembly. *J. Bacteriol.* **2005**, *187*, 6536LP–6544LP. [CrossRef]

168. Hamoen, L.W.; Meile, J.-C.; De Jong, W.; Noirot, P.; Errington, J. SepF, a novel FtsZ-interacting protein required for a late step in cell division. *Mol. Microbiol.* **2006**, *59*, 989–999. [CrossRef]

169. Vitha, S.; Froehlich, J.E.; Koksharova, O.; Pyke, K.A.; van Erp, H.; Osteryoung, K.W. ARC6 is a J-domain plastid division protein and an evolutionary descendant of the cyanobacterial cell division protein Ftn2. *Plant Cell* **2003**, *15*, 1918–1933. [CrossRef] [PubMed]

170. Singh, J.K.; Makde, R.D.; Kumar, V.; Panda, D. SepF increases the assembly and bundling of FtsZ polymers and stabilizes FtsZ protofilaments by binding along its length. *J. Biol. Chem.* **2008**, *283*, 31116–31124. [CrossRef] [PubMed]

171. MacCready, J.S.; Hakim, P.; Young, E.J.; Hu, L.; Liu, J.; Osteryoung, K.W.; Vecchiarelli, A.G.; Ducat, D.C. Protein gradients on the nucleoid position the carbon-fixing organelles of cyanobacteria. *elife* **2018**, *7*, 1–33. [CrossRef] [PubMed]

172. Springstein, B.L.; Arévalo, S.; Helbig, A.O.; Herrero, A.; Stucken, K.; Flores, E.; Dagan, T. A novel septal protein of multicellular heterocystous cyanobacteria is associated with the divisome. *Mol. Microbiol.* **2020**, *113*, 1140–1154. [CrossRef] [PubMed]

173. Marbouty, M.; Saguez, C.; Chauvat, F. The cyanobacterial cell division factor Ftn6 contains an N-terminal DnaD-like domain. *BMC Struct. Biol.* **2009**, *9*, 1–7. [CrossRef]

174. Rudolf, M.; Tetik, N.; Ramos-León, F.; Flinner, N.; Ngo, G.; Stevanovic, M.; Burnat, M.; Pernil, R.; Flores, E.; Schleiff, E. The peptidoglycan-binding protein SjcF1 influences septal junction function and channel formation in the filamentous cyanobacterium Anabaena. *MBio* **2015**, *6*, e00376–15. [CrossRef]

175. Zheng, Z.; Omairi-Nasser, A.; Li, X.; Dong, C.; Lin, Y.; Haselkorn, R.; Zhao, J. An amidase is required for proper intercellular communication in the filamentous cyanobacterium Anabaena sp. PCC 7120. *Proc. Natl. Acad. Sci. USA* **2017**, *114*, E1405–E1412. [CrossRef]

176. D'Ulisse, V.; Fagioli, M.; Ghelardini, P.; Paolozzi, L. Three functional subdomains of the Escherichia coli FtsQ protein are involved in its interaction with the other division proteins. *Microbiology* **2007**, *153*, 124–138. [CrossRef]

177. Mariscal, V.; Nürnberg, D.J.; Herrero, A.; Mullineaux, C.W.; Flores, E. Overexpression of SepJ alters septal morphology and heterocyst pattern regulated by diffusible signals in Anabaena. *Mol. Microbiol.* **2016**, *101*, 968–981. [CrossRef]

178. Karimova, G.; Dautin, N.; Ladant, D. Interaction network among Escherichia coli membrane proteins involved in cell division as revealed by bacterial two-hybrid analysis. *J. Bacteriol.* **2005**, *187*, 2233LP–2243LP. [CrossRef]

179. Cho, H. The role of cytoskeletal elements in shaping bacterial cells. *J. Microbiol. Biotechnol.* **2015**, *25*, 307–316. [CrossRef] [PubMed]

180. Szwedziak, P.; Ghosal, D. FtsZ-ring architecture and its control by MinCD. In *Prokaryotic Cytoskeletons: Filamentous Protein Polymers Active in the Cytoplasm of Bacterial and Archaeal Cells*; Löwe, J., Amos, L.A., Eds.; Springer: Berlin/Heidelberg, Germany, 2017; pp. 213–244.

181. Mukherjee, A.; Cao, C.; Lutkenhaus, J. Inhibition of FtsZ polymerization by SulA, an inhibitor of septation in Escherichia coli. *Proc. Natl. Acad. Sci. USA* **1998**, *95*, 2885LP–2890LP. [CrossRef] [PubMed]

182. Bi, E.; Lutkenhaus, J. Cell division inhibitors SulA and MinCD prevent formation of the FtsZ ring. *J. Bacteriol.* **1993**, *175*, 1118LP–1125LP. [CrossRef] [PubMed]

183. Raynaud, C.; Cassier-Chauvat, C.; Perennes, C.; Bergounioux, C. An Arabidopsis homolog of the bacterial cell division inhibitor SulA is involved in plastid division. *Plant Cell* **2004**, *16*, 1801–1811. [CrossRef] [PubMed]

184. Mukherjee, A.; Lutkenhaus, J. Dynamic assembly of FtsZ regulated by GTP hydrolysis. *EMBO J.* **1998**, *17*, 462–469.e6. [CrossRef]

185. Murray, S.M.; Howard, M. Center finding in E. coli and the role of mathematical modeling: Past, present and future. *J. Mol. Biol.* **2019**, *431*, 928–938. [CrossRef]

186. Jordan, A.; Chandler, J.; MacCready, J.S.; Huang, J.; Osteryoung, K.W.; Ducat, D.C. Engineering cyanobacterial cell morphology for enhanced recovery and processing of biomass. *Appl. Environ. Microbiol.* **2017**, *83*, 1–13. [CrossRef]

187. Liao, Y.; Rust, M.J. The min oscillator defines sites of asymmetric cell division in cyanobacteria during stress recovery. *Cell Syst.* **2018**, *7*, 471–481.e6. [CrossRef]

188. Fuchs, E.; Weber, K. Intermediate Filaments: Structure, dynamics, function and disease. *Annu. Rev. Biochem.* **1994**, *63*, 345–382. [CrossRef]

189. Herrmann, H.; Aebi, U. Intermediate filaments: Structure and assembly. *Cold Spring Harb. Perspect. Biol.* **2016**, *8*, a018242. [CrossRef]

190. Sundararajan, K.; Goley, E.D. Cytoskeletal proteins in caulobacter crescentus: Spatial orchestrators of cell cycle progression, development, and cell shape. *Subcell. Biochem.* **2017**, *84*, 103–137. [PubMed]

191. Van Teeseling, M.C.F.; de Pedro, M.A.; Cava, F. Determinants of bacterial morphology: From fundamentals to possibilities for antimicrobial targeting. *Front. Microbiol.* **2017**, *8*, 1–18. [CrossRef] [PubMed]

192. Hurme, R.; Namork, E.; Nurmiaho-Lassila, E.L.; Rhen, M. Intermediate filament-like network formed in vitro by a bacterial coiled coil protein. *J. Biol. Chem.* **1994**, *269*, 10675–10682.

193. Kelemen, G.H. Intermediate filaments supporting cell shape and growth in bacteria. In *Prokaryotic Cytoskeletons: Filamentous Protein Polymers Active in the Cytoplasm of Bacterial and Archaeal Cells*; Löwe, J., Amos, L.A., Eds.; Springer: Berlin/Heidelberg, Germany, 2017; pp. 161–211.

194. Erickson, H.P. Evolution of the cytoskeleton. *Bioessays* **2007**, *29*, 668–677. [CrossRef]

195. Wickstead, B.; Gull, K. The evolution of the cytoskeleton. *J. Cell Biol.* **2011**, *194*, 513–525. [CrossRef]

196. Fiuza, M.; Letek, M.; Leiba, J.; Villadangos, A.F.; Vaquera, J.; Zanella-Cléon, I.; Mateos, L.M.; Molle, V.; Gil, J.A. Phosphorylation of a novel cytoskeletal protein (RsmP) regulates rod-shaped morphology in Corynebacterium glutamicum. *J. Biol. Chem.* **2010**, *285*, 29387–29397. [CrossRef]

197. Bagchi, S.; Tomenius, H.; Belova, L.M.; Ausmees, N. Intermediate filament-like proteins in bacteria and a cytoskeletal function in Streptomyces. *Mol. Microbiol.* **2008**, *70*, 1037–1050. [PubMed]

198. Holmes, N.A.; Walshaw, J.; Leggett, R.M.; Thibessard, A.; Dalton, K.A.; Gillespie, M.D.; Hemmings, A.M.; Gust, B.; Kelemen, G.H. Coiled-coil protein Scy is a key component of a multiprotein assembly controlling polarized growth in Streptomyces. *Proc. Natl. Acad. Sci. USA* **2013**, *110*, E397–E406. [CrossRef]

199. Fröjd, M.J.; Flärdh, K. Apical assemblies of intermediate filament-like protein FilP are highly dynamic and affect polar growth determinant DivIVA in Streptomyces venezuelae. *Mol. Microbiol.* **2019**, *112*, 47–61. [CrossRef]

200. England, P.; Bourhy, P.; Picardeau, M.; Saint Girons, I.; Mazouni, K.; Pehau-Arnaudet, G. The scc spirochetal coiled-coil protein forms helix-like filaments and binds to nucleic acids generating nucleoprotein structures. *J. Bacteriol.* **2005**, *188*, 469–476.

201. Yang, R.; Bartle, S.; Otto, R.; Rogers, M.; Plamann, L.; Hartzell, P.L.; Stassinopoulos, A. AglZ is a filament-forming coiled-coil protein required for adventurous gliding motility of Myxococcus xanthus. *J. Bacteriol.* **2004**, *186*, 6168–6178. [CrossRef] [PubMed]

202. Specht, M.; Schätzle, S.; Graumann, P.L.; Waidner, B. Helicobacter pylori possesses four coiled-coil-rich proteins that form extended filamentous structures and control cell shape and motility. *J. Bacteriol.* **2011**, *193*, 4523–4530. [CrossRef] [PubMed]

203. Köster, S.; Weitz, D.A.; Goldman, R.D.; Aebi, U.; Herrmann, H. Intermediate filament mechanics in vitro and in the cell: From coiled coils to filaments, fibers and networks. *Curr. Opin. Cell Biol.* **2015**, *32*, 82–91. [CrossRef] [PubMed]

204. Springstein, B.L.; Woehle, C.; Weissenbach, J.; Helbig, A.O.; Dagan, T.; Stucken, K. Identification and characterization of novel filament-forming proteins in cyanobacteria. *Sci. Rep.* **2020**, *10*, 1894. [CrossRef]

205. Scanlan, D.J.; Ostrowski, M.; Mazard, S.; Dufresne, A.; Garczarek, L.; Hess, W.R.; Post, A.F.; Hagemann, M.; Paulsen, I.; Partensky, F. Ecological genomics of marine picocyanobacteria. *Microbiol. Mol. Biol. Rev.* **2009**, *73*, 249–299. [CrossRef]

206. Cho, Y.W.; Gonzales, A.; Harwood, T.V.; Huynh, J.; Hwang, Y.; Park, J.S.; Trieu, A.Q.; Italia, P.; Pallipuram, V.K.; Risser, D.D. Dynamic localization of HmpF regulates type IV pilus activity and directional motility in the filamentous cyanobacterium Nostoc punctiforme. *Mol. Microbiol.* **2017**, *106*, 252–265. [CrossRef]

207. Bhaya, D.; Takahashi, A.; Shahi, P.; Arthur, R. Novel motility mutants of synechocystis strain PCC 6803 generated by in vitro transposon mutagenesis. *J. Bacteriol.* **2001**, *183*, 1–5. [CrossRef]

208. Herrmann, H.; Aebi, U. Intermediate filaments: Molecular structure, assembly mechanism, and integration into functionally distinct intracellular scaffolds. *Annu. Rev. Biochem.* **2004**, *73*, 749–789. [CrossRef]

209. Šmarda, J.; Šmajs, D.; Komrska, J.; Krzyžánek, V. S-layers on cell walls of cyanobacteria. *Micron* **2002**, *33*, 257–277. [CrossRef]

210. Cohen, S.E.; McKnight, B.M.; Golden, S.S. Roles for ClpXP in regulating the circadian clock in Synechococcus elongatus. *Proc. Natl. Acad. Sci. USA* **2018**, *115*, E7805–E7813. [CrossRef]

211. Springstein, B.L.; Nürnberg, D.J.; Woehle, C.; Weissenbach, J.; Theune, M.L.; Helbig, A.O.; Maldener, I.; Dagan, T.; Stucken, K. Two novel heteropolymer-forming proteins maintain the multicellular shape of the cyanobacterium Anabaena sp. PCC 7120. *FEBS J.* **2020**. [CrossRef] [PubMed]

212. Lupas, A.; Van Dyke, M.; Stock, J. Predicting coiled coils from protein sequences. *Science* **1991**, *252*, 1162–1164. [CrossRef] [PubMed]

213. Bailey-Watts, A.E.; Bindloss, M.E. Freshwater primary production by a blue–green alga of bacterial size. *Nature* **1968**, *220*, 1344–1345. [CrossRef]

214. Bisalputra, T.; Oakley, B.R.; Walker, D.C.; Shields, C.M. Microtubular complexes in blue-green algae. *Protoplasma* **1975**, *86*, 19–28. [CrossRef]

215. Jensen, T.E.; Ayala, R.P. The fine structure of striated microtubules and sleeve bodies in several species of Anabaena. *J. Ultrasruct. Res.* **1976**, *57*, 185–193. [CrossRef]

216. Bermudes, D.; Hinkle, G.; Margulis, L. Do prokaryotes contain microtubules? *Microbiol. Rev.* **1994**, *58*, 387–400. [CrossRef]

217. Rast, A.; Schaffer, M.; Albert, S.; Wan, W.; Pfeffer, S.; Beck, F.; Plitzko, J.M.; Nickelsen, J.; Engel, B.D. Biogenic regions of cyanobacterial thylakoids form contact sites with the plasma membrane. *Nat. Plants* **2019**, *5*, 436–446. [CrossRef]

218. Jensen, T.E.; Ayala, R.P. Microtubule-like inclusions in isolates of the blue-green bacteria Anabaena and Nostoc. *Cytologia* **1980**, *45*, 315–326. [CrossRef]

219. Basler, M.; Pilhofer, M.; Henderson, G.P.; Jensen, G.J.; Mekalanos, J.J. Type VI secretion requires a dynamic contractile phage tail-like structure. *Nature* **2012**, *483*, 182–186. [CrossRef]

220. Porta, D.; Rippka, R.; Hernández-Mariné, M. Unusual ultrastructural features in three strains of Cyanothece (cyanobacteria). *Arch. Microbiol.* **2000**, *173*, 154–163. [CrossRef]

221. Medeiros, J.M.; Böck, D.; Weiss, G.L.; Kooger, R.; Wepf, R.A.; Pilhofer, M. Robust workflow and instrumentation for cryo-focused ion beam milling of samples for electron cryotomography. *Ultramicroscopy* **2018**, *190*, 1–11. [CrossRef] [PubMed]

Review

Small but Smart: On the Diverse Role of Small Proteins in the Regulation of Cyanobacterial Metabolism

Fabian Brandenburg and Stephan Klähn *

Helmholtz Centre for Environmental Research—UFZ, 04318 Leipzig, Germany; fabian.brandenburg@ufz.de
* Correspondence: stephan.klaehn@ufz.de; Tel.: +49-341-235-4787

Received: 28 October 2020; Accepted: 26 November 2020; Published: 1 December 2020

Abstract: Over the past few decades, bioengineered cyanobacteria have become a major focus of research for the production of energy carriers and high value chemical compounds. Besides improvements in cultivation routines and reactor technology, the integral understanding of the regulation of metabolic fluxes is the key to designing production strains that are able to compete with established industrial processes. In cyanobacteria, many enzymes and metabolic pathways are regulated differently compared to other bacteria. For instance, while glutamine synthetase in proteobacteria is mainly regulated by covalent enzyme modifications, the same enzyme in cyanobacteria is controlled by the interaction with unique small proteins. Other prominent examples, such as the small protein CP12 which controls the Calvin–Benson cycle, indicate that the regulation of enzymes and/or pathways via the attachment of small proteins might be a widespread mechanism in cyanobacteria. Accordingly, this review highlights the diverse role of small proteins in the control of cyanobacterial metabolism, focusing on well-studied examples as well as those most recently described. Moreover, it will discuss their potential to implement metabolic engineering strategies in order to make cyanobacteria more definable for biotechnological applications.

Keywords: small proteins; metabolic regulation; biotechnology

---

1. Introduction

In nature, proteins are one of the most versatile classes of biological compounds. They serve multiple purposes as structural components, enzymes, membrane transporters, signaling molecules or regulatory factors. Given their tremendous variability in fulfilling tasks in all aspects of life, it is not surprising that proteins come in a manifold of sizes and shapes. For example, they can be single domain proteins or a part of huge protein complexes. The biggest so-far known example that is not part of a multiunit structure is the protein Titin, which is part of vertebrate muscles [1]. Depending on the splice variant, Titin has a size of 27,000–35,000 amino acids and contains over 300 domains [2]. On the contrary, the protein Tal, which was found in *Drosophila melanogaster* is composed of only 11 amino acids [3]. Albeit being so small, it is involved in controlling gene expression and tissue folding and hence, is the shortest functional protein described so far.

It is known that the mean protein length of bacteria is 40–60% shorter than of eukaryotes [4]. Moreover, it was found that up to 16% of all proteins in a prokaryotic organisms might be actually smaller than 100 amino acids [5]. Consequently, more and more studies are suggesting that likely hundreds of small proteins are synthesized in bacterial cells and serve important structural and regulatory functions [6]. Of course, some of these small proteins are known for decades and well-studied, such as, for example, thioredoxins, which play important roles as antioxidants in almost all organisms, not only prokaryotes [7–9]. However, genes encoding small proteins are likely to be overlooked even in modern genome annotations because the minimal cutoff for small open reading frames is typically set to

100 amino acids [10,11]. In turn, this indicates the existence of a whole, unexplored universe of small proteins to be discovered in bacteria, which is especially exemplified by the phylum of cyanobacteria.

Cyanobacteria are the only prokaryotes performing oxygenic photosynthesis. To conduct and maintain their complex photosynthetic machinery, which is composed of several functionally related protein complexes, cyanobacteria use a plethora of small proteins. Some examples like Psb27 have been shown to be important in photosystem II (PSII) repair [12], while others like PetP are involved in stress adaptation of the photosynthetic electron transport chain [13]. In fact, more than 10 proteins smaller than 50 amino acids have been characterized to be important for the function of PSII alone [14,15]. Additionally, 293 candidate-genes for proteins smaller than 80 amino acids have been identified in the cyanobacterial model organism *Synechocystis* sp. PCC 6803 (hereafter *Synechocystis*), indicating that cyanobacteria provide a paradigm for the utilization of small proteins and hence, the functional characterization of bacterial micro-proteomes [16].

This review highlights further prominent examples of small proteins in cyanobacteria beyond the photosynthetic apparatus, i.e., those exercising a regulatory function related to primary metabolism. However, in the literature, different definitions for the term 'small protein' exist. Some studies limit the term to proteins ≤85 amino acids [17], while others also include proteins up to a size of 200 amino acids [18]. Some authors also use the term 'microproteins' which is typically defined as proteins up to a size of 80 amino acids [16]. In this review, we did not set a specific cut-off for the size of considered proteins, but focused on those candidates that regulate metabolic pathways via protein–protein interaction, among which various truly small proteins of only a few kDa are found. Finally, the potential of small proteins as an add-on for the currently existing molecular toolbox for metabolic engineering in cyanobacteria is discussed.

2. Light Regulation of the Calvin-Benson Cycle by the Small Protein CP12

In cyanobacteria, the Calvin–Benson cycle (CB) is the central pathway for the generation of biomass as it fixes CO_2 by using energy (ATP) and reduction equivalents (NADPH) derived from the photosynthetic electron transport chain. In darkness however, cyanobacteria need to oxidize carbohydrates to cover their needs of ATP and reductive power. For a long time, it was believed that the Embden–Meyerhof–Parnas pathway (glycolysis) and the oxidative pentose phosphate pathway (OPP) are the only pathways for carbohydrate oxidation in cyanobacteria. Only quite recently it could be shown that the Entner–Doudoroff (ED) pathway is also active [19]. However, switching from a phototrophic to a heterotrophic mode cannot be achieved by simply activating the respective pathways. Both glycolysis and the OPP share several intermediates with the CB and thus may form a futile cycle when operated at the same time [19]. In order to prevent this, these pathways as well as the CB respond to several signals that trigger a regulation of their activities. For the CB, these signals are thioredoxin, pH, the levels of magnesium and certain metabolites such as fructose 6-phosphate and sedoheptulose 7-phosphate [20]. Some of these signals are transmitted into the regulation of enzyme activity by protein-protein interactions. In plants, the key enzyme of the CB, ribulose-1,5-bisphosphate-carboxylase/-oxygenase (RuBisCO) is directly regulated by an enzyme called RuBisCO activase with a chaperone-like function [21]. However, this does not appear to be a general mechanism among cyanobacteria, as the respective enzyme could be found only in a few species [21].

On the contrary, the small 'chloroplast protein of 12 kDa' (CP12—74 aa, 8.3 kDA in *Synechocystis*) seems to be universally distributed among organisms performing oxygenic photosynthesis. CP12 presents an additional layer of regulation [22]. Under dark conditions, the protein forms a supramolecular complex with the enzymes glyceraldehyde-3-phosphate dehydrogenase (GAPDH) and phosphoribulokinase (PRK) and thereby inhibits their activity [23,24] (Figure 1). Both enzymes are important regulatory points of the CB because they act at the branching points of the CB and OPP. In an ATP-consuming step PRK produces the RuBisCO substrate ribulose 1,5-bisphosphate from the OPP intermediate ribulose 5-phosphate. GAPDH uses NADPH to produce glyceraldehyde 3-phosphate, which is the main exit point of the CB and also part of the OPP. Inhibition of these two enzymes therefore helps the cell to preserve energy, when the photosynthetic light reaction is not active.

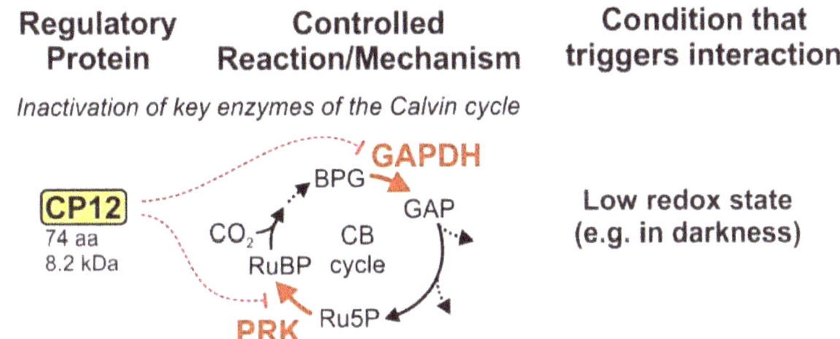

Figure 1. Function of the small protein CP12 in cyanobacteria. CP12 inhibits the CB cycle via complex formation with two key enzymes, PRK and GAPDH. The complex formation is initiated by a low redox status of the cell (e.g., low availability of reduction equivalents) and dinucleotide availability signals (e.g., changes in the NADP(H)/NAD(H) ratio) like they occur during transition from light to darkness. The protein size is given for the model strain *Synechocystis*. BPG—1,3-bisphopshoglycerate, GAP—glyceraldehyde 3-phosphate, Ru5P—ribose 5-phosphate, RuBP—ribulose 1,5 bisphosphate, GAPDH—glyceraldehyde 3-phosphate dehydrogenase, PRK—phosporibulokinase, CB cycle—Calvin–Benson cycle.

CP12 contains four conserved cysteine residues, two at each end. Under oxidizing conditions these cysteine residues form two terminal loops, which enable complex formation with first GADPH and then PRK [25–27]. Binding of GAPDH leads to a conformational change of CP12, which results in an extensive negative charge potential on its molecular surface. This negative charge potential mediates the binding of its N-terminal loop with PRK [25]. The formed complex drastically reduces the activity of both enzymes [28] The complex formation is modulated by the ratio of NADP(H) to NAD(H), which decreases to almost 50% upon transition from light to darkness, and the redox status of the cell [27,29]. Compared to wild-type (WT), CP12-deficient cells are unable to regulate the CB and hence, grow slower under fluctuating light conditions, while there is no difference under continuous light conditions [29]. Interestingly, cyanophages infecting marine picocyanobacteria of the genera *Prochlorococcus* and *Synechococcus* have been shown to express functional CP12 in their host cells, likely to shut down the CB and use the host's production of NADPH to fuel their own deoxynucleotide biosynthesis for replication [30].

3. Control of Glutamine Synthetase by Proteinaceous Inactivating Factors Unique to Cyanobacteria

Besides light and CO_2, nitrogen is another important environmental factor determining cyanobacterial growth. While some cyanobacteria are able to fix dinitrogen gas [31,32] most strains rely on the uptake of reduced nitrogen sources from their environment. Cyanobacteria can utilize a variety of nitrogen sources such as nitrate, nitrite, ammonium, urea, cyanate and some amino acids (such as arginine, glutamine and glutamate) [33–37]. Nevertheless, ammonium is preferred due to a lower energy demand for its assimilation compared to other nitrogen sources [35,38]. The manifold of nitrogen sources as well as their changing availability requires a well-orchestrated regulatory network of nitrogen metabolism in cyanobacteria.

Assimilated nitrate and nitrite are reduced inside the cell and the resulting ammonium is incorporated into carbon skeletons via glutamate dehydrogenase and the glutamine synthetase/glutamate synthase cycle (GS/GOGAT) [35]. The key enzyme GS is well known to be regulated by reversible adenylylation in several bacterial species [39]. By contrast, the cyanobacterial GS is regulated by interaction with small proteins, the so-called GS inactivating factors (IFs) two of which have been identified in *Synechocystis*: IF7

(65 aa, 7.5 kDa) and IF17 (149 aa, 16.7 kDa) [40]. Both IFs are synthesized under nitrogen-rich conditions and specifically bind to GS causing complete enzyme inactivation (Figure 2).

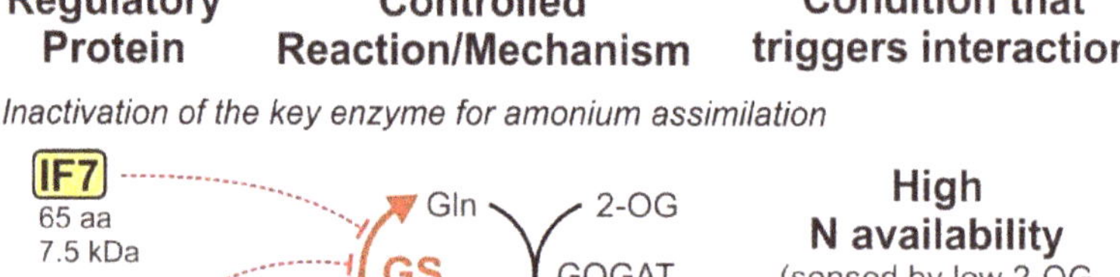

Figure 2. Inactivation of cyanobacterial GS by the interaction with small proteins. Most cyanobacteria harbor two homologous inactivating factors: IF7 and IF17, encoded by the genes *gifA* and *gifB*. The interaction with GS does not require a metabolic signal, hence IF7 and IF17 synthesis is tightly regulated at the transcriptional level by NtcA, which is further assured by regulatory RNAs acting at the post-transcriptional level. IF synthesis is stimulated by low 2-OG levels as well as high Gln levels (see text). Protein sizes are given for *Synechocystis* in which the proteins have initially been discovered. GS—glutamine synthetase, GOGAT—glutamine oxoglutarate aminotransferase (glutamate synthase), Gln—glutamine, Glu—glutamate, 2-OG—2-oxoglutarate.

GS activity is exclusively regulated by the abundance of IF7 and IF17 in the cell [40]. Consequently, their synthesis is target of tight control mechanisms, at the transcriptional and post-transcriptional level. For instance, the corresponding genes, *gifA* and *gifB* are repressed by NtcA [41], a universal transcriptional regulator of nitrogen assimilation in cyanobacteria, which is active and binds DNA under low-nitrogen conditions [42]. Accordingly, *gifA* and *gifB* expression rapidly increases in response to ammonium upshifts. Moreover, IF7 synthesis is negatively regulated by the small RNA NsiR4, which interacts with the *gifA* mRNA and interferes with its translation [43]. In addition, another unique RNA-dependent mechanism has evolved, namely a glutamine riboswitch, which is present in the 5′UTR of the *gifB* transcript. It tightly controls IF17 synthesis in response to a glutamine threshold that is passed when GS activity, i.e., glutamine synthesis exceeds a certain level [44]. The peculiarities of the complex regulation of GS by direct interaction with the small proteins IF7 and IF17 are also reviewed elsewhere in more detail [45]. However, it should be noted that although IFs are unique to cyanobacteria, GS regulation by small proteins is not restricted to this group *per se*. For instance, GS activity is stimulated by complex formation with the 23 aa peptide sP26 and further modulated in a 2-oxoglutarate (2-OG) dependent manner by the 114 aa protein GlnK$_1$ in the archaeal model *Methanosarcina mazei* [46,47]. Both proteins are not related to the cyanobacterial IFs, which greatly exemplifies how widespread and versatile regulatory mechanisms by small proteins are even when targeting the same enzyme.

4. Control of the Key Enzyme of Arginine Synthesis by Direct Interaction with the PII Protein

The PII signaling protein, a homolog to the aforementioned GlnK$_1$, fulfills important regulatory functions and hence, is widely distributed, i.e., present in archaea, bacteria and chloroplasts of plants [48]. With a size of 112 aa and 12.25 kDa in *Synechocystis* it can be defined as a small protein and will be highlighted even though it is also present in other bacteria. In cyanobacteria, the PII protein namely has distinctive regulatory functions. Unlike other bacterial phyla, cyanobacteria possess only one copy of PII [49]. Among other functions, the cyanobacterial PII protein regulates arginine synthesis by binding to the *N*-acetyl-*L*-glutamate kinase (NAGK), which catalyzes the second, rate limiting step of arginine synthesis from glutamate [50]. In addition to its incorporation into proteins, arginine is important as precursor for the nitrogen storage compound cyanophycin, which is a copolymer of aspartate and arginine [42,51]. NAGK is subject to strong feedback inhibition by arginine [52–54]. However, complex formation with PII prevents feedback inhibition of NAGK by arginine and thus strongly enhances the activity of the enzyme [55] (Figure 3).

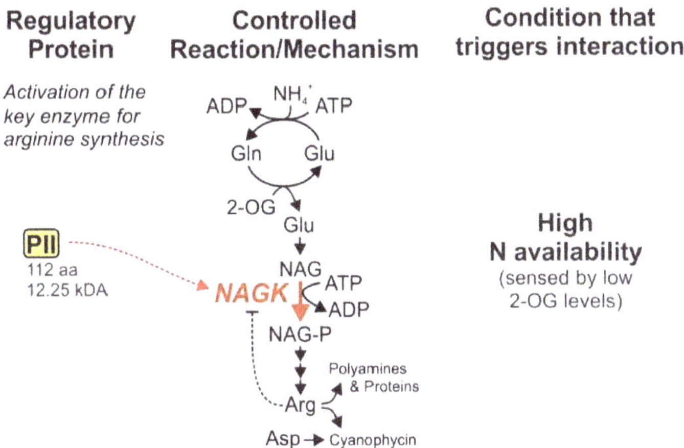

Figure 3. Regulation of N-acetyl-*L*-glutamate kinase (NAGK) by complex formation with the PII protein. NAGK is target of feedback inhibition by arginine, which is however minimized when PII interacts with NAGK. Thereby PII interaction enhances the flux through the rate-limiting step of arginine synthesis, which also impacts synthesis of the N storage compound cyanophycin. We chose NAGK as a prime example for the regulation of enzymes by direct interaction with the PII protein. Nevertheless, it should be noted that PII interacts with various other proteins including enzymes [56]. NAG—N-acetyl-L-glutamate, NAG-P—NAC phosphate, Arg—arginine, Asp—aspartate, Gln—glutamine, Glu—glutamate, 2-OG—2-oxoglutarate.

The activity of PII itself is regulated by phosphorylation, which is in contrast to heterotrophic bacteria, where PII proteins are controlled by uridylylation [38,57]. The phosphorylation status of PII correlates with the nitrogen status of the cell, i.e., fully phosphorylated PII protein is present in cells grown under nitrogen depleted conditions [38,58]. In contrast, the PII protein is completely dephosphorylated in cells grown in the presence of their preferred nitrogen source (ammonium) but gets phosphorylated when cells are using less preferred nitrogen sources like nitrate or nitrite [59]. The phosphorylation status of PII is dependent on the binding of both ATP and 2-OG which lead to a conformational change of PII that is recognized by the modifying enzymes PII-P phosphatase and PII kinase [60–62]. Here, 2-OG serves as a proxy for the nitrogen status of the cell, because due to the activity of the GS/GOGAT cycle the level of 2-OG correlates well with the nitrogen amount that is externally available for the cell [56,60]. Its function as a key regulatory protein is underlined by the fact that recombinant strains of *Synechococcus elongatus* PCC 7942 lacking functional PII are unable to adapt to changing environmental conditions, in particular to changes in nitrogen availability and changes from ammonium to other nitrogen sources, but also to changes in CO_2 concentrations and light intensities [58].

Furthermore, unphosphorylated PII is hypothesized to regulate the uptake of nitrate and nitrite at the post-translational level [49,63]. Likely the PII protein binds directly to the transport protein and thereby inhibits transport [64,65]. In fact, it has been shown that PII is involved in the uptake of ammonium by interaction with the ammonium permease Amt1, inhibits the uptake of nitrate by interaction with the NrtC and NrtD subunits of the nitrate/nitrite transporter NrtABCD, and interacts with the UrtE subunit of the urea transporter UrtABCDE [66]. Additionally, the uptake of bicarbonate is altered in PII-knockout mutants of *Synechococcus elongatus* PCC 7942 [67] and no uptake of methylammonium could be measured in PII-deficient cells of *Synechococcus elongatus* PCC 7942 [58].

In addition, multiple other interaction partners of PII have been identified like the enzyme acetyl-CoA carboxylase (ACCase) [68] or the membrane protein PamA [69]. Some of these interaction partners are small proteins themselves, which in turn exercise metabolic control (see below).

5. PII as an Antagonist for the Interaction of Small Proteins with Key Factors that Control Metabolic Fluxes

Most recently, two independent studies identified another interesting example for the regulation of metabolic processes by a small protein in cyanobacteria [70,71]. Both studies suggested different names for the same protein, namely carbon flow regulator A (CfrA, Muro-Pastor et al., 2020) and PII-interacting regulator of carbon metabolism (PirC, Orthwein et al. 2020), hence we refer to both names in this case. The protein is highly conserved and can be found in almost all cyanobacterial species [70,71]. Free CfrA/PirC (112 aa, 12.27 kDa in *Synechocystis*) binds and inhibits 3-phosphoglycerate mutase (PGAM), an enzyme whose activity is directing carbon flux from the CB cycle towards lower glycolysis [71] (Figure 4). Hence, recombinant strains showing overproduction of CfrA/PirC accumulated excessive amounts of glycogen, while respective knockout mutants were unable to accumulate glycogen even under nitrogen depletion [70]. Instead, those strains accumulated polyhydroxybutyrate (PHB), another carbon storage compound of cyanobacteria, which derives from acetyl-CoA, i.e., reactions downstream of the one that is catalyzed by PGAM [71]. Remarkably, both studies reported complex formation of CfrA/PirC with the PII protein, which is tuned by the 2-OG level [70,71]. Thereby, PII and its interaction with CfrA/PirC determines whether or not PGAM is inhibited by interacting with the small protein. Like other bacterial groups, cyanobacteria sense the nitrogen status via the levels of 2-OG and one of the main routes of newly fixed CO_2 is the synthesis of 2-OG for the assimilation of nitrogen via the GS-GOGAT cycle [59,72]. In the presence of low 2-OG levels, which signals a sufficient nitrogen status, CfrA/PirC interacts with PII. The PII-CfrA/PirC complex disassembles in presence of high 2-OG levels (e.g., under nitrogen limitation) and the released small protein then inhibits PGAM via protein-protein interaction.

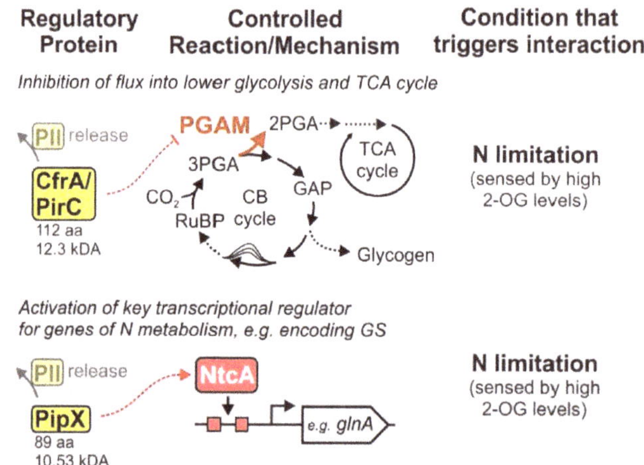

Figure 4. PII modulates and controls the interaction of further small proteins with key elements of cyanobacterial metabolism. Under conditions of sufficient nitrogen both, PipX and CfrA/PirC, form a complex with PII. Under nitrogen limitation, PII releases the small proteins that in turn interact with other target proteins thereby controlling their activity, e.g., enhancing DNA binding affinity of NtcA or inhibiting 3-phosphoglycerate mutase (PGAM).

A similar interplay between PII and a small protein refers to the activity of the transcription factor NtcA that regulates genes majorly encoding elements of nitrogen metabolism in cyanobacteria (for a review see [42]). The small protein PipX (89 aa, 10.53 kDa in *Synechocystis*) interacts with either NtcA or PII depending on the nitrogen status of the cell, sensed via 2-OG [73]. Under nitrogen sufficient conditions, i.e., low 2-OG levels, PipX is bound to PII and NtcA is inactive. In contrast, under nitrogen limiting conditions, PipX is released and interacts with NtcA. Thereby it functions as a coactivator required to enhance 2-OG-dependent binding of NtcA to its recognition sequence [73]. Albeit not exclusively, NtcA commonly acts as transcriptional activator [74] and hence, DNA binding that is triggered by 2-OG and PipX, induces sufficient expression of nitrogen assimilatory genes upon nitrogen limitation, e.g., the *glnA* gene encoding GS (Figure 4).

The structural basis for different binding preferences of PipX has also been investigated [75,76]. However, just like PII, PipX might have even more targets and functions than currently known. For instance, binding of PipX to PII facilitates the extension of the C-terminal region of PipX and thus allow for different interaction partners than the unbound PipX protein [77]. Moreover, it was found that under nitrogen-starving conditions, only 25% of the present PipX in *Synechococcus elongatus* PCC 7942 are needed to bind 100% of the present NtcA leaving 75% of the PipX protein unbound or potentially bound to unknown targets [77]. For instance, it has been shown that under nitrogen-sufficient conditions PII-bound PipX interacts with the GntR-like regulator PlmA [78]. Nevertheless, for the distinctive features of the cyanobacterial PipX protein we refer to a recent review [79].

In addition to the control that PII executes via direct interaction with enzymes such as NAGK, it appears as an antagonist of further small proteins that fulfill crucial regulatory functions in cyanobacterial metabolism. The two given examples are maybe only the tip of the iceberg and hence, further small proteins could function in a similar way. This is supported by the fact that another small protein, encoded by the *ssr0692* gene in *Synechocystis* has been copurified with PII previously [66]. Very recently, it was shown that Ssr0692 (51 aa, 5.8 kDa) is required to balance the synthesis of arginine and several other key amino acids under fluctuating N conditions. Indeed, the protein was confirmed to interact with PII and was hence named PII-interacting regulator of arginine synthesis (PirA, [80]).

6. Further Examples of Small Protein Regulators Affecting the Activity of Enzymes or Transporters

In addition to the PII protein, which is highly conserved in structure and function between different species, there are multiple so-called PII-like proteins that lack multiple consensus sequences and show rather low sequence identity [81]. One example is the protein GlnK, which is also involved in the control of nitrogen assimilation and GS regulation in various prokaryotes [47,82,83]. In cyanobacteria, the PII-like protein SbtB (104 aa, 11.51 kDa in *Synechocystis*) has been shown to interact with the bicarbonate transporter SbtA, which thereby likely has an inhibitory effect [84,85] (Figure 5). Structural analysis of SbtB revealed that increasing concentrations of ATP are one signal initiating the release of SbtB from SbtA [86]. This example shows that in addition to the complex regulatory network that has been unraveled for PII over the past decades, there might be a plethora of PII-like proteins, which could also serve important regulatory functions. The question remains if these proteins have a similar manifold of interaction partners, and with this, an equivalent regulatory importance that has been shown for PII.

Another very recent example for a small protein regulator affecting enzyme activity is AcnSP (44 aa, 5.09 kDa in *Synechocystis*), which is also the smallest protein presented in this review. AcnSP was proposed to stimulate the activity of its binding partner aconitase (Acn, Figure 5), and thereby impacts carbon flux towards the oxidative part of the tricarboxylic acid (TCA) cycle [87]. While the structural mechanism of protein–protein interaction and the exact biological function of AcnSP require further investigation, a potential link to high light adaptation has been already revealed [87].

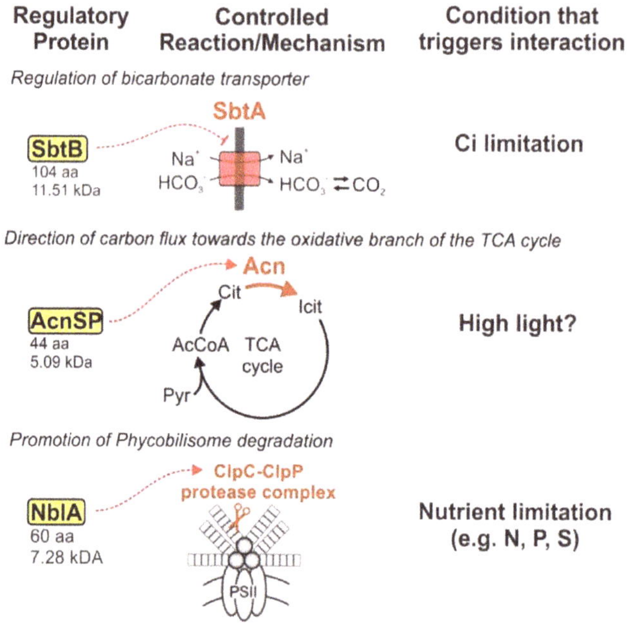

Figure 5. Further examples of small protein regulators affecting the activity of enzymes or transporters by direct interaction. The given protein sizes refer to *Synechocystis*. AcCoA—acetyl-CoA, Acn—aconitase, Ci—inorganic carbon, Cit—citrate, Icit—Isocitrate, Pyr—pyruvate, PSII—photosystem II, TCA cycle—tricarboxylic acid cycle.

Upon nutrient limitation, a common reaction is the degradation of the phycobilisomes, the major light-harvesting complexes of cyanobacteria. Several small proteins are involved in the process. Knockout of the corresponding genes leads to a nonbleaching (nbl) phenotype upon nitrogen starvation. The small protein NblA (60 aa, 7.28 kDa) was the first discovered protein involved in the process. The transcription of *nblA* increases moderately in phosphorous-deprived cells, but shows a much stronger response in sulfur- or nitrogen-deprived cells, leading to a partial or complete degradation of phycobilisomes, respectively [88]. Analysis of the crystal structure of NblA revealed that the protein is present as a homodimer and interacts via the C-terminus with the phycobilisomes [89]. Further studies revealed that in addition to the phycobilisome, NblA binds ClpC at the same time [90]. ClpC is part of the ClpC•ClpP protease complex. The interaction of NblA with ClpC initiates protein degradation by guiding the protease machinery to the phycobilisome [90] (Figure 5). Another small protein named NblB (233 aa, 25.51 kDa) functions analog to NblA and knockouts of *nblB* are also unable to to degrade phycobilisomes under nutrient limiting conditions [91]. In contrast to NblA, NblB is expressed similarly in both cells facing nutrient limitation and nutrient-sufficient conditions [91], indicating another, yet unknown, layer of regulation for this protein. Interestingly, recent results show that NblA and NblB not only facilitate degradation of phycobilisomes, but also binding and rearrangement of chromophores to phycobilisomes [92]. Thus, NblA and NblB might play an important role for short-term adjustments of the photocomplexes to optimize light harvesting under changing light conditions in natural environments.

Most recently, an additional small protein termed NblD (66 aa, 7.09 kDa) was described [93]. The gene locus for NblD was first discovered in a transcriptome study as a highly transcribed but not annotated region and determined as transcriptional unit (TU) 728 [94]. Later, it was discovered that TU728 accumulates upon nitrogen limitation and encodes a small protein, which was named

NsiR6 [16]. Studies with knockout strains revealed a role in phycobilisome degradation similar to NblA and the β-subunit of phycocyanin as interaction target. Accordingly, it was named NblD. However, the exact function of NblD and the interplay with the other Nbl-proteins remains to be unraveled at this point [93].

7. The Potential of Small Proteins for Metabolic Engineering and Biotechnological Applications

Interest in cyanobacteria as host organisms for biotechnological applications has increased steadily over the past decade [95–99]. To date, more than 20 chemicals have been synthesized in cyanobacteria directly from CO_2, e.g., 1,2-propanediol, cyclohexanol, ethanol, isobutyraldehyde, isobutanol, 1-butanol, isoprene, ethylene, hexoses, cellulose, mannitol, lactic acid and fatty acids [96,98,100–102]. Theoretically, every chemical that can be produced by heterotrophic bacteria can also be produced photoautotrophically using cyanobacteria. Thereby, a direct production based on photosynthetic CO_2 fixation is hypothesized to be beneficial over the intermediate CO_2 fixation into biomass, required as source for a heterotrophic production process [98,103].

The growing interest in the biotechnological utilization of cyanobacteria comes with an increasing demand for advanced molecular tools for the genetic engineering of cyanobacteria, which lack far behind the toolset that is available, e.g., for *E. coli* [104,105]. Besides tools for the predictable control of gene expression like promoters [106], this also includes ways to reroute the endogenous carbon flux towards the desired reaction, e.g., blocking of competing pathways or the synthesis of storage compounds [107]. Typically, this is achieved by knocking out the gene encoding the respective synthesis, for example by insertional inactivation with an antibiotic resistance cassette. One major drawback of knockout mutants is the permanent loss of a specific gene function, which might be disadvantageous under certain conditions, in particular in the long-term under natural, i.e., steadily fluctuating conditions. For instance, knockout mutants of glycogen biosynthesis show impaired growth under fluctuating light conditions [108]. This might be problematic, for example, in outdoor bioreactors during phases of biomass production, in which product formation is not the primary interest. Another approach to control gene regulation in bioengineered strains is RNA silencing by small antisense RNA (asRNA)—a technique that has been established also in cyanobacteria as an alternative method for permanent knockouts [109,110]. In comparison to permanent knockouts, asRNA constructs can be controlled, for example, by an inducible promoter independently or in the same way as the respective genes for the production process. However, RNA silencing approaches can be prone to off-target effects and genetic instability of the asRNA construct in subsequent generations [111]. The latter is especially critical in photoautotrophic bacterial systems with short generation times and ideally long or even continuous cultivation intervals.

As pointed out in this review, small proteins are able to sense and implement different signals, bind several interaction partners at the same time and thereby play important regulatory roles in diverse metabolic pathways. Hence, small proteins might pose an interesting addition to the molecular tool set of cyanobacteria. In fact, a few examples can already be found in literature. For instance, the aforementioned CP12 protein has been used to engineer carbon metabolism of the cyanobacterium *Synechococcus elongatus* PCC 7942, i.e., to improve CO_2 fixation and, together with other modifications, to increase the production of 2,3-butanediol [112]. Similarly, the carbon flow regulator CfrA/PirC together with two other modifications has been utilized to increase the production yield of the plastic alternative polyhydroxybutyrate (PHB) in *Synechocystis* from 15 to 63% per cell dry weight (CDW) and even to 81% with additional feeding of acetate [113]. In addition, targeted engineering of existing regulatory proteins may allow the generation of variants exhibiting different binding characteristic compared to the native protein. The principle has already been demonstrated for variants of the PII protein. For instance, variants mimicking either the phosphorylated or the unphosphorylated state of the protein resulted in strains with different nitrogen uptake characteristics compared to WT cells [63]. Moreover, other variants led to a constitutive interaction with NAGK, accompanied by enhanced arginine synthesis and cyanophycin accumulation [114]. Altogether, the greater chemical diversity of proteins compared to small pieces of RNA may result in a control mechanism that is less prone to off targets and more stable over several generations.

Author Contributions: Conceptualization, S.K.; original draft preparation, F.B.; review and editing, F.B. and S.K. All authors have read and agreed to the published version of the manuscript.

Funding: The work was financially supported by the Deutsche Forschungsgemeinschaft (DFG; Grant KL 3114/2-1 to S.K.). We acknowledge the use of the facilities of the Centre for Biocatalysis (MiKat) at the Helmholtz Centre for Environmental Research. The Helmholtz Centre for Environmental Research is supported by the European Regional Development Funds (EFRE, Europe funds Saxony) and the Helmholtz Association.

Conflicts of Interest: The authors declare no conflict of interest.

References

1. Labeit, S.; Kolmerer, B. Titins: Giant proteins in charge of muscle ultrastructure and elasticity. *Science* **1995**, *270*, 293–296. [CrossRef] [PubMed]
2. Opitz, C.A.; Kulke, M.; Leake, M.C.; Neagoe, C.; Hinssen, H.; Hajjar, R.J.; Linke, W.A. Damped elastic recoil of the titin spring in myofibrils of human myocardium. *Proc. Natl. Acad. Sci. USA* **2003**, *100*, 12688–12693. [CrossRef] [PubMed]
3. Galindo, M.I.; Pueyo, J.I.; Fouix, S.; Bishop, S.A.; Couso, J.P. Peptides encoded by short ORFs control development and define a new eukaryotic gene family. *PLoS Biol.* **2007**, *5*, 1052–1062. [CrossRef] [PubMed]
4. Zhang, J. Protein-length distributions for the three domains of life. *Trends Genet.* **2000**, *16*, 107–109. [CrossRef]
5. Miravet-Verde, S.; Ferrar, T.; Espadas-García, G.; Mazzolini, R.; Gharrab, A.; Sabido, E.; Serrano, L.; Lluch-Senar, M. Unraveling the hidden universe of small proteins in bacterial genomes. *Mol. Syst. Biol.* **2019**, *15*, 1–17. [CrossRef] [PubMed]
6. Storz, G.; Wolf, Y.I.; Ramamurthi, K.S. Small proteins can no Longer be ignored. *Annu. Rev. Biochem.* **2014**, *83*, 753–777. [CrossRef] [PubMed]
7. Holmgren, A. Thioredoxin and glutaredoxin systems. *Methods Enzymol.* **1989**, *264*, 286–296. [CrossRef]
8. Arnér, E.S.J.; Holmgren, A. Physiological functions of thioredoxin and thioredoxin reductase. *Eur. J. Biochem.* **2000**, *267*, 6102–6109. [CrossRef] [PubMed]
9. Lu, J.; Holmgren, A. The thioredoxin antioxidant system. *Free Radic. Biol. Med.* **2014**, *66*, 75–87. [CrossRef]
10. Su, M.; Ling, Y.; Yu, J.; Wu, J.; Xiao, J. Small proteins: Untapped area of potential biological importance. *Front. Genet.* **2013**, *4*, 286. [CrossRef]
11. Kliemt, J.; Soppa, J. Diverse functions of small RNAs (sRNAs) in halophilic Archaea: From non-coding regulatory sRNAs to microprotein-Encoding sRNAs. In *RNA Metabolism and Gene Expression in Archaea*, 1st ed.; Clouet-d'Orval, B., Ed.; Springer: Dordrecht, The Netherlands, 2017; Volume 5, pp. 225–242.
12. Grasse, N.; Mamedov, F.; Becker, K.; Styring, S.; Rögner, M.; Nowaczyk, M.M. Role of novel dimeric photosystem II (PSII)-Psb27 protein complex in PSII repair. *J. Biol. Chem.* **2011**, *286*, 29548–29555. [CrossRef]
13. Rexroth, S.; Rexroth, D.; Veit, S.; Plohnke, N.; Cormann, K.U.; Nowaczyk, M.M.; Rögner, M. Functional characterization of the small regulatory subunit PetP from the cytochrome b6f complex in *Thermosynechococcus elongatus*. *Plant Cell* **2014**, *26*, 3435–3448. [CrossRef] [PubMed]
14. Kashino, Y.; Lauber, W.M.; Carroll, J.A.; Wang, Q.; Whitmarsh, J.; Satoh, K.; Pakrasi, H.B. Proteomic analysis of a highly active photosystem II preparation from the cyanobacterium *Synechocystis* sp. PCC 6803 reveals the presence of novel polypeptides. *Biochemistry* **2002**, *41*, 8004–8012. [CrossRef] [PubMed]
15. Guskov, A.; Kern, J.; Gabdulkhakov, A.; Broser, M.; Zouni, A.; Saenger, W. Cyanobacterial photosystem II at 2.9-Å resolution and the role of quinones, lipids, channels and chloride. *Nat. Struct. Mol. Biol.* **2009**, *16*, 334–342. [CrossRef] [PubMed]
16. Baumgartner, D.; Kopf, M.; Klähn, S.; Steglich, C.; Hess, W.R. Small proteins in cyanobacteria provide a paradigm for the functional analysis of the bacterial micro-proteome. *BMC Microbiol.* **2016**, *16*, 1–15. [CrossRef] [PubMed]
17. Zuber, P. A peptide profile of the *Bacillus subtilis* genome. *Peptides* **2001**, *22*, 1555–1577. [CrossRef]
18. Yang, X.; Tschaplinski, T.J.; Hurst, G.B.; Jawdy, S.; Abraham, P.E.; Lankford, P.K.; Adams, R.M.; Shah, M.B.; Hettich, R.L.; Lindquist, E.; et al. Discovery and annotation of small proteins using genomics, proteomics, and computational approaches. *Genome Res.* **2011**, *21*, 634–641. [CrossRef]

19. Chen, X.; Schreiber, K.; Appel, J.; Makowka, A.; Fähnrich, B.; Roettger, M.; Hajirezaei, M.R.; Sönnichsen, F.D.; Schönheit, P.; Martin, W.F.; et al. The Entner-Doudoroff pathway is an overlooked glycolytic route in cyanobacteria and plants. *Proc. Natl. Acad. Sci. USA* **2016**, *113*, 5441–5446. [CrossRef]

20. Stitt, M. Metabolic Regulation of Photosynthesis. In *Photosynthesis and the Environment*, 1st ed.; Baker, N.A., Ed.; Springer: Dordrecht, The Netherlands, 2006; Volume 5, pp. 151–190.

21. Portis, A.R. Rubisco activase—Rubisco's catalytic chaperone. *Photosynth. Res.* **2003**, *75*, 11–27. [CrossRef]

22. Groben, R.; Kaloudas, D.; Raines, C.A.; Offmann, B.; Maberly, S.C.; Gontero, B. Comparative sequence analysis of CP12, a small protein involved in the formation of a Calvin cycle complex in photosynthetic organisms. *Photosynth. Res.* **2010**, *103*, 183–194. [CrossRef]

23. Pohlmeyer, K.; Paap, B.K.; Soll, J.; Wedel, N. CP12: A small nuclear-encoded chloroplast protein provides novel insights into higher-plant GAPDH evolution. *Plant Mol. Biol.* **1996**, *32*, 969–978. [CrossRef]

24. Wedel, N.; Soll, J.; Paap, B.K. CP12 provides a new mode of light regulation of Calvin cycle activity in higher plants. *Proc. Natl. Acad. Sci. USA* **1997**, *94*, 10479–10484. [CrossRef] [PubMed]

25. Matsumura, H.; Kai, A.; Maeda, T.; Tamoi, M.; Satoh, A.; Tamura, H.; Hirose, M.; Ogawa, T.; Kizu, N.; Wadano, A.; et al. Structure basis for the regulation of glyceraldehyde-3-phosphate dehydrogenase activity via the intrinsically disordered protein CP12. *Structure* **2011**, *19*, 1846–1854. [CrossRef] [PubMed]

26. Reichmann, D.; Jakob, U. The roles of conditional disorder in redox proteins. *Curr. Opin. Struct. Biol.* **2013**, *23*, 436–442. [CrossRef] [PubMed]

27. McFarlane, C.R.; Shah, N.R.; Kabasakal, B.V.; Echeverria, B.; Cotton, C.A.R.; Bubeck, D.; Murray, J.W. Structural basis of light-induced redox regulation in the Calvin–Benson cycle in cyanobacteria. *Proc. Natl. Acad. Sci. USA* **2019**, *116*, 20984–20990. [CrossRef] [PubMed]

28. Graciet, E.; Lebreton, S.; Camadro, J.M.; Gontero, B. Characterization of native and recombinant A4 glyceraldehyde 3-phosphate dehydrogenase: Kinetic evidence for conformation changes upon association with the small protein CP12. *Eur. J. Biochem.* **2003**, *270*, 129–136. [CrossRef]

29. Tamoi, M.; Miyazaki, T.; Fukamizo, T.; Shigeoka, S. The Calvin cycle in cyanobacteria is regulated by CP12 via the NAD(H)/NADP(H) ratio under light/dark conditions. *Plant J.* **2005**, *42*, 504–513. [CrossRef]

30. Thompson, L.R.; Zeng, Q.; Kelly, L.; Huang, K.H.; Singer, A.U.; Stubbe, J.A.; Chisholm, S.W. Phage auxiliary metabolic genes and the redirection of cyanobacterial host carbon metabolism. *Proc. Natl. Acad. Sci. USA* **2011**, *108*. [CrossRef]

31. Esteves-Ferreira, A.A.; Cavalcanti, J.H.F.; Vaz, M.G.M.V.; Alvarenga, L.V.; Nunes-Nesi, A.; Araújo, W.L. Cyanobacterial nitrogenases: Phylogenetic diversity, regulation and functional predictions. *Genet. Mol. Biol.* **2017**, *40*, 261–275. [CrossRef] [PubMed]

32. Zehr, J.P. Nitrogen fixation by marine cyanobacteria. *Trends Microbiol.* **2011**, *19*, 162–173. [CrossRef]

33. Quintero, M.J.; Montesinos, M.L.; Herrero, A.; Flores, E. Identification of genes encoding amino acid permeases by inactivation of selected ORFs from the *Synechocystis* genomic sequence. *Genome Res.* **2001**, *11*, 2034–2040. [CrossRef]

34. Valladares, A.; Montesinos, M.L.; Herrero, A.; Flores, E. An ABC-type, high-affinity urea permease identified in cyanobacteria. *Mol. Microbiol.* **2002**, *43*, 703–715. [CrossRef] [PubMed]

35. Flores, E.; Frías, J.E.; Rubio, L.M.; Herrero, A. Photosynthetic nitrate assimilation in cyanobacteria. *Photosynth. Res.* **2005**, *83*, 117–133. [CrossRef]

36. Muro-Pastor, M.I.; Reyes, J.C.; Florencio, F.J. Ammonium assimilation in cyanobacteria. *Photosynth. Res.* **2005**, *83*, 135–150. [CrossRef] [PubMed]

37. Kamennaya, N.A.; Chernihovsky, M.; Post, A.F. The cyanate utilization capacity of marine unicellular Cyanobacteria. *Limnol. Oceanogr.* **2008**, *53*, 2485–2494. [CrossRef]

38. Forchhammer, K.; Tandeau De Marsac, N. The P(II) protein in the cyanobacterium *Synechococcus* sp. strain PCC 7942 is modified by serine phosphorylation and signals the cellular N-status. *J. Bacteriol.* **1994**, *176*, 84–91. [CrossRef] [PubMed]

39. Merrick, M.J.; Edwards, R.A. Nitrogen control in bacteria. *Microbiol. Rev.* **1995**, *59*, 604–622. [CrossRef]

40. García-Domínguez, M.; Reyes, J.C.; Florencio, F.J. Glutamine synthetase inactivation by protein-protein interaction. *Proc. Natl. Acad. Sci. USA* **1999**, *96*, 7161–7166. [CrossRef]

41. García-Domínguez, M.; Reyes, J.C.; Florencio, F.J. NtcA represses transcription of *gifA* and *gifB*, genes that encode inhibitors of glutamine synthetase type I from *Synechocystis* sp. PCC 6803. *Mol. Microbiol.* **2000**, *35*, 1192–1201. [CrossRef]

42. Herrero, A.; Muro-Pastor, A.M.; Flores, E. Nitrogen control in cyanobacteria. *J. Bacteriol.* **2001**, *183*, 411–425. [CrossRef]

43. Klähn, S.; Schaal, C.; Georg, J.; Baumgartner, D.; Knippen, G.; Hagemann, M.; Muro-Pastor, A.M.; Hess, W.R. The sRNA NsiR4 is involved in nitrogen assimilation control in cyanobacteria by targeting glutamine synthetase inactivating factor IF7. *Proc. Natl. Acad. Sci. USA* **2015**, *112*, E6243–E6252. [CrossRef]

44. Klähn, S.; Bolay, P.; Wright, P.R.; Atilho, R.M.; Brewer, K.I.; Hagemann, M.; Breaker, R.R.; Hess, W.R. A glutamine riboswitch is a key element for the regulation of glutamine synthetase in cyanobacteria. *Nucleic Acids Res.* **2018**, *46*, 10082–10094. [CrossRef]

45. Bolay, P.; Muro-Pastor, M.I.; Florencio, F.J.; Klähn, S. The distinctive regulation of cyanobacterial glutamine synthetase. *Life* **2018**, *8*, 52. [CrossRef]

46. Gutt, M.; Jordan, B.; Weidenbach, K.; Gudzuhn, M.; Kiessling, C.; Cassidy, L.; Helbig, A.; Tholey, A.; Pyper, D.; Schwalbe, H.; et al. Small protein 26 interacts and enhances glutamine synthetase activity in *Methanosarcina mazei*. *bioRxiv* **2020**. [CrossRef]

47. Ehlers, C.; Weidenbach, K.; Veit, K.; Forchhammer, K.; Schmitz, R.A. Unique mechanistic features of post-translational regulation of glutamine synthetase activity in *Methanosarcina mazei* strain Göl in response to nitrogen availability. *Mol. Microbiol.* **2005**, *55*, 1841–1854. [CrossRef]

48. Huergo, L.F.; Chandra, G.; Merrick, M. PII signal transduction proteins: Nitrogen regulation and beyond. *FEMS Microbiol. Rev.* **2013**, *37*, 251–283. [CrossRef] [PubMed]

49. Lee, H.M.; Flores, E.; Herrero, A.; Houmard, J.; Tandeau De Marsac, N. A role for the signal transduction protein P(II) in the control of nitrate/nitrite uptake in a cyanobacterium. *FEBS Lett.* **1998**, *427*, 291–295. [CrossRef]

50. Heinrich, A.; Maheswaran, M.; Ruppert, U.; Forchhammer, K. The *Synechococcus elongatus* PII signal transduction protein controls arginine synthesis by complex formation with N-acetyl-L-glutamate kinase. *Mol. Microbiol.* **2004**, *52*, 1303–1314. [CrossRef]

51. Simon, R.D. Cyanophycin Granules from the Blue-Green Alga *Anabaena cylindrica*: A Reserve Material Consisting of Copolymers of Aspartic Acid and Arginine. *Proc. Natl. Acad. Sci. USA* **1971**, *68*, 265–267. [CrossRef]

52. Pauwels, K.; Abadjieva, A.; Hilven, P.; Stankiewicz, A.; Crabeel, M. The N-acetylglutamate synthase/N-acetylglutamate kinase metabolon of *Saccharomyces cerevisiae* allows coordinated feedback regulation of the first two steps in arginine biosynthesis. *Eur. J. Biochem.* **2003**, *270*, 1014–1024. [CrossRef]

53. Caldovic, L.; Tuchman, M. N -Acetylglutamate and its changing role through evolution. *Biochem J.* **2003**, *290*, 279–290. [CrossRef] [PubMed]

54. Cunin, R.; Glansdorff, N.; Pierard, A.; Stalon, V. Biosynthesis and metabolism of arginine in bacteria. *Microbiol. Rev.* **1986**, *50*, 314–352. [CrossRef] [PubMed]

55. Maheswaran, M.; Urbanke, C.; Forchhammer, K. Complex formation and catalytic activation by the PII signaling protein of N-acetyl-L-glutamate kinase from *Synechococcus elongatus* strain PCC 7942. *J. Biol. Chem.* **2004**, *279*, 55202–55210. [CrossRef] [PubMed]

56. Forchhammer, K. PII signal transducers: Novel functional and structural insights. *Trends Microbiol.* **2008**, *16*, 65–72. [CrossRef] [PubMed]

57. Maheswaran, M.; Forchhammer, K. Carbon-source-dependent nitrogen regulation in *Escherichia coli* is mediated through glutamine-dependent GlnB signalling. *Microbiology* **2003**, *149*, 2163–2172. [CrossRef] [PubMed]

58. Forchhammer, K.; Tandeau De Marsac, N. Functional analysis of the phosphoprotein P(II) (*glnB* gene product) in the cyanobacterium *Synechococcus* sp. strain PCC 7942. *J. Bacteriol.* **1995**, *177*, 2033–2040. [CrossRef] [PubMed]

59. Forchhammer, K. Global carbon/nitrogen control by PII signal transduction in cyanobacteria: From signals to targets. *FEMS Microbiol. Rev.* **2004**, *28*, 319–333. [CrossRef]

60. Forchhammer, K. The PII protein in *Synechococcus* PCC 7942 senses and signals 2-oxoglutarate under ATP-replete conditions. In *The Phototrophic Prokaryotes*, 1st ed.; Peschek, G.A., Löffelhardt, W., Schmetterer, G., Eds.; Springer: Berlin/Heidelberg, Germany, 1999; pp. 549–553.

61. Forchhammer, K.; De Marsac, N.T. Phosphorylation of the P(II) protein (*glnB* gene product) in the cyanobacterium *Synechococcus* sp. strain PCC 7942: Analysis of in vitro kinase activity. *J. Bacteriol.* **1995**, *177*, 5812–5817. [CrossRef]

62. Kloft, N.; Rasch, G.; Forchhammer, K. Protein phosphatase PphA from *Synechocystis* sp. PCC 6803: The physiological framework of PII-P dephosphorylation. *Microbiology* **2005**, *151*, 1275–1283. [CrossRef]

63. Lee, H.M.; Flores, E.; Forchhammer, K.; Herrero, A.; Tandeau De Marsac, N. Phosphorylation of the signal transducer P(II) protein and an additional effector are required for the P(II)-mediated regulation of nitrate and nitrite uptake in the cyanobacterium *Synechococcus* sp. PCC 7942. *Eur. J. Biochem.* **2000**, *267*, 591–600. [CrossRef]

64. Heinrich, A.; Woyda, K.; Brauburger, K.; Meiss, G.; Detsch, C.; Stülke, J.; Forchhammer, K. Interaction of the membrane-bound GlnK-AmtB complex with the master regulator of nitrogen metabolism TnrA in *Bacillus subtilis*. *J. Biol. Chem.* **2006**, *281*, 34909–34917. [CrossRef] [PubMed]

65. Radchenko, M.V.; Thornton, J.; Merrick, M. Control of AmtB-GlnK complex formation by intracellular levels of ATP, ADP, and 2-oxoglutarate. *J. Biol. Chem.* **2010**, *285*, 31037–31045. [CrossRef] [PubMed]

66. Watzer, B.; Spät, P.; Neumann, N.; Koch, M.; Sobotka, R.; MacEk, B.; Hennrich, O.; Forchhammer, K. The signal transduction protein PII controls ammonium, nitrate and urea uptake in cyanobacteria. *Front. Microbiol.* **2019**, *10*, 1–20. [CrossRef] [PubMed]

67. Hisbergues, M.; Jeanjean, R.; Joset, F.; Tandeau De Marsac, N.; Bédu, S. Protein PII regulates both inorganic carbon and nitrate uptake and is modified by a redox signal in *Synechocystis* PCC 6803. *FEBS Lett.* **1999**, *463*, 216–220. [CrossRef]

68. Hauf, W.; Schmid, K.; Gerhardt, E.C.M.; Huergo, L.F.; Forchhammer, K. Interaction of the nitrogen regulatory protein GlnB (PII) with biotin carboxyl carrier protein (BCCP) controls acetyl-CoA levels in the cyanobacterium *Synechocystis* sp. PCC 6803. *Front. Microbiol.* **2016**, *7*, 1–14. [CrossRef]

69. Osanai, T.; Sato, S.; Tabata, S.; Tanaka, K. Identification of PamA as a PII-binding membrane protein important in nitrogen-related and sugar-catabolic gene expression in *Synechocystis* sp. PCC 6803. *J. Biol. Chem.* **2005**, *280*, 34684–34690. [CrossRef]

70. Muro-Pastor, M.I.; Cutillas-Farray, Á.; Pérez-Rodríguez, L.; Pérez-Saavedra, J.; Vega-de Armas, A.; Paredes, A.; Robles-Rengel, R.; Florencio, F.J. CfrA, a novel carbon flow regulator, adapts carbon metabolism to nitrogen deficiency in cyanobacteria. *Plant Physiol.* **2020**. [CrossRef]

71. Orthwein, T.; Scholl, J.; Spät, P.; Lucius, S.; Koch, M.; Macek, B.; Hagemann, M.; Forchhammer, K. The Novel PII-Interacting Regulator PirC (Sll0944) Identifies 3-Phosphoglycerate Mutase (PGAM) as Central Control Point of Carbon Storage Metabolism in Cyanobacteria. *bioRxiv* **2020**. [CrossRef]

72. Leigh, J.A.; Dodsworth, J.A. Nitrogen regulation in bacteria and archaea. *Annu. Rev. Microbiol.* **2007**, *61*, 349–377. [CrossRef]

73. Espinosa, J.; Forchhammer, K.; Burillo, S.; Contreras, A. Interaction network in cyanobacterial nitrogen regulation: PipX, a protein that interacts in a 2-oxoglutarate dependent manner with PII and NtcA. *Mol. Microbiol.* **2006**, *61*, 457–469. [CrossRef]

74. Giner-Lamia, J.; Robles-Rengel, R.; Hernández-Prieto, M.A.; Isabel Muro-Pastor, M.; Florencio, F.J.; Futschik, M.E. Identification of the direct regulon of NtcA during early acclimation to nitrogen starvation in the cyanobacterium *Synechocystis* sp. PCC 6803. *Nucleic Acids Res.* **2017**, *45*, 11800–11820. [CrossRef] [PubMed]

75. Llácer, J.L.; Espinosa, J.; Castells, M.A.; Contreras, A.; Forchhammer, K.; Rubio, V. Structural basis for the regulation of NtcA-dependent transcription by proteins PipX and PII. *Proc. Natl. Acad. Sci. USA* **2010**, *107*, 15397–15402. [CrossRef] [PubMed]

76. Zhao, M.-X.; Jiang, Y.-L.; He, Y.-X.; Chen, Y.-F.; Teng, Y.-B.; Chen, Y.; Zhang, C.-C.; Zhou, C.-Z. Structural basis for the allosteric control of the global transcription factor NtcA by the nitrogen starvation signal 2-oxoglutarate. *Proc. Natl. Acad. Sci. USA* **2010**, *107*, 12487–12492. [CrossRef]

77. Forcada-Nadal, A.; Llácer, J.L.; Contreras, A.; Marco-Marín, C.; Rubio, V. The PII-NAGK-PipX-NtcA regulatory axis of cyanobacteria: A tale of changing partners, allosteric effectors and non-covalent interactions. *Front. Mol. Biosci.* **2018**, *5*, 1–18. [CrossRef]

78. Labella, J.I.; Obrebska, A.; Espinosa, J.; Salinas, P.; Forcada-Nadal, A.; Tremiño, L.; Rubio, V.; Contreras, A. Expanding the cyanobacterial nitrogen regulatory network: The GntR-like regulator PlmA interacts with the PII-PipX complex. *Front. Microbiol.* **2016**, *7*, 1–17. [CrossRef]

79. Labella, J.I.; Cantos, R.; Salinas, P.; Espinosa, J.; Contreras, A. Distinctive features of PipX, a unique signaling protein of cyanobacteria. *Life* **2020**, *10*, 79. [CrossRef] [PubMed]

80. Bolay, P.; Muro-pastor, M.I.; Rozbeh, R.; Timm, S.; Hagemann, M.; Florencio, F.J.; Forchhammer, K.; Klähn, S. The novel PII-interacting protein PirA regulates flux into the cyanobacterial ornithine-ammonia cycle. *bioRxiv* **2020**. [CrossRef]

81. Forchhammer, K.; Lüddecke, J. Sensory properties of the PII signalling protein family. *FEBS J.* **2016**, *283*, 425–437. [CrossRef]

82. Van Heeswijk, W.C.; Hoving, S.; Molenaar, D.; Stegeman, B.; Kahn, D.; Westerhoff, H.V. An alternative P(II) protein in the regulation of glutamine synthetase in *Escherichia coli*. *Mol. Microbiol.* **1996**, *21*, 133–146. [CrossRef]

83. Atkinson, M.R.; Ninfa, A.J. Role of the GlnK signal transduction protein in the regulation of nitrogen assimilation in *Escherichia coli*. *Mol. Microbiol.* **1998**, *29*, 431–447. [CrossRef]

84. Du, J.; Förster, B.; Rourke, L.; Howitt, S.M.; Price, G.D. Characterisation of cyanobacterial bicarbonate transporters in *E. coli* shows that SbtA homologs are functional in this heterologous expression system. *PLoS ONE* **2014**, *9*, e115905. [CrossRef] [PubMed]

85. Selim, K.A.; Haase, F.; Hartmann, M.D.; Hagemann, M.; Forchhammer, K. PII-like signaling protein SbtB links cAMP sensing with cyanobacterial inorganic carbon response. *Proc. Natl. Acad. Sci. USA* **2018**, *115*, E4861–E4869. [CrossRef] [PubMed]

86. Kaczmarski, J.A.; Hong, N.S.; Mukherjee, B.; Wey, L.T.; Rourke, L.; Förster, B.; Peat, T.S.; Price, G.D.; Jackson, C.J. Structural basis for the allosteric regulation of the SbtA bicarbonate transporter by the PII-like protein, SbtB, from *Cyanobium* sp. PCC7001. *Biochemistry* **2019**, *58*, 5030–5039. [CrossRef] [PubMed]

87. de Alvarenga, L.V.; Hess, W.R.; Hagemann, M. AcnSP—A novel small protein regulator of aconitase activity in the cyanobacterium *Synechocystis* sp. PCC 6803. *Front. Microbiol.* **2020**, *11*, 1–12. [CrossRef]

88. Collier, J.L.; Grossman, A.R. A small polypeptide triggers complete degradation of light-harvesting phycobiliproteins in nutrient-deprived cyanobacteria. *EMBO J.* **1994**, *13*, 1039–1047. [CrossRef] [PubMed]

89. Bienert, R.; Baier, K.; Volkmer, R.; Lockau, W.; Heinemann, U. Crystal structure of NblA from *Anabaena* sp. PCC 7120, a small protein playing a key role in phycobilisome degradation. *J. Biol. Chem.* **2006**, *281*, 5216–5223. [CrossRef] [PubMed]

90. Karradt, A.; Sobanski, J.; Mattow, J.; Lockau, W.; Baier, K. NblA, a key protein of phycobilisome degradation, interacts with ClpC, a HSP100 chaperone partner of a cyanobacterial Clp protease. *J. Biol. Chem.* **2008**, *283*, 32394–32403. [CrossRef]

91. Dolganov, N.; Grossman, A.R. A polypeptide with similarity to phycocyanin α-subunit phycocyanobilin lyase involved in degradation of phycobilisomes. *J. Bacteriol.* **1999**, *181*, 610–617. [CrossRef]

92. Hu, P.P.; Hou, J.Y.; Xu, Y.L.; Niu, N.N.; Zhao, C.; Lu, L.; Zhou, M.; Scheer, H.; Zhao, K.H. The role of lyases, NblA and NblB proteins and bilin chromophore transfer in restructuring the cyanobacterial light-harvesting complex. *Plant J.* **2020**, *102*, 529–540. [CrossRef]

93. Krauspe, V.; Fahrner, M.; Spät, P.; Steglich, C.; Frankenberg-Dinkel, N.; Macek, B.; Schilling, O.; Hess, W.R. Discovery of a novel small protein factor involved in the coordinated degradation of phycobilisomes in cyanobacteria. *bioRxiv* **2020**. [CrossRef]

94. Kopf, M.; Klähn, S.; Scholz, I.; Matthiessen, J.K.F.; Hess, W.R.; Voß, B. Comparative analysis of the primary transcriptome of *Synechocystis* sp. PCC 6803. *DNA Res.* **2014**, *21*, 527–539. [CrossRef]

95. Hitchcock, A.; Hunter, C.N.; Canniffe, D.P. Progress and challenges in engineering cyanobacteria as chassis for light-driven biotechnology. *Microb. Biotechnol.* **2020**, *13*, 363–367. [CrossRef] [PubMed]

96. Savakis, P.; Hellingwerf, K.J. Engineering cyanobacteria for direct biofuel production from CO_2. *Curr. Opin. Biotechnol.* **2015**, *33*, 8–14. [CrossRef] [PubMed]

97. Hagemann, M.; Hess, W.R. Systems and synthetic biology for the biotechnological application of cyanobacteria. *Curr. Opin. Biotechnol.* **2018**, *49*, 94–99. [CrossRef] [PubMed]

98. Ducat, D.C.; Way, J.C.; Silver, P.A. Engineering cyanobacteria to generate high-value products. *Trends Biotechnol.* **2011**, *29*, 95–103. [CrossRef] [PubMed]

99. Abed, R.M.M.; Dobretsov, S.; Sudesh, K. Applications of cyanobacteria in biotechnology. *J. Appl. Microbiol.* **2009**, *106*, 1–12. [CrossRef] [PubMed]

100. Rosgaard, L.; de Porcellinis, A.J.; Jacobsen, J.H.; Frigaard, N.U.; Sakuragi, Y. Bioengineering of carbon fixation, biofuels, and biochemicals in cyanobacteria and plants. *J. Biotechnol.* **2012**, *162*, 134–147. [CrossRef] [PubMed]

101. David, C.; Schmid, A.; Adrian, L.; Wilde, A.; Bühler, K. Production of 1,2-propanediol in photoautotrophic *Synechocystis* is linked to glycogen turn-over. *Biotechnol. Bioeng.* **2018**, *115*, 300–311. [CrossRef] [PubMed]

102. Hoschek, A.; Toepel, J.; Hochkeppel, A.; Karande, R.; Bühler, B.; Schmid, A. Light-dependent and aeration-independent gram-scale hydroxylation of cyclohexane to cyclohexanol by CYP450 harboring *Synechocystis* sp. PCC 6803. *Biotechnol. J.* **2019**, *14*, 1–10. [CrossRef]

103. Van Dam, J.E.G.; De Klerk-Engels, B.; Struik, P.C.; Rabbinge, R. Securing renewable resource supplies for changing market demands in a bio-based economy. *Ind. Crops Prod.* **2005**, *21*, 129–144. [CrossRef]

104. Wang, B.; Wang, J.; Meldrum, D.R. Application of synthetic biology in cyanobacteria and algae. *Front. Microbiol.* **2012**, *3*, 1–15. [CrossRef] [PubMed]

105. Markley, A.L.; Begemann, M.B.; Clarke, R.E.; Gordon, G.C.; Pfleger, B.F. Synthetic biology toolbox for controlling gene expression in the cyanobacterium *Synechococcus* sp. strain PCC 7002. *ACS Synth. Biol.* **2015**, *4*, 595–603. [CrossRef] [PubMed]

106. Englund, E.; Liang, F.; Lindberg, P. Evaluation of promoters and ribosome binding sites for biotechnological applications in the unicellular cyanobacterium *Synechocystis* sp. PCC 6803. *Sci. Rep.* **2016**, *6*, 1–12. [CrossRef]

107. Zhou, J.; Zhu, T.; Cai, Z.; Li, Y. From cyanochemicals to cyanofactories: A review and perspective. *Microb. Cell Fact.* **2016**, *15*, 1–9. [CrossRef] [PubMed]

108. Gründel, M.; Scheunemann, R.; Lockau, W.; Zilliges, Y. Impaired glycogen synthesis causes metabolic overflow reactions and affects stress responses in the cyanobacterium *Synechocystis* sp. PCC 6803. *Microbiology* **2012**, *158*, 3032–3043. [CrossRef]

109. Srivastava, A.; Brilisauer, K.; Rai, A.K.; Ballal, A.; Forchhammer, K.; Tripathi, A.K. Down-regulation of the alternative sigma factor SIGJ confers a photoprotective phenotype to *Anabaena* PCC 7120. *Plant Cell Physiol.* **2017**, *58*, 287–297. [CrossRef]

110. Tailor, V.; Ballal, A. Novel molecular insights into the function and the antioxidative stress response of a Peroxiredoxin Q protein from cyanobacteria. *Free Radic. Biol. Med.* **2017**, *106*, 278–287. [CrossRef]

111. Frizzi, A.; Huang, S. Tapping RNA silencing pathways for plant biotechnology. *Plant Biotechnol. J.* **2010**, *8*, 655–677. [CrossRef]

112. Kanno, M.; Carroll, A.L.; Atsumi, S. Global metabolic rewiring for improved CO_2 fixation and chemical production in cyanobacteria. *Nat. Commun.* **2017**, *8*, 1–11. [CrossRef]

113. Koch, M.; Bruckmoser, J.; Scholl, J.; Hauf, W.; Rieger, B.; Forchhammer, K. Maximizing PHB content in *Synechocystis* sp. PCC 6803: Development of a new photosynthetic overproduction strain. *bioRxiv* **2020**. [CrossRef]

114. Watzer, B.; Engelbrecht, A.; Hauf, W.; Stahl, M.; Maldener, I.; Forchhammer, K. Metabolic pathway engineering using the central signal processor PII. *Microb. Cell Fact.* **2015**, *14*, 1–12. [CrossRef] [PubMed]

Review

Distinctive Features of PipX, a Unique Signaling Protein of Cyanobacteria

Jose I. Labella, Raquel Cantos, Paloma Salinas, Javier Espinosa and Asunción Contreras *

Dpto. Fisiología, Genética y Microbiología, Universidad de Alicante, 03690 Alicante, Spain;
ls.joseignacio@ua.es (J.I.L.); raquel.cantos@ua.es (R.C.); paloma.salinas@ua.es (P.S.); javier.espinosa@ua.es (J.E.)
* Correspondence: contrera@ua.es

Received: 10 May 2020; Accepted: 26 May 2020; Published: 28 May 2020

Abstract: PipX is a unique cyanobacterial protein identified by its ability to bind to PII and NtcA, two key regulators involved in the integration of signals of the nitrogen/carbon and energy status, with a tremendous impact on nitrogen assimilation and gene expression in cyanobacteria. PipX provides a mechanistic link between PII, the most widely distributed signaling protein, and NtcA, a global transcriptional regulator of cyanobacteria. PII, required for cell survival unless PipX is inactivated or down-regulated, functions by protein–protein interactions with transcriptional regulators, transporters, and enzymes. In addition, PipX appears to be involved in a wider signaling network, supported by the following observations: (i) PII–PipX complexes interact with PlmA, an as yet poorly characterized transcriptional regulator also restricted to cyanobacteria; (ii) the *pipX* gene is functionally connected with *pipY*, a gene encoding a universally conserved pyridoxal phosphate binding protein (PLPBP) involved in vitamin B6 and amino acid homeostasis, whose loss-of-function mutations cause B6-dependent epilepsy in humans, and (iii) *pipX* is part of a relatively robust, six-node synteny network that includes *pipY* and four additional genes that might also be functionally connected with *pipX*. In this overview, we propose that the study of the protein–protein interaction and synteny networks involving PipX would contribute to understanding the peculiarities and idiosyncrasy of signaling pathways that are conserved in cyanobacteria.

Keywords: cyanobacteria; signal transduction; nitrogen regulation; interaction network; synteny network

---

1. Introduction

Cyanobacteria, phototrophic organisms that perform oxygenic photosynthesis, constitute an ecologically important phylum that is responsible for the evolution of the oxygenic atmosphere, and are the main contributors to marine primary production [1]. Their photosynthetic lifestyle and ease of cultivation make them ideal production systems for a number of high-value compounds, including biofuels [2]. However, cyanobacteria have developed sophisticated systems to maintain the homeostasis of carbon/nitrogen (reviewed by [3,4]), the two most abundant nutrient elements in all living forms; therefore, understanding the regulatory mechanisms affecting their metabolic balance is of paramount importance from the biotechnological as well as the environmental points of view.

Cyanobacteria can use different nitrogen sources that are then converted into ammonium and incorporated via the glutamine synthetase-glutamate synthase (GS-GOGAT) cycle into carbon skeleton 2-oxoglutarate (2-OG) for the biosynthesis of amino acids and other N-containing compounds. The 2-OG, a universal indicator of the intracellular carbon-to-nitrogen balance [5,6], appears to be particularly suitable for this role in cyanobacteria, because the lack of 2-OG dehydrogenase results in the accumulation of 2-OG during nitrogen starvation [7]. Recently, a role as an antioxidant agent involved in reactive oxygen species (ROS) homeostasis has also been proposed for 2-OG in cyanobacteria [8].

In bacteria and plants, 2-OG is sensed by the widely distributed signal transduction protein PII, which is encoded by *glnB* and is a homotrimer with one binding site per subunit for 2-OG. PII regulates the activity of proteins implicated in nitrogen metabolism by direct protein–protein interactions [3]. The first two PII receptors in cyanobacteria [9,10] were identified using the yeast two-hybrid system [11] to search for proteins interacting with PII in *Synechococcus elongatus* PCC7942 (hereafter *S. elongatus*). One of the identified proteins was the enzyme N-acetyl-glutamate-kinase (NAGK), which catalyzes a key regulatory step in the biosynthesis of arginine that is stimulated by PII in cyanobacteria and plant chloroplasts [10,12,13]. The other one was a small and previously unknown protein of 89 amino acids that was named PipX (PII interacting protein X). PipX was also found as prey in yeast two-hybrid searches with NtcA, the global transcriptional regulator involved in nitrogen assimilation in cyanobacteria [14,15]. Since PII and NtcA are 2-OG receptors, PipX appeared to be a novel component of the nitrogen signal transduction pathway in this phylum. Subsequent work confirmed the role of PipX as a regulatory link between PII and NtcA, and unraveled the functional and structural details of the PipX–PII and PipX–NtcA interactions [16–21]. Although most of the significant advances relate to *S. elongatus*, experimental and *in silico* evidence supports conservation of the same nitrogen-related regulatory interactions of PipX in the cyanobacterial phylum [21–28].

Additional "guilty by association" approaches have extended the physical and functional networks of PipX. Yeast three-hybrid searches with PipX–PII as bait resulted in the identification of the cyanobacterial transcriptional regulator PlmA as an interacting protein [29], while co-expression and synteny approaches functionally connected PipX with PipY, a conserved pyridoxal phosphate-binding protein involved in amino/keto acid and pyridoxal phosphate (PLP) homeostasis [30–32].

2. The Complex Nitrogen Signaling Network of Cyanobacteria

2.1. PII and PipX as Dynamic Hubs of an Extended Protein Interaction Network

PII and PipX mediate protein–protein interactions with regulatory targets that include transcriptional regulators, enzymes, and transporters involved in nitrogen and carbon assimilation, forming an ever-growing and dynamic interaction network that largely depends on PII effectors. Thus, the levels of 2-OG and the ATP/ADP ratio are the main intracellular signal molecules determining protein–protein interactions [33]. In addition, this complex network is necessarily affected by the relative abundance of the different components. In *S. elongatus*, estimations of the numbers of chains of the relevant proteins can be obtained from massive proteomic studies [34]. In a comprehensive scheme, Figure 1A illustrates the nitrogen interaction network and the position of PII and PipX as regulatory hubs, integrating information on the relative abundance of the protein components and the levels of signal molecules determining protein–protein interactions. For the sake of simplicity, only protein components whose interactions with the hubs have been characterized to a certain extent (discussed in this section) are included in the illustration. The relative abundance of PipY, considered a member of the corresponding PipX–PII regulatory network (see below), is also illustrated for comparison.

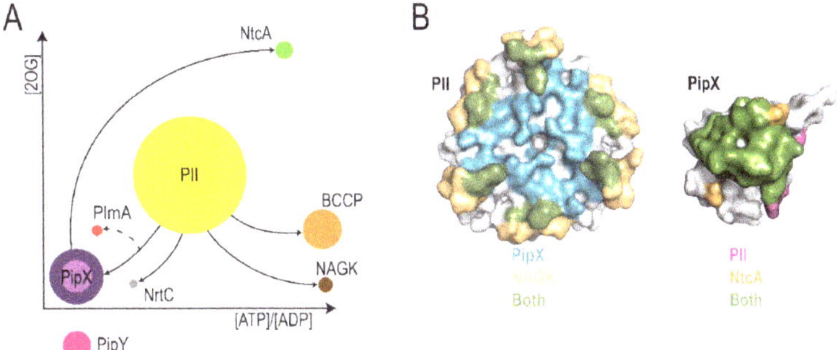

Figure 1. The PipX interaction network. (**A**) PII and PII PipX interactions, according to the concentration spectra of 2-OG and ATP/ADP ratios. Molecular players are illustrated as circles whose sizes, drawn to scale, refer to the number of protein molecules, according to [34], taking into account their quaternary structure: 20.078 PII trimers, 4.560 PipX monomers, and 1.520 PipX trimers (dark- and light-colored circles, respectively), as well as 510 NtcA dimers, 200 PlmA dimers, 160 NrtC monomers, 359 NAGK hexamers, and 2409 biotin carboxyl carrier protein (BCCP) monomers. The arrows go from hub to target proteins, with PipX being considered as a PII target. The position of the arrowheads (towards target proteins) indicates the conditions favoring the corresponding complexes, one of which is a ternary complex (PII–PipX–PlmA). A dashed arrow is used for PlmA interaction with PipX–PII, which has not been characterized at the molecular level but should be favored by relatively low 2-OG levels and low ATP/ADP ratios. PipY (1.275 monomers), for which no protein–protein interactions are known, is shown outside the graphical representation. (**B**) The three-dimensional (3D) structures of PII (left) and PipX (right), with surfaces colored according to the area of interaction, with NAGK (yellow) and PipX (blue) for PII, and with PII (purple) and NtcA (yellow) for PipX. (Adapted from [35] (**A**) and [16] (**B**)).

PII perceives metabolic information by the competitive binding of ATP or ADP and by the synergistic binding of ATP and 2-OG [36]. The PII trimer has three binding sites for ATP/ADP (in some species AMP) and 2-OG. PII binds to NAGK, stimulating its activity and promoting nitrogen storage as arginine in cyanobacteria and plants [10,37–39]. When abundant, 2-OG binds to MgATP-complexed PII, triggering conformational changes that prevent the interaction of PII with either NAGK or PipX [9,37]. In the absence of 2-OG, only the ATP/ADP ratio and concentration of ADP directs the competitive interaction of PII with these targets in vitro. PipX increases the affinity of PII for ADP, and, conversely, the interaction between PII and PipX is highly sensitive to fluctuations in the ATP/ADP ratio [40]. Since the same surface of PII binds either NAGK or PipX (Figure 1B), these proteins do not form ternary complexes. Although competition between NAGK and PipX for PII binding can be observed in vitro [40,41], the great excess of PII over these two binding partners and the presence of additional actors makes competition a less likely scenario in vivo.

PII also binds to the biotin carboxyl carrier protein (BCCP) of acetyl-CoA carboxylase (ACCase), inhibiting its ability to control acetyl-CoA levels in organisms in which PII is present [42]. PII-dependent inhibition of nitrate transport, known to occur after the addition of ammonium to nitrate-containing cultures or the transfer of cultures to darkness, requires interaction of PII with the NrtD and NrtC subunits of the nitrate transporter (NRT) [28,43–45]. Other recently discovered PII receptors in *Synechocystis* sp. PCC 6803 are the ammonium (Amt1) and urea (UrtE subunit involved) transporters and two proteins of as yet unknown functions (Sll0944/DUF1830/_0891 and Ssr0692) [43].

2-OG stimulates complex formation between the global transcriptional regulator NtcA, a CRP-like protein, and PipX [9], as well as the binding of NtcA to target sites [46] and transcription activation in vitro [47]. The PipX–NtcA complex consists of one active (2-OG bound) NtcA dimer and two PipX molecules. Each NtcA subunit binds one PipX molecule in such a way that it stabilizes the active NtcA conformation and probably helps to recruit RNA polymerase without providing extra DNA contacts [16,48]. Comparative studies of cyanobacterial NtcA and CRP proteins and their interactions with DNA and effectors (2-OG and cAMP) provided additional details of NtcA–PipX functional interactions [22].

Importantly, PipX provides a mechanistic link between PII signaling and gene expression, depending on NtcA, which controls a large regulon in response to nitrogen limitation [23,49]. PipX uses the same surface from its N-terminal domain to bind to either 2-OG-bound NtcA (Figure 1B), stimulating DNA binding and transcriptional activity, or to 2-OG-free PII, to form PipX–PII complexes [16,21]. Here the relative abundance of the excess PII over PipX provides a predictable scenario of competition with NtcA for PipX in vivo, at least under physiological conditions in which the affinity of PipX for NtcA is not optimal.

PII-PipX complexes interact with the transcriptional factor PlmA [29], suggesting a role of nitrogen regulators in the transcriptional control of the yet unknown PlmA regulon. The main contacts of PlmA with PII-PipX complexes appear restricted to surface-exposed elements of PipX, specifically one residue in its Tudor Like Domain (TLD/KOW) and the C-terminal helices, which acquire an open conformation when bound to PII [50]. Therefore, the PipX determinants for binding to PlmA appear be very different from those involved in PipX–NtcA complexes.

A comprehensive summary of relevant complexes and interactions involving PipX and PII proteins are illustrated in Figure 2. The reader is referred to [33] for additional structural information and details on complex formation that are omitted in this review.

2.2. Role of PII and PipX in Cyanobacterial Survival

While genetic inactivation of *pipX* appears to have little impact in *S. elongatus* survival under standard laboratory conditions [17,51], genetic inactivation of *glnB* is not viable in a wild-type background. Unsuccessful attempts to completely segregate *glnB* null alleles thus indicate that the PII protein is essential for survival, a finding in close agreement with the importance of *glnB* genes in many other microorganisms, where inactivation leads to severe growth defects or lethality. Interestingly, the metabolic basis of *glnB* deficiency in the studied microorganisms seems to be diverse (discussed in [18]). In cyanobacteria, the essentiality of PII appears to be related with the importance of PipX–PII complex formation, presumably to counteract PipX functions. In line with this, a small reduction in PipX levels suppresses the need for PII for the survival of *S. elongatus* [18–20]. The importance of PII for cell survival is even more pronounced in nitrogen-rich media [51], conditions in which the inhibitory effect of 2-OG on PipX–PII interactions would not take place.

It is worth noting that cyanobacterial genomes always contain at least as many copies of *glnB* as of *pipX*, with duplications of *pipX* correlating with duplications of *glnB* [52], in line with the idea that a relatively high ratio of PII over PipX is required to counteract unwanted interactions with low-affinity PipX partners. In *Synechococcus* WH5701, a cyanobacterium with two PipX and two PII-like proteins, differences in affinities between PII and PipX paralogs and their binding partners—PipX-I, PipX-II, GlnB-A, GlnB-B, NAGK, or NtcA—presumably increases their regulatory potential. Therefore, by integrating multiple signaling pathways, PII and PipX are likely to play currently unknown roles in adaptation-to-environment situations faced by cyanobacteria.

The basis of the phenomenon that we have called "PipX toxicity" in the absence of PII has been explored by mutational analyses, which initially took advantage of point mutations, either identified as spontaneous suppressors in *glnB* strains or generated in the course of our investigations on PipX determinants for interactions with PII or NtcA. The question of whether PipX toxicity is due to over-activation of the NtcA regulon when there is not enough PII to prevent PipX binding to NtcA has been explored by subjecting *pipX* mutant derivatives to co-activation assays (reporting from NtcA activated promoters *glnB* and *glnN*) and to transcriptomic analysis (in the case of two gain-of-functions mutations in the contexts of NtcA coactivation and PipX toxicity) [19,20,49]. However, a cause–effect relationship between PipX toxicity and the over-expression of NtcA gene targets could not be established [20], and it is thus possible that both over-expression of NtcA gene targets and interactions of PipX with a third partner may contribute to PipX toxicity. In the later context, genetic and transcriptomic analyses have also suggested additional regulatory targets of PipX [49].

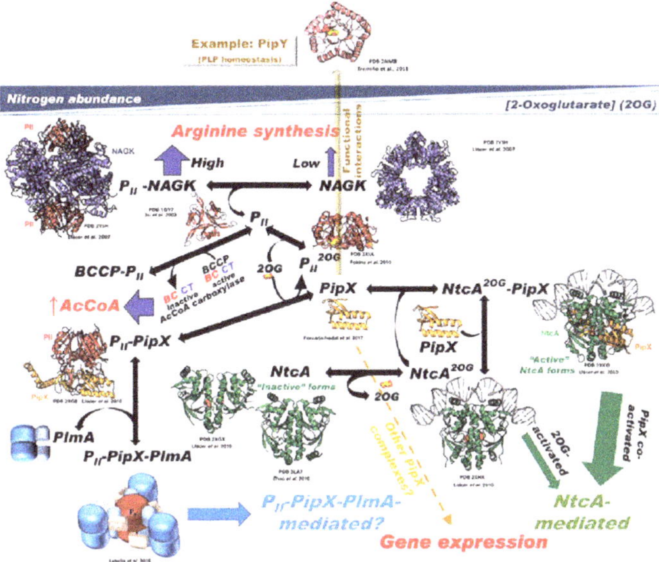

Figure 2. Summary of the PII–PipX–NtcA network of *S. elongatus*. The network illustrates its different elements and complexes depending on nitrogen abundance (inversely related to 2-OG level), as well as the structures of the macromolecules and complexes formed (when known). For PlmA (dimer in darker and lighter blue hues for its dimerization and DNA-binding domains, respectively) and its complex, the architectural coarse model proposed [29] is shown, with the C-terminal helices of PipX (schematized in the extended conformation) illustrated as pink-colored and the two PII molecules in dark red. The DNA complexed with NtcA and with NtcA-PipX is modeled from the structure of DNA–CRP [16], since no DNA–NtcA structure has been reported. BCCP is the biotin carboxyl carrier protein of bacterial acetyl CoA carboxylase (abbreviated AcCoA carboxylase); the other two components of this enzyme, biotin carboxylase and carboxyl transferase, are abbreviated BC and CT, respectively. No structural model of BCC has been shown, because the structure of this component has not been determined in *S. elongatus*, and also because the structures of this protein from other bacteria lack a disordered, 77-residue N-terminal portion that could be highly relevant for interaction with PII. The broken yellow arrow highlights the possibility of further PipX interactions not mediated by NtcA or PII–PlmA, resulting in changes in gene expression [49]. The solid, semi-transparent, yellowish arrow emerging perpendicularly from the flat network symbolizes the possibility of functional interactions of PipX not mediated by physical contacts between the macromolecules involved in the interaction, giving as an example the functional interaction with PipY. Its position outside the network tries to express this different type of interaction (relative to the physical contacts shown in the remainder of the network), as well as to place it outside the field of 2-OG concentrations (taken from [33]).

Interestingly, PII and PipX display distinct localization patterns during diurnal cycles, co-localizing into foci located at the periphery and cell poles during dark periods (Figure 3), a process dependent on energy levels [35]. These dynamic regulatory interactions would play circadian-independent roles in the attenuation of transcriptional activity and other functions during dark periods, while facilitating the return to the essential, and energy costly, light-dependent or light-induced activities.

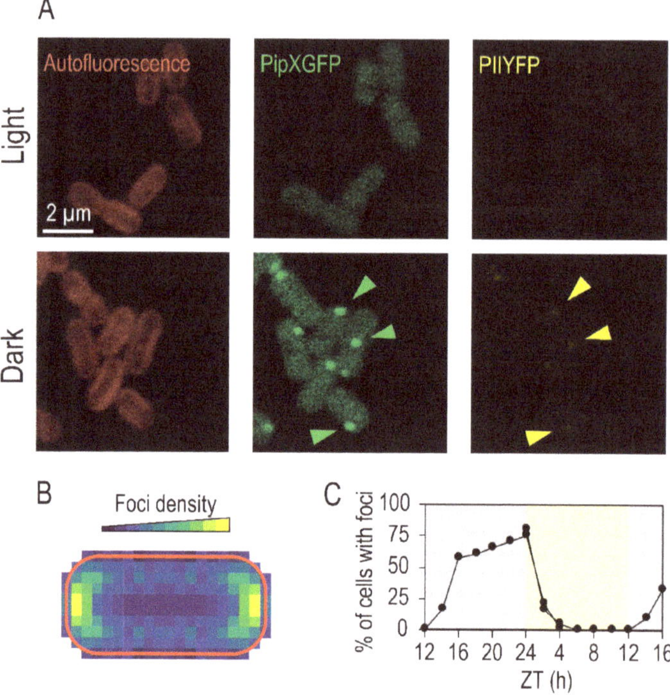

Figure 3. PipXGFP and PIIYFP fusion proteins localize into foci in darkness. (**A**) *S. elongatus* cells were imaged from cultures grown with nitrate in the light or dark (8 h). The autofluorescence (red), PipXGFP (green), and PIIYFP (yellow) signals are shown as indicated. (**B**) Heatmap of PipXGFP distribution relative to the cell autofluorescence border (in red). Colors represent the density of foci at each cellular position. (**C**) Percentage of the cell population that showed PipXGFP foci in diurnal conditions (light and dark conditions corresponding to yellow and gray regions, respectively). Cells were entrained in light–dark cycles and sampled every 2 hours in either the light (ZT 0–12) or dark (ZT 12–24). ZT (Zeitgeber time) refers to the time relative to "lights on" (adapted from [35]).

2.3. PipX Role in Gene Expression and Interactions with the Unique Transcriptional Regulators NtcA and PlmA

In cyanobacteria, multiple metabolic and developmental processes are induced by nitrogen starvation. NtcA, the global regulator for nitrogen control, activates genes involved in nitrogen assimilation, heterocyst differentiation, and acclimation to nitrogen starvation [14,15,53–56]. The interaction between PipX and NtcA is known to be relevant under nitrogen limitation for the activation of NtcA-dependent genes in *S. elongatus* and *Anabaena* sp. Strain PCC 7120 [9,17,24,26], and presumably in *Prochlorococcus* [27].

From a combination of genetic, transcriptomic, and multivariate analyses, we previously obtained groups of genes differentially regulated by PipX that have improved the definition of the consensus NtcA binding motif for *S. elongatus* and provided further insights into the function of NtcA–PipX complexes. Importantly, additional groups of genes that are differentially regulated by PipX suggested the involvement of PipX in NtcA-independent regulatory pathways, indicating that PipX is involved in a much wider interaction network affecting nitrogen assimilation, translation, and photosynthesis [49].

PlmA, the other transcriptional regulator interacting with PipX (as part of PipX–PII complexes) belongs to the GntR super-family of transcriptional regulators [57]. It is involved in photosystem stoichiometry in *Synechocystis* sp. PCC 6803 [58], plasmid maintenance in *Anabaena* sp. strain PCC

7120 [59], and in regulation of the highly conserved cyanobacterial sRNA YFR2 in marine picocyanobacteria [60]. Recently, it has been shown that in *Synechocystis* sp. PCC 6803, PlmA is reduced by the thioredoxin TrxM [61], suggesting that its function may depend on the redox status.

NtcA and PlmA belong to large families of transcriptional regulators (CRP and GntR). Both are universally present in cyanobacteria, and their presence is also restricted to this group (Figure 4), as is the case with PipX. Therefore, the three proteins are hallmarks of cyanobacteria [29,57,59]. In contrast to the already abundant structural and functional details about NtcA, very little is still known about PlmA, but its unique and exclusive distribution within cyanobacteria suggest that PlmA functions are important to cope with regulatory signals and metabolites that are relevant in the context of photosynthesis.

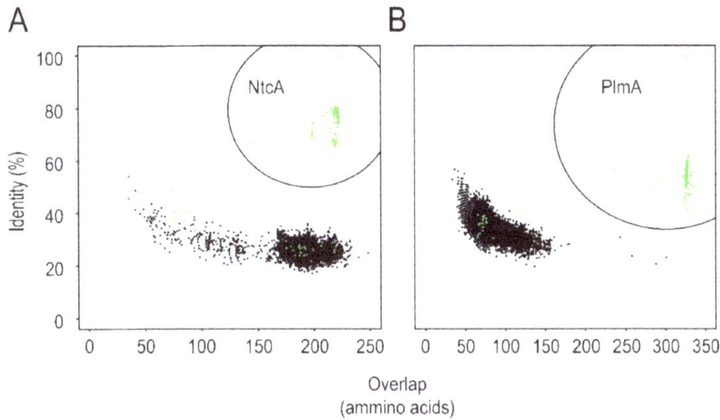

Figure 4. The NtcA (**A**) and PlmA (**B**) subfamilies. Bidirectional blast queries with *S. elongatus* PlmA and NtcA sequences as queries resulted in 33.903 and 102 738 sequences homologous to NtcA and PlmA, respectively, from the Refseq genomic bacterial database. Horizontal and vertical axes represent, respectively, the amino acid overlap and identity of each hit. Green dots, accounting for 285 (in **A**) and 264 (in **B**) hits, were retrieved from cyanobacterial genomes. Those dots corresponding to cyanobacterial NtcA and PlmA orthologs (238 and 234 hits, respectively) are encircled (adapted from [29]).

3. Regulatory Connections between PipX and the Conserved PLP-Binding Protein PipY

3.1. The Tight Link between pipX and pipY Genes in Cyanobacteria

Inspection of cyanobacterial genomes [30,32] revealed a tight connection between *pipX* and its downstream gene, named *pipY*, encoding a member of the widely distributed family of pyridoxal phosphate (PLP) binding proteins (PLPBP; COG0325), which control vitamin B6 and amino acid cell homeostasis. In *S. elongatus*, where the two genes form an operon, PipX increases expression of either *pipY* or a reporter gene occupying the *pipY* locus [31], thus suggesting the importance of the PipX–PipY balance. On the other hand, and given the relatively low number of genes that form part of polycistronic units (55% for *S. elongatus*), it is significant that, in almost 80% of the cyanobacterial genomes analyzed, *pipX* genes were found adjacent to *pipY* genes (always upstream and in the same orientation), presumably as part of bicistronic *pipXY* operons. In addition, tight co-regulation and even translational coupling can be inferred by the relatively short or non-existent intergenic distances found between contiguous *pipX* and *pipY* coding sequences, strongly suggesting a functional interaction between PipX and PipY in most, if not all, cyanobacteria. When not adjacent, as in *Synechocystis* strains, *pipX* sequences constitute monocistronic operons. Figure 5 summarizes *in silico* evidence for the tight link between *pipX* and *pipY* genes found in cyanobacteria genomes.

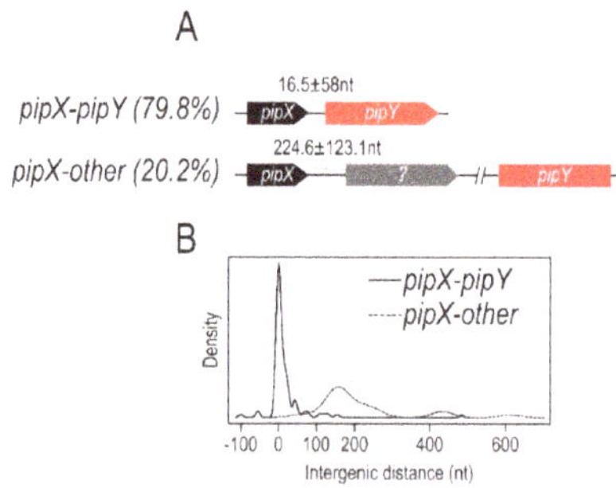

Figure 5. *pipX–pipY* and other gene arrangements. The two possible situations concerning *pipX* and *pipX* genes found in cyanobacterial genomes are represented: *pipX* (black) is followed by *pipY* (red), always in the same orientation (as illustrated by the direction of the arrows), or by apparently non-related and diverse genes (illustrated by a grey arrow). For each situation, the corresponding percentage found in cyanobacterial genomes is shown in brackets. The intergenic distance between *pipX* and downstream genes in cyanobacterial genomes is shown as the mean and standard deviation (**A**) and density curve (**B**). (Adapted from [31]).

An open question is whether PipX and PipY proteins physically interact in cyanobacteria, which is so far unsupported by yeast two-hybrid assays [30]. However, a putative interaction between PipX and PipY may require another factor(s) present in *S. elongatus* but not in yeast nuclei, and thus PipX–PipY complex formation should not be ruled out yet.

3.2. Common Structural and Functional Features between PipY and Other PLPBP Members

Structures of PipY homologs from phylogenetically distant organisms, such as yeast (Ybl036c) and cyanobacteria (PipY), have already been experimentally solved or modelled (human PLPHP), and a remarkable conservation is found when both protein sequence and three-dimensional (3D) structures are compared [62–64]. PLPBP structures reveal a single domain monomer, folded as the TIM barrel of type III-fold PLP enzymes, with the PLP cofactor solvent exposed. These structures all have an α-helical extension of the C-terminal β-strand binding the phosphate of PLP, which could act as a trigger for PLP exchange (Figure 6) [62,64].

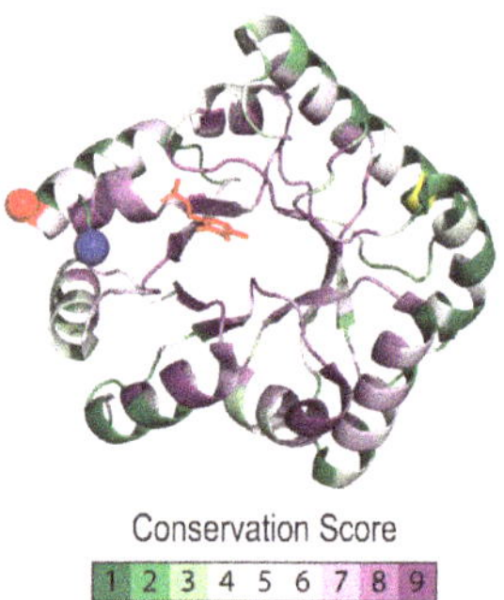

Figure 6. Pyridoxal phosphate binding protein (PLPBP) residue conservation and distinctive structural features. Conservation scores were calculated using CONSURF [65] with COG0325 alignment and a phylogenetic tree from the EGGNOG database as query. The color code illustrates lowest to highest conservation in a green to purple gradient, with yellow for non-informative residues, plotted over the Synpcc7942_2060 (PDB 5NM8) structure. The N-terminal and C-terminal ends are indicated with a red and blue sphere, respectively. The PLP molecule is colored in red.

Although our initial interest in PipY stems from its predicted involvement in the cyanobacterial nitrogen regulatory network, the idea that PLPBP members may perform the same basic functions in the context of amino acid and PLP homeostasis in all types of cells has gained strength during the course of our studies. It was concluded that PipY performs the same basic functions inferred for YggS (the *E. coli* homolog), and PLPHP and is therefore a bona fide member of the PLPBP family. In particular, *S. elongatus pipY* mutants [30], like *E. coli yggS* mutants, showed sensitivity to pyridoxine (PN) and an imbalance of the amino/keto acid pools. *pipY* mutants also showed sensitivity to antibiotics targeting essential PLP holoenzymes. Importantly, distantly-related PLPBP members were able to rescue species (or even strain)-specific defects, such as Val overproduction and the PN sensitivity phenotypes of *E. coli yggS* mutants. These two *E. coli yggS* mutant phenotypes were respectively rescued by heterologous expression of orthologs from bacilli (YlmE), yeast (Ybl036c), and humans [66]; and of orthologs from plants (*Zea mays, Arabidopsis thaliana*), yeast, and humans [67]. In humans, mutations affecting protein levels or PLP-binding at the PLPHP causes vitamin B6-dependent epilepsy [64,67,68].

Synthetic lethality between PLPBP and PLP holoenzymes, previously reported for *E. coli yggS* and either *glyA* [69,70] or *serA* [71], has also been found between *pipY* and *cysK*, the two most abundant PLP-binding proteins in *S. elongatus* [30]. Synthetic lethality probably reflects the common involvement of the corresponding protein pairs in amino acid and PLP homeostasis, but may also indicate that any relatively abundant PLP-binding protein could fulfill a role as a PLP reservoir, thus helping to cope with PLP toxicity. In this context, it is worth emphasizing that the cofactor in PLPBP is solvent-exposed, and thus is best suited for a role as a PLP delivering modules for essential apo-enzymes in cells.

While the mechanistic details affecting PLPBP-mediated PLP homeostasis and those concerning the regulatory connections between PipY and PipX remain to be elucidated, the apparent recruitment of PipY into the 2-OG-dependent nitrogen interaction network, and its connection with PipX in cyanobacteria (see below), provides a unique opportunity to investigate the functions of a bona fide PLPBP member in the context of a relatively characterized signaling network in a bacterial model system.

3.3. The Intriguing Connection between PipY and the Cyanobacterial Nitrogen Network

Two experimental lines of evidence place PipX and PipY within the same regulatory pathway: the two proteins contribute to *S. elongatus* resistance to PLP-targeting antibiotics D-cycloserine and β-chloro-D-alanine, and affect expression of a common set of transcripts.

In particular, transcriptomic analysis with single and double *pipX* and *pipY* null mutants has revealed the implication of both PipX and PipY in the control of NtcA activated transcripts, where PipY plays a positive role [30]. In the context of NtcA-independent transcripts, PipX and PipY could have similar or opposite effects, suggesting a rather complex regulation [30].

In addition, transcriptomic analysis with *pipX* null and gain-of-function mutant derivatives was consistent with PipX having a role as a repressor of many photosynthesis- and translation-related genes independent of NtcA [49]. The model emerging from those analyses was that of PipX implicated in the fine-tuning of different regulons, of which only the NtcA one was anticipated.

To explain the positive role of PipY on the levels of NtcA-activated transcripts, it is tempting to propose that PipY increases the levels of 2-OG in *S. elongatus* cells—that is, PipY activity would alter nitrogen signaling in cyanobacteria by affecting the levels of 2-OG, a possibility worth considering given the importance of PLPBP proteins in maintaining the balance of the amino/keto acid pools. Similarly, the expression of NtcA-independent transcripts co-regulated by PipX–PipY may also respond to amino/keto acids, or to PLP-related compounds, via a yet-unidentified factor. Here it is tempting to speculate that the transcriptional regulator PlmA may be involved. Figure 7 illustrates our current working model, integrating PipY and PlmA into the PipX regulatory network.

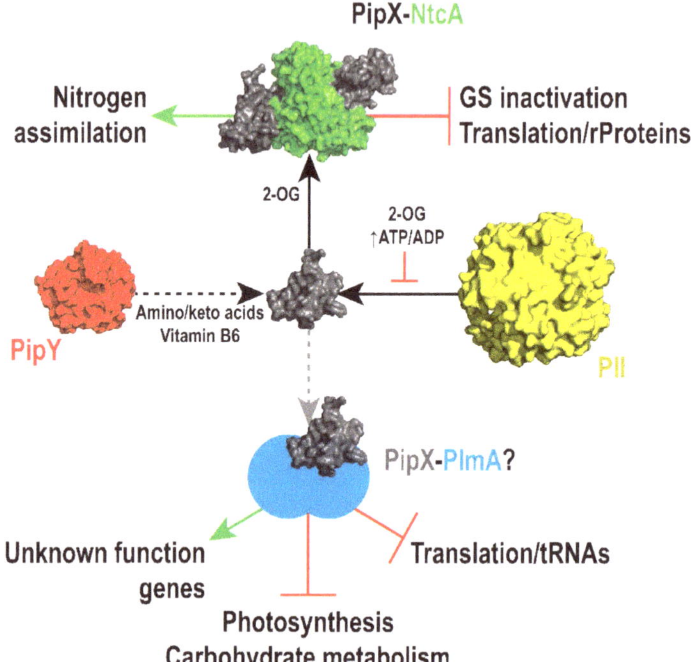

Figure 7. Model of PipX function in transcriptional regulation. PipX, as the NtcA co-regulator, is involved in the (2-OG dependent) activation of multiple operons for nitrogen assimilation, as well as in the repression of translation-related genes and inhibitors of key nitrogen assimilation genes (glutamine synthetase). PipX also works independently of NtcA in the regulation of key processes, including translation, photosynthesis, and carbohydrate metabolism, presumably in a complex with at least one other transcriptional regulator, likely PlmA, whose interaction with PipX depends on PII, and would be regulated by 2-OG and ATP/ADP. PipY may affect all PipX regulatory interactions by altering the levels of amino/keto acids and vitamin B6. Positive and negative regulation are depicted by green arrows and red lines, respectively. Black solid arrows illustrate the formation of complexes driven by the indicated regulatory signals. Dotted arrows indicate putative regulatory interactions by PipY or by hypothetical complex(es) (depicted in a lighter color arrow) that may involve PlmA.

4. Unknown Functions of PipX and the Synteny Approach

4.1. PipX and NusG Family Proteins Share a Domain Involved in Operon Polarity

While investigating regulatory connections between *pipX* and *pipY* genes, we became aware of a rather intriguing function of PipX: polarity suppression or the enhancement of *pipY* expression, specifically in *cis* [31]. Most intriguing was the finding that the TLD/KOW structural domain of PipX is shared by NusG proteins, since NusG paralogs typified by RfaH are non-essential, operon-specific bacterial factors involved in the upregulation of horizontally-acquired genes, normally located downstream within the same operon [72–74]. Not underestimating the fact that the NusG paralogs act in *trans*, PipX plays the same role over the downstream gene *pipY*, opening the question of whether PipX can also use the TLD/KOW domain to interact with the translation machinery (Figure 8A), as NusG proteins do. On the one hand, key NusG residues contacting ribosomal protein S10 are conserved between PipX and RfaH [31], suggesting the possibility of a similar interaction between PipX and S10. On the other hand, the C-terminal domain of PipX contains a basic arginine-rich patch (Figure 8B) that provides interaction determinants in non-canonical RNA binding proteins [75], as it is compatible with a putative role of PipX in RNA binding. It is therefore tempting to speculate that PipX may also be involved in the regulation of gene expression at the level of translation.

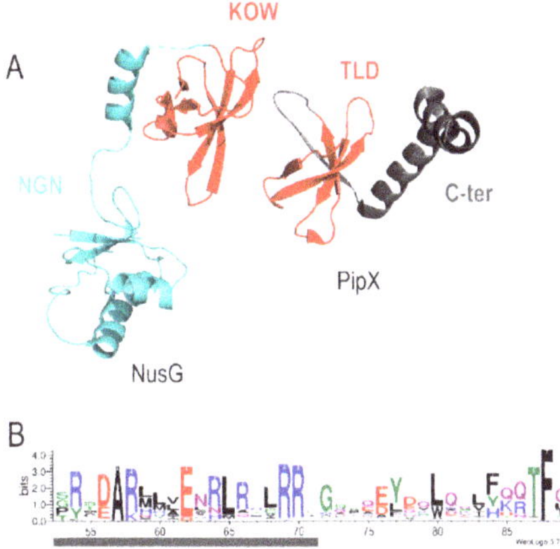

Figure 8. Structural features of PipX, discussed in the context of a possible role in translation regulation. (**A**) Whole length protein structures of NusG from *E. coli* (PDB 5TBZ:J) and PipX from *S. elongatus* (PDB 2XG8:D). TLD/KOW domains are shown in red, the N-terminal domain of NusG (NGN) in blue, and the C-terminal domain of PipX in grey. (**B**) WebLogo of the C-terminal domain of PipX. The first alpha helix containing an arginine-rich patch is underlined in grey (adapted from [31]).

4.2. The PipX Synteny Network

As previously mentioned, the discovery of PipX was due to the implementation of the "guilty by association" principle implicit in protein–protein interaction screenings. The same principle, applied in the context of gene linkage, led us to connect PipY with the cyanobacterial nitrogen regulation network. Given that in this phylum, most signaling proteins are encoded in monocistronic units, we have recently taken this "guilty by association" principle a step further, to look for genes that, independent of their operon structures, are closely associated with PipX or PipY in cyanobacterial genomes, and may thus be functionally connected. For this we turned to Cyanobacterial Linked Genome [32], a database generated on the basis of conservation of gene neighborhoods across cyanobacterial species, accessed through an interactive platform. The default outcome places PipX and PipY as part of a relatively robust six-node network (Figure 9), raising questions on the possible functional connections amongst them. Because one of the network nodes is EngA (YphC/Der/YfgK), a GTPase involved in ribosome assembly [76,77] that could play a role in the coordination of photosynthesis activity with ribosome function [78,79], current work is now focused in investigating its functional connections with PipX.

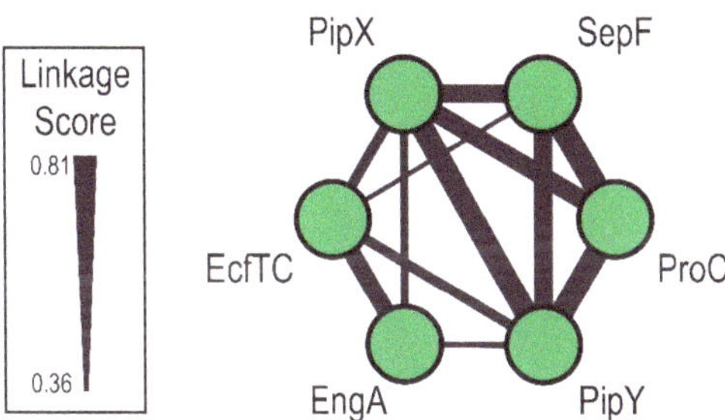

Figure 9. The PipX synteny network. Proteins are represented as nodes/circles, and syntenic relationships as lines whose width is proportional to the corresponding linkage score between nodes (adapted from [32]).

5. Conclusions

Cyanobacteria have developed sophisticated mechanisms to adapt their metabolic processes to important environmental challenges, like those imposed by the succession of days and nights. However, the regulatory mechanisms and molecular details behind the versatility and environmental adaptations of cyanobacteria are largely unknown, and the study of unique proteins like PipX, which is restricted to this phylum, provides an opportunity to unravel some of them. Particularly challenging is the identification of the biological processes and environmental contexts in which a rather promiscuous protein like PipX participates, a task that has been fueled by "guilty by association" approaches, in particular protein–protein interactions and genetic linkage. In this context, bioinformatic tools like the recently developed Cyanobacterial Linked Genome should also help to provide working hypotheses in the context of other proteins that may be unique to cyanobacteria.

Author Contributions: Conceptualization, A.C and J.I.L.; writing—original draft preparation, A.C.; writing—review and editing, J.I.L., R.C., P.S., J.E., and A.C.; funding acquisition, A.C. All authors have read and agreed to the published version of the manuscript.

Funding: Generalitat Valenciana: PROMETEO/2017/129; Universidad de Alicante: VIGROB-126/19; UAUSTI18 and FPUUA59 fellowship to J.I.L.

Acknowledgments: We would like to thank V. Rubio and R. Dixon for critical reading of the manuscript.

Conflicts of Interest: The authors declare no conflict of interest.

References

1. Blank, C.E.; Sánchez-Baracaldo, P. Timing of morphological and ecological innovations in the cyanobacteria—A key to understanding the rise in atmospheric oxygen. *Geobiology* **2010**, *8*, 1–23. [CrossRef] [PubMed]

2. Khan, S.; Fu, P. Biotechnological perspectives on algae: A viable option for next generation biofuels. *Curr. Opin. Biotechnol.* **2020**, *62*, 146–152. [CrossRef] [PubMed]

3. Forchhammer, K.; Selim, K.A. Carbon/nitrogen homeostasis control in cyanobacteria. *FEMS Microbiol. Rev.* **2019**, *44*, 33–53. [CrossRef]

4. Zhang, C.-C.; Zhou, C.-Z.; Burnap, R.L.; Peng, L. Carbon/Nitrogen Metabolic Balance: Lessons from Cyanobacteria. *Trends Plant Sci.* **2018**, *23*, 1116–1130. [CrossRef] [PubMed]

5. Senior, P.J. Regulation of nitrogen metabolism in *Escherichia coli* and *Klebsiella aerogenes*: Studies with the continuous culture technique. *J. Bacteriol.* **1975**, *123*, 407–418. [CrossRef] [PubMed]

6. Huergo, L.F.; Dixon, R. The Emergence of 2-Oxoglutarate as a Master Regulator Metabolite. *Microbiol. Mol. Biol. Rev.* **2015**, *79*, 419–435. [CrossRef]

7. Stanier, R.Y.; Bazine, G.C. Phototrophic Prokaryotes: The Cyanobacteria. *Annu. Rev. Microbiol.* **1977**, *31*, 225–274. [CrossRef]

8. Robles-Rengel, R.; Florencio, F.J.; Muro-Pastor, M.I. Redox interference in nitrogen status via oxidative stress is mediated by 2-oxoglutarate in cyanobacteria. *New Phytol.* **2019**, *224*, 216–228. [CrossRef]

9. Espinosa, J.; Forchhammer, K.; Burillo, S.; Contreras, A. Interaction network in cyanobacterial nitrogen regulation: PipX, a protein that interacts in a 2-oxoglutarate dependent manner with PII and NtcA. *Mol. Microbiol.* **2006**, *61*, 457–469. [CrossRef]

10. Burillo, S.; Luque, I.; Fuentes, I.; Contreras, A. Interactions between the nitrogen signal transduction protein PII and N-acetyl glutamate kinase in organisms that perform oxygenic photosynthesis. *J. Bacteriol.* **2004**, *186*, 3346–3354. [CrossRef]

11. Fields, S.; Song, O.K. A novel genetic system to detect protein-protein interactions. *Nature* **1989**, *340*, 245–246. [CrossRef] [PubMed]

12. Chellamuthu, V.R.; Alva, V.; Forchhammer, K. From cyanobacteria to plants: Conservation of PII functions during plastid evolution. *Planta* **2013**, *237*, 451–462. [CrossRef] [PubMed]

13. Heinrich, A.; Maheswaran, M.; Ruppert, U.; Forchhammer, K. The *Synechococcus elongatus* PII signal transduction protein controls arginine synthesis by complex formation with N-acetyl-L-glutamate kinase. *Mol. Microbiol.* **2004**, *52*, 1303–1314. [CrossRef] [PubMed]

14. Herrero, A.; Muro-Pastor, A.M.; Flores, E. Nitrogen Control in Cyanobacteria. *J. Bacteriol.* **2001**, *183*, 411–425. [CrossRef] [PubMed]

15. Esteves-Ferreira, A.A.; Inaba, M.; Fort, A.; Araújo, W.L.; Sulpice, R. Nitrogen metabolism in cyanobacteria: Metabolic and molecular control, growth consequences and biotechnological applications. *Crit. Rev. Microbiol.* **2018**, *44*, 541–560. [CrossRef]

16. Llácer, J.L.; Espinosa, J.; Castells, M.A.; Contreras, A.; Forchhammer, K.; Rubio, V. Structural basis for the regulation of NtcA-dependent transcription by proteins PipX and PII. *Proc. Natl. Acad. Sci. USA* **2010**, *107*, 15397–15402. [CrossRef]

17. Espinosa, J.; Forchhammer, K.; Contreras, A. Role of the *Synechococcus* PCC 7942 nitrogen regulator protein PipX in NtcA-controlled processes. *Microbiology* **2007**, *153*, 711–718. [CrossRef]

18. Espinosa, J.; Castells, M.A.; Laichoubi, K.B.; Contreras, A. Mutations at *pipX* suppress lethality of PII-deficient mutants of *Synechococcus elongatus* PCC 7942. *J. Bacteriol.* **2009**, *191*, 4863–4869. [CrossRef]

19. Espinosa, J.; Castells, M.A.; Laichoubi, K.B.; Forchhammer, K.; Contreras, A. Effects of spontaneous mutations in PipX functions and regulatory complexes on the cyanobacterium *Synechococcus elongatus* strain PCC 7942. *Microbiology* **2010**, *156*, 1517–1526. [CrossRef]

20. Laichoubi, K.B.; Espinosa, J.; Castells, M.A.; Contreras, A. Mutational analysis of the cyanobacterial nitrogen regulator PipX. *PLoS ONE* **2012**, *7*, e35845. [CrossRef]

21. Zhao, M.X.; Jiang, Y.L.; Xu, B.Y.; Chen, Y.; Zhang, C.C.; Zhou, C.Z. Crystal structure of the cyanobacterial signal transduction protein PII in complex with PipX. *J. Mol. Biol.* **2010**, *402*, 552–559. [CrossRef] [PubMed]

22. Forcada-Nadal, A.; Forchhammer, K.; Rubio, V. SPR analysis of promoter binding of *Synechocystis* PCC6803 transcription factors NtcA and CRP suggests cross-talk and sheds light on regulation by effector molecules. *FEBS Lett.* **2018**, *592*, 2378. [CrossRef] [PubMed]

23. Giner-Lamia, J.; Robles-Rengel, R.; Hernández-Prieto, M.A.; Isabel Muro-Pastor, M.; Florencio, F.J.; Futschik, M.E. Identification of the direct regulon of NtcA during early acclimation to nitrogen starvation in the cyanobacterium *Synechocystis* sp. PCC 6803. *Nucleic Acids Res.* **2017**, *45*, 11800–11820. [CrossRef]

24. Valladares, A.; Rodríguez, V.; Camargo, S.; Martínez-Noël, G.M.A.; Herrero, A.; Luque, I. Specific role of the cyanobacterial *pipX* factor in the heterocysts of *Anabaena* sp. strain PCC 7120. *J. Bacteriol.* **2011**, *193*, 1172–1182. [CrossRef] [PubMed]

25. Chen, H.L.; Bernard, C.S.; Hubert, P.; My, L.; Zhang, C.C. Fluorescence resonance energy transfer based on interaction of PII and PipX proteins provides a robust and specific biosensor for 2-oxoglutarate, a central metabolite and a signalling molecule. *FEBS J.* **2014**, 1742–4658. [CrossRef] [PubMed]

26. Camargo, S.; Valladares, A.; Forchhammer, K.; Herrero, A. Effects of PipX on NtcA-dependent promoters and characterization of the cox3 promoter region in the heterocyst-forming cyanobacterium *Anabaena* sp. PCC 7120. *FEBS Lett.* **2014**, *588*, 1787–1794. [CrossRef] [PubMed]

27. Domínguez-Martín, M.A.; López-Lozano, A.; Clavería-Gimeno, R.; Velázquez-Campoy, A.; Seidel, G.; Burkovski, A.; Díez, J.; García-Fernández, J.M. Differential NtcA responsiveness to 2-oxoglutarate underlies the diversity of C/N balance regulation in *Prochlorococcus*. *Front. Microbiol.* **2018**, *8*, 2641. [CrossRef]

28. Ohashi, Y.; Shi, W.; Takatani, N.; Aichi, M.; Maeda, S.; Watanabe, S.; Yoshikawa, H.; Omata, T. Regulation of nitrate assimilation in cyanobacteria. *J. Exp. Bot.* **2011**, *62*, 1411–1424. [CrossRef]

29. Labella, J.I.; Obrebska, A.; Espinosa, J.; Salinas, P.; Forcada-Nadal, A.; Tremiño, L.; Rubio, V.; Contreras, A. Expanding the Cyanobacterial Nitrogen Regulatory Network: The GntR-Like Regulator PlmA Interacts with the PII-PipX Complex. *Front. Microbiol.* **2016**, *7*, 1677. [CrossRef]

30. Labella, J.I.; Cantos, R.; Espinosa, J.; Forcada-Nadal, A.; Rubio, V.; Contreras, A. PipY, a Member of the Conserved COG0325 Family of PLP-Binding Proteins, Expands the Cyanobacterial Nitrogen Regulatory Network. *Front. Microbiol.* **2017**, *8*, 1244. [CrossRef]

31. Cantos, R.; Labella, J.I.; Espinosa, J.; Contreras, A. The nitrogen regulator PipX acts in cis to prevent operon polarity. *Environ. Microbiol. Rep.* **2019**, *11*, 495–507. [CrossRef]

32. Labella, J.I.; Llop, A.; Contreras, A. The default cyanobacterial linked genome: An interactive platform based on cyanobacterial linkage networks to assist functional genomics. *FEBS Lett.* **2020**, *10*, 1873–3468. [CrossRef] [PubMed]

33. Forcada-Nadal, A.; Llácer, J.L.; Contreras, A.; Marco-Marín, C.; Rubio, V. The PII-NAGK-PipX-NtcA regulatory axis of cyanobacteria: A tale of changing partners, allosteric effectors and non-covalent interactions. *Front. Mol. Biosci.* **2018**, *5*, 91. [CrossRef] [PubMed]

34. Guerreiro, A.C.L.; Benevento, M.; Lehmann, R.; Van Breukelen, B.; Post, H.; Giansanti, P.; Altelaar, A.F.M.; Axmann, I.M.; Heck, A.J.R. Daily rhythms in the cyanobacterium *Synechococcus elongatus* probed by high-resolution mass spectrometry-based proteomics reveals a small defined set of cyclic proteins. *Mol. Cell. Proteom.* **2014**, *13*, 2042–2055. [CrossRef]

35. Espinosa, J.; Labella, J.I.; Cantos, R.; Contreras, A. Energy drives the dynamic localization of cyanobacterial nitrogen regulators during diurnal cycles. *Environ. Microbiol.* **2018**, *20*, 1240–1252. [CrossRef] [PubMed]

36. Kamberov, E.S.; Atkinson, M.R.; Ninfa, A.J. The *Escherichia coli* PII signal transduction protein is activated upon binding 2-ketoglutarate and ATP. *J. Biol. Chem.* **1995**, *270*, 17797–17807. [CrossRef]

37. Llácer, J.L.; Contreras, A.; Forchhammer, K.; Marco-Marín, C.; Gil-Ortiz, F.; Maldonado, R.; Fita, I.; Rubio, V. The crystal structure of the complex of PII and acetylglutamate kinase reveals how PII controls the storage of nitrogen as arginine. *Proc. Natl. Acad. Sci. USA* **2007**, *104*, 17644–17649. [CrossRef]

38. Llácer, J.L.; Fita, I.; Rubio, V. Arginine and nitrogen storage. *Curr. Opin. Struct. Biol.* **2008**, *18*, 673–681. [CrossRef] [PubMed]

39. Selim, K.A.; Ermilova, E.; Forchhammer, K. From Cyanobacteria to Archaeplastida: New evolutionary insights into PII signaling in the plant kingdom. *New Phytol.* **2020**. [CrossRef]

40. Zeth, K.; Fokina, O.; Forchhammer, K. Structural basis and target-specific modulation of ADP sensing by the *Synechococcus elongatus* PII signaling protein. *J. Biol. Chem.* **2014**, *298*, 8960–8972. [CrossRef]

41. Lüddecke, J.; Forchhammer, K. Energy sensing versus 2-oxoglutarate dependent ATPase switch in the control of *Synechococcus* PII interaction with its targets NAGK and PipX. *PLoS ONE* **2015**, *10*, e0137114. [CrossRef] [PubMed]

42. Hauf, W.; Schmid, K.; Gerhardt, E.C.M.; Huergo, L.F.; Forchhammer, K. Interaction of the nitrogen regulatory protein GlnB (PII) with biotin carboxyl carrier protein (BCCP) controls acetyl-Coa levels in the cyanobacterium *Synechocystis* sp. PCC 6803. *Front. Microbiol.* **2016**, *7*, 1700. [CrossRef] [PubMed]

43. Watzer, B.; Spät, P.; Neumann, N.; Koch, M.; Sobotka, R.; MacEk, B.; Hennrich, O.; Forchhammer, K. The signal transduction protein PII controls ammonium, nitrate and urea uptake in cyanobacteria. *Front. Microbiol.* **2019**, *10*, 1428. [CrossRef] [PubMed]

44. Lee, H.-M.; Flores, E.; Herrero, A.; Houmard, J.; Tandeau de Marsac, N. A role for the signal transduction protein PII in the control of nitrate/nitrite uptake in a cyanobacterium. *FEBS Lett.* **1998**, *427*, 291–295. [CrossRef]

45. Kobayashi, M.; Rodríguez, R.; Lara, C.; Omata, T. Involvement of the C-terminal domain of an ATP-binding subunit in the regulation of the ABC-type nitrate/nitrite transporter of the cyanobacterium *Synechococcus* sp. Strain PCC 7942. *J. Biol. Chem.* **1997**, *272*, 27197–27201. [CrossRef] [PubMed]

46. Vázquez-Bermúdez, M.F.; Herrero, A.; Flores, E. 2-Oxoglutarate increases the binding affinity of the NtcA (nitrogen control) transcription factor for the *Synechococcus glnA* promoter. *FEBS Lett.* **2002**, *512*, 71–74. [CrossRef]

47. Tanigawa, R.; Shirokane, M.; Maeda, S.I.; Omata, T.; Tanaka, K.; Takahashi, H. Transcriptional activation of NtcA-dependent promoters of *Synechococcus* sp. PCC 7942 by 2-oxoglutarate *in vitro*. *Proc. Natl. Acad. Sci. USA* **2002**, *99*, 4251–4255. [CrossRef]

48. Zhao, M.X.; Jiang, Y.L.; He, Y.X.; Chen, Y.F.; Teng, Y.B.; Chen, Y.; Zhang, C.C.; Zhou, C.Z. Structural basis for the allosteric control of the global transcription factor NtcA by the nitrogen starvation signal 2-oxoglutarate. *Proc. Natl. Acad. Sci. USA* **2010**, *107*, 12487–12492. [CrossRef]

49. Espinosa, J.; Rodríguez-Mateos, F.; Salinas, P.; Lanza, V.F.; Dixon, R.; De La Cruz, F.; Contreras, A. PipX, the coactivator of NtcA, is a global regulator in cyanobacteria. *Proc. Natl. Acad. Sci. USA* **2014**, *111*, E2423–E2430. [CrossRef]

50. Forcada-Nadal, A.; Palomino-Schätzlein, M.; Neira, J.L.; Pineda-Lucena, A.; Rubio, V. The PipX Protein, When Not Bound to Its Targets, Has Its Signaling C-Terminal Helix in a Flexed Conformation. *Biochemistry* **2017**, *56*, 3211–3224. [CrossRef]

51. Chang, Y.; Takatani, N.; Aichi, M.; Maeda, S.I.; Omata, T. Evaluation of the effects of PII deficiency and the toxicity of PipX on growth characteristics of the PII-less mutant of the cyanobacterium *Synechococcus elongatus*. *Plant Cell Physiol.* **2013**, *45*, 1504–1514. [CrossRef] [PubMed]

52. Laichoubi, K.B.; Beez, S.; Espinosa, J.; Forchhammer, K.; Contreras, A. The nitrogen interaction network in *Synechococcus* WH5701, a cyanobacterium with two PipX and two P II-like proteins. *Microbiology* **2011**, *157*, 1220–1228. [CrossRef] [PubMed]

53. Luque, I.; Zabulon, G.; Contreras, A.; Houmard, J. Convergence of two global transcriptional regulators on nitrogen induction of the stress-acclimation gene *nblA* in the cyanobacterium *Synechococcus* sp. PCC 7942. *Mol. Microbiol.* **2001**, *41*, 937–947. [CrossRef] [PubMed]

54. Sauer, J.; Dirmeier, U.; Forchhammer, K. The *Synechococcus* strain PCC 7942 *glnN* product (Glutamine Synthetase III) helps recovery from prolonged nitrogen chlorosis. *J. Bacteriol.* **2000**, *182*, 5615–5619. [CrossRef]

55. Luque, I.; Contreras, A.; Zabulon, G.; Herrero, A.; Houmard, J. Expression of the glutamyl-tRNA synthetase gene from the cyanobacterium *Synechococcus* sp. PCC 7942 depends on nitrogen availability and the global regulator NtcA. *Mol. Microbiol.* **2002**, *41*, 937–947. [CrossRef]

56. Herrero, A.; Flores, E. Genetic responses to carbon and nitrogen availability in *Anabaena*. *Environ. Microbiol.* **2019**, *21*, 1–17. [CrossRef]

57. Hoskisson, P.A.; Rigali, S. Chapter 1 Variation in Form and Function. In *Advances in Applied Microbiology*; Academic Press: Cambridge, MA, USA, 2009; pp. 1–22. ISBN 9780123748249.

58. Fujimori, T.; Higuchi, M.; Sato, H.; Aiba, H.; Muramatsu, M.; Hihara, Y.; Sonoike, K. The Mutant of sll1961, Which Encodes a Putative Transcriptional Regulator, Has a Defect in Regulation of Photosystem Stoichiometry in the Cyanobacterium *Synechocystis* sp. PCC 6803. *Plant Physiol.* **2005**, *139*, 408–416. [CrossRef]

59. Lee, M.H.; Scherer, M.; Rigali, S.; Golden, J.W. PlmA, a new member of the GntR family, has plasmid maintenance functions in *Anabaena* sp. strain PCC 7120. *J. Bacteriol.* **2003**, *185*, 4315–4325. [CrossRef]

60. Lambrecht, S.J.; Wahlig, J.M.L.; Steglich, C. The GntR family transcriptional regulator PMM1637 regulates the highly conserved cyanobacterial sRNA Yfr2 in marine picocyanobacteria. *DNA Res.* **2018**, *28*, 489–497. [CrossRef]

61. Kujirai, J.; Nanba, S.; Kadowaki, T.; Oka, Y.; Nishiyama, Y.; Hayashi, Y.; Arai, M.; Hihara, Y. Interaction of the GntR-family transcription factor Sll1961 with thioredoxin in the cyanobacterium *Synechocystis* sp. PCC 6803. *Sci. Rep.* **2018**, *8*, 6666. [CrossRef]

62. Tremiño, L.; Forcada-Nadal, A.; Contreras, A.; Rubio, V. Studies on cyanobacterial protein PipY shed light on structure, potential functions, and vitamin B6-dependent epilepsy. *FEBS Lett.* **2017**, *591*, 3431–3442. [CrossRef]

63. Eswaramoorthy, S.; Gerchman, S.; Graziano, V.; Kycia, H.; Studier, F.W.; Swaminathan, S. Structure of a yeast hypothetical protein selected by a structural genomics approach. *Acta Crystallogr Sect. D Biol. Crystallogr.* **2003**, *95*, 127–135. [CrossRef]

64. Tremiño, L.; Forcada-Nadal, A.; Rubio, V. Insight into vitamin B6-dependent epilepsy due to PLPBP (previously PROSC) missense mutations. *Hum. Mutat.* **2018**. [CrossRef]

65. Ashkenazy, H.; Abadi, S.; Martz, E.; Chay, O.; Mayrose, I.; Pupko, T.; Ben-Tal, N. ConSurf 2016: an improved methodology to estimate and visualize evolutionary conservation in macromolecules. *Nucleic Acids Res.* **2016**. [CrossRef]

66. Ito, T.; Iimori, J.; Takayama, S.; Moriyama, A.; Yamauchi, A.; Hemmi, H.; Yoshimura, T. Conserved pyridoxal protein that regulates Ile and Val metabolism. *J. Bacteriol.* **2013**, *195*, 5439–5449. [CrossRef]

67. Darin, N.; Reid, E.; Prunetti, L.; Samuelsson, L.; Husain, R.A.; Wilson, M.; El Yacoubi, B.; Footitt, E.; Chong, W.K.; Wilson, L.C.; et al. Mutations in PROSC Disrupt Cellular Pyridoxal Phosphate Homeostasis and Cause Vitamin-B6-Dependent Epilepsy. *Am. J. Hum. Genet.* **2016**, *99*, 1325–1337. [CrossRef]

68. Plecko, B.; Zweier, M.; Begemann, A.; Mathis, D.; Schmitt, B.; Striano, P.; Baethmann, M.; Vari, M.S.; Beccaria, F.; Zara, F.; et al. Confirmation of mutations in PROSC as a novel cause of vitamin B6-dependent epilepsy. *J. Med. Genet.* **2017**, *54*, 809–814. [CrossRef] [PubMed]

69. Nichols, R.J.; Sen, S.; Choo, Y.J.; Beltrao, P.; Zietek, M.; Chaba, R.; Lee, S.; Kazmierczak, K.M.; Lee, K.J.; Wong, A.; et al. Phenotypic landscape of a bacterial cell. *Cell* **2011**, *156*, 1493–1506. [CrossRef] [PubMed]

70. Prunetti, L.; El Yacoubi, B.; Schiavon, C.R.; Kirkpatrick, E.; Huang, L.; Bailly, M.; El Badawi-Sidhu, M.; Harrison, K.; Gregory, J.F.; Fiehn, O.; et al. Evidence that COG0325 proteins are involved in PLP homeostasis. *Microbiology (UK)* **2016**, *99*, 1325–1337. [CrossRef] [PubMed]

71. Ito, T.; Hori, R.; Hemmi, H.; Downs, D.M.; Yoshimura, T. Inhibition of glycine cleavage system by pyridoxine 5′-phosphate causes synthetic lethality in *glyA yggS* and *serA yggS* in *Escherichia coli*. *Mol. Microbiol.* **2020**, *113*, 270–284. [CrossRef]

72. Belogurov, G.A.; Mooney, R.A.; Svetlov, V.; Landick, R.; Artsimovitch, I. Functional specialization of transcription elongation factors. *EMBO J.* **2009**, *28*, 112–122. [CrossRef]

73. Goodson, J.R.; Klupt, S.; Zhang, C.; Straight, P.; Winkler, W.C. LoaP is a broadly conserved antiterminator protein that regulates antibiotic gene clusters in *Bacillus amyloliquefaciens*. *Nat. Microbiol.* **2017**, *2*, 17003. [CrossRef]

74. Chatzidaki-Livanis, M.; Coyne, M.J.; Comstock, L.E. A family of transcriptional antitermination factors necessary for synthesis of the capsular polysaccharides of *Bacteroides fragilis*. *J. Bacteriol.* **2009**, *191*, 7288–7295. [CrossRef]

75. Järvelin, A.I.; Noerenberg, M.; Davis, I.; Castello, A. The new (dis)order in RNA regulation. *Cell Commun. Signal.* **2016**, *14*, 9. [CrossRef]

76. Hwang, J.; Inouye, M. The tandem GTPase, Der, is essential for the biogenesis of 50S ribosomal subunits in *Escherichia coli*. *Mol. Microbiol.* **2006**, *61*, 1660–1672. [CrossRef]

77. Bharat, A.; Brown, E.D. Phenotypic investigations of the depletion of EngA in *Escherichia coli* are consistent with a role in ribosome biogenesis. *FEMS Microbiol. Lett.* **2014**, *353*, 26–32. [CrossRef]

78. Kato, Y.; Hyodo, K.; Sakamoto, W. The photosystem II repair cycle requires FTSH turnover through the ENGA GtPase. *Plant Physiol.* **2018**, *178*, 596–611. [CrossRef]

79. Jeon, Y.; Ahn, C.S.; Jung, H.J.; Kang, H.; Park, G.T.; Choi, Y.; Hwang, J.; Pai, H.S. DER containing two consecutive GTP-binding domains plays an essential role in chloroplast ribosomal RNA processing and ribosome biogenesis in higher plants. *J. Exp. Bot.* **2014**, *65*, 117–130. [CrossRef]

Review

Stress Signaling in Cyanobacteria: A Mechanistic Overview

Raphaël Rachedi, Maryline Foglino and Amel Latifi *

CNRS, Laboratoire de Chimie Bactérienne LCB, Aix Marseille University, IMM, 13009 Marseille, France; rrachedi@imm.cnrs.fr (R.R.); foglino@imm.cnrs.fr (M.F.)
* Correspondence: latifi@imm.cnrs.fr; Tel.: +33-4-91-16-41-88

Received: 22 October 2020; Accepted: 25 November 2020; Published: 26 November 2020

Abstract: Cyanobacteria are highly diverse, widely distributed photosynthetic bacteria inhabiting various environments ranging from deserts to the cryosphere. Throughout this range of niches, they have to cope with various stresses and kinds of deprivation which threaten their growth and viability. In order to adapt to these stresses and survive, they have developed several global adaptive responses which modulate the patterns of gene expression and the cellular functions at work. Sigma factors, two-component systems, transcriptional regulators and small regulatory RNAs acting either separately or collectively, for example, induce appropriate cyanobacterial stress responses. The aim of this review is to summarize our current knowledge about the diversity of the sensors and regulators involved in the perception and transduction of light, oxidative and thermal stresses, and nutrient starvation responses. The studies discussed here point to the fact that various stresses affecting the photosynthetic capacity are transduced by common mechanisms.

Keywords: cyanobacteria; gene expression; signaling; stress

---

1. Introduction

The domain of bacteria includes an ancient, monophyletic phylum of organisms called cyanobacteria which are able to undergo oxygenic photosynthesis. Their phototrophic metabolism makes them leading players in the biosphere because of their impact on the global carbon and nitrogen cycles—they are thought to account for 25% of the global primary production [1,2], and in view of the N_2-fixing ability of some strains, they are held to be the main source of combined nitrogen in the marine environment [3]. They therefore play an important role in the fields of agriculture, aquatic ecology and environmental protection [4]. In addition, due to the great progress made in the field of genetic manipulations and the recent emergence of synthetic biology, cyanobacteria are now being applied successfully in many biotechnological processes such as bioremediation, high-value secondary metabolite synthesis, and biofuel (including ethanol and hydrogen) production [4,5]. From their early emergence up to the present day, cyanobacteria have succeeded in colonizing a wide range of aquatic to terrestrial ecological niches [6]. This impressively wide pattern of distribution is due to their ability to cope with many kinds of starvation and stress, such as nutrient deprivation, light and temperature fluctuations, and oxidative, thermal and osmotic stresses. In response to environmental changes of various kinds, the ability to trigger and coordinate suitable adaptive mechanisms depends on the ability of these bacteria to rapidly sense the physical stimuli present and to appropriately transduce the signals perceived into gene expression modulation processes.

The molecular mechanisms developed by cyanobacteria for adapting to stress conditions have been studied in detail, in several strains. In addition to the publications included in this Special Issue, these aspects have been addressed in the following reviews [7–11]. The aim of the present review is to sum up the latest knowledge available on the perception of and regulatory mechanisms involved in stress transduction in cyanobacteria. DNA microarray technology, and proteome analysis combined with systematic mutagenesis, has made it possible to identify several regulators involved in the stress response of the unicellular cyanobacterium model *Synechocystis* PCC 6803 (called *Synechocystis* hereafter), as reviewed in [12]. Here, we give an update of this knowledge and summarize the recent progress made in studies on stress regulation in *Synechocystis* and other cyanobacterial strains.

2. Light Stress

Since cyanobacteria use solar energy for their growth, the perception of light and the physiological changes that occur in response to light variations are important adaptative responses that they have to orchestrate. They are equipped with light-absorbing antenna called phycobilisomes, which are part of the main systems of acclimation to light [7]. In particular, several strains are able to vary the composition of their phycobilisomes to respond to changes in the quality of light, based on a mechanism known as complementary chromatic acclimation (CCA) [13,14]. In addition, motile strains use a process of phototaxis to move towards the optimal light conditions required for their photosynthesis [15]. CCA and phototaxis are advantageous adaptative responses that do not induce any general stress responses. They will therefore not be discussed here.

In addition to qualitative changes in light, cyanobacteria can also be exposed in their environment to quantitative daily changes. Irradiances far above the light saturation level of the photosynthetic machinery are harmful, as they induce the photoinhibition and photo destruction of the photosystems [16]. In addition, when challenged by high light (HL) stress, photosynthetic organisms generate reactive oxygen species which are deleterious to all the macromolecules in the organism (see below). The adaptive responses to HL and oxidative stresses are therefore tightly bound together, which makes it difficult to identify the signal transduction systems specific to each type of condition. Short-term acclimation to strong light can be achieved by quenching excess light energy and redistributing between the photosystems the excitation energy required [17]. By contrast, long-term adaptation often requires a particular process of modulation of the patterns of gene expression. The transcription of a number of genes is induced in response to HL stress, and the promoters of several of these genes harbor a conserved tandem sequence known as the HL regulatory region (HLRR) [18–20]. The use of this sequence in DNA/protein interaction assays has been a decisive step towards identifying the regulatory mechanisms involved in the transmission of light stress. This is how the RpaB protein was found to be the master HL stress regulator in *Synechocystis* PCC 6803 and *Synechococcus elongatous* PCC 7942 (called *Synechococcus* hereafter) [20–23]. The *rpaB* gene, which is largely conserved in cyanobacterial genomes, encodes a response regulator protein belonging to the two-component systems. Genetic and biochemical experiments have shown that RpaB acts as a repressor of HL-induced genes during growth under normal light [24–26]. In *Synechococcus*, the histidine kinase NblS has been found to be the sensor partner of RpaB in the HL transduction signal [24]. NblS and its orthologue Hik33 (DpsA) in *Synechocytis* are the most highly conserved histidine kinases in the genomes of cyanobacteria [27]. The role of NblS/Hik33 in the response to HL was actually established long before that of RpaB [28,29], but establishing the proof of its direct involvement in this response has been a rather tedious task due to the fact that this kinase is a pleiotropic regulator involved in the transduction of multiple stresses (see below). In *Synechococcus*, a second response regulator (SsrA) is phosphorylated by NblS. *ssrA* gene expression is induced by HL under the control of RpaB. Once it has been produced, SsrA might quench the phosphotransfer from NblS to RpaB [24,26]. In addition, the kinase activities of NblS and Hik33 have been found to be stimulated through their interactions with SipA (NblS-interacting protein A) [30]. Since the *sipA* gene is also conserved in cyanobacterial genomes, the data derived from *Synechocystis* and *Synechococcus* might well apply to other strains. The NblS pathway involved in HL control is therefore a hierarchical cascade where two interfering response regulators (RpaB and SipA) cooperate, and in which the activity of the sensor is finely tuned to ensure optimal acclimation to HL in cyanobacteria (Figure 1).

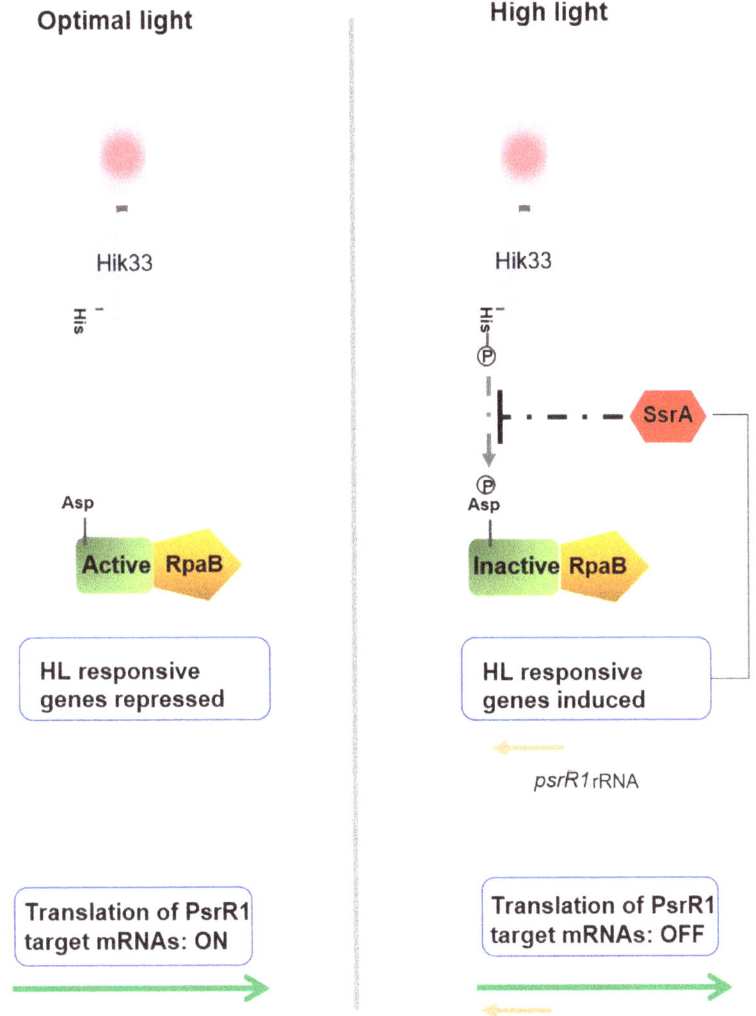

Figure 1. Regulation of high light stress acclimation.

Multiple *rpoD* genes encoding sigma factors have been identified in the genomes of cyanobacteria, and it has been suggested that many of them may act only under specific growth conditions [31]. In *Synechococcus*, the expression of the *rpoD3* gene is induced in response to HL stress under the control of RpaB, and the *rpoD3* deletion mutant is unable to survive this type of stress [23]. RpoD3 (as well as its orthologues in other cyanobacteria) might therefore be the specific alternative sigma factor enabling the polymerase to transcribe HL-induced genes.

In addition to the control process exerted at the initiation of transcription, described above, the regulation of gene expression in response to HL also occurs at the post-transcriptional level. In *Synechocystis*, a small regulatory RNA (sRNA) called PsrR1 (photosynthesis regulatory RNA1), which is conserved in the cyanobacterial genomes, has been found to be expressed in response to HL stress under the control of RpaB [32]. Based on computational and genetic findings, it has been established that PsrR1 regulates several photosynthetic genes negatively by interacting with their ribosome binding sites, and thus inhibiting the translation of their transcripts [33]. The existence of this negative regulation process is consistent with the fact that phycobilisome and photosystem reduction is one of the main physiological processes of adaptation to HL [9]. A second sRNA (RblR) accumulates under HL conditions in *Synechocystis*. RblR acts as an anti-sense RNA to the *rbcL* mRNA encoding the large subunit of the Rubisco [34]. The idea that this RNA may enhance carbon assimilation has been suggested, based on the phenotypes of mutant strains over- or under-expressing the *rblR* gene [34], but how exactly the regulation of the *rblR* transcription process induced in response to HL stress is achieved has not yet been established. In the marine cyanobacterium *Synechococcus* sp. WH7803, the expression of six non-coding RNA genes is regulated in response to HL, several of which possess photosynthesis genes as predictive targets [35]. All these data provide strong evidence that non-coding RNAs play an important role in the regulation of genes involved in the HL stress response. Elucidating the identity of all the non-coding RNA-target genes, the molecular mechanisms involved and how they are integrated into the global process of acclimation to HL is certainly the most challenging perspective ahead of us in this field.

The response regulator RpaB represses gene transcription under normal light conditions. In response to high light (HL) stress, RpaB is phosphorylated, resulting in its inactivation and the subsequent induction of HL-inducible genes. The Hik33/NblS (DspA) kinase is involved in the perception of HL stress. The putative phosphorylation of RpaB by Hik33 is thought to be inhibited by the SsrA protein. The *psrR1* gene belonging to the RpaB regulon expresses an sRNA that inhibits the translation of several genes under HL stress conditions.

3. Oxidative Stress

Reactive oxygen species (ROS), such as the singlet oxygen species ($^1O_2$), the hydroxyl radical (OH$^\bullet$), the superoxide anion (O_2^-) and hydrogen peroxide (H_2O_2), are by-products of aerobic metabolism, and in photosynthetic organisms they are mainly produced when the intensity of the light collected by the photosystems is greater than the rate of electron consumption [36]. Cyanobacteria, like all living organisms, have developed several defense systems which reduce and eliminate some of these ROS before they can react with biomolecules and oxidize them. The state of imbalance where the level of ROS exceeds the amount of antioxidant molecules is called oxidative stress [8,36,37]. No enzymatic defense mechanisms exist for $^1O_2$ and OH$^\bullet$, and so the cellular responses to these species do not result in any modulation of gene expression. The regulation of oxidative stress therefore consists of the transduction of superoxide and peroxide signals.

Superoxide anions resulting from the photoreduction of oxygen can be converted into H_2O_2 by the metalloenzymes superoxide dismutases (SOD), which are thought to constitute the main antioxidant defense system against O_2^- [37]. In the multicellular cyanobacterium *Nostoc* (*Anabaena*) PCC 7120, the transcription of the *sodB* gene encoding FeSOD is under the direct negative control of the transcriptional factor CalA (cyanobacterial AbrB like) [38]. AbrB homologues, which are present in all the cyanobacterial genomes [39], have been found to regulate several metabolic pathways, such as the carbon fixation, nitrogen uptake and hydrogen oxidation pathways [40–44]. Since CalA is essential in *Nostoc*, obtaining a strain deleted from *calA* is not possible. The negative effect on the transcription of *sodB* was therefore analyzed by overexpressing the *calA* gene [38]. The question as to how CalA perceives the superoxide anion has not yet been answered. Interestingly, in *Synechocytis*, the transcription of the *sodB* gene was reported to be repressed by the transcriptional factor PrqR, and this control was found to be indirect [45]. If CalA also represses the transcription of *sodB* in *Synechocystis*, the regulatory scheme responsible for superoxide signaling may be based on the fact that PrqR perceives the signal and transduces it via CalA (Figure 2A).

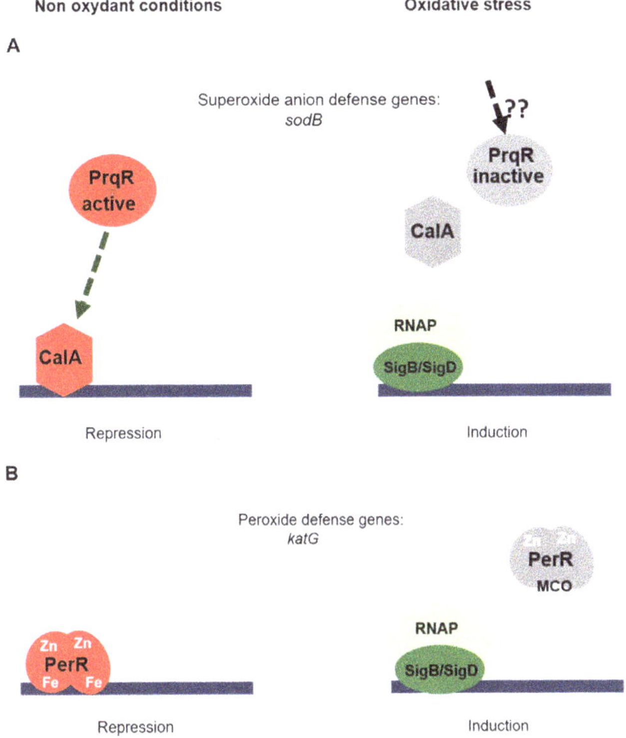

Figure 2. Regulation of oxidative stress responses. (**A**) Superoxide anion stress. In the absence of stress, the PrqR sensor might phosphorylate the response regulator CalA, which therefore represses the gene transcription process. In the presence of superoxide, PrqR is inactivated by a hitherto unknown mechanism. The Group 2 sigma factors SigB and/or SigD associate(s) with the RNA polymerase, resulting in the initiation of the transcription process. (**B**) Peroxide stress. The metalloregulator PerR represses gene transcription in the absence of peroxide. Under oxidative conditions, metal catalyzed oxidation (MCO) inactivates PerR, thus inducing gene transcription.

Several studies have converged in designating PerR as the main specific regulator of the response to peroxide in cyanobacteria [46–49]. PerR is a zinc-metalloprotein, in which either Fe^{2+} or Mn^{2+} ligated to two conserved His and Asp residues act as corepressors. The activity of PerR is regulated by a metal-catalyzed oxidation (MCO) process—the metal present in the catalytic center reacts with peroxide in a Fenton-type chemistry process which leads to the irreversible inactivation of the repressor and the induction of the target genes, including peroxidase and catalase encoding genes [50]. Based on structural modeling studies, it has been suggested that the binding of PerR to its target genes may occur via multimers of PerR-protein interacting with the AT-rich repeats present in the promoters of the repressed genes [51]. The involvement of PerR in the repression of the transcription of peroxide-related genes in cyanobacteria has been described in *Synechocystis* and *Nostoc*, and metal oxidation has been found to contribute to the action of PerR in *Nostoc* [46]. In addition, the overexpression of PerR has been found to affect the composition and the stability of the photosynthetic machinery and the division process in *Nostoc* [52]. This link between PerR, photosynthesis and cell division might explain why *perR* is an essential gene in this strain.

In addition to the regulatory effects mentioned above, the response of *Synechocytis* to oxidative stress resulting from HL or peroxide treatment has also been found to rely on Group 2 alternative sigma factors, namely SigB and SigD factors [53]. A strain lacking all the Group 2 sigma factors was unable to sustain its growth when challenged by oxidative stress, although this ability was rescued by the introduction of the *sigB* or *sigD* gene. In addition, RNA polymerase holoenzyme associated with either SigB or SigD accumulates in response to peroxide stress [53]. The signaling of oxidative stress is therefore based on transcriptional regulators (CalA, PrqR, PerR) combined with the programming of the RNA polymerase with dedicated alternative sigma factors (Figure 2B).

4. Salt Stress

Salinity, defined as the total inorganic ion concentration in the environment, is one of the main changing abiotic factors that cyanobacteria have to cope with in both aquatic and terrestrial niches [10,54]. A high ion concentration in the medium results in an osmotic loss of water and a concomitant increase in some inorganic ions which can be deleterious to the macromolecules in the cell. By contrast, at low salinity levels, water largely flows into the cell, resulting in lysis due to high turgor pressure. Even if the acclimation to changing salinity levels is of two kinds, depending on the salt concentration in the surrounding medium, salt stress nomenclature is often attributed to high salinity conditions. Cyanobacteria, like many other non-halophilic prokaryotes, respond to salt stress by accumulating small organic solutes (often in the form of sugars) and monitoring the active export of inorganic ions. This strategy has therefore been called "the salt-out-strategy" [55]. The organic solutes in question have a low molecular mass and do not interfere with the cell metabolism, which explains why they are referred to as compatible solutes. In cyanobacteria, the strain-specific salt tolerance level is correlated with the nature of the main compatible solute produced [55].

Many studies in which it was proposed to elucidate how *Synechocystis* adapt to salt stress (reviewed in [10,56,57]) have shown that the cellular responses involved are dynamic processes, and that they occur at various levels, including the allosteric regulation of the activity of several transporters and enzymes, as well as a global change in the process of gene transcription, in which the molecular mechanisms involved are not totally known. The exception here is the regulation of the transcription of genes involved in the synthesis of the organic solutes produced by *Synechocystis*, namely sucrose and glycosylglycerol. The synthesis of sucrose involves two enzymes: the sucrose phosphate phosphatase (Spp) and the sucrose phosphate synthase (SpsA), which is the rate limiting enzyme [58]. Upon exposure to salt stress, the activity of the SpsA enzyme and the transcription level of the *spsA* gene are both enhanced [59]. Under normal salinity conditions, the expression of *spsA* is repressed by the Rre39 response regulation [57], but since this is an orphan regulator (i.e., the cognate histidine kinase has not yet been identified), the question as to how the "high salinity" signal is transduced to Rre39 has not yet been elucidated. The synthesis of glycosylglycerol is a two-step reaction involving the glycosylglycerol phosphate synthase (GgpS) and the glycosylglycerol phosphate phosphatase (GgpP) enzymes [60]. Under salt stress conditions, the production of glycosylglycerol is enhanced by the induction of *ggpS* gene transcription and by an increase in the activity of the GgpS enzyme [60]. The Group 2 sigma factor SigF seems to be the specific sigma factor responsible for *ggpS* transcription, as a *sigF* mutant is unable to adapt to salt stress and shows significantly low accumulation rates of *ggpS* transcripts [61,62]. Upstream of the *ggpS* gene, the small regulatory gene *ggpR* has been identified, the deletion of which increases the *ggpS* mRNA levels under normal salinity conditions, which suggests that this gene may act as a repressor of *ggpS* expression [63]. As the GgpR protein does not contain any typical DNA binding motifs, it is questionable whether GgpR acts after binding directly to the *ggpS* promoter. The *ggpR* gene is not conserved in any other cyanobacterial genomes, but interestingly, the synteny between *ggpS* and another small gene is observed in many genomes, which means that the possible involvement of the corresponding protein in the regulation of *ggpS* expression cannot be ruled out [63]. In addition to this specific regulation, the expression of the *ggpS* gene has been shown to be controlled by the global transcriptional regulator LexA in *Synechocystis*—RNA seq analyses have shown that the transcript level of the *ggpS* gene is more than 10-fold higher in a strain lacking the *lexA* gene compared to the wild type strain, which suggests a negative control of LexA upon *ggpS* transcription. This control is likely direct, as a gel shift approach has indicated that LexA is able to interact with the promoter of *ggpS* [64]. The LexA protein has recently been proposed to be dephosphorylated on a Serine residue in response to salt stress [65]. This result might suggest that LexA represses the transcription of salt-induced genes under repressive conditions, and that their induction in response to salt stress requires the modulation of LexA activity together with the action of GgpR (Figure 3A).

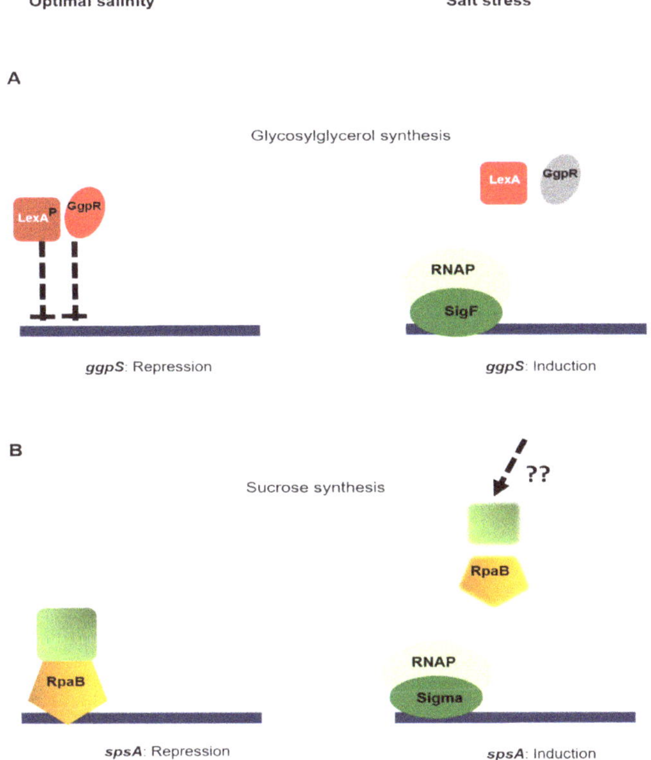

Figure 3. Regulation of salt stress responses. (**A**) the transcription of the *spsA* gene involved in the synthesis of sucrose is repressed by RpaB. The cognate sensor involved has not yet been identified. (**B**) the transcription of the *ggpS* gene involved in the synthesis of glycosylglycerol is subjected to the negative control of GgpR and LexA. Under salt stress conditions, the presence of the SigF sigma factor enables the RNA polymerase to initiate the transcription process. The dephosphorylation of LexA inhibits its action. The mechanism underlying GgpR inhibition is not known.

Sucrose is the only osmolyte produced in *Nostoc* in response to salt stress, and in addition to being a compatible solute, it is involved in the nitrogen fixation process [66,67]. The synthesis of sucrose involves two sucrose phosphate synthases (SpsA and SpsB) and a sucrose phosphate phosphatase (SppA) [68]. Sucrose can also be formed by the reversible action of the synthases SusA and SusB, but as they are thought to preferentially catalyze the degradation of sucrose in vivo, they will not be further discussed here [66]. The increase observed in the sucrose accumulation rates in response to salt is due to higher Spps activity and the induction of the transcription of the *spsA* gene [69]. Salt induction of *spsA* was abolished in a mutant strain lacking the response regulator OrrA, which indicates that this transcriptional regulator may play a positive role in the process of sucrose synthesis [70]. The *orrA* gene has been detected in a genetic screen set up for the identification of salt-induced genes in *Nostoc* [71]; the signaling pathway leading from salt perception to OrrA activation still remains to be determined (Figure 3B).

In conclusion, the dynamics of compatible solute accumulation and the regulation of the enzymatic activities involved have both been thoroughly documented in cyanobacteria, but this is far from being the case as far as the regulatory mechanisms involved in gene expression are concerned. Since the nature of the main solute(s) produced and the ability to adapt to high salt levels vary among cyanobacteria, the possibility cannot be ruled out that the regulatory mechanisms involved may also differ from one strain to another.

5. Heat Shock

The optimal growth temperatures for cyanobacteria cover a large range, as this phylum includes several strains inhabiting extreme environments, from hotspring to cryosphere environments [72]. However, with the exception of these extremophile members, most mesophilic strains are sensitive to temperature fluctuations, and the processes of photosynthesis and nitrogen fixation are both inhibited by heat [73,74]; the molecular responses to heat shock are therefore crucial. Like many other organisms, cyanobacteria induce the expression of heat shock genes (*hsp*) in response to temperature upshifts [74–77]. Many of the HSPs are molecular chaperones or proteases playing a major role in proteostasis, such as the Hsp60 members (mainly GroEL and GroES), which are the most abundant HSPs produced in cyanobacteria after the occurrence of temperature upshifts.

In many bacteria, the *hsp* gene promoters contain a highly conserved 9-bp inverted repeat sequence, which is required for the heat-induction process [78]. This sequence, which has been called the controlling inverted repeat of chaperone expression (CIRCE), is the binding site of the HrcA repressor, which is also highly conserved in many bacteria [79]. HrcA is a dimeric transcriptional regulator that undergoes denaturation upon being subjected to temperature upshifts. The subsequent synthesis of GroEL regenerates the HrcA dimer, thus restoring the repression of the *hsps* genes. Chaperone activity therefore acts as the molecular heat shock sensor [80]. The *hrcA* gene is widely conserved in the cyanobacterial genomes, and its role in the repression of *hsps* genes expression has been documented in both *Synechocystis* and *Nostoc* [74,75,81]. In both strains, the deletion of the *hrcA* gene results in the constitutive expression of *hsps* genes, which is consistent with the idea that a negative control may be exerted by HrcA on heat shock genes. Microarray studies, in which the global gene response of a *Synechocystis* mutant strain deleted from the *hrcA* gene was compared with that of the wild type strain, have shown that HrcA might also regulate some genes devoid of the CIRCE element that do not belong to the Hsp60 family [75,81], which suggests that HrcA might be a global regulator of gene expression, or alternatively that the latter effect might be due to the constitutive expression of the *groEL* and *groES* genes in this mutant. Interestingly, in both *Synechocystis* and *Nostoc*, the expression of some *hsps* genes in the *hrcA*-deleted strain was not fully derepressed, since a small induction was still observed in response to a temperature upshift, which indicates that their expression is regulated by another mechanism in addition to HrcA [74,75].

A fast response to temperature upshifts is ensured by the mechanism controlling the translation of some *hsps* mRNA. This mechanism involves specific sequences located in the 5′ untranslated region of the mRNA, which change their conformation in response to heat shock [82,83]. At low temperatures, the secondary structure they generate encompasses the ribosome binding site, which affects the translation efficiency. These riboswiches, which are known as thermometer RNAs (or thermosensor RNAs), are present in many bacteria [82,83], including cyanobacteria. The *hsp17* gene of *Synechocytis* harbors a rather small 5′ untranslated region which has been found to act like a typical thermosensor [82,84], and similar cis-acting riboregulatory RNAs have been identified in *hsp* genes in *Anabaena variabilis*, *Nostoc* and the thermophilic cyanobacterium *Thermosynechococcus elongatus* [85].

Another regulation system of the *hsp* genes which occurs in response to heat shock consists of reprogramming the RNA polymerase core enzyme with the appropriate alternative sigma factor. In *Synechocystis*, the sigma factors SigB and SigD play an important role in high-temperature responses; the growth of a double *sigBsigD* mutant is much more severely impaired at 43 degrees than that of the simple mutants [75,86]. Interestingly, a protein (SinA) interacting with the principal sigma factor

(RpoD1) and playing a role in heat shock responses has been recently identified in *Synechococcus* [87]. The RNA polymerase–RpoD–SinA complex was dissociated after a temperature upshift, a *sinA*-deletion mutant was unable to sustain its growth at 40 °C, and the induction profile of *hspA* gene was affected in the mutant. All in all, these data point to the conclusion that SinA may play a role in the replacement of RpoD1 by the heat stress-specific sigma factor. The finding that homologs of SinA were present in 361 genomes out of the 367 analyzed suggests that the function of this protein may be widely conserved among the members of the cyanobacterial phylum [87] (Figure 4).

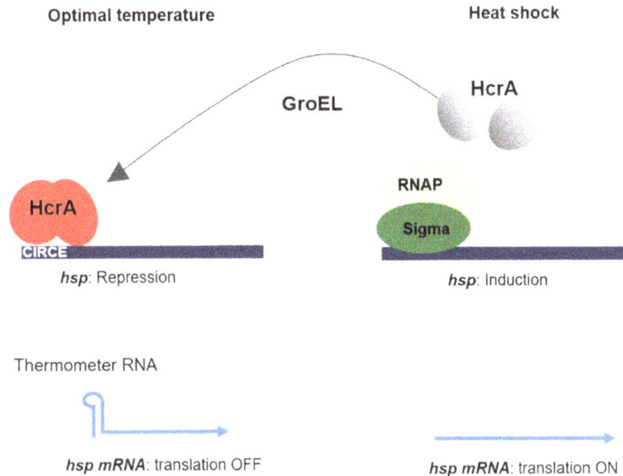

Figure 4. Heat shock response regulation. Under optimal growth conditions, the transcription of the *hsp* genes is repressed by HrcA, which binds to a conserved sequence called CIRCE. Upon undergoing heat shock, HrcA is inactivated and *hsp* transcription is induced. The GroEL chaperone facilitates the refolding of HrcA at the release of the stress, thus acting as the molecular sensor of the signal. Some of the *hsp* mRNAs carry an untranslated sequence forming a secondary (thermometer) structure sequestering the ribosome binding site under normal growth temperature conditions. Upon undergoing heat shock, this secondary structure dissociates, which makes it possible for the translation process to proceed.

6. Cold Stress

Mesophilic bacteria are often challenged by temperature downshifts to below their optimum growth temperature, which results in a decrease in the fluidity of the cell membranes and in the efficiency of the transcription and translation processes due to the abnormal stabilization of secondary structures in the DNA and RNA [88] The activities of the ribosomes and those of protein foldases are also impaired [88]. Bacteria respond by inducing the production of proteins called CSPs (cold shock-induced proteins) that serve to enhance the transcription and translation processes by acting on secondary nucleic structures; these are RNA binding proteins which affect the transcription and translation processes at low temperature via their RNA chaperoning function, and RNA helicases which stimulate the degradation or translation of RNA at low temperatures [88]. The second main response is the induction of desaturated fatty acids which counterbalance the loss of membrane fluidity [88]. The expression of the *csps* genes, which has been most closely studied in *Escherichia coli*, is regulated at the transcription, mRNA stabilization, and translation levels [89]. In *Nostoc*, the RNA-helicase encoding gene *chrC* is specifically induced in response to temperature downshifts, and its regulation was found to occur at several levels, including the transcription, mRNA stability and translation levels, but the exact molecular mechanisms involved have not yet been determined [90]. The ribosomal protein S2 has also been found to be continuously phosphorylated in *Nostoc* during exposure to cold stress, resulting in the downregulation of the translation process, with the exception of cold stress-induced mRNA [91]. In *Synechocystis*, the expression of about half of the cold-induced genes is controlled by the transmembrane histidine kinase Hik33 [92], and depends on the fluidity of the membrane [93]. Hik33 kinase also controls the responses to oxidative, osmotic and salt stress (see below), which suggests the possible existence of a common signal triggering gene induction in response to various stresses. It has been suggested that the oxidation status of the quinone pool, which was found to vary depending on the membrane fluidity during cold stress exposure, may be the common response signal to stressors affecting the membrane fluidity [94].

7. Nutrient Starvation

Among the multiple environmental stresses that cyanobacteria encounter, nutrient depletion is often a limiting factor for their growth. Like most bacterial species, cyanobacteria do not form typical dormant spores, but are nevertheless able to survive long periods of nutrient starvation. How these starved cells manage to survive and how they resume their metabolic activities once the nutrients are available de novo is one of the most intriguing questions being addressed today (see for examples [95–97]).

7.1. Carbone Starvation

Being autotrophic organisms, cyanobacteria use inorganic carbon (CO_2 and bicarbonate) as a carbon source for their growth [2]. They are able to acclimate to limited carbon conditions by optimizing their carbon fixation activity through a mechanism that concentrates CO_2 near the ribulose biphosphate carboxylase/oxygenase enzyme (RubisCo) [98,99]. This mechanism (known as CCM for carbon concentrating mechanism) includes carboxysomes and transporters for the internalization of dissolved inorganic carbon [100]. Two transcriptional regulators (CmpR and NdhR) belonging to the LysR family are the main specific factors ensuring gene expression modulation in response to carbon starvation [11]. CmpR has been identified in *Synechocystis,* and has been shown to activate the transcription of the *cmp* operon that encodes a bicarbonate transporter in response to low carbon conditions [101]. The activity of CmpR has been found to be stimulated by the binding of 2-phosphoglycolate (2-PG), which is generated from the oxygenation of ribulose-1,5-bisphosphate (RuBP) by the oxygenase activity of Rubisco under low CO_2 conditions [102]. 2-PG is a toxic by-product of the RubisCO oxygenase reaction that is metabolized by the essential photorespiration process [99]. Structural studies indicated that CmpR acts as a dimeric protein with one molecule of RuBP bound between two monomers [103], but the association between CmpR and 2-PG has not been structurally demonstrated yet. In *Nostoc*, transcription of the *cmpR* gene is positively regulated in response to combined nitrogen starvation, making this regulator a factor connecting nitrogen and carbon metabolisms [104]. The involvement of NdhR (also called CcmR) in the control of the response to the low carbon condition has been established in *Synechocystis* for the first time [105]. NdhR represses the transcription of a large set of genes, including the *ndh3* operon encoding NAD(P)H dehydrogenase (*ndh*) subunits [105], the *sbtA/B* genes encoding the Na^+/HCO_3^- symporter, and genes belonging to the *ndh-I3* operon that encodes for the high-affinity CO_2 uptake system subunits [106,107]. NdhR has also been found to regulate all the identified low-carbon-inducible genes in the coastal strain *Synechococcus* PCC 7002 [108]. The negative control of gene expression by NdhR can be thus considered as the main adaptive mechanism in carbon limitation. Like CmpR, NdhR is submitted to allosteric regulation. In vitro analysis has shown that the binding of NAD^+ and 2-oxoglutarate (2-OG), accumulating under carbon-sufficient conditions, enhances the NdhR DNA binding activity [109]. Further progress in the understanding of the allosteric regulation of NdhR has been achieved recently [110]. A combination of in vivo and structural approaches has confirmed that the role of 2-OG acts as co-repressor, and established 2-PG as the inducer (co-repressor) as it inhibits NdhR binding to DNA [110]. 2-OG is one of the two signaling molecules of nitrogen starvation (see below), and therefore NdhR coordinates the responses to nitrogen and carbon status in cyanobacteria (Figure 5A). In addition to the control exerted by CmpR and NdhR, the AbrB-type regulator cyAbrB2 has been shown to adapt carbon and nitrogen metabolisms to the photosynthetic activity in *Synechocystis* [111], and LexA has been found to be required for surviving low carbon conditions [112].

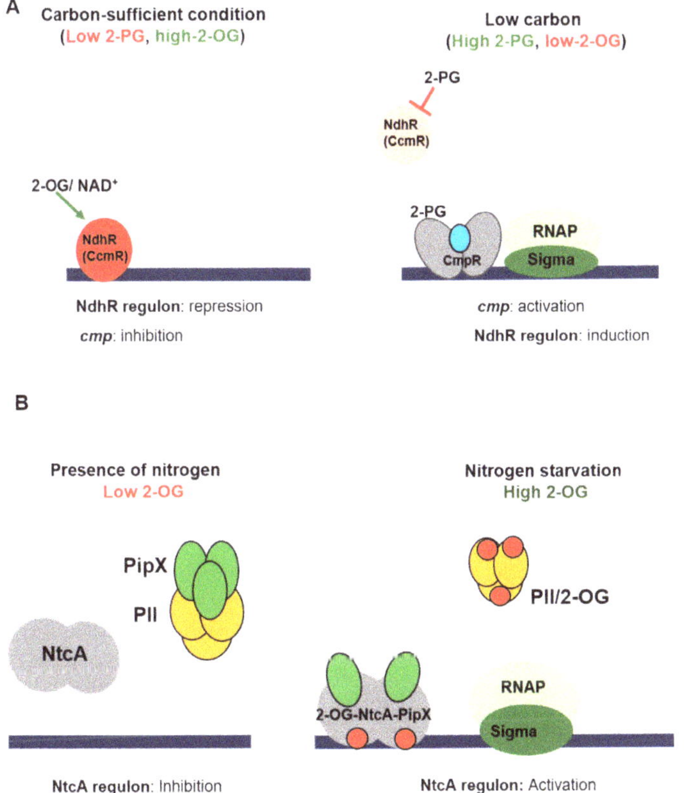

Figure 5. Regulation of carbon and nitrogen acclimations. (**A**) Low carbon response. NdhR (CcmR) is the main transcriptional regulator of the low carbon response. The activity of NdhR is submitted to a dual allosteric regulation where 2-OG and NAD$^+$ act as co-activators and 2-PG as co-repressor. When inorganic carbon level is sufficient, the carboxylase activity of Rubisco is advantaged and 2-PG concentration is low. The levels of 2-OG and NAD$^+$ are high, and NdhR is active and represses the transcription of target genes. Under carbon starvation, the oxygenase activity of Rubisco generates 2-PG that binds to NdhR and inhibits its activity, leading to the induction of CO$_2$ concentrating mechanism (CCM). The *cmp* operon, which is also part of CCM, is under the positive control of CmpR. 2-PG and RuBP (1,5 ribulose biphosphate, blue dot) act as co-activators of CmpR. 2-PG and RubP are generated by the oxygenase activity of RubisCo under low carbon conditions. The low carbon response is also regulated by LexA and AbrB and submitted to post-transcriptional regulations (see text for details). (**B**) Nitrogen starvation. The NtcA protein, the main transcriptional factor of nitrogen-induced genes, is activated by PipX. The intracellular level of 2-OG (red circles), which reflects the nitrogen status of the cell, is perceived by PII and NtcA. Under replete nitrogen conditions, the 2-OG level is low, PipX interacts with PII, and NtcA is inactive. When nitrogen is a limiting factor, 2-OG accumulates, and PipX changes its partner and interacts with NtcA. The 2-OG-NtcA-PipX complex regulates the gene transcription process. The post-transcriptional regulation mechanisms involved in nitrogen starvation signaling are not presented in the figure.

Interestingly, the transcription of several genes expressing sRNAs is modulated according to the availability of inorganic carbon in *Synechocystis* (CsiR1, Ncr0700, NcR1200, SyR12) [107,113]. As this response has been found to be maintained in a strain lacking the *ndHR* gene, a post-transcriptional regulation acting independently from NdhR may contribute to the acclimation to carbon starvation.

In addition to CCM induction, the expression of the genes encoding the flavodiiron proteins Flv2 and Flv4, acting as electron valves to protect the photosystems from oxidation [114], is also highly induced after a shift to low carbon [115]. The transcript of the *flv4-2* gene has been shown to be negatively regulated by its anti-sens RNA As1_flv4 [115]. This control of Flv protein production extends the post-transcriptional regulation to mechanisms adjusting photosynthesis to carbon availability.

7.2. Nitrogen Starvation

Cyanobacteria survive prolonged nitrogen starvation by decreasing their central metabolism and by degrading their photosynthetic apparatus, resulting in a loss of autofluorescence and cell bleaching, a state called chlorosis (Reviewed in [116]). Chlorosis is a highly orchestrated process which starts with the synthesis of the Clp-protease adaptor protein NblA [117]. Since the regulation of the *nblA* gene has recently been reviewed [118], we will not discuss this topic here, but rather focus on specific signaling pathways involved in the transduction of nitrogen deficiency.

Although cyanobacteria are able to assimilate a number of combined nitrogen compounds, including ammonium, nitrate, nitrite and urea, the preferred one is ammonium [119–121]. The metabolism of various compounds therefore starts with their intracellular assimilation to ammonium, which is then incorporated into the carbon skeleton of 2-OG via the glutamine synthetase–glutamate synthase (GS-GOGAT) pathway, giving glutamate [122]. The fact that 2-OG is an intermediate of the TCA cycle means that the processes of nitrogen and carbon assimilation are interconnected. A state of combined nitrogen deficiency therefore leads to inhibition of the GS-GOGAT cycle, and ultimately to 2-OG accumulation, which provides us with a useful indicator of the nitrogen status of the cell [119].

Non-diazotrophic cyanobacterial strains have to cope with nitrogen deficiency, and this adaptive response depends on the ability to perceive the state of starvation and to modulate the pattern of gene expression accordingly in order to use alternative nitrogen sources. The facultatively nitrogen-fixing cyanobacteria also have to perceive the state of combined nitrogen depletion in order to induce the genetic program enabling them to shift their metabolism towards the reduction of atmospheric nitrogen. In both cases, 2-OG has been found to act as a molecular sensor of nitrogen deficiency [120]. We will focus below on the response to nitrogen starvation in non-fixing cyanobacteria, as it is only in these organisms that this situation constitutes a stress [116]. The transduction of the 2-OG signal, which has been intensively studied in unicellular freshwater strains (mainly *Synechocystis* and *Synechococcus*), involves several factors, including the transcriptional regulator NtcA, which is thought to be the main one involved. The NtcA protein is a member of the CRP family of transcription regulators, and deletion of the *ntcA* gene impairs the ability of the strains to grow on nitrogen sources other than ammonium, which is consistent with the finding that NtcA activates the transcription of genes required for the assimilation of nitrogen sources, such as nitrate [120,123]. The activity of NtcA is modulated depending on the nitrogen source and its concentration: it is induced in the absence of ammonium and under conditions where the 2-OG level is high (which corresponds to low levels of nitrogen) [124]. A dimer of NtcA has the ability to bind to 2-OG, and structural studies have established that the binding of the effector induces a conformational change that enhances the DNA-binding activity [125,126]. PipX, a protein present only in cyanobacteria, acts as a coactivator of NtcA [127]. Biochemical studies have shown that PipX enhances the affinity of NtcA for promoters and the effective affinity of NtcA for 2-OG [128]. In addition to NtcA, the second sensor at work in 2-OG signaling is the protein PII, which is encoded by the *glnB* gene present in all the cyanobacterial genomes. PII proteins constitute one of the largest and most widely distributed families of signal transduction factors, all the members of which are able to bind to ATP/ADP in addition to 2-OG (for a recent review on PII, see [129]). Via protein–protein interactions, they control the activity of target proteins in response to cellular ATP/ADP levels and the 2-OG status, thus creating a link between the carbon and nitrogen metabolisms [129]. When the combined nitrogen source is abundant (and the intracellular 2-OG level is low), PII-ADP binds to PipX and NtcA is mainly present in the apo form, whereas under low combined nitrogen levels (and high 2-OG concentrations) PII interacts with 2-OG and ATP, inhibiting its interactions with PipX. At the same time, the binding of 2-OG to NtcA favors its interaction with PipX, resulting in the enhancement of the transcriptional activation of the genes involved in nitrogen assimilation [126,127]. The role played by PII and its protein partners in the control of the nitrogen/carbon balance in cyanobacteria has been intensively discussed over the last few years. For further information on this subject, readers can consult the following reviews [121,130,131] (Figure 5B).

The regulation of nitrogen assimilation in cyanobacteria also occurs at a post-transcriptional level through two distinct mechanisms: riboswiches and sRNA. (i) Comparative genome analyses designed for riboswitch probing have identified the presence of two RNA motifs (*glnA* and Downstream peptide) present only in cyanobacterial genomes and metagenomic sequences [132]. Shortly after, these two RNA motifs have been shown to specifically bind glutamine [133]. In a study comparing 60 cyanobacterial genomes, these RNA motifs have been found to be frequently present in the gene *gifB* that encodes the IF17 protein involved in the inactivation of the glutamine synthetase [134]. In this study, it has been proposed to rename the *glnA* and Downstream peptide as glutamine type 1 and glutamine type 2 riboswiches, respectively, which will avoid the confusion between the *glnA* RNA motif and the *glnA* gene encoding glutamine synthetase. The RNA motif located at the 5′ UTR of the *gifB* of *Synechocystis* has been proven to act as a riboswich since its interaction with glutamine increased protein synthesis in vivo [134]. These data unearthed the role of glutamine as a signaling molecule, specifically in cyanobacteria. (ii) Transcriptome and ChIP seq analyses have shown that several genes expressing small regulatory RNA are responsive to nitrogen starvation [113]. The transcription of two of them (NsrR1 and NsRi4) is under the control of NtcA in *Nostoc* and *Synechocystis* [135–137]. In *Nostoc*, NsrR1 (nitrogen repressed RNA1) has been shown to regulate negatively the expression of a gene required for diazotrophic growth [135], and to be involved in the expression of the *nblA* gene [135], which points to a role of this RNA in the control of nitrogen metabolism and chlorosis in this bacterium. NsiR4 (nitrogen stress-induced RNA 4) has been shown to be required for the negative regulation of the *gifA* gene encoding the glutamine synthetase inactivation factor in *Synechocystis* and *Nostoc* [136]. Since genes expressing NsiR4 are widely conserved in the genomes of cyanobacteria [136], the control of nitrogen assimilation by this snRNA is likely to also be conserved.

7.3. Phosphate Starvation

Phosphate deficiency affects photosynthetic activity, cell growth, phospholipid and nucleotide synthesis, and cell growth. When starved of phosphorus, cyanobacteria induce the expression of specific genes (known as the Pho regulon), which enhances phosphate uptake [138–140], triggers a process of alkaline phosphatase synthesis releasing phosphorus from several components [141–143], and decreasing the phospholipid levels present in the membrane via a remodeling process [144]. The Pho regulon has been found in *Synechocystis* [145,146] and *Synechococcus* [147,148] to be under the exclusive positive control of the SphS/SphR two-component system. The signal transduction mechanism performed by the SphS/SphR system has been studied in *Synechocystis*; the fact that the deletion of the extended N-terminal extremity of the SphS kinase sensor abolished the activation of the Pho regulon suggests that this sequence is required for sensing the phosphorus levels [149]. The activity of the SphS/SpHR system is negatively regulated by the SphU protein, probably by interacting with and inactivating the transmitter domain of SpHS [150]. Gel mobility shift assays have shown the existence of a conserved sequence in the promoters of genes belonging to the Pho regulon, known as the 'Pho box', which is required for the activation of transcription by SphR [146]. The Pho regulon has been predicted to exist in 19 cyanobacterial strains, and interestingly, the loss of SphS/ SphR was observed in the genomes of 3 of them known to inhabit phosphate-rich niches [151]. Whether the need to adapt to phosphate deficiency has been lost in the course of evolution or whether a regulator other than SphR is involved in these strains still remains to be elucidated (Figure 6A).

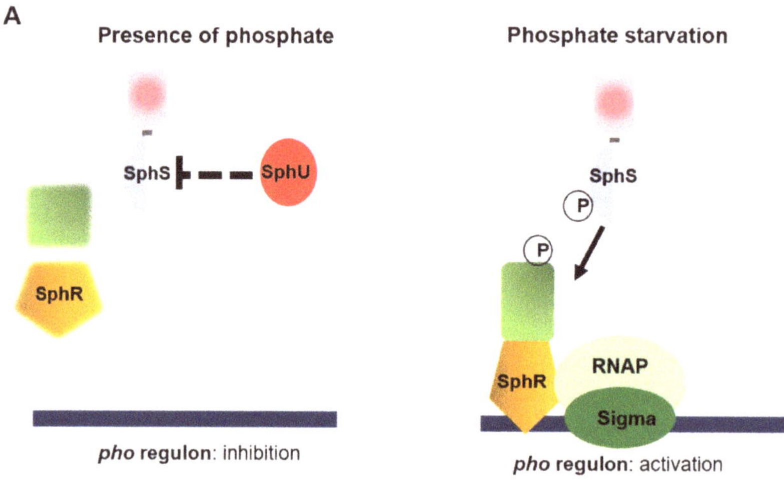

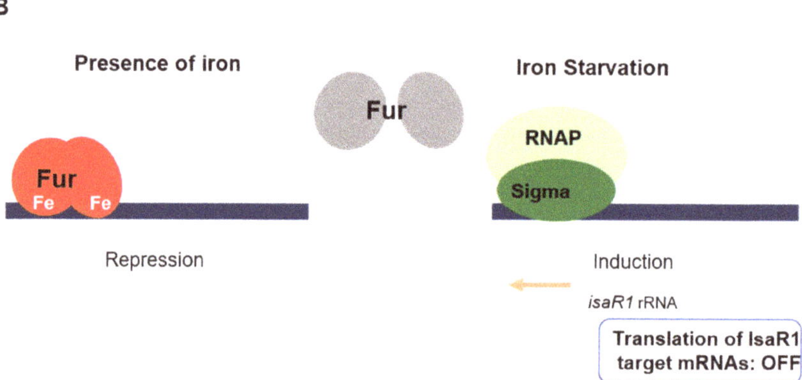

Figure 6. Regulation of phosphate and iron acclimations. (**A**) Phosphate starvation. The group of genes induced in response to phosphate starvation is called "the *pho* regulon". In response to phosphate limitation, the transcription of these genes is activated by the TCS formed by the kinase SphS and the response regulator SphR. Under replete phosphate conditions, the activity of SphS is inhibited by the protein SphU, presumably by interacting with the transmitter domain of SphS. (**B**) Iron starvation. The Fur repressor is a metalloprotein in which iron serves as a cofactor. When iron becomes limited, it is inactivated by the shift of Fur to its apoform, and the transcription process is thus induced. In addition, the translation of several genes is inhibited by the sRNA isaR1, which is expressed under iron starvation conditions.

7.4. Iron Starvation

Iron plays the role of cofactor in the case of several essential proteins, but free iron is rarely available in nature, which makes it an important limiting factor for bacterial and phytoplankton growth in various environments [152]. Iron homeostasis is tightly regulated to prevent both starvation and excess, which lead to oxidative stress in cyanobacteria [8]. The ferric uptake regulator (Fur) is the main transcriptional regulator of the genes involved in iron homeostasis in most bacteria [153]. Fur acts as a repressor and an iron sensor; at higher iron concentrations, it binds to Fe^{2+}, dimerizes, and binds to target promoters in a conserved sequence termed "Fur-box". During iron starvation, the release of Fe^{2+} inactivates Fur and cancels the repression exerted by this regulator [154]. Fur homologues are widely distributed in cyanobacterial genomes, and the involvement of Fur in their adaptation to iron starvation has been investigated in some model freshwater strains (*Nostoc, Synechococcus, Synechocytis*) [155–157]. Interestingly, the latter studies have shown that the *fur* gene is essential in cyanobacteria, which suggests that it is required for some essential processes in addition to iron-response control [155–157]. Studies using a "transcript-depletion" strategy have shown that in *Nostoc*, Fur controls the expression of genes involved in several processes, including exopolysaccharide biosynthesis, phycobilisome degradation, chlorophyll catabolism, nitrogen fixation, and exopolysaccharide biosynthesis [158]. Several functions must be inhibited in response to iron starvation, but how could this control be exerted since Fur is inactivated under these conditions? A recent study on *Synechocytis* has yielded a clue to understanding the molecular basis of this homeostasis [159]. The gene expressing the small regulatory RNA Isar1 (Iron Stress-Activated RNA 1) is repressed by Fur. Isar1 accumulates in response to iron starvation and controls at a post-transcriptional level the expression of several genes involved in central cellular processes (photosynthesis, (iron–sulfur) cluster biogenesis, citrate cycle and tetrapyrrole biogenesis [159]. Homologs of IsaR1 are largely conserved in the genomes of cyanobacteria [159], and the involvement of this riboregulator in the control of photosynthesis via iron homeostasis might be conserved in the cyanobacterial phylum (Figure 6B).

8. Multiple Stresses Sharing Common Sensors and/or Transducers

All the environmental stresses discussed above generally decrease the maximum photosynthetic capacity of cyanobacteria, resulting in the hyper-reduction of the electron flow and ultimately in a decrease in the anabolism. It is therefore often difficult to distinguish between the specific effect of a particular stress and its indirect impact through a change in the redox state of the cell. This distinction could be made in the case of nitrogen deficiency in *Synechoccocus* [160]. In this study, nitrogen deficiency was mimicked using an inhibitor of the glutamine synthetase. This treatment resulted in the induction of the *glnB* gene (specific to nitrogen deficiency), and of the *nblA* gene (chlorosis marker). Interestingly, the addition of nitrate suppressed the induction of *nblA* without significantly affecting the expression of *glnB*. By acting as an electron sink, nitrate decreased the hyper-reduction of the photosystems and probably modified the redox state, which might therefore be considered as the specific signal triggering *nblA* [160].

Genetic screening for mutations impacting the adaptive responses to several stresses at work has led to the identification of the sensor Hik33/NblS (DspA) and the cognate response regulators as leading players in the transduction of the stress signals (reviewed in [12]). Photosynthetic redox stress can be assumed to be the actual signal that is perceived by Hik33/NblS/(DspA) kinase during exposure to various stresses. As mentioned above, the redox state of the quinone pool reflects the fluidity status of the membrane [94]. The redox-sensitive transcriptional regulator PedR has been found to be reduced by thioredoxin and to be inactivated under HL conditions, which shows the existence of a relationship between gene expression and photosynthetic activity [161]. Gene regulation through PedR can thus be expected to respond to various stresses that affect the photosynthetic abilities of these bacteria. In addition to the redox signal that mediates a pleiotropic transduction pathway, more specific regulatory cross-talk occurs between some stress responses.

9. Conclusions

Given the ecological role of cyanobacteria, their wide pattern of distribution and their versatile metabolism, data on the stress responses at work in these bacteria are relevant to many fields, including industrial biotechnological applications. A thorough knowledge of the regulatory networks mediating stress responses is a prerequisite for circumventing inhibitory mechanisms in order to maintain the growth of these microorganisms, even under the unfavorable conditions that frequently occur during large-scale production processes.

Author Contributions: Conceptualization: A.L.; writing and editing: R.R., M.F., A.L.; funding acquisition: A.L. All authors have read and agreed to the revised version of the manuscript.

Funding: R.R. has a fellowship from the Région SUD and CNRS. Work in A.L. group is funded by the "Agence Nationale de la Recherche" ANR program 18-CE05-0029.

Conflicts of Interest: The authors declare that they have no conflicts of interest to declare. The funders had no role in the design of the study; in the collection, analyses, or interpretation of data; in the writing of the manuscript, or in the decision to publish the results.

References

1. Flombaum, P.; Gallegos, J.L.; Gordillo, R.A.; Rincon, J.; Zabala, L.L.; Jiao, N.; Karl, D.M.; Li, W.K.; Lomas, M.W.; Veneziano, D.; et al. Present and future global distributions of the marine Cyanobacteria Prochlorococcus and Synechococcus. *Proc. Natl. Acad. Sci. USA* **2013**, *110*, 9824–9829. [CrossRef] [PubMed]

2. Veaudor, T.; Blanc-Garin, V.; Chenebault, C.; Diaz-Santos, E.; Sassi, J.F.; Cassier-Chauvat, C.; Chauvat, F. Recent Advances in the Photoautotrophic Metabolism of Cyanobacteria: Biotechnological Implications. *Life* **2020**, *10*, 71. [CrossRef]

3. Karl, D.L.; Tupas, L.R.; Dore, J.; Christian, J.; Hebel, D. The role of nitrogen fixation in biogeochemical cycling in the subtropical North Pacific Ocean. *Nature* **1997**, *388*, 533–538. [CrossRef]

4. Singh, J.S.; Kumar, A.; Rai, A.N.; Singh, D.P. Cyanobacteria: A Precious Bio-resource in Agriculture, Ecosystem, and Environmental Sustainability. *Front. Microbiol.* **2016**, *7*, 529. [CrossRef] [PubMed]

5. Santos-Merino, M.; Singh, A.K.; Ducat, D.C. New Applications of Synthetic Biology Tools for Cyanobacterial Metabolic Engineering. *Front. Bioeng. Biotechnol.* **2019**, *7*, 33. [CrossRef] [PubMed]

6. Moreira, C.; Vasconcelos, V.; Antunes, A. Phylogeny and biogeography of cyanobacteria and their produced toxins. *Mar. Drugs* **2013**, *11*, 4350–4369. [CrossRef]

7. Wiltbank, L.B.; Kehoe, D.M. Diverse light responses of cyanobacteria mediated by phytochrome superfamily photoreceptors. *Nat. Rev. Microbiol.* **2019**, *17*, 37–50. [CrossRef]

8. Latifi, A.; Ruiz, M.; Zhang, C.C. Oxidative stress in cyanobacteria. *FEMS Microbiol. Rev.* **2009**, *33*, 258–278. [CrossRef]

9. Muramatsu, M.; Hihara, Y. Acclimation to high-light conditions in cyanobacteria: From gene expression to physiological responses. *J. Plant Res.* **2012**, *125*, 11–39. [CrossRef]

10. Hagemann, M. Molecular biology of cyanobacterial salt acclimation. *FEMS Microbiol. Rev.* **2011**, *35*, 87–123. [CrossRef]

11. Burnap, R.L.; Hagemann, M.; Kaplan, A. Regulation of CO_2 Concentrating Mechanism in Cyanobacteria. *Life* **2015**, *5*, 348–371. [CrossRef] [PubMed]

12. Los, D.A.; Zorina, A.; Sinetova, M.; Kryazhov, S.; Mironov, K.; Zinchenko, V.V. Stress sensors and signal transducers in cyanobacteria. *Sensors* **2010**, *10*, 2386–2415. [CrossRef] [PubMed]

13. Montgomery, B.L. Sensing the light: Photoreceptive systems and signal transduction in cyanobacteria. *Mol. Microbiol.* **2007**, *64*, 16–27. [CrossRef] [PubMed]

14. Sanfilippo, J.E.; Garczarek, L.; Partensky, F.; Kehoe, D.M. Chromatic Acclimation in Cyanobacteria: A Diverse and Widespread Process for Optimizing Photosynthesis. *Annu. Rev. Microbiol.* **2019**, *73*, 407–433. [CrossRef]

15. Schuergers, N.; Mullineaux, C.W.; Wilde, A. Cyanobacteria in motion. *Curr. Opin. Plant Biol.* **2017**, *37*, 109–115. [CrossRef]

16. Nixon, P.J.; Michoux, F.; Yu, J.; Boehm, M.; Komenda, J. Recent advances in understanding the assembly and repair of photosystem II. *Ann. Bot.* **2010**, *106*, 1–16. [CrossRef]

17. Karapetyan, N.V. Protective dissipation of excess absorbed energy by photosynthetic apparatus of cyanobacteria: Role of antenna terminal emitters. *Photosynth. Res.* **2008**, *97*, 195–204. [CrossRef]

18. Hihara, Y.; Kamei, A.; Kanehisa, M.; Kaplan, A.; Ikeuchi, M. DNA microarray analysis of cyanobacterial gene expression during acclimation to high light. *Plant. Cell* **2001**, *13*, 793–806. [CrossRef]

19. Eriksson, J.; Salih, G.F.; Ghebramedhin, H.; Jansson, C. Deletion mutagenesis of the 5′ psbA2 region in Synechocystis 6803: Identification of a putative cis element involved in photoregulation. *Mol. Cell Biol. Res. Commun.* **2000**, *3*, 292–298. [CrossRef]

20. Kappell, A.D.; van Waasbergen, L.G. The response regulator RpaB binds the high light regulatory 1 sequence upstream of the high-light-inducible hliB gene from the cyanobacterium Synechocystis PCC 6803. *Arch. Microbiol.* **2007**, *187*, 337–342. [CrossRef]

21. Kappell, A.D.; Bhaya, D.; van Waasbergen, L.G. Negative control of the high light-inducible hliA gene and implications for the activities of the NblS sensor kinase in the cyanobacterium Synechococcus elongatus strain PCC 7942. *Arch. Microbiol.* **2006**, *186*, 403–413. [CrossRef] [PubMed]

22. Hanaoka, M.; Tanaka, K. Dynamics of RpaB-promoter interaction during high light stress, revealed by chromatin immunoprecipitation (ChIP) analysis in Synechococcus elongatus PCC 7942. *Plant J. Cell Mol. Biol.* **2008**, *56*, 327–335. [CrossRef] [PubMed]

23. Seki, A.; Hanaoka, M.; Akimoto, Y.; Masuda, S.; Iwasaki, H.; Tanaka, K. Induction of a group 2 sigma factor, RPOD3, by high light and the underlying mechanism in Synechococcus elongatus PCC 7942. *J. Biol. Chem.* **2007**, *282*, 36887–36894. [CrossRef] [PubMed]

24. Kato, H.; Kubo, T.; Hayashi, M.; Kobayashi, I.; Yagasaki, T.; Chibazakura, T.; Watanabe, S.; Yoshikawa, H. Interactions between histidine kinase NblS and the response regulators RpaB and SrrA are involved in the bleaching process of the cyanobacterium Synechococcus elongatus PCC 7942. *Plant Cell Physiol.* **2011**, *52*, 2115–2122. [CrossRef] [PubMed]

25. Moronta-Barrios, F.; Espinosa, J.; Contreras, A. In vivo features of signal transduction by the essential response regulator RpaB from Synechococcus elongatus PCC 7942. *Microbiology* **2012**, *158*, 1229–1237. [CrossRef]

26. Lopez-Redondo, M.L.; Moronta, F.; Salinas, P.; Espinosa, J.; Cantos, R.; Dixon, R.; Marina, A.; Contreras, A. Environmental control of phosphorylation pathways in a branched two component system. *Mol. Microbiol.* **2010**, *78*, 475–489. [CrossRef]

27. Ashby, M.K.; Houmard, J. Cyanobacterial two-component proteins: Structure, diversity, distribution, and evolution. *Microbiol. Mol. Biol. Rev.* **2006**, *70*, 472–509. [CrossRef]

28. van Waasbergen, L.G.; Dolganov, N.; Grossman, A.R. nblS, a gene involved in controlling photosynthesis-related gene expression during high light and nutrient stress in Synechococcus elongatus PCC 7942. *J. Bacteriol.* **2002**, *184*, 2481–2490. [CrossRef]

29. Hsiao, H.Y.; He, Q.; Van Waasbergen, L.G.; Grossman, A.R. Control of photosynthetic and high-light-responsive genes by the histidine kinase DspA: Negative and positive regulation and interactions between signal transduction pathways. *J. Bacteriol.* **2004**, *186*, 3882–3888. [CrossRef]

30. Salinas, P.; Ruiz, D.; Cantos, R.; Lopez-Redondo, M.L.; Marina, A.; Contreras, A. The regulatory factor SipA provides a link between NblS and NblR signal transduction pathways in the cyanobacterium *Synechococcus* sp. PCC 7942. *Mol. Microbiol.* **2007**, *66*, 1607–1619. [CrossRef]

31. Osanai, T.; Ikeuchi, M.; Tanaka, K. Group 2 sigma factors in cyanobacteria. *Physiol. Plant* **2008**, *133*, 490–506. [CrossRef] [PubMed]

32. Kadowaki, T.; Nagayama, R.; Georg, J.; Nishiyama, Y.; Wilde, A.; Hess, W.R.; Hihara, Y. A Feed-Forward Loop Consisting of the Response Regulator RpaB and the Small RNA PsrR1 Controls Light Acclimation of Photosystem I Gene Expression in the Cyanobacterium *Synechocystis* sp. PCC 6803. *Plant Cell Physiol.* **2016**, *57*, 813–823. [CrossRef]

33. Georg, J.; Dienst, D.; Schurgers, N.; Wallner, T.; Kopp, D.; Stazic, D.; Kuchmina, E.; Klahn, S.; Lokstein, H.; Hess, W.R.; et al. The small regulatory RNA SyR1/PsrR1 controls photosynthetic functions in cyanobacteria. *Plant Cell* **2014**, *26*, 3661–3679. [CrossRef] [PubMed]

34. Hu, J.; Li, T.; Xu, W.; Zhan, J.; Chen, H.; He, C.; Wang, Q. Small Antisense RNA RblR Positively Regulates RuBisCo in *Synechocystis* sp. PCC 6803. *Front. Microbiol.* **2017**, *8*, 231. [CrossRef]

35. Gierga, G.; Voss, B.; Hess, W.R. Non-coding RNAs in marine Synechococcus and their regulation under environmentally relevant stress conditions. *ISME J.* **2012**, *6*, 1544–1557. [CrossRef] [PubMed]

36. Imlay, J.A. Pathways of oxidative damage. *Annu. Rev. Microbiol.* **2003**, *57*, 395–418. [CrossRef]
37. Johnson, L.A.; Hug, L.A. Distribution of reactive oxygen species defense mechanisms across domain bacteria. *Free Radic. Biol. Med.* **2019**, *140*, 93–102. [CrossRef]
38. Agervald, A.; Baebprasert, W.; Zhang, X.; Incharoensakdi, A.; Lindblad, P.; Stensjo, K. The CyAbrB transcription factor CalA regulates the iron superoxide dismutase in *Nostoc* sp. strain PCC 7120. *Environ. Microbiol.* **2010**, *12*, 2826–2837. [CrossRef]
39. Larsson, J.; Nylander, J.A.; Bergman, B. Genome fluctuations in cyanobacteria reflect evolutionary, developmental and adaptive traits. *BMC Evol. Biol.* **2011**, *11*, 187. [CrossRef]
40. Lieman-Hurwitz, J.; Haimovich, M.; Shalev-Malul, G.; Ishii, A.; Hihara, Y.; Gaathon, A.; Lebendiker, M.; Kaplan, A. A cyanobacterial AbrB-like protein affects the apparent photosynthetic affinity for CO_2 by modulating low-CO_2-induced gene expression. *Environ. Microbiol.* **2009**, *11*, 927–936. [CrossRef] [PubMed]
41. Agervald, A.; Zhang, X.; Stensjo, K.; Devine, E.; Lindblad, P. CalA, a cyanobacterial AbrB protein, interacts with the upstream region of hypC and acts as a repressor of its transcription in the cyanobacterium *Nostoc* sp. strain PCC 7120. *Appl. Environ. Microbiol.* **2010**, *76*, 880–890. [CrossRef] [PubMed]
42. Dutheil, J.; Saenkham, P.; Sakr, S.; Leplat, C.; Ortega-Ramos, M.; Bottin, H.; Cournac, L.; Cassier-Chauvat, C.; Chauvat, F. The AbrB2 autorepressor, expressed from an atypical promoter, represses the hydrogenase operon to regulate hydrogen production in Synechocystis strain PCC6803. *J. Bacteriol.* **2012**, *194*, 5423–5433. [CrossRef] [PubMed]
43. Yamauchi, Y.; Kaniya, Y.; Kaneko, Y.; Hihara, Y. Physiological roles of the cyAbrB transcriptional regulator pair Sll0822 and Sll0359 in *Synechocystis* sp. strain PCC 6803. *J. Bacteriol.* **2011**, *193*, 3702–3709. [CrossRef] [PubMed]
44. Oliveira, P.; Lindblad, P. An AbrB-Like protein regulates the expression of the bidirectional hydrogenase in *Synechocystis* sp. strain PCC 6803. *J. Bacteriol.* **2008**, *190*, 1011–1019. [CrossRef] [PubMed]
45. Khan, R.I.; Wang, Y.; Afrin, S.; Wang, B.; Liu, Y.; Zhang, X.; Chen, L.; Zhang, W.; He, L.; Ma, G. Transcriptional regulator PrqR plays a negative role in glucose metabolism and oxidative stress acclimation in *Synechocystis* sp. PCC 6803. *Sci. Rep.* **2016**, *6*, 32507. [CrossRef]
46. Yingping, F.; Lemeille, S.; Talla, E.; Janicki, A.; Denis, Y.; Zhang, C.C.; Latifi, A. Unravelling the cross-talk between iron starvation and oxidative stress responses highlights the key role of PerR (alr0957) in peroxide signalling in the cyanobacterium Nostoc PCC 7120. *Environ. Microbiol. Rep.* **2014**, *6*, 468–475. [CrossRef]
47. Li, H.; Singh, A.K.; McIntyre, L.M.; Sherman, L.A. Differential gene expression in response to hydrogen peroxide and the putative PerR regulon of *Synechocystis* sp. strain PCC 6803. *J. Bacteriol.* **2004**, *186*, 3331–3345. [CrossRef]
48. Kobayashi, M.; Ishizuka, T.; Katayama, M.; Kanehisa, M.; Bhattacharyya-Pakrasi, M.; Pakrasi, H.B.; Ikeuchi, M. Response to oxidative stress involves a novel peroxiredoxin gene in the unicellular cyanobacterium *Synechocystis* sp. PCC 6803. *Plant Cell Physiol.* **2004**, *45*, 290–299. [CrossRef]
49. Houot, L.; Floutier, M.; Marteyn, B.; Michaut, M.; Picciocchi, A.; Legrain, P.; Aude, J.C.; Cassier-Chauvat, C.; Chauvat, F. Cadmium triggers an integrated reprogramming of the metabolism of Synechocystis PCC6803, under the control of the Slr1738 regulator. *BMC Genom.* **2007**, *8*, 350. [CrossRef]
50. Lee, J.W.; Helmann, J.D. The PerR transcription factor senses H_2O_2 by metal-catalysed histidine oxidation. *Nature* **2006**, *440*, 363–367. [CrossRef]
51. Garcin, P.; Delalande, O.; Zhang, J.Y.; Cassier-Chauvat, C.; Chauvat, F.; Boulard, Y. A transcriptional-switch model for Slr1738-controlled gene expression in the cyanobacterium Synechocystis. *BMC Struct. Biol.* **2012**, *12*, 1. [CrossRef] [PubMed]
52. Sevilla, E.; Sarasa-Buisan, C.; Gonzalez, A.; Cases, R.; Kufryk, G.; Peleato, M.L.; Fillat, M.F. Regulation by FurC in Anabaena Links the Oxidative Stress Response to Photosynthetic Metabolism. *Plant Cell Physiol.* **2019**, *60*, 1778–1789. [CrossRef] [PubMed]
53. Koskinen, S.; Hakkila, K.; Kurkela, J.; Tyystjarvi, E.; Tyystjarvi, T. Inactivation of group 2 sigma factors upregulates production of transcription and translation machineries in the cyanobacterium *Synechocystis* sp. PCC 6803. *Sci. Rep.* **2018**, *8*, 10305. [CrossRef] [PubMed]
54. Allakhverdiev, S.I.; Murata, N. Salt stress inhibits photosystems II and I in cyanobacteria. *Photosynth. Res.* **2008**, *98*, 529–539. [CrossRef]
55. Reed, R.H.; Stewart, W.D.P. Osmotic adjustment and organic solute accumulation in unicellular cyanobacteria from freshwater and marine habitats. *Mar. Biol.* **1985**, *88*, 1–9. [CrossRef]

56. Pade, N.; Hagemann, M. Salt acclimation of cyanobacteria and their application in biotechnology. *Life* **2014**, *5*, 25–49. [CrossRef]

57. Chen, L.; Wu, L.; Zhu, Y.; Song, Z.; Wang, J.; Zhang, W. An orphan two-component response regulator Slr1588 involves salt tolerance by directly regulating synthesis of compatible solutes in photosynthetic *Synechocystis* sp. PCC 6803. *Mol. Biosyst.* **2014**, *10*, 1765–1774. [CrossRef]

58. Curatti, L.; Folco, E.; Desplats, P.; Abratti, G.; Limones, V.; Herrera-Estrella, L.; Salerno, G. Sucrose-phosphate synthase from *Synechocystis* sp. strain PCC 6803: Identification of the spsA gene and characterization of the enzyme expressed in Escherichia coli. *J. Bacteriol.* **1998**, *180*, 6776–6779. [CrossRef]

59. Kirsch, F.; Klahn, S.; Hagemann, M. Salt-Regulated Accumulation of the Compatible Solutes Sucrose and Glucosylglycerol in Cyanobacteria and Its Biotechnological Potential. *Front. Microbiol.* **2019**, *10*, 2139. [CrossRef]

60. Hagemann, M.; Schoor, A.; Jeanjean, R.; Zuther, E.; Joset, F. The stpA gene form *Synechocystis* sp. strain PCC 6803 encodes the glucosylglycerol-phosphate phosphatase involved in cyanobacterial osmotic response to salt shock. *J. Bacteriol.* **1997**, *179*, 1727–1733. [CrossRef]

61. Huckauf, J.; Nomura, C.; Forchhammer, K.; Hagemann, M. Stress responses of *Synechocystis* sp. strain PCC 6803 mutants impaired in genes encoding putative alternative sigma factors. *Microbiology* **2000**, *146 Pt 11*, 2877–2889. [CrossRef]

62. Marin, K.; Huckauf, J.; Fulda, S.; Hagemann, M. Salt-dependent expression of glucosylglycerol-phosphate synthase, involved in osmolyte synthesis in the cyanobacterium *Synechocystis* sp. strain PCC 6803. *J. Bacteriol.* **2002**, *184*, 2870–2877. [CrossRef] [PubMed]

63. Klahn, S.; Hohne, A.; Simon, E.; Hagemann, M. The gene ssl3076 encodes a protein mediating the salt-induced expression of ggpS for the biosynthesis of the compatible solute glucosylglycerol in *Synechocystis* sp. strain PCC 6803. *J. Bacteriol.* **2010**, *192*, 4403–4412. [CrossRef]

64. Kizawa, A.; Kawahara, A.; Takimura, Y.; Nishiyama, Y.; Hihara, Y. RNA-seq Profiling Reveals Novel Target Genes of LexA in the Cyanobacterium *Synechocystis* sp. PCC 6803. *Front. Microbiol.* **2016**, *7*, 193. [CrossRef]

65. Takashima, K.; Nagao, S.; Kizawa, A.; Suzuki, T.; Dohmae, N.; Hihara, Y. The role of transcriptional repressor activity of LexA in salt-stress responses of the cyanobacterium *Synechocystis* sp. PCC 6803. *Sci. Rep.* **2020**, *10*, 17393. [CrossRef] [PubMed]

66. Curatti, L.; Flores, E.; Salerno, G. Sucrose is involved in the diazotrophic metabolism of the heterocyst-forming cyanobacterium *Anabaena* sp. *FEBS Lett.* **2002**, *513*, 175–178. [CrossRef]

67. Lopez-Igual, R.; Flores, E.; Herrero, A. Inactivation of a heterocyst-specific invertase indicates a principal role of sucrose catabolism in heterocysts of *Anabaena* sp. *J. Bacteriol.* **2010**, *192*, 5526–5533. [CrossRef]

68. Vargas, W.; Cumino, A.; Salerno, G.L. Cyanobacterial alkaline/neutral invertases. Origin of sucrose hydrolysis in the plant cytosol? *Planta* **2003**, *216*, 951–960. [CrossRef]

69. Desplats, P.; Folco, E.; Salerno, G.L. Sucrose may play an additional role to that of an osmolyte in *Synechocystis* sp. PCC 6803 salt-shocked cells. *Plant. Physiol. Biochem.* **2005**, *43*, 133–138. [CrossRef]

70. Ehira, S.; Kimura, S.; Miyazaki, S.; Ohmori, M. Sucrose synthesis in the nitrogen-fixing cyanobacterium *Anabaena* sp. strain PCC 7120 is controlled by the two-component response regulator OrrA. *Appl. Environ. Microbiol.* **2014**, *80*, 5672–5679. [CrossRef]

71. Schwartz, S.H.; Black, T.A.; Jager, K.; Panoff, J.M.; Wolk, C.P. Regulation of an osmoticum-responsive gene in *Anabaena* sp. strain PCC 7120. *J. Bacteriol.* **1998**, *180*, 6332–6337. [CrossRef] [PubMed]

72. Chrismas, N.A.; Anesio, A.M.; Sanchez-Baracaldo, P. Multiple adaptations to polar and alpine environments within cyanobacteria: A phylogenomic and Bayesian approach. *Front. Microbiol.* **2015**, *6*, 1070. [CrossRef] [PubMed]

73. Allakhverdiev, S.I.; Hayashi, H.; Nishiyama, Y.; Ivanov, A.G.; Aliev, J.A.; Klimov, V.V.; Murata, N.; Carpentier, R. Glycinebetaine protects the D1/D2/Cytb559 complex of photosystem II against photo-induced and heat-induced inactivation. *J. Plant Physiol.* **2003**, *160*, 41–49. [CrossRef] [PubMed]

74. Rajaram, H.; Kumar Apte, S. Heat-shock response and its contribution to thermotolerance of the nitrogen-fixing cyanobacterium *Anabaena* sp. strain L-31. *Arch. Microbiol.* **2003**, *179*, 423–429. [CrossRef] [PubMed]

75. Singh, A.K.; Summerfield, T.C.; Li, H.; Sherman, L.A. The heat shock response in the cyanobacterium *Synechocystis* sp. Strain PCC 6803 and regulation of gene expression by HrcA and SigB. *Arch. Microbiol.* **2006**, *186*, 273–286. [CrossRef] [PubMed]

76. Borbely, G.; Suranyi, G.; Korcz, A.; Palfi, Z. Effect of heat shock on protein synthesis in the cyanobacterium *Synechococcus* sp. strain PCC 6301. *J. Bacteriol.* **1985**, *161*, 1125–1130. [CrossRef]

77. Suzuki, I.; Simon, W.J.; Slabas, A.R. The heat shock response of *Synechocystis* sp. PCC 6803 analysed by transcriptomics and proteomics. *J. Exp. Bot.* **2006**, *57*, 1573–1578. [CrossRef]

78. Zuber, U.; Schumann, W. CIRCE, a novel heat shock element involved in regulation of heat shock operon dnaK of Bacillus subtilis. *J. Bacteriol.* **1994**, *176*, 1359–1363. [CrossRef]

79. Schulz, A.; Schumann, W. hrcA, the first gene of the Bacillus subtilis dnaK operon encodes a negative regulator of class I heat shock genes. *J. Bacteriol.* **1996**, *178*, 1088–1093. [CrossRef]

80. Schumann, W. Regulation of bacterial heat shock stimulons. *Cell Stress Chaperones* **2016**, *21*, 959–968. [CrossRef]

81. Nakamoto, H.; Suzuki, M.; Kojima, K. Targeted inactivation of the hrcA repressor gene in cyanobacteria. *FEBS Lett.* **2003**, *549*, 57–62. [CrossRef]

82. Kortmann, J.; Narberhaus, F. Bacterial RNA thermometers: Molecular zippers and switches. *Nat. Rev. Microbiol.* **2012**, *10*, 255–265. [CrossRef] [PubMed]

83. Krajewski, S.S.; Narberhaus, F. Temperature-driven differential gene expression by RNA thermosensors. *Biochim. Biophys. Acta* **2014**, *1839*, 978–988. [CrossRef] [PubMed]

84. Wagner, D.; Rinnenthal, J.; Narberhaus, F.; Schwalbe, H. Mechanistic insights into temperature-dependent regulation of the simple cyanobacterial hsp17 RNA thermometer at base-pair resolution. *Nucleic Acids Res.* **2015**, *43*, 5572–5585. [CrossRef] [PubMed]

85. Cimdins, A.; Klinkert, B.; Aschke-Sonnenborn, U.; Kaiser, F.M.; Kortmann, J.; Narberhaus, F. Translational control of small heat shock genes in mesophilic and thermophilic cyanobacteria by RNA thermometers. *RNA Biol.* **2014**, *11*, 594–608. [CrossRef]

86. Tuominen, I.; Pollari, M.; Tyystjarvi, E.; Tyystjarvi, T. The SigB sigma factor mediates high-temperature responses in the cyanobacterium *Synechocystis* sp. PCC6803. *FEBS Lett.* **2006**, *580*, 319–323. [CrossRef]

87. Hasegawa, H.; Tsurumaki, T.; Kobayashi, I.; Imamura, S.; Tanaka, K. Identification and analysis of a principal sigma factor interacting protein SinA, essential for growth at high temperatures in a cyanobacterium Synechococcus elongatus PCC 7942. *J. Gen. Appl. Microbiol.* **2020**, *66*, 66–72. [CrossRef]

88. Phadtare, S. Recent developments in bacterial cold-shock response. *Curr. Issues Mol. Biol.* **2004**, *6*, 125–136.

89. Phadtare, S.; Inouye, M. Genome-wide transcriptional analysis of the cold shock response in wild-type and cold-sensitive, quadruple-csp-deletion strains of Escherichia coli. *J. Bacteriol.* **2004**, *186*, 7007–7014. [CrossRef]

90. Chamot, D.; Owttrim, G.W. Regulation of cold shock-induced RNA helicase gene expression in the cyanobacterium *Anabaena* sp. strain PCC 7120. *J. Bacteriol.* **2000**, *182*, 1251–1256. [CrossRef]

91. El-Fahmawi, B.; Owttrim, G.W. Cold-stress-altered phosphorylation of EF-Tu in the cyanobacterium *Anabaena* sp. strain PCC 7120. *Can. J. Microbiol.* **2007**, *53*, 551–558. [CrossRef] [PubMed]

92. Mikami, K.; Kanesaki, Y.; Suzuki, I.; Murata, N. The histidine kinase Hik33 perceives osmotic stress and cold stress in *Synechocystis* sp. PCC 6803. *Mol. Microbiol.* **2002**, *46*, 905–915. [CrossRef] [PubMed]

93. Mironov, K.S.; Sidorov, R.A.; Trofimova, M.S.; Bedbenov, V.S.; Tsydendambaev, V.D.; Allakhverdiev, S.I.; Los, D.A. Light-dependent cold-induced fatty acid unsaturation, changes in membrane fluidity, and alterations in gene expression in Synechocystis. *Biochim. Biophys. Acta* **2012**, *1817*, 1352–1359. [CrossRef] [PubMed]

94. Maksimov, E.G.; Mironov, K.S.; Trofimova, M.S.; Nechaeva, N.L.; Todorenko, D.A.; Klementiev, K.E.; Tsoraev, G.V.; Tyutyaev, E.V.; Zorina, A.A.; Feduraev, P.V.; et al. Membrane fluidity controls redox-regulated cold stress responses in cyanobacteria. *Photosynth. Res.* **2017**, *133*, 215–223. [CrossRef] [PubMed]

95. Klotz, A.; Georg, J.; Bucinska, L.; Watanabe, S.; Reimann, V.; Januszewski, W.; Sobotka, R.; Jendrossek, D.; Hess, W.R.; Forchhammer, K. Awakening of a Dormant Cyanobacterium from Nitrogen Chlorosis Reveals a Genetically Determined Program. *Curr. Biol. CB* **2016**, *26*, 2862–2872. [CrossRef] [PubMed]

96. Roth-Rosenberg, D.; Aharonovich, D.; Luzzatto-Knaan, T.; Vogts, A.; Zoccarato, L.; Eigemann, F.; Nago, N.; Grossart, H.P.; Voss, M.; Sher, D. Prochlorococcus Cells Rely on Microbial Interactions Rather than on Chlorotic Resting Stages to Survive Long-Term Nutrient Starvation. *mBio* **2020**, *11*. [CrossRef]

97. Sauer, J.; Schreiber, U.; Schmid, R.; Volker, U.; Forchhammer, K. Nitrogen starvation-induced chlorosis in Synechococcus PCC 7942. Low-level photosynthesis as a mechanism of long-term survival. *Plant Physiol.* **2001**, *126*, 233–243. [CrossRef]

98. Kaplan, A.; Reinhold, L. CO_2 Concentrating Mechanisms in Photosynthetic Microorganisms. *Annu. Rev. Plant Physiol. Plant Mol. Biol.* **1999**, *50*, 539–570. [CrossRef]

99. Hagemann, M.; Kern, R.; Maurino, V.G.; Hanson, D.T.; Weber, A.P.; Sage, R.F.; Bauwe, H. Evolution of photorespiration from cyanobacteria to land plants, considering protein phylogenies and acquisition of carbon concentrating mechanisms. *J. Exp. Bot.* **2016**, *67*, 2963–2976. [CrossRef]

100. Kerfeld, C.A.; Melnicki, M.R. Assembly, function and evolution of cyanobacterial carboxysomes. *Curr. Opin. Plant. Biol.* **2016**, *31*, 66–75. [CrossRef]

101. Omata, T.; Gohta, S.; Takahashi, Y.; Harano, Y.; Maeda, S. Involvement of a CbbR homolog in low CO_2-induced activation of the bicarbonate transporter operon in cyanobacteria. *J. Bacteriol.* **2001**, *183*, 1891–1898. [CrossRef]

102. Nishimura, T.; Takahashi, Y.; Yamaguchi, O.; Suzuki, H.; Maeda, S.; Omata, T. Mechanism of low CO_2-induced activation of the cmp bicarbonate transporter operon by a LysR family protein in the cyanobacterium Synechococcus elongatus strain PCC 7942. *Mol. Microbiol.* **2008**, *68*, 98–109. [CrossRef] [PubMed]

103. Mahounga, D.M.; Sun, H.; Jiang, Y.L. Crystal structure of the effector-binding domain of Synechococcus elongatus CmpR in complex with ribulose 1,5-bisphosphate. *Acta Crystallogr. F Struct. Biol. Commun.* **2018**, *74*, 506–511. [CrossRef] [PubMed]

104. Lopez-Igual, R.; Picossi, S.; Lopez-Garrido, J.; Flores, E.; Herrero, A. N and C control of ABC-type bicarbonate transporter Cmp and its LysR-type transcriptional regulator CmpR in a heterocyst-forming cyanobacterium, *Anabaena* sp. *Environ. Microbiol.* **2012**, *14*, 1035–1048. [CrossRef]

105. Figge, R.M.; Cassier-Chauvat, C.; Chauvat, F.; Cerff, R. Characterization and analysis of an NAD(P)H dehydrogenase transcriptional regulator critical for the survival of cyanobacteria facing inorganic carbon starvation and osmotic stress. *Mol. Microbiol.* **2001**, *39*, 455–468. [CrossRef] [PubMed]

106. Wang, H.L.; Postier, B.L.; Burnap, R.L. Alterations in global patterns of gene expression in *Synechocystis* sp. PCC 6803 in response to inorganic carbon limitation and the inactivation of ndhR, a LysR family regulator. *J. Biol. Chem.* **2004**, *279*, 5739–5751. [CrossRef] [PubMed]

107. Klahn, S.; Orf, I.; Schwarz, D.; Matthiessen, J.K.; Kopka, J.; Hess, W.R.; Hagemann, M. Integrated Transcriptomic and Metabolomic Characterization of the Low-Carbon Response Using an ndhR Mutant of *Synechocystis* sp. PCC 6803. *Plant Physiol.* **2015**, *169*, 1540–1556. [CrossRef]

108. Woodger, F.J.; Bryant, D.A.; Price, G.D. Transcriptional regulation of the CO_2-concentrating mechanism in a euryhaline, coastal marine cyanobacterium, *Synechococcus* sp. Strain PCC 7002: Role of NdhR/CcmR. *J. Bacteriol.* **2007**, *189*, 3335–3347. [CrossRef]

109. Daley, S.M.; Kappell, A.D.; Carrick, M.J.; Burnap, R.L. Regulation of the cyanobacterial CO_2-concentrating mechanism involves internal sensing of NADP+ and alpha-ketogutarate levels by transcription factor CcmR. *PLoS ONE* **2012**, *7*, e41286. [CrossRef]

110. Jiang, Y.L.; Wang, X.P.; Sun, H.; Han, S.J.; Li, W.F.; Cui, N.; Lin, G.M.; Zhang, J.Y.; Cheng, W.; Cao, D.D.; et al. Coordinating carbon and nitrogen metabolic signaling through the cyanobacterial global repressor NdhR. *Proc. Natl. Acad. Sci. USA* **2018**, *115*, 403–408. [CrossRef]

111. Orf, I.; Schwarz, D.; Kaplan, A.; Kopka, J.; Hess, W.R.; Hagemann, M.; Klahn, S. CyAbrB2 Contributes to the Transcriptional Regulation of Low CO_2 Acclimation in *Synechocystis* sp. PCC 6803. *Plant Cell Physiol.* **2016**, *57*, 2232–2243. [CrossRef] [PubMed]

112. Domain, F.; Houot, L.; Chauvat, F.; Cassier-Chauvat, C. Function and regulation of the cyanobacterial genes lexA, recA and ruvB: LexA is critical to the survival of cells facing inorganic carbon starvation. *Mol. Microbiol.* **2004**, *53*, 65–80. [CrossRef]

113. Muro-Pastor, A.M.; Hess, W.R. Regulatory RNA at the crossroads of carbon and nitrogen metabolism in photosynthetic cyanobacteria. *Biochim Biophys Acta Gene Regul. Mech.* **2020**, *1863*, 194477. [CrossRef] [PubMed]

114. Zhang, P.; Eisenhut, M.; Brandt, A.M.; Carmel, D.; Silen, H.M.; Vass, I.; Allahverdiyeva, Y.; Salminen, T.A.; Aro, E.M. Operon flv4-flv2 provides cyanobacterial photosystem II with flexibility of electron transfer. *Plant Cell* **2012**, *24*, 1952–1971. [CrossRef] [PubMed]

115. Eisenhut, M.; Georg, J.; Klahn, S.; Sakurai, I.; Mustila, H.; Zhang, P.; Hess, W.R.; Aro, E.M. The antisense RNA As1_flv4 in the Cyanobacterium *Synechocystis* sp. PCC 6803 prevents premature expression of the flv4-2 operon upon shift in inorganic carbon supply. *J. Biol. Chem.* **2012**, *287*, 33153–33162. [CrossRef]

116. Schwarz, R.; Forchhammer, K. Acclimation of unicellular cyanobacteria to macronutrient deficiency: Emergence of a complex network of cellular responses. *Microbiology* **2005**, *151*, 2503–2514. [CrossRef]

117. Collier, J.L.; Grossman, A.R. A small polypeptide triggers complete degradation of light-harvesting phycobiliproteins in nutrient-deprived cyanobacteria. *EMBO J.* **1994**, *13*, 1039–1047. [CrossRef]

118. Forchhammer, K.; Schwarz, R. Nitrogen chlorosis in unicellular cyanobacteria - a developmental program for surviving nitrogen deprivation. *Environ. Microbiol.* **2019**, *21*, 1173–1184. [CrossRef]

119. Muro-Pastor, M.I.; Reyes, J.C.; Florencio, F.J. Ammonium assimilation in cyanobacteria. *Photosynth. Res.* **2005**, *83*, 135–150. [CrossRef]

120. Herrero, A.; Muro-Pastor, A.M.; Flores, E. Nitrogen control in cyanobacteria. *J. Bacteriol.* **2001**, *183*, 411–425. [CrossRef]

121. Forchhammer, K.; Selim, K.A. Carbon/nitrogen homeostasis control in cyanobacteria. *FEMS Microbiol. Rev.* **2020**, *44*, 33–53. [CrossRef] [PubMed]

122. Meeks, J.C.; Wolk, C.P.; Lockau, W.; Schilling, N.; Shaffer, P.W.; Chien, W.S. Pathways of assimilation of [13N]N2 and 13NH4+ by cyanobacteria with and without heterocysts. *J. Bacteriol.* **1978**, *134*, 125–130. [CrossRef]

123. Vega-Palas, M.A.; Flores, E.; Herrero, A. NtcA, a global nitrogen regulator from the cyanobacterium Synechococcus that belongs to the Crp family of bacterial regulators. *Mol. Microbiol.* **1992**, *6*, 1853–1859. [CrossRef] [PubMed]

124. Vazquez-Bermudez, M.F.; Herrero, A.; Flores, E. 2-Oxoglutarate increases the binding affinity of the NtcA (nitrogen control) transcription factor for the Synechococcus glnA promoter. *FEBS Lett.* **2002**, *512*, 71–74. [CrossRef]

125. Zhao, M.X.; Jiang, Y.L.; He, Y.X.; Chen, Y.F.; Teng, Y.B.; Chen, Y.; Zhang, C.C.; Zhou, C.Z. Structural basis for the allosteric control of the global transcription factor NtcA by the nitrogen starvation signal 2-oxoglutarate. *Proc. Natl. Acad. Sci. USA* **2010**, *107*, 12487–12492. [CrossRef] [PubMed]

126. Llacer, J.L.; Espinosa, J.; Castells, M.A.; Contreras, A.; Forchhammer, K.; Rubio, V. Structural basis for the regulation of NtcA-dependent transcription by proteins PipX and PII. *Proc. Natl. Acad. Sci. USA* **2010**, *107*, 15397–15402. [CrossRef]

127. Espinosa, J.; Forchhammer, K.; Burillo, S.; Contreras, A. Interaction network in cyanobacterial nitrogen regulation: PipX, a protein that interacts in a 2-oxoglutarate dependent manner with PII and NtcA. *Mol. Microbiol.* **2006**, *61*, 457–469. [CrossRef]

128. Forcada-Nadal, A.; Forchhammer, K.; Rubio, V. SPR analysis of promoter binding of Synechocystis PCC6803 transcription factors NtcA and CRP suggests cross-talk and sheds light on regulation by effector molecules. *FEBS Lett.* **2014**, *588*, 2270–2276. [CrossRef]

129. Forchhammer, K.; Luddecke, J. Sensory properties of the PII signalling protein family. *FEBS J.* **2016**, *283*, 425–437. [CrossRef]

130. Zhang, C.C.; Zhou, C.Z.; Burnap, R.L.; Peng, L. Carbon/Nitrogen Metabolic Balance: Lessons from Cyanobacteria. *Trends Plant Sci.* **2018**, *23*, 1116–1130. [CrossRef]

131. Labella, J.I.; Cantos, R.; Salinas, P.; Espinosa, J.; Contreras, A. Distinctive Features of PipX, a Unique Signaling Protein of Cyanobacteria. *Life* **2020**, *10*, 79. [CrossRef] [PubMed]

132. Weinberg, Z.; Wang, J.X.; Bogue, J.; Yang, J.; Corbino, K.; Moy, R.H.; Breaker, R.R. Comparative genomics reveals 104 candidate structured RNAs from bacteria, archaea, and their metagenomes. *Genome Biol.* **2010**, *11*, R31. [CrossRef]

133. Ames, T.D.; Breaker, R.R. Bacterial aptamers that selectively bind glutamine. *RNA Biol.* **2011**, *8*, 82–89. [CrossRef] [PubMed]

134. Klahn, S.; Bolay, P.; Wright, P.R.; Atilho, R.M.; Brewer, K.I.; Hagemann, M.; Breaker, R.R.; Hess, W.R. A glutamine riboswitch is a key element for the regulation of glutamine synthetase in cyanobacteria. *Nucleic Acids Res.* **2018**, *46*, 10082–10094. [CrossRef] [PubMed]

135. Alvarez-Escribano, I.; Brenes-Alvarez, M.; Olmedo-Verd, E.; Vioque, A.; Muro-Pastor, A.M. The Nitrogen Stress-Repressed sRNA NsrR1 Regulates Expression of all1871, a Gene Required for Diazotrophic Growth in *Nostoc* sp. PCC 7120. *Life* **2020**, *10*, 54. [CrossRef] [PubMed]

136. Klahn, S.; Schaal, C.; Georg, J.; Baumgartner, D.; Knippen, G.; Hagemann, M.; Muro-Pastor, A.M.; Hess, W.R. The sRNA NsiR4 is involved in nitrogen assimilation control in cyanobacteria by targeting glutamine synthetase inactivating factor IF7. *Proc. Natl. Acad. Sci. USA* **2015**, *112*, E6243–E6252. [CrossRef] [PubMed]

137. Alvarez-Escribano, I.; Vioque, A.; Muro-Pastor, A.M. NsrR1, a Nitrogen Stress-Repressed sRNA, Contributes to the Regulation of nblA in *Nostoc* sp. PCC 7120. *Front. Microbiol.* **2018**, *9*, 2267. [CrossRef]

138. Hudek, L.; Premachandra, D.; Webster, W.A.; Brau, L. Role of Phosphate Transport System Component PstB1 in Phosphate Internalization by Nostoc punctiforme. *Appl. Environ. Microbiol.* **2016**, *82*, 6344–6356. [CrossRef]

139. Pitt, F.D.; Mazard, S.; Humphreys, L.; Scanlan, D.J. Functional characterization of *Synechocystis* sp. strain PCC 6803 pst1 and pst2 gene clusters reveals a novel strategy for phosphate uptake in a freshwater cyanobacterium. *J. Bacteriol.* **2010**, *192*, 3512–3523. [CrossRef]

140. Falkner, R.; Priewasser, M.; Falkner, G. Information Processing by Cyanobacteria during Adaptation to Environmental Phosphate Fluctuations. *Plant Signal. Behav.* **2006**, *1*, 212–220. [CrossRef]

141. Ray, J.M.; Bhaya, D.; Block, M.A.; Grossman, A.R. Isolation, transcription, and inactivation of the gene for an atypical alkaline phosphatase of *Synechococcus* sp. strain PCC 7942. *J. Bacteriol.* **1991**, *173*, 4297–4309. [CrossRef] [PubMed]

142. Wagner, K.U.; Masepohl, B.; Pistorius, E.K. The cyanobacterium *Synechococcus* sp. strain PCC 7942 contains a second alkaline phosphatase encoded by phoV. *Microbiology* **1995**, *141 Pt 12*, 3049–3058. [CrossRef]

143. Sebastian, M.; Ammerman, J.W. The alkaline phosphatase PhoX is more widely distributed in marine bacteria than the classical PhoA. *ISME J.* **2009**, *3*, 563–572. [CrossRef] [PubMed]

144. Peng, Z.; Feng, L.; Wang, X.; Miao, X. Adaptation of *Synechococcus* sp. PCC 7942 to phosphate starvation by glycolipid accumulation and membrane lipid remodeling. *Biochim. Biophys. Acta Mol. Cell Biol. Lipids* **2019**, *1864*, 158522. [CrossRef] [PubMed]

145. Hirani, T.A.; Suzuki, I.; Murata, N.; Hayashi, H.; Eaton-Rye, J.J. Characterization of a two-component signal transduction system involved in the induction of alkaline phosphatase under phosphate-limiting conditions in *Synechocystis* sp. PCC 6803. *Plant Mol. Biol.* **2001**, *45*, 133–144. [CrossRef]

146. Suzuki, S.; Ferjani, A.; Suzuki, I.; Murata, N. The SphS-SphR two component system is the exclusive sensor for the induction of gene expression in response to phosphate limitation in synechocystis. *J. Biol. Chem.* **2004**, *279*, 13234–13240. [CrossRef]

147. Nagaya, M.; Aiba, H.; Mizuno, T. The sphR product, a two-component system response regulator protein, regulates phosphate assimilation in *Synechococcus* sp. strain PCC 7942 by binding to two sites upstream from the phoA promoter. *J. Bacteriol.* **1994**, *176*, 2210–2215. [CrossRef]

148. Aiba, H.; Nagaya, M.; Mizuno, T. Sensor and regulator proteins from the cyanobacterium Synechococcus species PCC7942 that belong to the bacterial signal-transduction protein families: Implication in the adaptive response to phosphate limitation. *Mol. Microbiol.* **1993**, *8*, 81–91. [CrossRef]

149. Burut-Archanai, S.; Incharoensakdi, A.; Eaton-Rye, J.J. The extended N-terminal region of SphS is required for detection of external phosphate levels in *Synechocystis* sp. PCC 6803. *Biochem. Biophys. Res. Commun.* **2009**, *378*, 383–388. [CrossRef]

150. Juntarajumnong, W.; Hirani, T.A.; Simpson, J.M.; Incharoensakdi, A.; Eaton-Rye, J.J. Phosphate sensing in *Synechocystis* sp. PCC 6803: SphU and the SphS-SphR two-component regulatory system. *Arch. Microbiol.* **2007**, *188*, 389–402. [CrossRef]

151. Su, Z.; Olman, V.; Xu, Y. Computational prediction of Pho regulons in cyanobacteria. *BMC Genom.* **2007**, *8*, 156. [CrossRef] [PubMed]

152. Wandersman, C.; Delepelaire, P. Bacterial iron sources: From siderophores to hemophores. *Annu. Rev. Microbiol.* **2004**, *58*, 611–647. [CrossRef] [PubMed]

153. Hantke, K. Iron and metal regulation in bacteria. *Curr. Opin. Microbiol.* **2001**, *4*, 172–177. [CrossRef]

154. Pecqueur, L.; D'Autreaux, B.; Dupuy, J.; Nicolet, Y.; Jacquamet, L.; Brutscher, B.; Michaud-Soret, I.; Bersch, B. Structural changes of Escherichia coli ferric uptake regulator during metal-dependent dimerization and activation explored by NMR and X-ray crystallography. *J. Biol. Chem.* **2006**, *281*, 21286–21295. [CrossRef] [PubMed]

155. Ghassemian, M.; Straus, N.A. Fur regulates the expression of iron-stress genes in the cyanobacterium *Synechococcus* sp. strain PCC 7942. *Microbiology* **1996**, *142 Pt 6*, 1469–1476. [CrossRef]

156. Kunert, A.; Vinnemeier, J.; Erdmann, N.; Hagemann, M. Repression by Fur is not the main mechanism controlling the iron-inducible isiAB operon in the cyanobacterium *Synechocystis* sp. PCC 6803. *FEMS Microbiol. Lett.* **2003**, *227*, 255–262. [CrossRef]

157. Gonzalez, A.; Bes, M.T.; Valladares, A.; Peleato, M.L.; Fillat, M.F. FurA is the master regulator of iron homeostasis and modulates the expression of tetrapyrrole biosynthesis genes in *Anabaena* sp. PCC 7120. *Environ. Microbiol.* **2012**, *14*, 3175–3187. [CrossRef]

158. Gonzalez, A.; Bes, M.T.; Peleato, M.L.; Fillat, M.F. Expanding the Role of FurA as Essential Global Regulator in Cyanobacteria. *PLoS ONE* **2016**, *11*, e0151384. [CrossRef]

159. Georg, J.; Kostova, G.; Vuorijoki, L.; Schon, V.; Kadowaki, T.; Huokko, T.; Baumgartner, D.; Muller, M.; Klahn, S.; Allahverdiyeva, Y.; et al. Acclimation of Oxygenic Photosynthesis to Iron Starvation Is Controlled by the sRNA IsaR1. *Curr. Biol. CB* **2017**, *27*, 1425–1436 e1427. [CrossRef]

160. Klotz, A.; Reinhold, E.; Doello, S.; Forchhammer, K. Nitrogen Starvation Acclimation in Synechococcus elongatus: Redox-Control and the Role of Nitrate Reduction as an Electron Sink. *Life* **2015**, *5*, 888–904. [CrossRef]

161. Horiuchi, M.; Nakamura, K.; Kojima, K.; Nishiyama, Y.; Hatakeyama, W.; Hisabori, T.; Hihara, Y. The PedR transcriptional regulator interacts with thioredoxin to connect photosynthesis with gene expression in cyanobacteria. *Biochem. J.* **2010**, *431*, 135–140. [CrossRef] [PubMed]

Publisher's Note: MDPI stays neutral with regard to jurisdictional claims in published maps and institutional affiliations.

Review

The Role of the Cyanobacterial Type IV Pilus Machinery in Finding and Maintaining a Favourable Environment

Fabian D. Conradi [1], Conrad W. Mullineaux [1] and Annegret Wilde [2],*

[1]   School of Biological and Chemical Sciences, Queen Mary University of London, London E1 4NS, UK; f.d.conradi@qmul.ac.uk (F.D.C.); c.mullineaux@qmul.ac.uk (C.W.M.)

[2]   Institute of Biology III, University of Freiburg, Schänzlestr. 1, 79104 Freiburg; Germany

*   Correspondence: annegret.wilde@biologie.uni-freiburg.de

Received: 21 September 2020; Accepted: 21 October 2020; Published: 23 October 2020

Abstract: Type IV pili (T4P) are proteinaceous filaments found on the cell surface of many prokaryotic organisms and convey twitching motility through their extension/retraction cycles, moving cells across surfaces. In cyanobacteria, twitching motility is the sole mode of motility properly characterised to date and is the means by which cells perform phototaxis, the movement towards and away from directional light sources. The wavelength and intensity of the light source determine the direction of movement and, sometimes in concert with nutrient conditions, act as signals for some cyanobacteria to form mucoid multicellular assemblages. Formation of such aggregates or flocs represents an acclimation strategy to unfavourable environmental conditions and stresses, such as harmful light conditions or predation. T4P are also involved in natural transformation by exogenous DNA, secretion processes, and in cellular adaptation and survival strategies, further cementing the role of cell surface appendages. In this way, cyanobacteria are finely tuned by external stimuli to either escape unfavourable environmental conditions via phototaxis, exchange genetic material, and to modify their surroundings to fit their needs by forming multicellular assemblies.

Keywords: Type IV pili; cyanobacteria; phototaxis; flocculation; competence; *Synechocystis*

---

1. The Type IV Pilus Machinery Conveys Twitching Motility to Cyanobacteria

Cyanobacteria are found in a wide variety of ecological niches, ranging from polar latitudes [1,2] to desert soil crusts [3]. In marine environments, a small number of cyanobacterial genera are so abundant that they account for a substantial portion of marine primary productivity [4]. Despite the differences in preferred niches and cellular morphology, many species of cyanobacteria take a remarkably similar approach in adapting to environmental changes. Cyanobacteria are not known to possess flagella and instead, most motile cyanobacteria rely on T4P to convey twitching motility across surfaces, allowing them to move towards favourable environments or to escape unfavourable environments. A well-studied example of cyanobacterial twitching motility is found in *Synechocystis* sp. PCC 6803 (hereafter *Synechocystis*), but other instances have been reported and characterised in both single-celled [5] and filamentous cyanobacteria [6], which largely match the *Synechocystis* T4P machinery in component genes and operon structure [7].

T4P are protein filaments extended from a membrane-spanning pore complex. In *Synechocystis*, the components of the T4P complex have been largely identified by homology with the Type IVa pilus (T4aP) systems of the heterotrophic Gram-negative bacteria *Myxococcus xanthus* [8] and *Pseudomonas aeruginosa* [9], in contrast to the Type IVb pilus, which mainly differs by the length and the amino acid sequence of the pilin proteins [10]. The filament, composed of pilin proteins characterised by their cleavage motif [11], is extended by the secretion ATPase PilB and passes through the membrane at the inner membrane platform protein PilC and the outer membrane pore PilQ. PilC and PilQ are connected by a set of accessory proteins. The structure and function of the cyanobacterial T4P apparatus have been reviewed in [7]. A schematic representation of the T4P complex is given in Figure 1.

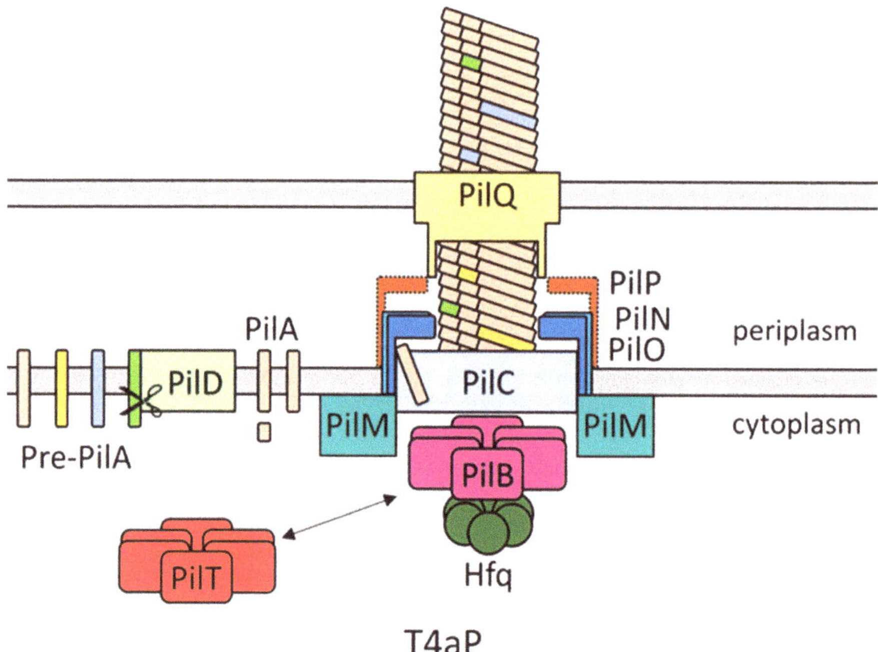

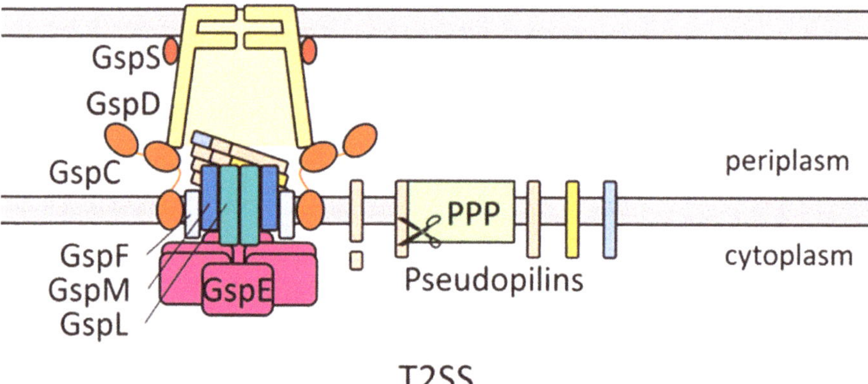

Figure 1. Schematic representation of the *Synechocystis* Type IVa pilus (T4aP) apparatus adapted from *Myxococcus xanthus* structure and nomenclature [12] and of the type II secretion system (T2SS) based on structural data from [13–15] using Escherichia component nomenclature. Colours denote proteins fulfilling homologous functions between the two systems. Dotted outline of PilP denotes a lack of experimental data confirming the in silico identification by Taton et al. [16].

Electron micrographs show two types of cell appendages on the *Synechocystis* surface, which have been termed thick pili and thin pili for their respective diameters. Thick *Synechocystis* pili have a diameter of 6–8 nm [8], matching the diameter of T4P found in heterotrophic bacteria [17]. Thick pili are essential for twitching motility and deletion mutants of the *pilA1* gene encoding the major pilin, a homologue of *P. aeruginosa* PilA, lose both thick pili and twitching motility. It is currently unclear whether PilA1 is the only pilin incorporated in *Synechocystis* thick pili. In contrast, thin pili are 2–3 nm in diameter and unable to convey twitching motility by themselves, as Δ*pilA1* mutants retain thin pili [8]. The nature and subunit composition of *Synechocystis* thin pili have not been solved so far. However, recent structures of *Thermus thermophilus* wide and narrow pili have shown that T4P machinery can produce structurally distinct pili depending on the type of pilin subunit incorporated [18]. Besides PilA1, *Synechocystis* contains a complement of ten other known PilA variants termed minor pilins [11]. The filamentous cyanobacterium *Nostoc punctiforme* also contains multiple PilA-like proteins and many of the components of the *N. punctiforme* T4P machinery have been identified according to *Synechocystis* annotations [19].

Pilins are processed by the PilD peptidase, which cleaves the N-terminal signal peptide. The mature pilins are then polymerised by the action of the hexameric ATPase PilB [20], thought to rotate during ATP hydrolysis, rotating PilC in turn and incorporating pilin subunits [21]. The retraction motor PilT, also a hexameric ATPase, rotates in the opposite direction to PilB, depolymerising the filament during pilus retraction [21]. The polymerised filament exits the outer membrane through PilQ. A set of accessory proteins have been identified in *Synechocystis* by homology with heterotrophic T4P systems, termed PilM, PilN, and PilO [9]. A PilP homologue has recently been identified in *Synechococcus elongatus* PCC 7942 [16]. Taton et al. also suggest that several other cyanobacteria, including *Synechocystis*, encode PilP homologues, though the sequence conservation is less pronounced [16]. As shown in Figure 1, the alignment complex composed of the PilMNOP accessory proteins forms a set of two rings in the periplasmic space [12] and links the T4P apparatus components in the inner and outer membranes. A recent study also indicated that many marine picocyanobacterial strains may contain a T4P apparatus homologous to *S. elongatus* PCC 7942 [22].

In contrast to single celled motile cyanobacteria like *Synechocystis* or *Synechococcus elongatus* UTEX 3055, vegetative filaments of many filamentous cyanobacteria are not inherently motile. Instead, they differentiate into specialised motile filaments termed hormogonia. While vegetative filaments of *N. punctiforme* are non-piliated, cells that are part of hormogonia show abundant surface piliation [19] arrayed in rings at the cell poles [23] and are capable of twitching motility using T4P [6]. In contrast to *Synechocystis*, however, only thick pili have been detected in *Nostoc* hormogonia [19].

2. Twitching Motility Enables Cyanobacteria to Seek out Favourable and Escape Unfavourable Environments

Cyanobacteria couple twitching motility conveyed by the T4P apparatus with environmental sensing and engage in tactic behaviour—the movement towards or away from light sources (phototaxis) or chemical gradients (chemotaxis). Phototaxis allows cyanobacteria to move towards environments that provide sufficient photosynthetically active light, avoiding higher light intensities or shorter wavelengths of light that might damage the photosynthetic apparatus [24]. Chemotaxis has been reported for some species of filamentous cyanobacteria such as *N. punctiforme* in response to plant hosts [25], which provide opportunity for symbiotic lifestyles, and *Oscillatoria* towards CO_2 [26]. Twitching motility-based phototaxis, in contrast, is a much more common feature in cyanobacteria, being observed and studied in single-celled cyanobacteria such as *Synechocystis* [27] and *S. elongatus* UTEX 3055 [5] as well as filamentous cyanobacteria [23]. The means by which phototaxis is performed, however, varies between cyanobacteria. Filamentous cyanobacteria like *N. punctiforme* perform both chemotaxis and phototaxis by moving up and down gradients, the motile hormogonia moving along the filament axis and controlling the rates of reversal in order to achieve net motion towards the attractant [6]. When applied to phototaxis, this behaviour is termed scotophobia [28]. Light perception in *Phormidium uncinatum*, another filamentous cyanobacterium, depends on decreases in light intensity at the leading end of the filament and increases in light intensity at the lagging end, though the mechanism for this spatial comparison is unclear [29]. Despite the prevalence of movement aligned with the filament axis, partial illumination of *Anabaena variabilis* trichomes has produced motility towards a light source placed to the side of the filament, indicating some ability to reorient the trichome and thus, the direction of movement [30]. Until recently, the mechanism by which motile filamentous cyanobacteria such as *N. punctiforme*, *Oscillatoria* [31], or *Anabaena variabilis* [32] move was under debate, with other hypotheses besides T4P including polysaccharide extrusion as the driving force of hormogonium motility [33]. However, it has become clear that at least *N. punctiforme* hormogonia move using a T4P system with the pili exerting a pulling force [6], as is common for twitching motility. The hormogonium polysaccharides extruded by *N. punctiforme* nonetheless support motility across surfaces [23] as has also been suggested for *Synechocystis* [34–36]. The hormogonium achieves the coordinated T4P action required for this behaviour by dynamically localising the partial coiled-coil-rich protein HmpF, encoded by a gene in the chemotaxis-like *hmp* gene cluster, to the forward cell pole [6]. Cho et al. showed that HmpF is essential for hormogonium motility and phototaxis as well as cell piliation in *N. punctiforme* [6].

In contrast to the bias towards a light source in *N. punctiforme*, *Synechocystis* has been shown to sense light directly by using the cell body as a microlens [37,38]. Although several control elements which are likely to be involved in downstream signal transmission have been identified, the exact signal transduction mechanism is unclear so far. The lensing of light in *Synechocystis* enables the cells to perform true phototaxis, moving directly towards light sources instead of performing biased random walks as in chemotaxis. *Synechocystis* has pili distributed around the entire cell perimeter and dynamically localises the extension motor PilB1 in the direction of movement [39], whereas *N. punctiforme* do not show dynamic PilB localisation [23]. This feature enables *Synechocystis* to move in any direction rather than being constrained in the direction to the cell or filament axis.

The regulation of twitching motility in response to external stimuli is achieved by sets of receptors and downstream regulators. In *Synechocystis*, while light lensing through the cell body provides the directionality to motility, movement towards or away from the light source is determined by the action of several photoreceptors. In particular, excess blue and UV light cause lower growth rates in *Synechocystis* [40], induce non-photochemical quenching in photosynthetic organisms indicating photoinhibition [41], oxidative stress [42], and various other deleterious effects in different cyanobacteria [43]. Many *Synechocystis* photoreceptors thus perceive short-wavelength light and *Synechocystis* cells use this information to either cease movement (blue light) [44] or reverse movement direction and move away from the stimulus (UV-A light) [45]. Stimuli from longer-wavelength light are also sensed, such as high-intensity red light leading to no movement or even negative phototaxis [46,47]. Cyanobacteriochromes (CBCR), Light-oxygen-voltage (LOV) photoreceptors, and Sensor of Blue Light using Flavin adenine dinucleotide (BLUF) photoreceptors sense the wavelength and intensity of incident light by photoconversion of bound bilin [48], flavin mononucleotide [49], or flavin adenine dinucleotide [50] chromophores, respectively.

The CBCR PixJ1 controls the direction of motility depending on the ratio of blue/green light [51] and mutants deficient in PixJ1 perform negative phototaxis (movement away from the light source) in environments where wt *Synechocystis* move towards the light source [52]. PixJ1 is part of the chemotaxis-like system *tax1* and carries a methyl-accepting chemotaxis protein (MCP)-like domain [51] and the downstream CheY-like response regulator PixG contains a PATAN domain thought to interact with the T4P apparatus [28,53]. PixJ homologues have been identified in various cyanobacteria, including *Nostoc punctiforme* ATCC 29133 [54], *Thermosynechococcus elongatus* BP-1 [55], *Anabaena* sp. PCC 7120 [56], and *Synechococcus elongatus* UTEX 3055 [5]. The photoconversion of the different PixJ homologues is variable across a wide range including red/green [57] and blue/green photocycles [51]. However, only some of the species confirmed to carry PixJ homologues have been shown to use them in phototaxis [5,54], partially owed to the general loss of motility in several cyanobacterial isolates. UirS, another *Synechocystis* CBCR, contains a UV/green photoconvertible GAF domain and causes motility reversal under UV illumination via its downstream effector, the PATAN domain containing response regulator LsiR [45]. Although the mechanism of action by which LsiR affects the T4P apparatus is not currently known, the dependence of negative phototaxis under UV-A illumination on both *uirS* and *lsiR* indicates their interaction with the T4P machinery either directly or indirectly. A third photoreceptor system which is involved in controlling phototaxis in *Synechocystis* is the PixD–PixE complex [58]. PixD is a BLUF domain photoreceptor which binds the PATAN domain response regulator PixE [59,60]. It is hypothesised that upon blue light exposure, PixE dissociates from PixD, binds to PilB1 and reverses the direction of movement, resulting in negative phototaxis [61]. The *Synechocystis* chemotaxis-like system *tax3* for which the signal input is unknown, leads to loss of thick pili and motility entirely when disrupted [52,62]. This system also contains a PATAN domain CheY-like response regulator, implying that PATAN domains could link external signals with the T4P apparatus to control its function or localization. The photoreceptor Cph2 is also implicated in *Synechocystis* motility. Mutants deficient in Cph2 show phototaxis towards blue light, whereas wt *Synechocystis* are non-motile in the same conditions [44]. In contrast to PixJ1, Cph2 transmits downstream signals through the production of the second messenger cyclic di-GMP (c-di-GMP) via its C-terminal GGDEF domain, which becomes active when blue light intensity is high relative to green light intensity [63]. Elevated c-di-GMP levels are commonly associated with reduced motility and increased sessility (reviewed in [64]). The blue/green light-dependent activity of Cph2 also leads to a host of pilus- and cell surface-related transcriptional changes [65], indicating multiple modes of action of Cph2-based c-di-GMP signalling. Furthermore, the N-terminus of Cph2 contains a red/far-red light sensing dual GAF domain module. The c-di-GMP synthesizing enzyme Slr1143 interacts with Cph2 and modulates motility under high-intensity red light by a so far unknown mechanism [46]. In *S. elongatus* PCC 7942, the LOV domain-based blue light receptor SL2 shows phosphodiesterase activity, breaking down c-di-GMP in the dark. This process unexpectedly accelerates upon blue illumination [66]. As *S. elongatus* PCC 7942 is non-motile, however, the effect of SL2-based photoperception on motility cannot be assessed, although the closely related and motile *Synechococcus elongatus* UTEX 3055 strain [5] may enable studies of the involvement of photoreceptors in *Synechococcus* phototaxis and biofilm formation. The mechanisms of many of the photoreceptors mentioned here have recently been reviewed in [28].

Cyanobacteria thus move towards light sources until one or more photoreceptors detect unfavourable light conditions, such as high-intensity red, blue, or UV light, which provoke a change in movement direction or a cessation of motility. In this way, cyanobacteria can position themselves in an optimal light environment. A model of this process in a biofilm context is shown in Figure 2. A biofilm provides a substrate for T4P to latch onto during twitching motility, making it a more likely environment for applying such a model compared to planktonic culture. The variety of chromophores, photosensing domains, and downstream signalling mechanisms with which cyanobacteria regulate the direction and extent of phototaxis via the T4P apparatus enable a high degree of adaptation to fluctuating environmental light conditions.

Figure 2. Model of movement of single-celled cyanobacteria in a multi-species phototrophic microbial mat. Black arrows indicate twitching motility in the direction of the arrow. The relative attenuation of different wavelengths of light in phototrophic communities (adapted from [67]) is represented by the colour gradients of the respective downward arrows, with purple representing UV-A radiation. The distinct light environments at different depths in the mat can activate or inactivate motility and trigger switches from positive to negative phototaxis and vice versa. This leads to an accumulation of the cyanobacteria at a depth where there is a favourable light environment.

3. Many Species of Cyanobacteria Form Large-Scale Multicellular Assemblages

While phototaxis and chemotaxis enable cyanobacteria to seek out favourable environments, many species of cyanobacteria are also known to form multicellular assemblies, allowing cyanobacteria to create their own niches. Aggregate formation in bacteria usually involves the secretion of various substances such as polysaccharides, proteins, and nucleic acids, which together form extracellular polymeric substances (EPS). Examples of aggregate formation include the well-documented blooms of floating *Microcystis aeruginosa* colonies, a unicellular freshwater cyanobacterium showing extensive cell surface piliation in liquid medium and on agar surfaces [68]. Nakasugi and Neilan (2005) present electron micrographs which indicate connections between cells that are likely made by T4P [68]. Colonies of different *Microcystis* species vary in size and growth rate with nutrient concentrations and growth medium pH [69]. Ma et al. (2014) suggest the smaller colony sizes and higher growth rate observed in certain high N and high P conditions represent a better utilisation of the available nutrients by *Microcystis* [69]. Nutrient starvation in bacterial aggregates has also been observed in heterotrophs [70] and *Synechocystis* [71] and slow mass-transfer into aggregate interiors is one of the drawbacks of communal lifestyles, leading to steep gradients of important nutrients such as CO_2 [72]. Similarly to *M. aeruginosa*, *Synechocystis* cell growth increased and lower aggregation was observed when the extracellular nutrient concentration was raised [71].

Synechocystis cells form floating aggregates termed flocs [71,73], although these flocs show a less dense colony morphology than their *Microcystis* counterparts. They are distinguished from biofilms in their lack of attachment to a substratum. Formation of these filamentous structures is dependent on some T4P components and the string-like appearance of *Synechocystis* flocs (shown in Figure 3b,c) might be a strategy to minimise nutrient limitation by increasing the surface-area-to-volume ratio [71]. Despite the disadvantages of nutrient limitation experienced in colonial lifestyles, lower diffusion rates through EPS also benefit the cells contained in it. Heterotrophic bacteria such as *P. aeruginosa* [74] and *Klebsiella pneumoniae* [75] show significantly increased resilience to many antibiotics in intact biofilms compared to planktonic cultures. Although the non-infectious cyanobacteria are less likely to encounter antibiotic treatment, the same principle of EPS shielding cells from harmful extraneous effects has been extended to salt and metal ion toxicity [76–78], phage infection [79], and predation by some (though not all) grazers [80]. Work on the *Pseudomonas* genus has shown that functional, retractile T4P are also required for infection by a number of bacteriophages [81–83]. Although this has not been confirmed to date in cyanobacteria to our knowledge, particularly as no known phages exist for the frequently studied *Synechocystis*, it may provide an additional layer of phage protection in multicellular aggregate contexts where a high intracellular c-di-GMP level down-regulates T4P dynamics.

The flocculation process in *Synechocystis* is known to be dependent on light colour via Cph2-based c-di-GMP signalling, increasing flocculation in blue light relative to green light [71]. *Synechocystis* biofilm formation has likewise been shown to be stimulated by blue light, likely via c-di-GMP [84]. Similar wavelength-dependent effects have been described for aggregation in *Thermosynechococcus vulcanus*, where the blue/green photoconvertible CBCR SesA induces aggregation in blue/UV light [85,86]. The SesB and SesC CBCRs in turn suppress *T. vulcanus* aggregation in red or green light [86]. Aggregate formation in *Synechocystis* and *Thermosynechococcus* may thus be, among other functions, a protective measure against short-wavelength light, both by increased light attenuation in dense parts of cyanobacterial aggregates [72] and by secretion of photoprotective extracellular matrix components as seen in various cyanobacteria [43,87,88]. It has been proposed that particularly the regulation of aggregation in response to the ratio of blue to green light by photoreceptors like Cph2 or the SesABC system might also serve to sense cell shading and regulate the size of aggregates, as green light penetrates deeper into cyanobacterial aggregates than blue light [71,89]. Enomoto and Ikeuchi [89] have shown that *T. vulcanus* aggregation is dependent on initial culture density as would be expected for such a system. A proposed model of this process is shown in Figure 3a. Some species of filamentous cyanobacteria are known to move vertically within phototrophic mats with diurnal cycles and in response to different light intensities [67,90] and wavelengths [91]. Cell shading sensors like blue/green photoconvertible photoreceptors of *Synechocystis* and *Thermosynechococcus* [89] and direct light direction sensing in *Synechocystis* may also enable cyanobacteria in complex communities to determine their location within the community [37]. However, to our knowledge, little evidence exists of unicellular cyanobacterial genera taking part in such migrations to date. Figure 2 shows a speculative model of such migrations according to light gradients for single cellular cyanobacteria in a multi-species phototrophic mat. In this model, cyanobacteria migrate in the direction of incident light towards the surface of the biofilm until blue light intensity relative to green light intensity is sufficient to block further motility. Cyanobacteria near the surface of the biofilm would in turn migrate away from UV-A radiation, which is strongly attenuated by the biofilm [91]. In this way, a zone might form in phototrophic mats in which cyanobacteria accumulate, with precise depth likely depending on the species. Cyanobacteria could thus control their light environment even within communities. This model comes with the caveat of a lack of spatiotemporal data on motility and c-di-GMP levels in aggregates of unicellular cyanobacteria.

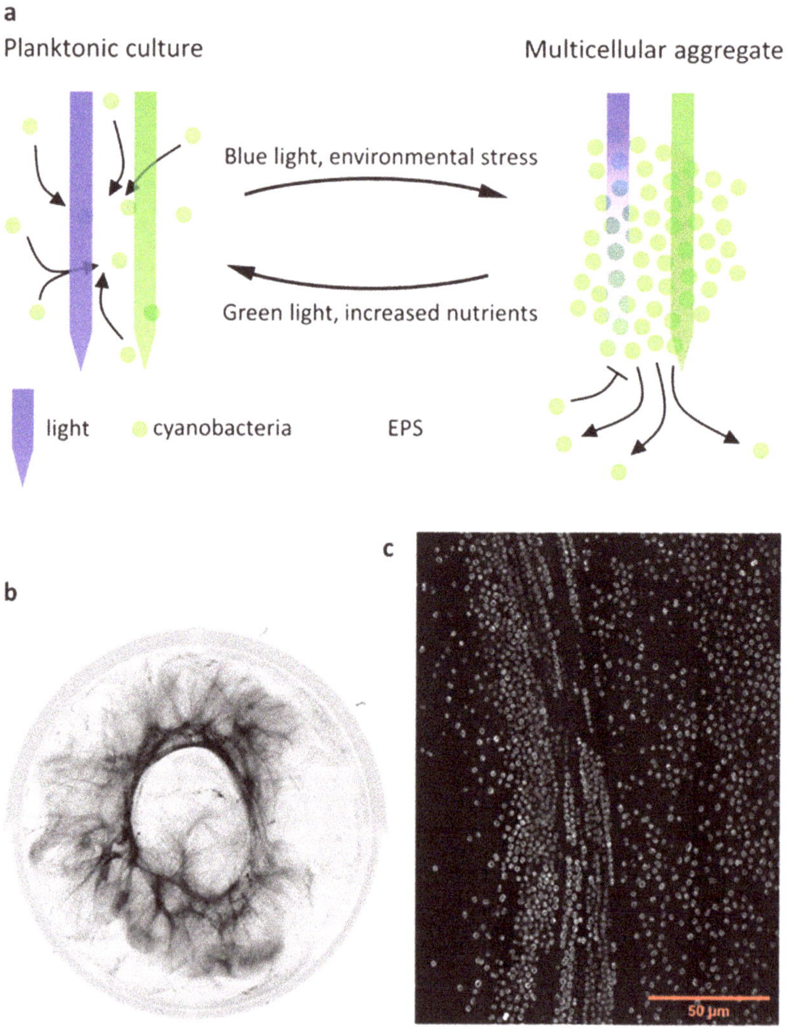

Figure 3. (**a**) Model of dynamic aggregation in *Synechocystis* and the *Thermosynechococcus* genus depending on external factors such as light penetration. Small arrows denote aggregation or the dissolution of aggregates. (**b**) Example of *Synechocystis* wt flocculation of a liquid culture in a 6-well plate (imaged vertically) with autofluorescence displayed in inverted greyscale. (**c**) Confocal micrograph of a *Synechocystis* floc, displaying autofluorescence in greyscale.

Another example of unicellular cyanobacteria forming multicellular communities is *S. elongatus* PCC 7942 biofilm formation. Intriguingly, *S. elongatus* PCC 7942 possesses an autoinhibitory system controlling biofilm formation in a manner dependent on a PilB homologue and PilC [92]. Schatz et al. (2013) found that when the T4P system was disrupted at the point of the motor ATPase or PilC, the usually planktonic *S. elongatus* PCC 7942 formed biofilms at the bottom of culture vessels [92].

4. The T4P Apparatus Has Structural and Secretory Roles in Cyanobacterial Community Formation

Many roles of T4P in biofilm development in heterotrophic bacteria have been established, including patterning [93,94] and surface sensing [95,96]. In cyanobacteria, in contrast, research into T4P involvement in multicellular aggregate formation has been more limited. In *Synechocystis*, thick pili (via deletion of *pilA1*) and associated twitching motility were found to be non-essential for flocculation, whereas PilB1, the RNA chaperone Hfq, PilC, and the minor pilins of the *pilA9-slr2019* operon were crucial for flocculation and mutants in any of the respective genes lost flocculation entirely [71,73]. It has been suggested that some minor pilins may be incorporated into T4P as has been found in *P. aeruginosa* [97], with thin pili visible during electron microscopy potentially being composed of a different set of pilins. Neuhaus et al. recently found that *Thermus thermophilus* produces pili with different diameters that vary in pilin composition [18]. *Synechocystis* thin pili may thus likewise be (partially) composed of minor pilins and play an important role in flocculation.

T4P have been implicated in mechanosensing in several species of heterotrophic bacteria, sensing surface contact via downstream cAMP and c-di-GMP signalling and leading to enhanced surface colonisation [95,96,98]. There is currently no evidence that a similar process occurs in *Synechocystis* or other cyanobacteria, and neither PilA1 nor PilT1 being required for flocculation indicates that if mechanosensing via retractile T4P exists in *Synechocystis*, it is not essential for flocculation.

Biofilm formation in *S. elongatus* PCC 7942 suggests an additional, secretory role for the T4P apparatus [99]. Inactivation of the T4P apparatus leads to a biofilm phenotype in this cyanobacterium. Notably, wt conditioned medium was able to restore planktonic growth in these mutants, indicating that a small secreted factor inhibits biofilm formation in wt *S. elongatus* PCC 7942 [92]. In absence of this putative factor, short peptidase-processed proteins are secreted by *S. elongatus* PCC 7942 and support biofilm formation [100]. Although quorum sensing systems are widespread in proteobacteria and often essential for biofilm formation [101], limited examples of quorum sensing systems exist in cyanobacteria to date [102]. The suppression of *S. elongatus* PCC 7942 biofilm formation by secreted factors may represent a non-traditional quorum sensing system, suggesting quorum sensing may be more common in cyanobacteria than previously thought.

The type II secretion system (T2SS) is closely related to the T4P apparatus, with many T2SS protein components showing functions equivalent to their T4P counterpart. A schematic comparison between the architecture of T4P apparatus and the T2SS is shown in Figure 1. Although the *S. elongatus* PCC 7942 secretion ATPase deleted by Schatz et al. might be a homologue of either GspE or PilB [92], bioinformatic analyses have suggested that cyanobacteria do not contain a T2SS and the *S. elongatus* PCC 7942 system is rather a T4P system spread over multiple loci [103,104]. Denise et al. found that there are only two species (both *Gloeobacter*) containing generic type IV filament systems in the cyanobacterial clade and no dedicated T2SS at all, while type IVa pilus systems were found in 79 genomes, including various filamentous and unicellular cyanobacterial species [104]. Considering the ambiguity in assignment of the PilB-type ATPase responsible for the suppression of biofilm formation in *S. elongatus* PCC 7942, the impact of the *S. elongatus* PCC 7942 PilB homologue on biofilm formation and general protein secretion [99] suggests a secretory role of the T4P apparatus in biofilm formation in this cyanobacterium. Involvement of the T4P apparatus in secretion is not without precedent in cyanobacteria. Secretion of heterologous proteins in *Synechococcus elongatus* and other cyanobacteria has been suggested to proceed via the T4P apparatus by an unknown mechanism [105,106]. Furthermore, accumulation of extracellular PilA in *N. punctiforme* is connected with polysaccharide secretion via HmpF, PilB, and PilQ [6]. Cho et al.

find that, particularly, loss of PilB and PilQ drastically reduces released polysaccharides and suggest polysaccharide export through the T4P machinery [6], though regulation via mechanosensing or similar T4P-mediated signalling mechanisms leading to downstream regulation of polysaccharide export by second messenger signalling could also explain the observations. Recently, several new genes have been identified in *N. punctiforme* encoding putative hormogonium polysaccharide (hps)-producing proteins, including homologues of Wza, Wzc, and Wzy [107]. Zuniga et al. also showed a connection between PilA secretion and the *hps* genes, with PilA secretion and motility depending strongly on several identified genes, although the mechanism of this interaction remains unclear [107]. In *Synechocystis*, polysaccharide export is substantially dependent on homologues of Wzm/Wzt [108] and Wza/Wzc [78], which have been found to be important in cell–cell and cell–surface adherence and cell buoyancy in *Synechocystis*, respectively. However, in contrast to *N. punctiforme*, no connection with the T4P machinery is known to date. This suggests that polysaccharide secretion may be dependent on T4P in some but not all species of cyanobacteria. Moreover, it has been shown that the homologue of the RNA chaperone Hfq binds to PilB1 in *Synechocystis* and that its correct localization at the pilus base is important for its function. Inactivation of *hfq* leads to non-motile cells and to changes in transcript accumulation, which are similar in *pilB1* and *pilC* mutant strains [109]. Hfq binds to a specific C-terminal domain of PilB1, which is found only in cyanobacterial assembly ATPases of T4P. Therefore, it is tempting to speculate that in general, in cyanobacterial mutants which do not form functional T4P, secondary effects occur, for example due to the incorrectly localised Hfq protein.

5. Regulation of the T4P Machinery

Although pilus motors can be directly regulated by c-di-GMP as found for the *Vibrio cholerae* Msh pilus [110], several other avenues of regulation exist, particularly at the transcriptional level. The second messenger cyclic 3′-5′-AMP (c-AMP) is induced by blue light in *Synechocystis* via the adenylate cyclase Cya1 [111], which is responsible for a large portion of intracellular cAMP production [112]. Which blue light-receptor is involved in activating Cya1 is currently unclear. Furthermore, *cya1* is downregulated by elevated bicarbonate levels [113], showing the diverse regulatory inputs in cAMP signalling. Cya1 is known to be crucial in *Synechocystis* phototaxis, with cells unable to form the characteristic finger-like projections in *cya1* deletion mutants [112,114]. Sycrp1 and Sycrp2 are the *Synechocystis* versions of the typical cAMP binding transcriptional regulator Crp known from many bacteria [115]. They propagate the Cya1 signal downstream, regulating among others the *pilA9-slr2019* operon and leading to a loss of cell surface piliation when deleted [116]. The downregulation of the *pilA9-slr2019* operon in Δ*sycrp2* mutants and the putative cooperation between Sycrp1 and Sycrp2 observed by Song et al. are sufficient to explain the impairment of motility in Δ*cya1*, Δ*sycrp1* [115], and Δ*sycrp2* [65,116] mutants. The abundance of *pilA11* mRNA and protein is further regulated by the PilR antisense RNA. PilR, however, does not regulate *pilA9*, *pilA10*, or *slr2018* expression, all part of the same operon as *pilA11*, showing that regulation of minor pilins can be highly specific [117].

Other species of cyanobacteria contain adenylate cyclases showing homology to *Synechocystis cya1* [112]. Though the connection to cell piliation in those bacteria is not as well established as in *Synechocystis*, cAMP signalling is involved in stress responses in other cyanobacteria such as desiccation tolerance in *Anabaena sp.* PCC 7120 [118], and mat formation in *Spirulina platensis*, both during extracellular cAMP addition [119] and when intracellular cAMP phosphodiesterases were inhibited [120]. Uptake of extracellular cAMP and compensation for low intracellular cAMP have also been observed in *Synechocystis* motility [114].

Several cyanobacteria have been confirmed to contain SigF, a stress response-related group 3 sigma factor, including many community-forming or filamentous species [121]. Although Imamura and Asayama found SigF in all cyanobacteria investigated, only *Synechocystis* and *N. punctiforme* SigF function has been characterised to date [121]. *Synechocystis* SigF is known to directly regulate the transcription of the *pilA1-pilA2* operon [122]. Deletion of *sigF* in *Synechocystis* correspondingly leads to a reduction in *pilA1-pilA2* mRNA and a loss of thick pili and phototactic motility [123,124].

Furthermore, Flores et al. found that *sigF* deletion significantly alters the exoproteome, with many proteins being present in reduced quantities or not at all [124]. Although this may indicate a connection between T4P action and protein secretion, as has been found in *S. elongatus* PCC 7942, Flores et al. also point out the large number of proteins of unknown function regulated by SigF, so definite conclusions on protein export will require further characterisation [124]. The *Synechocystis* Δ*sigF* mutant also shows an enhanced degree of cell sedimentation and flocculation [124], both in keeping with the drastic increase in exopolysaccharide production and change in monosaccharide composition. The increase in flocculation despite the downregulation of the *pilA1-pilA2* operon confirms that the major pilin PilA1 is not required in *Synechocystis* flocculation. Similarly, Miranda et al. have found *Synechocystis* to flocculate in batch cultures when the gene *slr1783*, coding for a monooxygenase, was deleted, leading to increased exopolysaccharide content [125]. The enhanced cellular aggregation in exopolysaccharide-overproducing mutants poses the question of the relative importance of T4P and exopolysaccharides in enabling *Synechocystis* flocculation depending on external conditions.

Similarly to *Synechocystis*, *N. punctiforme* SigF regulates *pilA*, which is very highly expressed in hormogonia, with deletion of *sigF* leading to almost total loss of *pilA* mRNA [126]. Intriguingly, all other T4P components investigated by Gonzalez et al. showed strong regulation by SigJ, rather than SigF [126]. The *Synechocystis sigF* deletion mutant showed no differential regulation of other components of the T4P [124], indicating that regulation of the major pilin specifically may be a common feature of cyanobacterial SigF.

6. Cyanobacterial Natural Competence Requires T4P

The uptake of exogenous DNA, known as natural competence, allows cells to adapt to environmental conditions through the exchange of situationally useful genes. Competence is known to depend on functional T4P in several bacteria [81,127], with DNA binding by T4P likely occurring at the pilus tip [128,129].

In *Synechocystis*, a close interplay between T4P and competence has also become evident. Yoshihara et al. have shown that many core components of the T4P apparatus are required for transformability in addition to their role in motility [9]. They also identified the *comA* gene, which, when deleted, causes a complete loss of transformability but only has a moderate impact on thick pili abundance and deletion mutants retain their motility [9]. Likewise, Nakasugi et al. identified a homologue of the *comF* gene in *Synechocystis*, which also leads to loss of competence but does not disrupt surface piliation when deleted [130]. ComF deletion, however, leads to a loss of motility [130]. This suggests a correlation between transformability and motility which was not apparent in the study by Yoshihara et al. on the effects of *comA* deletion. However, Yoshihara et al. assessed motility only from colony morphology, which may be less reliable than the colony motility assay method employed by Nakasugi et al, which is now more widely favoured. Nonetheless, these findings suggest that *Synechocystis* has DNA binding proteins which are required for natural transformability and which also strongly influence pilus function. Intriguingly, Nakasugi et al. also report the formation of filamentous aggregates in liquid cultures of *comF* deletion mutants matching the structures described in recent work on *Synechocystis* flocculation [71]. They suggest that the increased bundling of pili in the Δ*comF* mutant may be the cause of this flocculating phenotype, indicating that competence genes may be important factors in controlling *Synechocystis* T4P morphology and multicellularity [71].

Similar dependence of natural transformation on parts of the T4P apparatus has been observed in *S. elongatus* PCC 7942, including T4P motor proteins, *comEA/comEC/comF* homologues, and *sigF*, showing that many similar factors are involved as in *Synechocystis* [16]. Intriguingly, *S. elongatus* PCC 7942 is non-motile but naturally competent, asking questions about the cause of its deficiency in phototaxis, the mechanism of its competence, or both. Taton et al. also discovered a strong dependence of natural transformation on the circadian clock, likely via circadian control of the T4P machinery [16]. However, not all strains of *Synechococcus* are naturally competent, indicating some variability within the genus [131].

Although T4P in other naturally competent cyanobacteria have been suggested to be important in natural transformation, for example in *Microcystis aeruginosa* PCC 7806 [132], *Phormidium lacuna* [133], and *Thermosynechococcus elongatus* BP-1 [134], no direct link between T4P and competence in these organisms has been made to date. Recent in silico approaches, however, have shown that many cyanobacteria share high sequence similarity in their T4P genes with naturally competent cyanobacteria [16,133], suggesting that T4P may enable natural transformation in many more cyanobacterial genera than have been confirmed experimentally.

7. Concluding Remarks

Several unanswered questions on the involvement of T4P in cyanobacterial motility and multicellularity remain. Particularly mechanosensing, emerging as a frequent feature among heterotrophic bacteria, has not yet been demonstrated in cyanobacteria. Aggregation into biofilms, flocs, and microbial mats is a more common strategy in cyanobacteria than previously thought, particularly among laboratory strains such as *Synechocystis* or *S. elongatus* PCC 7942. The latter strain provides an example of the dangers of relying on strains that exhibit potentially non-representative phenotypes with the isolation of *S. elongatus* UTEX 3055, a very close relative of *S. elongatus* PCC 7942, indicating that freshwater *Synechococcus* may be motile and natively community-forming. *Thermosynechococcus vulcanus* NIES-2134 (RKN) and *Thermosynechococcus elongatus* BP-1 similarly are very closely related [135] and yet show divergent aggregation phenotype, with *T. elongatus* BP-1 being deficient in aggregation [136], illustrating the importance of selecting appropriate strains for research. The examples of *S. elongatus* PCC 7942 and *N. punctiforme* have demonstrated that much is left to be understood about the secretory roles of the T4P apparatus in cyanobacteria, providing potential avenues of cell–cell communication which have so far been largely missing in the phylum.

The loss of motility in laboratory strains as a result of prolonged cultivation in particular is further exemplified by the microevolution of *Synechocystis* laboratory strains [137]. We have likewise observed in the past that at least some strains of the non-motile Kazusa branch of *Synechocystis* (ATC27184) do not flocculate. It therefore seems prudent to be mindful of such pitfalls given the large areas of cyanobacterial physiology that are yet to be thoroughly explored.

Research in the last few years has shown that cyanobacteria are capable of complex and co-operative behaviour. Much remains to be learned about these behaviours and the survival advantages that they may confer in the natural environment.

Author Contributions: F.D.C. prepared the original draft of the manuscript; C.W.M. and A.W. reviewed and edited the manuscript. All authors have read and agreed to the published version of the manuscript.

Funding: Research in this field in the authors' laboratories was funded by the Deutsche Forschungsgemeinschaft (DFG, German Science Foundation), grant number WI2014/8-1, A2-SFB1381 to A.W. and Leverhulme Trust research grant RPG-2020-054 to C.W.M. and F.D.C. The article processing charge was funded by the Baden-Wuerttemberg Ministry of Science, Research and Art and the University of Freiburg in the funding programme Open Access Publishing.

Conflicts of Interest: The authors declare no conflict of interest.

References

1. Vigneron, A.; Cruaud, P.; Mohit, V.; Martineau, M.-J.; Culley, A.I.; Lovejoy, C.; Vincent, W.F. Multiple strategies for light-harvesting, photoprotection, and carbon flow in high latitude microbial mats. *Front. Microbiol.* **2018**, *9*, 1–12. [CrossRef] [PubMed]

2. Chrismas, N.A.M.; Barker, G.; Anesio, A.M.; Sánchez-Baracaldo, P. Genomic mechanisms for cold tolerance and production of exopolysaccharides in the Arctic cyanobacterium *Phormidesmis priestleyi* BC1401. *BMC Genom.* **2016**, *17*, 533. [CrossRef] [PubMed]

3. Hagemann, M.; Henneberg, M.; Felde, V.J.M.N.L.; Berkowicz, S.M.; Raanan, H.; Pade, N.; Felix-Henningsen, P.; Kaplan, A. Cyanobacterial populations in biological soil crusts of the northwest Negev Desert, Israel-effects of local conditions and disturbance. *FEMS Microbiol. Ecol.* **2017**, *93*, 1–9. [CrossRef] [PubMed]

4. Flombaum, P.; Gallegos, J.L.; Gordillo, R.A.; Rincón, J.; Zabala, L.L.; Jiao, N.; Karl, D.M.; Li, W.K.W.; Lomas, M.W.; Veneziano, D.; et al. Present and future global distributions of the marine cyanobacteria *Prochlorococcus* and *Synechococcus*. *Proc. Natl. Acad. Sci. USA* **2013**, *110*, 9824–9829. [CrossRef]

5. Yang, Y.; Lam, V.; Adomako, M.; Simkovsky, R.; Jakob, A.; Rockwell, N.C.; Cohen, S.E.; Taton, A.; Wang, J.; Lagarias, J.C.; et al. Phototaxis in a wild isolate of the cyanobacterium *Synechococcus elongatus*. *Proc. Natl. Acad. Sci. USA* **2018**, *115*, E12378–E12387. [CrossRef]

6. Cho, Y.W.; Gonzales, A.; Harwood, T.V.; Huynh, J.; Hwang, Y.; Park, J.S.; Trieu, A.Q.; Italia, P.; Pallipuram, V.K.; Risser, D.D. Dynamic localization of HmpF regulates type IV pilus activity and directional motility in the filamentous cyanobacterium *Nostoc punctiforme*. *Mol. Microbiol.* **2017**, *106*, 252–265. [CrossRef]

7. Schuergers, N.; Wilde, A. Appendages of the Cyanobacterial Cell. *Life* **2015**, *5*, 700–715. [CrossRef]

8. Bhaya, D.; Bianco, N.R.; Bryant, D. Type IV pilus biogenesis and motility in the cyanobacterium *Synechocystis* sp. PCC6803. *Mol. Microbiol.* **2000**, *37*, 941–951. [CrossRef]

9. Yoshihara, S.; Geng, X.X.; Okamoto, S.; Yura, K.; Murata, T.; Go, M.; Ohmori, M.; Ikeuchi, M. Mutational analysis of genes involved in pilus structure, motility and transformation competency in the unicellular motile cyanobacterium *Synechocystis sp.* PCC 6803. *Plant Cell Physiol.* **2001**, *42*, 63–73. [CrossRef]

10. Roux, N.; Spagnolo, J.; De Bentzmann, S. Neglected but amazingly diverse type IVb pili. *Res. Microbiol.* **2012**, *163*, 659–673. [CrossRef]

11. Linhartová, M.; Bučinská, L.; Halada, P.; Ječmen, T.; Šetlík, J.; Komenda, J.; Sobotka, R. Accumulation of the Type IV prepilin triggers degradation of SecY and YidC and inhibits synthesis of Photosystem II proteins in the cyanobacterium *Synechocystis* PCC 6803. *Mol. Microbiol.* **2014**, *93*, 1207–1223. [CrossRef] [PubMed]

12. Chang, Y.W.; Rettberg, L.A.; Treuner-Lange, A.; Iwasa, J.; Søgaard-Andersen, L.; Jensen, G.J. Architecture of the type IVa pilus machine. *Science (80-)* **2016**, *351*, aad2001. [CrossRef]

13. Korotkov, K.V.; Sandkvist, M.; Hol, W.G.J. The type II secretion system: Biogenesis, molecular architecture and mechanism. *Nat. Rev. Microbiol.* **2012**, *10*, 336–351. [CrossRef] [PubMed]

14. Chernyatina, A.A.; Low, H.H. Core architecture of a bacterial type II secretion system. *Nat. Commun.* **2019**, *10*, 1–10. [CrossRef]

15. Hay, I.D.; Belousoff, M.J.; Dunstan, R.A.; Bamert, R.S.; Lithgow, T. Structure and membrane topography of the vibrio-type secretin complex from the type 2 secretion system of enteropathogenic *Escherichia coli*. *J. Bacteriol.* **2018**, *200*, 1–15. [CrossRef]

16. Taton, A.; Erikson, C.; Yang, Y.; Rubin, B.E.; Rifkin, S.A.; Golden, J.W.; Golden, S.S. The circadian clock and darkness control natural competence in cyanobacteria. *Nat. Commun.* **2020**, *11*, 1688. [CrossRef] [PubMed]

17. Craig, L.; Pique, M.E.; Tainer, J.A. Type IV pilus structure and bacterial pathogenicity. *Nat. Rev. Microbiol.* **2004**, *2*, 363–378. [CrossRef]

18. Neuhaus, A.; Selvaraj, M.; Salzer, R.; Langer, J.D.; Kruse, K.; Kirchner, L.; Sanders, K.; Daum, B.; Averhoff, B.; Gold, V.A.M. Cryo-electron microscopy reveals two distinct type IV pili assembled by the same bacterium. *Nat. Commun.* **2020**, *11*, 2231. [CrossRef]

19. Duggan, P.S.; Gottardello, P.; Adams, D.G. Molecular analysis of genes in *Nostoc punctiforme* involved in pilus biogenesis and plant infection. *J. Bacteriol.* **2007**, *189*, 4547–4551. [CrossRef]

20. Sukmana, A.; Yang, Z. The type IV pilus assembly motor PilB is a robust hexameric ATPase with complex kinetics. *Biochem. J.* **2018**, *475*, 1979–1993. [CrossRef]

21. McCallum, M.; Tammam, S.; Khan, A.; Burrows, L.L.; Lynne Howell, P. The molecular mechanism of the type IVa pilus motors. *Nat. Commun.* **2017**, *8*, 15091. [CrossRef] [PubMed]

22. del Mar Aguilo-Ferretjans, M.; Bosch, R.; Puxty, R.; Latva, M.; Zadjelovic, V.; Chhun, A.; Sousoni, D.; Polin, M.; Scanlan, D.; Christie-Oleza, J. Pili allow dominant marine cyanobacteria to avoid sinking and evade predation. *bioRxiv* **2020**. [CrossRef]

23. Khayatan, B.; Meeks, J.C.; Risser, D.D. Evidence that a modified type IV pilus-like system powers gliding motility and polysaccharide secretion in filamentous cyanobacteria. *Mol. Microbiol.* **2015**, *98*, 1021–1036. [CrossRef]

24. Rehman, A.U.; Cser, K.; Sass, L.; Vass, I. Characterization of singlet oxygen production and its involvement in photodamage of Photosystem II in the cyanobacterium *Synechocystis* PCC 6803 by histidine-mediated chemical trapping. *Biochim. Biophys. Acta Bioenerg.* **2013**, *1827*, 689–698. [CrossRef] [PubMed]

25. Campbell, E.L.; Meeks, J.C. Characteristics of hormogonia formation by symbiotic *Nostoc spp.* in response to the presence of *Anthoceros punctatus* or its extracellular products. *Appl. Environ. Microbiol.* **1989**, *55*, 125–131. [CrossRef] [PubMed]

26. Malin, G.; Walsby, A.E. Chemotaxis of a cyanobacterium on concentration gradients of carbon dioxide, bicarbonate and oxygen. *Microbiology* **1985**, *131*, 2643–2652. [CrossRef]

27. Choi, J.; Chungl, Y.; Moon, Y.; Kimtl, C.; Watanabe, M.; Song, P.; Joe, C.-O.; Bogorad, L.; Park, Y.M. Photomovement of the gliding cyanobacterium *Synechocystis sp.* PCC 6803. *Photochem. Photobiol.* **1999**, *70*, 95–102. [CrossRef]

28. Wilde, A.; Mullineaux, C.W. Light-controlled motility in prokaryotes and the problem of directional light perception. *FEMS Microbiol. Rev.* **2017**, *41*, 900–922. [CrossRef]

29. Gabai, V.L. A one-instant mechanism of phototaxis in the cyanobacterium *Phormidium uncinatum*. *FEMS Microbiol. Lett.* **1985**, *30*, 125–129. [CrossRef]

30. Nultsch, W.; Wenderoth, K. Partial Irradiation Experiments with *Anabaena variabilis* (Kütz). *Z. für Pflanzenphysiol.* **1983**, *111*, 1–7. [CrossRef]

31. Halfen, L.N.; Castenholz, R.W. Gliding motility in the blue-green alga *Oscillatoria princeps*. *J. Phycol.* **1971**, *7*, 133–145. [CrossRef]

32. Nultsch, W.; Schuchart, H.; Höhl, M. Investigations on the phototactic orientation of *Anabaena variabilis*. *Arch. Microbiol.* **1979**, *122*, 85–91. [CrossRef]

33. Hoiczyk, E.; Baumeister, W. The junctional pore complex, a prokaryotic secretion organelle, is the molecular motor underlying gliding motility in cyanobacteria. *Curr. Biol.* **1998**, *8*, 1161–1168. [CrossRef]

34. Varuni, P.; Menon, S.N.; Menon, G.I. Phototaxis as a collective phenomenon in cyanobacterial colonies. *Sci. Rep.* **2017**, *7*, 17799. [CrossRef]

35. Menon, S.N.; Varuni, P.; Menon, G.I. Information integration and collective motility in phototactic cyanobacteria. *PLoS Comput. Biol.* **2020**, *16*, 1–18. [CrossRef] [PubMed]

36. Wilde, A.; Mullineaux, C.W. Motility in cyanobacteria: Polysaccharide tracks and Type IV pilus motors. *Mol. Microbiol.* **2015**, *98*, 998–1001. [CrossRef] [PubMed]

37. Schuergers, N.; Lenn, T.; Kampmann, R.; Meissner, M.V.; Esteves, T.; Temerinac-Ott, M.; Korvink, J.G.; Lowe, A.R.; Mullineaux, C.W.; Wilde, A. Cyanobacteria use micro-optics to sense light direction. *eLife* **2016**, *5*, e12620. [CrossRef]

38. Nakane, D.; Nishizaka, T. Asymmetric distribution of type IV pili triggered by directional light in unicellular cyanobacteria. *Proc. Natl. Acad. Sci. USA* **2017**, *114*, 6593–6598. [CrossRef] [PubMed]

39. Schuergers, N.; Nürnberg, D.J.; Wallner, T.; Mullineaux, C.W.; Wilde, A. PilB localization correlates with the direction of twitching motility in the cyanobacterium *Synechocystis sp.* PCC 6803. *Microbiology (United Kingdom)* **2015**, *161*, 960–966. [CrossRef] [PubMed]

40. Luimstra, V.M.; Schuurmans, J.M.; Verschoor, A.M.; Hellingwerf, K.J.; Huisman, J.; Matthijs, H.C.P. Blue light reduces photosynthetic efficiency of cyanobacteria through an imbalance between photosystems I and II. *Photosynth. Res.* **2018**, *138*, 177–189. [CrossRef] [PubMed]

41. Vass, I. Molecular mechanisms of photodamage in the Photosystem II complex. *Biochim. Biophys. Acta Bioenerg.* **2012**, *1817*, 209–217. [CrossRef] [PubMed]

42. He, Y.Y.; Häder, D.P. Reactive oxygen species and UV-B: Effect on cyanobacteria. *Photochem. Photobiol. Sci.* **2002**, *1*, 729–736. [CrossRef] [PubMed]

43. Ehling-Schulz, M.; Scherer, S. UV protection in cyanobacteria. *Eur. J. Phycol.* **1999**, *34*, 329–338. [CrossRef]

44. Wilde, A.; Fiedler, B.; Börner, T. The cyanobacterial phytochrome Cph2 inhibits phototaxis towards blue light. *Mol. Microbiol.* **2002**, *44*, 981–988. [CrossRef] [PubMed]

45. Song, J.Y.; Cho, H.S.; Cho, J.-I.; Jeon, J.S.; Lagarias, J.C.; Park, Y. Il Near-UV cyanobacteriochrome signaling system elicits negative phototaxis in the cyanobacterium *Synechocystis sp.* PCC 6803. *Proc. Natl. Acad. Sci. USA* **2011**, *108*, 10780–10785. [CrossRef] [PubMed]

46. Angerer, V.; Schwenk, P.; Wallner, T.; Kaever, V.; Hiltbrunner, A.; Wilde, A. The protein Slr1143 is an active diguanylate cyclase in *Synechocystis sp.* PCC 6803 and interacts with the photoreceptor Cph2. *Microbiology* **2017**, *163*, 920–930. [CrossRef]

47. Ng, W.O.; Grossman, A.R.; Bhaya, D. Multiple light inputs control phototaxis in *Synechocystis sp.* Strain PCC6803. *J. Bacteriol.* **2003**, *185*, 1599–1607. [CrossRef]

48. Fushimi, K.; Narikawa, R. Cyanobacteriochromes: Photoreceptors covering the entire UV-to-visible spectrum. *Curr. Opin. Struct. Biol.* **2019**, *57*, 39–46. [CrossRef]

49. Pudasaini, A.; El-Arab, K.K.; Zoltowski, B.D. LOV-based optogenetic devices: Light-driven modules to impart photoregulated control of cellular signaling. *Front. Mol. Biosci.* **2015**, *2*, 1–15. [CrossRef]

50. Tanaka, K.; Nakasone, Y.; Okajima, K.; Ikeuchi, M.; Tokutomi, S.; Terazima, M. Light-induced conformational change and transient dissociation reaction of the BLUF photoreceptor *Synechocystis* PixD (Slr1694). *J. Mol. Biol.* **2011**, *409*, 773–785. [CrossRef]

51. Yoshihara, S.; Ikeuchi, M. Phototactic motility in the unicellular cyanobacterium *Synechocystis sp.* PCC 6803. *Photochem. Photobiol. Sci.* **2004**, *3*, 512–518. [CrossRef]

52. Bhaya, D.; Takahashi, A.; Grossman, A.R. Light regulation of type IV pilus-dependent motility by chemosensor-like elements in *Synechocystis* PCC6803. *Proc. Natl. Acad. Sci. USA* **2001**, *98*, 7540–7545. [CrossRef] [PubMed]

53. Makarova, K.S.; Koonin, E.V.; Haselkorn, R.; Galperin, M.Y. Cyanobacterial response regulator PatA contains a conserved N-terminal domain (PATAN) with an alpha-helical insertion. *Bioinformatics* **2006**, *22*, 1297–1301. [CrossRef]

54. Campbell, E.L.; Hagen, K.D.; Chen, R.; Risser, D.D.; Ferreira, D.P.; Meeks, J.C. Genetic analysis reveals the identity of the photoreceptor for phototaxis in hormogonium filaments of *Nostoc punctiforme*. *J. Bacteriol.* **2015**, *197*, 782–791. [CrossRef]

55. Ishizuka, T.; Shimada, T.; Okajima, K.; Yoshihara, S.; Ochiai, Y.; Katayama, M.; Ikeuchi, M. Characterization of cyanobacteriochrome TePixJ from a thermophilic cyanobacterium *Thermosynechococcus elongatus* strain BP-1. *Plant Cell Physiol.* **2006**, *47*, 1251–1261. [CrossRef]

56. Narikawa, R.; Fukushima, Y.; Ishizuka, T.; Itoh, S.; Ikeuchi, M. A novel photoactive GAF domain of cyanobacteriochrome AnPixJ that shows reversible green/red photoconversion. *J. Mol. Biol.* **2008**, *380*, 844–855. [CrossRef] [PubMed]

57. Rockwell, N.C.; Martin, S.S.; Lagarias, J.C. Red/green cyanobacteriochromes: Sensors of color and power. *Biochemistry* **2012**, *51*, 9667–9677. [CrossRef] [PubMed]

58. Masuda, S.; Ono, T.A. Biochemical characterization of the major adenylyl cyclase, Cya1, in the cyanobacterium *Synechocystis sp.* PCC 6803. *FEBS Lett.* **2004**, *577*, 255–258. [CrossRef] [PubMed]

59. Tanaka, K.; Nakasone, Y.; Okajima, K.; Ikeuchi, M.; Tokutomi, S.; Terazima, M. Time-resolved tracking of interprotein signal transduction: *Synechocystis* PixD-PixE complex as a sensor of light intensity. *J. Am. Chem. Soc.* **2012**, *134*, 8336–8339. [CrossRef]

60. Ren, S.; Sato, R.; Hasegawa, K.; Ohta, H.; Masuda, S. A predicted structure for the PixD-PixE complex determined by homology modeling, docking simulations, and a mutagenesis study. *Biochemistry* **2013**, *52*, 1272–1279. [CrossRef]

61. Jakob, A.; Nakamura, H.; Kobayashi, A.; Sugimoto, Y.; Wilde, A.; Masuda, S. The (PATAN)-CheY-like response regulator PixE interacts with the motor ATPase PilB1 to control negative phototaxis in the cyanobacterium *Synechocystis sp.* PCC 6803. *Plant Cell Physiol.* **2019**, *0*, 1–12. [CrossRef]

62. Yoshihara, S.; Geng, X.; Ikeuchi, M. *pilG* gene cluster and split *pilL* genes involved in pilus biogenesis, motility and genetic transformation in the cyanobacterium *Synechocystis sp.* PCC 6803. *Plant Cell Physiol.* **2002**, *43*, 513–521. [CrossRef]

63. Savakis, P.; De Causmaecker, S.; Angerer, V.; Ruppert, U.; Anders, K.; Essen, L.O.; Wilde, A. Light-induced alteration of c-di-GMP level controls motility of *Synechocystis sp.* PCC 6803. *Mol. Microbiol.* **2012**, *85*, 239–251. [CrossRef] [PubMed]

64. Römling, U.; Galperin, M.Y.; Gomelsky, M. Cyclic di-GMP: The First 25 Years of a Universal Bacterial Second Messenger. *Microbiol. Mol. Biol. Rev.* **2013**, *77*, 1–52. [CrossRef]

65. Wallner, T.; Pedroza, L.; Voigt, K.; Kaever, V.; Wilde, A. The cyanobacterial phytochrome 2 regulates the expression of motility-related genes through the second messenger cyclic di-GMP. *Photochem. Photobiol. Sci.* **2020**, 631–643. [CrossRef]

66. Cao, Z.; Livoti, E.; Losi, A.; Gärtner, W. A blue light-inducible phosphodiesterase activity in the cyanobacterium *Synechococcus elongatus*. *Photochem. Photobiol.* **2010**, *86*, 606–611. [CrossRef]

67. Lichtenberg, M.; Cartaxana, P.; Kühl, M. Vertical migration optimizes photosynthetic efficiency of motile cyanobacteria in a coastal mnicrobial mat. *Front. Mar. Sci.* **2020**, *7*, 1–13. [CrossRef]

68. Nakasugi, K.; Neilan, B.A. Identification of pilus-like structures and genes in *Microcystis aeruginosa* PCC7806. *Appl. Environ. Microbiol.* **2005**, *71*, 7621–7625. [CrossRef]

69. Ma, J.; Brookes, J.D.; Qin, B.; Paerl, H.W.; Gao, G.; Wu, P.; Zhang, W.; Deng, J.; Zhu, G.; Zhang, Y.; et al. Environmental factors controlling colony formation in blooms of the cyanobacteria *Microcystis spp.* in Lake Taihu, China. *Harmful Algae* **2014**, *31*, 136–142. [CrossRef]

70. Liu, J.; Prindle, A.; Humphries, J.; Gabalda-Sagarra, M.; Asally, M.; Lee, D.Y.D.; Ly, S.; Garcia-Ojalvo, J.; Süel, G.M. Metabolic co-dependence gives rise to collective oscillations within biofilms. *Nature* **2015**, *523*, 550–554. [CrossRef] [PubMed]

71. Conradi, F.D.; Zhou, R.; Oeser, S.; Schuergers, N.; Wilde, A.; Mullineaux, C.W. Factors Controlling Floc Formation and Structure in the Cyanobacterium *Synechocystis sp.* Strain PCC 6803. *J. Bacteriol.* **2019**, *201*, e00344-19. [CrossRef]

72. Kühl, M.; Fenchel, T. Bio-optical Characteristics and the Vertical Distribution of Photosynthetic Pigments and Photosynthesis in an Artificial Cyanobacterial Mat. *Microb. Ecol.* **2000**, *40*, 94–103. [CrossRef] [PubMed]

73. Allen, R.; Rittmann, B.E.; Curtiss, R. Axenic biofilm formation and aggregation by *Synechocystis sp.* strain PCC 6803 are induced by changes in nutrient concentration and require cell surface structures. *Appl. Environ. Microbiol.* **2019**, *85*, e02192-18. [CrossRef] [PubMed]

74. Walters III, M.C.; Roe, F.; Bugnicourt, A.; Franklin, M.J.; Stewart, P.S. Contributions of antibiotic penetration, oxygen limitation, and low metabolic activity to tolerance of *Pseudomonas aeruginosa* biofilms to ciprofloxacin and tobramycin. *Antimicrob. Agents Chemother.* **2003**, *47*, 317–323. [CrossRef]

75. Anderl, J.N.; Franklin, M.J.; Stewart, P.S. Role of antibiotic penetration limitation in *Klebsiella pneumoniae* biofilm resistance to ampicillin and ciprofloxacin. *Antimicrob. Agents Chemother.* **2000**, *44*, 1818–1824. [CrossRef]

76. Grujić, S.; Vasić, S.; Čomić, L.; Ostojić, A.; Radojević, I. Heavy metal tolerance and removal potential in mixed-species biofilm. *Water Sci. Technol.* **2017**, *76*, 806–812. [CrossRef]

77. Giner-Lamia, J.; Pereira, S.B.; Bovea-Marco, M.; Futschik, M.E.; Tamagnini, P.; Oliveira, P. Extracellular proteins: Novel key components of metal resistance in cyanobacteria? *Front. Microbiol.* **2016**, *7*, 1–8. [CrossRef]

78. Jittawuttipoka, T.; Planchon, M.; Spalla, O.; Benzerara, K.; Guyot, F.; Cassier-Chauvat, C.; Chauvat, F. Multidisciplinary evidences that *Synechocystis* PCC6803 exopolysaccharides operate in cell sedimentation and protection against salt and metal stresses. *PLoS ONE* **2013**, *8*, e55564. [CrossRef]

79. Testa, S.; Berger, S.; Piccardi, P.; Oechslin, F.; Resch, G.; Mitri, S. Spatial structure affects phage efficacy in infecting dual-strain biofilms of *Pseudomonas aeruginosa*. *Commun. Biol.* **2019**, *2*, 1–12. [CrossRef]

80. Seiler, C.; van Velzen, E.; Neu, T.R.; Gaedke, U.; Berendonk, T.U.; Weitere, M. Grazing resistance of bacterial biofilms: A matter of predators' feeding trait. *FEMS Microbiol. Ecol.* **2017**, *93*, 1–9. [CrossRef] [PubMed]

81. Graupner, S.; Weger, N.; Sohni, M.; Wackernagel, W. Requirement of novel competence genes pilT and pilU of *Pseudomonas stutzeri* for natural transformation and suppression of pilT deficiency by a hexahistidine tag on the type IV pilus protein PilAI. *J. Bacteriol.* **2001**, *183*, 4694–4701. [CrossRef] [PubMed]

82. Bradley, D.E.; Pitt, T.L. Pilus-dependence of four *Pseudomonas aeruginosa* bacteriophages with non-contractile tails. *J. Gen. Virol.* **1974**, *24*, 1–15. [CrossRef]

83. Romantschuk, M.; Bamford, D.H. Function of Pili in bacteriophage Ø6 penetration. *J. Gen. Virol.* **1985**, *66*, 2461–2469. [CrossRef]

84. Agostoni, M.; Waters, C.M.; Montgomery, B.L. Regulation of biofilm formation and cellular buoyancy through modulating intracellular cyclic di-GMP levels in engineered cyanobacteria. *Biotechnol. Bioeng.* **2016**, *113*, 311–319. [CrossRef]

85. Enomoto, G.; Nomura, R.; Shimada, T.; Narikawa, R.; Ikeuchi, M. Cyanobacteriochrome SesA is a diguanylate cyclase that induces cell aggregation in *Thermosynechococcus*. *J. Biol. Chem.* **2014**, *289*, 24801–24809. [CrossRef]

86. Enomoto, G.; Win, N.; Narikawa, R.; Ikeuchi, M. Three cyanobacteriochromes work together to form a light color-sensitive input system for c-di-GMP signaling of cell aggregation. *Proc. Natl. Acad. Sci. USA* **2015**, *112*, 8082–8087. [CrossRef]

87. Balskus, E.P.; Case, R.J.; Walsh, C.T. The biosynthesis of cyanobacterial sunscreen scytonemin in intertidal microbial mat communities. *FEMS Microbiol. Ecol.* **2011**, *77*, 322–332. [CrossRef]

88. Zhang, L.; Li, L.; Wu, Q. Protective effects of mycosporine-like amino acids of *Synechocystis sp.* PCC 6803 and their partial characterization. *J. Photochem. Photobiol. B Biol.* **2007**, *86*, 240–245. [CrossRef] [PubMed]

89. Enomoto, G.; Ikeuchi, M. Blue-/Green-light-responsive cyanobacteriochromes are cell shade sensors in red-light replete niches. *iScience* **2020**, *23*, 100936. [CrossRef] [PubMed]

90. Garcia-Pichel, F.; Mechling, M.; Castenholtz, R.W. Diel migrations of microorganisms within a benthic, hypersaline mat community. *Appl. Environ. Microbiol.* **1994**, *60*, 1500–1511. [CrossRef] [PubMed]

91. Bebout, B.M.; Garcia-Pichel, F. UV B-induced vertical migrations of cyanobacteria in a microbial mat. *Appl. Environ. Microbiol.* **1995**, *61*, 4215–4222. [CrossRef] [PubMed]

92. Schatz, D.; Nagar, E.; Sendersky, E.; Parnasa, R.; Zilberman, S.; Carmeli, S.; Mastai, Y.; Shimoni, E.; Klein, E.; Yeger, O.; et al. Self-suppression of biofilm formation in the cyanobacterium *Synechococcus elongatus*. *Environ. Microbiol.* **2013**, *15*, 1786–1794. [CrossRef] [PubMed]

93. Klausen, M.; Aaes-Jørgensen, A.; Molin, S.; Tolker-Nielsen, T. Involvement of bacterial migration in the development of complex multicellular structures in *Pseudomonas aeruginosa* biofilms. *Mol. Microbiol.* **2003**, *50*, 61–68. [CrossRef] [PubMed]

94. Klausen, M.; Heydorn, A.; Ragas, P.; Lambertsen, L.; Aaes-Jørgensen, A.; Molin, S.; Tolker-Nielsen, T. Biofilm formation by *Pseudomonas aeruginosa* wild type, flagella and type IV pili mutants. *Mol. Microbiol.* **2003**, *48*, 1511–1524. [CrossRef] [PubMed]

95. Persat, A.; Inclan, Y.F.; Engel, J.N.; Stone, H.A.; Gitai, Z. Type IV pili mechanochemically regulate virulence factors in *Pseudomonas aeruginosa*. *Proc. Natl. Acad. Sci. USA* **2015**, *112*, 7563–7568. [CrossRef]

96. Ellison, C.K.; Kan, J.; Dillard, R.S.; Kysela, D.T.; Ducret, A.; Berne, C.; Hampton, C.M.; Ke, Z.; Wright, E.R.; Biais, N.; et al. Obstruction of pilus retraction stimulates bacterial surface sensing. *Science (80-)* **2017**, *358*, 535–538. [CrossRef]

97. Giltner, C.L.; Habash, M.; Burrows, L.L. *Pseudomonas aeruginosa* minor pilins are incorporated into type IV Pili. *J. Mol. Biol.* **2010**, *398*, 444–461. [CrossRef]

98. Del Medico, L.; Cerletti, D.; Schächle, P.; Christen, M.; Christen, B. The type IV pilin PilA couples surface attachment and cell-cycle initiation in *Caulobacter crescentus*. *Proc. Natl. Acad. Sci. USA* **2020**, *117*, 9546–9553. [CrossRef]

99. Nagar, E.; Zilberman, S.; Sendersky, E.; Simkovsky, R.; Shimoni, E.; Gershtein, D.; Herzberg, M.; Golden, S.S.; Schwarz, R. Type 4 pili are dispensable for biofilm development in the cyanobacterium *Synechococcus elongatus*. *Environ. Microbiol.* **2017**, *19*, 2862–2872. [CrossRef]

100. Parnasa, R.; Nagar, E.; Sendersky, E.; Reich, Z.; Simkovsky, R.; Golden, S.; Schwarz, R. Small secreted proteins enable biofilm development in the cyanobacterium *Synechococcus elongatus*. *Sci. Rep.* **2016**, *6*, 32209. [CrossRef]

101. Parsek, M.R.; Greenberg, E.P. Sociomicrobiology: The connections between quorum sensing and biofilms. *Trends Microbiol.* **2005**, *13*, 27–33. [CrossRef] [PubMed]

102. Sharif, D.I.; Gallon, J.; Smith, C.J.; Dudley, E. Quorum sensing in Cyanobacteria: N-octanoyl-homoserine lactone release and response, by the epilithic colonial cyanobacterium *Gloeothece* PCC6909. *ISME J.* **2008**, *2*, 1171–1182. [CrossRef] [PubMed]

103. Abby, S.S.; Cury, J.; Guglielmini, J.; Néron, B.; Touchon, M.; Rocha, E.P.C. Identification of protein secretion systems in bacterial genomes. *Sci. Rep.* **2016**, *6*, 1–14. [CrossRef] [PubMed]

104. Denise, R.; Abby, S.S.; Rocha, E.P.C. Diversification of the type IV filament superfamily into machines for adhesion, protein secretion, DNA uptake, and motility. *PLoS Biol.* **2019**, *17*, e3000390. [CrossRef]

105. Russo, D.A.; Zedler, J.A.Z.; Wittmann, D.N.; Möllers, B.; Singh, R.K.; Batth, T.S.; Van Oort, B.; Olsen, J.V.; Bjerrum, M.J.; Jensen, P.E. Expression and secretion of a lytic polysaccharide monooxygenase by a fast-growing cyanobacterium. *Biotechnol. Biofuels* **2019**, *12*, 74. [CrossRef]

106. Russo, D.A.; Zedler, J.A.Z. Genomic insights into cyanobacterial protein translocation systems. *Biol. Chem.* **2020**. [CrossRef]

107. Zuniga, E.G.; Boateng, K.K.A.; Bui, N.U.; Kurnfuli, S.; Muthana, S.M.; Risser, D.D. Identification of a hormogonium polysaccharide-specific gene set conserved in filamentous cyanobacteria. *Mol. Microbiol.* **2020**, 1–12. [CrossRef]

108. Fisher, M.L.; Allen, R.; Luo, Y.; Curtiss, R. Export of extracellular polysaccharides modulates adherence of the Cyanobacterium *Synechocystis*. *PLoS ONE* **2013**, *8*, e74514. [CrossRef]

109. Schuergers, N.; Ruppert, U.; Watanabe, S.; Nürnberg, D.J.; Lochnit, G.; Dienst, D.; Mullineaux, C.W.; Wilde, A. Binding of the RNA chaperone Hfq to the type IV pilus base is crucial for its function in *Synechocystis sp.* PCC 6803. *Mol. Microbiol.* **2014**, *92*, 840–852. [CrossRef]

110. Wang, Y.C.; Chin, K.H.; Tu, Z.L.; He, J.; Jones, C.J.; Sanchez, D.Z.; Yildiz, F.H.; Galperin, M.Y.; Chou, S.H. Nucleotide binding by the widespread high-affinity cyclic di-GMP receptor MshEN domain. *Nat. Commun.* **2016**, *7*, 1–12. [CrossRef]

111. Terauchi, K.; Ohmori, M. Blue light stimulates cyanobacterial motility via a cAMP signal transduction system. *Mol. Microbiol.* **2004**, *52*, 303–309. [CrossRef] [PubMed]

112. Terauchi, K.; Ohmori, M. An adenylate cyclase, Cya1, regulates cell motility in the cyanobacterium *Synechocystis sp.* PCC 6803. *Plant Cell Physiol.* **1999**, *40*, 248–251. [CrossRef] [PubMed]

113. Masuda, S.; Ono, T.A. Adenylyl cyclase activity of Cya1 from the cyanobacterium *Synechocystis sp.* strain PCC 6803 is inhibited by bicarbonate. *J. Bacteriol.* **2005**, *187*, 5032–5035. [CrossRef] [PubMed]

114. Bhaya, D.; Nakasugi, K.; Fazeli, F.; Burriesci, M.S. Phototaxis and impaired motility in adenylyl cyclase and cyclase receptor protein mutants of *Synechocystis sp.* strain PCC 6803. *J. Bacteriol.* **2006**, *188*, 7306–7310. [CrossRef] [PubMed]

115. Yoshimura, H.; Yoshihara, S.; Okamoto, S.; Ikeuchi, M.; Ohmori, M. A cAMP receptor protein, SYCRP1, is responsible for the cell motility of *Synechocystis sp.* PCC 6803. *Plant Cell Physiol.* **2002**, *43*, 460–463. [CrossRef]

116. Song, W.-Y.; Zang, S.-S.; Li, Z.-K.; Dai, G.-Z.; Liu, K.; Chen, M.; Qiu, B.-S. Sycrp2 is essential for twitching motility in the cyanobacterium *Synechocystis sp.* Strain PCC 6803. *J. Bacteriol.* **2018**, *200*, 1–13. [CrossRef]

117. Hu, J.; Zhan, J.; Chen, H.; He, C.; Cang, H.; Wang, Q. The small regulatory antisense RNA PilR affects pilus formation and cell motility by negatively regulating *pilA11* in *Synechocystis sp.* PCC 6803. *Front. Microbiol.* **2018**, *9*, 786. [CrossRef]

118. Higo, A.; Ikeuchi, M.; Ohmori, M. cAMP regulates respiration and oxidative stress during rehydration in *Anabaena sp.* PCC 7120. *FEBS Lett.* **2008**, *582*, 1883–1888. [CrossRef]

119. Ohmori, K.; Hirose, M.; Ohmori, M. Function of cAMP as a mat-forming factor in the cyanobacterium *Spirulina platensis*. *Plant Cell Physiol.* **1992**, *33*, 21–25. [CrossRef]

120. Ohmori, K.; Hirose, M.; Ohmori, M. An increase in the intracellular concentration of cAMP triggers formation of an algal mat by the cyanobacterium *Spirulina platensis*. *Plant Cell Physiol.* **1993**, *34*, 169–171. [CrossRef]

121. Imamura, S.; Asayama, M. Sigma factors for cyanobacterial transcription. *Gene Regul. Syst. Biol.* **2009**, *3*, GRSB-S2090. [CrossRef] [PubMed]

122. Asayama, M.; Imamura, S. Stringent promoter recognition and autoregulation by the group 3 σ-factor SigF in the cyanobacterium *Synechocystis sp.* strain PCC 6803. *Nucleic Acids Res.* **2008**, *36*, 5297–5305. [CrossRef] [PubMed]

123. Bhaya, D.; Watanabe, N.; Ogawa, T.; Grossman, A.R. The role of an alternative sigma factor in motility and pilus formation in the cyanobacterium *Synechocystis sp.* strain PCC6803. *Proc. Natl. Acad. Sci. USA* **1999**, *96*, 3188–3193. [CrossRef] [PubMed]

124. Flores, C.; Santos, M.; Pereira, S.B.; Mota, R.; Rossi, F.; De Philippis, R.; Couto, N.; Karunakaran, E.; Wright, P.C.; Oliveira, P.; et al. The alternative sigma factor SigF is a key player in the control of secretion mechanisms in *Synechocystis sp.* PCC 6803. *Environ. Microbiol.* **2019**, *21*, 343–359. [CrossRef] [PubMed]

125. Miranda, H.; Immerzeel, P.; Gerber, L.; Hörnaeus, K.; Lind, S.B.; Pattanaik, B.; Lindberg, P.; Mamedov, F.; Lindblad, P. Sll1783, a monooxygenase associated with polysaccharide processing in the unicellular cyanobacterium *Synechocystis* PCC 6803. *Physiol. Plant.* **2017**, *161*, 182–195. [CrossRef]

126. Gonzalez, A.; Riley, K.W.; Harwood, T.V.; Zuniga, E.G.; Risser, D.D. A tripartite, hierarchical sigma factor cascade promotes hormogonium development in the filamentous cyanobacterium *Nostoc punctiforme*. *mSphere* **2019**, *4*, e00231-19. [CrossRef] [PubMed]

127. Ellison, C.K.; Dalia, T.N.; Vidal Ceballos, A.; Wang, J.C.Y.; Biais, N.; Brun, Y.V.; Dalia, A.B. Retraction of DNA-bound type IV competence pili initiates DNA uptake during natural transformation in *Vibrio cholerae*. *Nat. Microbiol.* **2018**, *3*, 773–780. [CrossRef]

128. Van Schaik, E.J.; Giltner, C.L.; Audette, G.F.; Keizer, D.W.; Slupsky, C.M.; Irvin, R.T. DNA binding: A novel function of *Pseudomonas aeruginosa* type IV pili. *J. Bacteriol.* **2005**, *187*, 1455–1464. [CrossRef]

129. Salleh, M.Z.; Karuppiah, V.; Snee, M.; Thistlethwaite, A.; Levy, C.W.; Knight, D.; Derrick, J.P. Structure and properties of a natural competence-associated pilin suggest a unique pilus Tip-associated DNA receptor. *MBio* **2019**, *10*, e00614-19. [CrossRef]

130. Nakasugi, K.; Svenson, C.J.; Neilan, B.A. The competence gene, *comF*, from *Synechocystis sp.* strain PCC 6803 is involved in natural transformation, phototactic motility and piliation. *Microbiology* **2006**, *152*, 3623–3631. [CrossRef]

131. Yu, J.; Liberton, M.; Cliften, P.F.; Head, R.D.; Jacobs, J.M.; Smith, R.D.; Koppenaal, D.W.; Brand, J.J.; Pakrasi, H.B. *Synechococcus elongatus* UTEX 2973, a fast growing cyanobacterial chassis for biosynthesis using light and CO_2. *Sci. Rep.* **2015**, *5*, 8132. [CrossRef] [PubMed]

132. Nakasugi, K.; Alexova, R.; Svenson, C.J.; Neilan, B.A. Functional Analysis of PilT from the toxic cyanobacterium *Microcystis aeruginosa* PCC 7806. *J. Bacteriol.* **2007**, *189*, 1689–1697. [CrossRef] [PubMed]

133. Nies, F.; Mielke, M.; Pochert, J.; Tilman, L. Natural transformation of the filamentous cyanobacterium *Phormidium lacuna*. *PLoS ONE* **2020**, *15*, e0234440. [CrossRef] [PubMed]

134. Iwai, M.; Katoh, H.; Katayama, M.; Ikeuchi, M. Improved genetic transformation of the thermophilic cyanobacterium, *Thermosynechococcus elongatus* BP-1. *Plant Cell Physiol.* **2004**, *45*, 171–175. [CrossRef]

135. Cheng, Y.I.; Chou, L.; Chiu, Y.F.; Hsueh, H.T.; Kuo, C.H.; Chu, H.A. Comparative genomic analysis of a novel strain of Taiwan hot-spring cyanobacterium *Thermosynechococcus sp.* CL-1. *Front. Microbiol.* **2020**, *11*, 82. [CrossRef]

136. Kawano, Y.; Saotome, T.; Ochiai, Y.; Katayama, M.; Narikawa, R.; Ikeuchi, M. Cellulose accumulation and a cellulose synthase gene are responsible for cell aggregation in the cyanobacterium *Thermosynechococcus vulcanus* RKN. *Plant Cell Physiol.* **2011**, *52*, 957–966. [CrossRef]

137. Trautmann, D.; Voß, B.; Wilde, A.; Al-Babili, S.; Hess, W.R. Microevolution in cyanobacteria: Re-sequencing a motile substrain of *Synechocystis sp.* PCC 6803. *DNA Res.* **2012**, *19*, 435–448. [CrossRef]

Publisher's Note: MDPI stays neutral with regard to jurisdictional claims in published maps and institutional affiliations.

Review

Function and Benefits of Natural Competence in Cyanobacteria: From Ecology to Targeted Manipulation

Alexandra M. Schirmacher, Sayali S. Hanamghar and Julie A. Z. Zedler *

Matthias Schleiden Institute for Genetics, Bioinformatics and Molecular Botany, Friedrich Schiller University Jena, 07743 Jena, Germany; alexandra.schirmacher@uni-jena.de (A.M.S.); sayali.hanamghar@uni-jena.de (S.S.H.)
* Correspondence: julie.zedler@uni-jena.de

Received: 27 September 2020; Accepted: 20 October 2020; Published: 22 October 2020

Abstract: Natural competence is the ability of a cell to actively take up and incorporate foreign DNA in its own genome. This trait is widespread and ecologically significant within the prokaryotic kingdom. Here we look at natural competence in cyanobacteria, a group of globally distributed oxygenic photosynthetic bacteria. Many cyanobacterial species appear to have the genetic potential to be naturally competent, however, this ability has only been demonstrated in a few species. Reasons for this might be due to a high variety of largely uncharacterised competence inducers and a lack of understanding the ecological context of natural competence in cyanobacteria. To shed light on these questions, we describe what is known about the molecular mechanisms of natural competence in cyanobacteria and analyse how widespread this trait might be based on available genomic datasets. Potential regulators of natural competence and what benefits or drawbacks may derive from taking up foreign DNA are discussed. Overall, many unknowns about natural competence in cyanobacteria remain to be unravelled. A better understanding of underlying mechanisms and how to manipulate these, can aid the implementation of cyanobacteria as sustainable production chassis.

Keywords: cyanobacteria; DNA uptake; DNA processing; type IV pili; T4P; *com* genes; *pil* genes; natural competence; transformation; genetic engineering

---

1. Introduction

Natural competence refers to the ability of prokaryotes to take up DNA from the environment and insert it into their own genome by homologous recombination. This prokaryotic trait is widespread within many phylogenetic taxa of Proteobacteria, Firmicutes, Chlorobi, Deinococcus-Thermus and Euryarchaeota [1]. The best-studied examples of natural competence are all heterotrophic organisms: *Bacillus subtilis, Streptococcus pneumoniae, Thermus thermophilus, Neisseria gonorrhoeae, Vibrio cholerae, Helicobacter pylori, Acinetobacter* spp. and *Haemophilus influenzae* [1,2]. In all known naturally competent bacteria, except for *Helicobacter pylori* [3,4], the first step of DNA uptake is mediated by type IV pili (T4P). In cyanobacteria, these are also referred to as thick pili in order to distinguish them from morphologically distinct thin pili [5]. T4P are multifunctional cellular appendages known to be involved in natural competence, twitching motility [6], predation [7], cell adhesion [8], biofilm formation [9–11], virulence [12] and secretion [13].

Natural competence is not restricted to heterotrophic bacteria, but also present in cyanobacteria. Cyanobacteria are oxygenic photosynthetic prokaryotes found in a wide range of environments, which have also attracted interest as production hosts due to their photoautotrophic growth regime and metabolic versatility. Despite their ecological and biotechnological significance and a large body of literature about these organisms, not much is known about the prevalence of natural competence (first described in [14]). Natural competence offers a simple and efficient method of transformation.

Understanding and utilising this trait would therefore also enable faster and easier exploitation of these organisms for applications. Here, we review the molecular mechanisms of DNA uptake and processing, the regulation of natural competence and its prevalence in cyanobacteria. We further discuss the implications of manipulating and exploiting this trait in engineering cyanobacteria.

2. The Molecular Basis of Natural Competence

Transformation via natural competence first involves DNA uptake, followed by DNA processing and homologous recombination. The first step of DNA uptake is mediated by binding to the T4P filament. Upon pilus retraction, the DNA is pulled along with the filament into the periplasmic space and translocated into the cytoplasm [1,15,16]. In cyanobacteria, experimental studies of DNA uptake mechanisms are lacking, however, information regarding the proteins involved is available. Table 1 gives an overview of the proteins and their assigned function in natural competence of cyanobacteria. An overview of how DNA uptake and processing might occur in *Synechocystis* sp. PCC 6803 (hereafter *Synechocystis*) is shown in Figure 1.

Table 1. Proteins known to be involved in natural competence of *Synechocystis*.

Protein	Gene Assignment	Function	Reference
ComA	*slr0197*	DNA translocation (DNA binding and nuclease activity?)	[17,18]
ComE	*sll1929*	DNA translocation (translocase activity?)	[17]
ComF	*slr0388*	DNA translocation (transitioning DNA uptake and homologous recombination?)	[19]
DprA	*slr1197*	DNA processing protein	[20–22]
Hfq	*ssr3341*	Pilus biogenesis	[23,24]
PilA1	*sll1694*	Filament formation (major pilin)	[5]
PilA2	*sll1695*	Filament formation (minor pilin)	[5,17]
PilB1	*slr0063*	Motor protein (polymerisation)	[17]
PilB2	*slr0079*	Unknown	[17]
PilC	*slr0162-slr0163*	Platform protein	[5]
PilD	*slr1120*	Prepilin peptidase	[5,17,25]
PilH	*slr1042*	Che-like response regulator pilus assembly	[26]
PilJ	*slr1044*	Che-like response regulator pilus assembly	[26]
PilL-C	*slr0322*	Che-like response regulator pilus assembly	[26]
PilM	*slr1274*	Pilus alignment complex	[17]
PilN	*slr1275*	Pilus alignment complex	[17]
PilO	*slr1276*	Pilus alignment complex	[17]
PilQ	*slr1277*	Secretin	[17]
PilT1	*slr0161*	Motor protein (depolymerisation)	[5,27]
RecA	*sll0569*	Homologous recombination	[20,21,28]

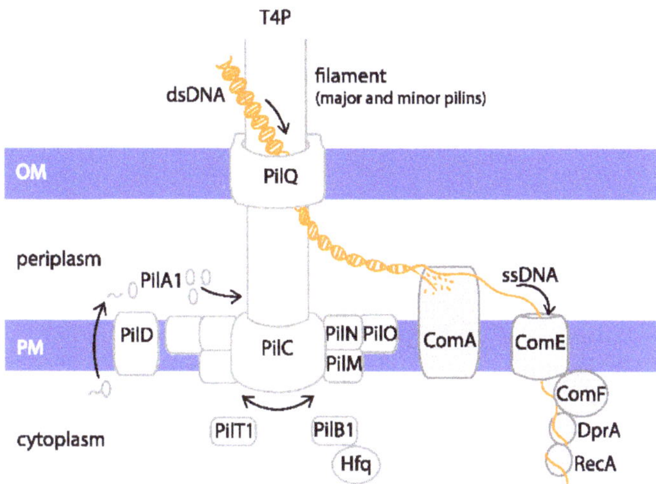

Figure 1. Overview of DNA uptake and processing machinery in *Synechocystis*. Double-stranded DNA (dsDNA) is taken up from the extracellular space through PilQ, the outer membrane (OM) pore of the T4P, by binding to the filament (consisting of major pilin PilA1 and minor pilins). ComA (putative DNA binding and endonuclease domains) is responsible for dsDNA processing into single stranded DNA (ssDNA), which is further translocated across the plasma membrane (PM) via ComE. In the cytoplasm, the ssDNA is further processed by ComF, DprA and RecA.

2.1. T4P Are Crucial for Natural Competence in Cyanobacteria

Cyanobacterial T4P have been shown to play a role in twitching motility [5,29], metal uptake [30,31], biofilm formation [32], flocculation [33] and also in natural competence [14,34]. The filament of T4P mainly consists of the major pilin, which in cyanobacteria, is referred to as PilA1 [5,15,17,35,36]. PilA1 is exported to the periplasm via the General Secretory (Sec) pathway [37]. The peptidase PilD cleaves off the Sec signal peptide and methylates PilA1 [38–40]. In addition to PilA1, further minor pilins of low abundance might be incorporated into the filament. In cyanobacteria, these are thought to assist with flocculation [33], motility [41,42] and also seem to play a role in natural competence. For instance, in *Synechocystis* a knockout of the gene encoding for the minor pilin PilA2 (*sll1695*) led to a transformation efficiency of only 52% compared to the wild type (WT) [17], while motility was not affected [5]. In heterotrophic bacteria, minor pilins were shown to have versatile functions such as ensuring correct pilus assembly, assisting in pilus adhesion and, interestingly, they might also be involved in DNA binding [15]. Where exactly the DNA is bound to the filament and if this is mediated by minor or major pilins is not certain. In *V. cholerae*, *Neisseria meningitidis* and *T. thermophilus* minor pilins are thought to be located at the tip of the filament [15,43]. However, other models favour the binding of DNA to the major pilin or to minor pilins along the filament [34]. A similar scenario could also apply to cyanobacteria. For instance, PilA2 and/or other minor pilins in *Synechocystis* could function as DNA binding proteins similarly to PilA3 in *Synechococcus elongatus* PCC 7942 (hereafter *S. elongatus*) [22]. However, this needs experimental validation. Additionally, it has been suggested that the outer membrane pore-forming secretin PilQ, through which the DNA is translocated into the periplasm, may directly be involved in DNA binding [44].

Further pilus subunits are known to affect natural competence of cyanobacteria, although their function in the T4P system has not been established. For instance, PilB2 seems to have an additional role in natural competence—a Δ*pilB2* mutant shows a reduced transformation efficiency at 37% of the WT while retaining motility [17]. In addition, two gene clusters that encode components of Che-like chemotaxis systems can also indirectly affect natural competence: A *pilH* mutant of *Synechocystis* was shown to accumulate T4P, but lost motility and competence was reduced to 28.6% of WT levels [26]. Similarily, *pilJ* and *pilL-C* mutants were non-motile, non-piliated and competence was reduced to 1.5% of the WT [26]. A *pilI* mutant showed a reduction in the number of T4P, whilst the *pilL-N* mutant was hyperpiliated. PilG does not seem to be involved in competence as a mutant retained transformability [26]. An overview of the different phenotypes of knockout mutants is shown in Figure 2.

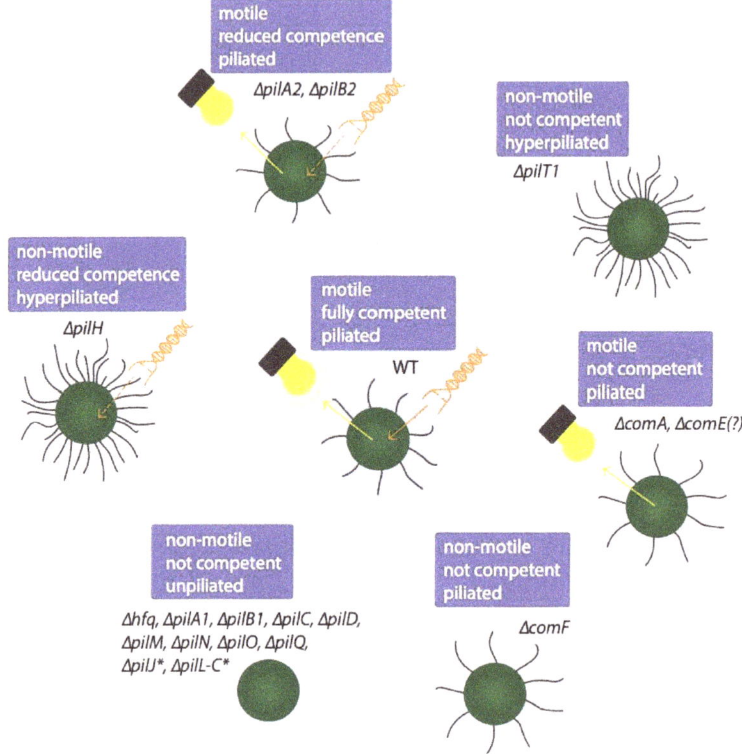

Figure 2. Phenotypes of *Synechocystis* mutants deficient in a single component of the natural competence machinery. Wild type (WT) cells are motile, piliated and competent (centre) while in Δ*pilH*, Δ*pilA2* and Δ*pilB2* mutants a reduced competence (~30 and 50%) is observed [5,17,26]. Other knockouts lead to strongly limited competence (denoted with an asterisk) or a complete loss of competence [5,17,19,22,23,26].

A recent study identified several other T4P-related genes as essential for natural competence in *S. elongatus* [22]: *pilA3* (Synpcc7942_2590), *pilW* (Synpcc7942_2591), *rntB* (Synpcc7942_2485), *rntA* (Synpcc7942_2486) and *sigF2* (Synpcc7942_1784). The function of *rntA* and *rntB* is not known, however, interestingly, they contain a type IV pilin-like signal peptide suggesting a role within the T4P. SigF2 is part of the regulatory network inducing natural competence. While *pilA3* and *pilW* are essential for natural competence in *S. elongatus*, these genes are not present in other naturally competent cyanobacteria [22]. Overexpression of *pilA3* also led to a higher transformation efficiency, supporting an essential role of minor pilins for DNA binding.

2.2. Com Proteins Mediate DNA Uptake and Processing

Once the DNA has been pulled into the periplasm, one DNA strand is degraded whilst the other is translocated further across the cytoplasmic membrane through the action of ComAEF [1,16,44]. In cyanobacteria, the exact function of each of these proteins is poorly understood. ComA is predicted to be plasma membrane (PM)-localised with periplasmic DNA-binding and endonuclease domains [18]. This protein was shown to be crucial for transformation in *Synechocystis* and dispensable for motility [17]. In *Synechocystis*, ComE has been identified as a homologue to ComEC from *B. subtilis* [17] where it forms a pore in the PM and is proposed to translocate one strand of DNA whilst degrading the other [1,16]. ComF was shown to play a crucial role in both phototaxis and transformation in *Synechocystis* [19]. The *S. pneumoniae* homologue ComFC was proposed to mediate the transition between DNA uptake and homologous recombination. It was shown to interact with DprA and to form a complex with ComFA, an ATPase that also binds to single-stranded DNA (ssDNA) [45]. Translocation of ssDNA across the PM is known to be ComFA-dependent in multiple gram-positive bacteria [1,16] but no homologue has been identified in cyanobacteria. Once the ssDNA reaches the cytoplasm, binding of DprA offers protection from nucleases and mediates the recruitment of the recombinase RecA to the ssDNA. RecA is then responsible for integration into the genome via homologous recombination [1,3,16,21]. Both proteins, DprA and RecA, are vital for homologous recombination. However, these mechanisms remain to be experimentally validated in cyanobacteria.

3. Versatile Factors Control Natural Competence

Natural competence in bacteria is generally not a constitutive function. Underlying induction and regulation mechanisms are highly diverse and complex [3,46]. For example, induction of competence can depend on growth phase [1,3,47,48], quorum sensing [47,49–51], biofilm formation [50,52,53], nutrient limitation [54], DNA damage [55,56], presence of antibiotics [56,57] or certain substrates [47,58–60]. For more detailed insights, the reader is directed to recent reviews [1,3,46,61,62].

In cyanobacteria, one of the competence-inducing factors seems to be the growth phase. *Synechococcus* sp. PCC 7002 and *Synechocystis* were shown to be most competent during the exponential phase, and their transformability is drastically reduced in the stationary phase [63,64]. On the other hand, competence of *S. elongatus* was reported to not depend on the growth phase [65]. Generally, it seems to be species-dependent if and how a specific growth phase influences competence. Despite a correlation of growth phase and level of transformability in several species, the growth phase itself might not be the actual inducer. Many factors, such as nutrient limitation or quorum sensing, correlate with the stationary phase and may thus be the true factor of varied transformability levels. Another trigger of natural competence in cyanobacteria could be changes in lifestyle. For instance, in the plant pathogen *Xylella fastidiosa*, cultivation in microfluidic chambers mimicking its natural environment and promoting biofilm formation was shown to induce competence [52]. Many cyanobacteria are capable of biofilm formation and a link between these factors could exist.

Recently, the circadian clock was identified as a major regulator of natural competence in *S. elongatus* [22]. Cells grown under a light-dark cycle showed the highest transformation efficiency at dusk. Expression of *comEA* is up-regulated at dusk or shortly before, and other genes relevant for competence (*sigF2*, *pilA3*, *rntA* and *dprA*) are induced by darkness. Interestingly, most T4P component genes were expressed in the morning. Consequently, cells were piliated during daytime, and numbers of pili decreased until the second half of the night [22].

There might be many other factors that impact natural competence in cyanobacteria, but these might be difficult to identify. For instance, heterocyst-forming species lead a complex lifestyle and excrete a large number of extracellular nucleases [66], making the characterisation of their competence systems difficult. Many filamentous cyanobacteria produce hormogonia, which are known to produce

T4P and to be motile [67]. The hormogonial state might be more suitable for the uptake and incorporation of exogenous DNA than the filamentous state as the new genetic information could be transferred during cell division. Thus, factors influencing and regulating natural competence in cyanobacteria are diverse and complex. It is likely that many more remain to be unravelled.

4. Natural Competence Might Be More Frequent among Cyanobacteria than Initially Anticipated

4.1. Experimental Evidence of Natural Competence in Cyanobacteria

The first record of natural competence in cyanobacteria is from 1970 for *Synechococcus* sp. PCC 7943 (previously *Anacystis nidulans* 602) [14]. Over the years, multiple members of the *Synechococcus* genus were found to be naturally competent: *S. elongatus* R2 [68], *S. elongatus* PCC 11801 [69] and *Synechococcus* sp. PCC 7002 [70]. It is unclear if *Synechococcus* sp. PCC 6301 is in fact naturally competent as there are many older records reporting it as such [63,71–75], although other studies claim the opposite [76,77]. This strain might have lost this trait over decades of lab cultivation or its competence machinery might be controlled by an unknown inducer. Most records of natural competence in cyanobacteria are found for the model organism *Synechocystis*, where it was first described in 1982 [33]. Apart from this, only few other, mostly unicellular, cyanobacteria are known to be naturally competent, including *Thermosynechococcus elongatus* BP-1 [78] and *Microcystis aeruginosa* PCC 7806 [79].

Records of natural competence in filamentous cyanobacteria are very scarce, suggesting it might not be common within this morphological group. In fact, filamentous cyanobacteria are widely regarded as being not naturally competent [80–82]. One case of natural competence was recorded for *Nostoc muscorum* in 1981 [83], and two more records of natural competence in this species exist from 1987 [84] and 1990 [85]. The latter study also reported the successful transformation of the filamentous *Anabaena doliolum* [85]. Since then, no more records of natural competence of *N. muscorum* or *A. doliolum* exist. After several decades without progress, two recent studies have demonstrated successful natural transformation in the filamentous *Phormidium lacuna* HE10DO [86] and the ramified *Chlorogloeopsis fritschii* PCC 6912 [87]. These recent findings may prompt investigations into other filamentous cyanobacteria and indicate that natural competence might be more common in this morphological group than the few literature records denote.

4.2. Genomics Give Insights into the Prevalence of Cyanobacterial Natural Competence

A bigger picture of how widespread natural competence is among cyanobacteria is still lacking. However, recent studies suggest a higher prevalence than experimental evidence has so far shown [22,86]. A total of 345 cyanobacterial genomes were recently analysed for their presence of the genes *pilA1*, *pilD*, *pilB*, *pilT*, *pilC*, *pilM*, *pilN*, *pilO* and *pilQ* and combined with data on *comA*, *comE* and *comF* from a preceding study [88,89]. However, *com* gene data are only available for 21% of the 345 again analysed species. Therefore, after curating this dataset by retaining only species where data for both *pil* and *com* genes were available, this resulted in 73 cyanobacterial genomes (Table S1, taxonomy retrieved from NCBI taxonomy browser, accessed 12 September 2020). An overview of the presence of *pil* genes in our cyanobacterial database is given in Table 2.

Table 2. Identification of *pil* (*pilA1*, *pilD*, *pilB*, *pilT*, *pilC*, *pilM*, *pilN*, *pilO*, *pilQ*) and *com* (*comA*, *comE*, *comF*) genes across cyanobacterial orders. Cells are highlighted in shades of green correlating to their number.

Order	Number of Species	Number of *pil* and *com* Genes Identified in the Genome												
		0	1	2	3	4	5	6	7	8	9	10	11	12
Chroococcales	10	0	0	0	0	0	0	0	0	0	0	0	2	8
Chroococcidiopsidales	1	0	0	0	0	0	0	0	0	0	0	0	0	1
Gloeobacterales	2	0	0	0	0	0	0	0	0	0	0	0	1	1
Nostocales	13	0	0	0	0	0	0	0	0	0	0	1	0	12
Oscillatoriales	8	0	0	0	0	0	0	0	0	0	0	0	1	7
Pleurocapsales	2	0	0	0	0	0	0	0	0	0	0	0	0	2
Synechococcales	37	2	3	0	1	1	2	0	2	1	0	2	8	15

These data show that 63% of the analysed species have at least one copy of each of the *pil* and *com* genes. This suggests that the majority of cyanobacteria have a full complement of T4P genes. 46 out of 48 of all species which have a full set of *pil* genes, also have a full set of *com* genes (Table S1). Conversely, among the 61 species which have all *com* genes, 15 lack at least one *pil* gene. The dataset contains two members of the basal order of Gloeobacterales: *Gloeobacter violaceus* PCC 7421 and *Gloeobacter kilaueensis* JS1. Both species contain all three *com* homologues and a full (*G. violaceus*) or almost full (*G. kilaueensis*) set of *pil* genes. The presence of these genes in the *Gloeobacter* lineage, that diverged early from all other cyanobacterial lineages [90], suggests an early emergence of these genes in cyanobacteria. Thus, natural competence might be a primary trait in cyanobacteria but was lost during evolution in some lineages.

Interestingly, a large proportion of cyanobacterial species lacking *pil* or *com* genes belong to the genus *Prochlorococcus* (Table S1). These are marine, free-living picocyanobacteria, typically with minimal genomes, abundant in subtropical and oligotrophic oceans [91]. Considering their unique characteristics, it is not surprising that members of this genus also lack the *pil* and *com* genes. However, there are a few exceptions: *Prochlorococcus marinus* MIT 9303 and MIT 9313, which both have all of the analysed genes apart from *pilM*. These two strains are very distinct from 'typical' *Prochlorococcus* strains as they have a larger cell and genome size and are the only *Prochlorococcus* species known to possess *dprA* [91] and an *hfq* homologue [24]. To the best of our knowledge, it has not been experimentally shown whether these strains are naturally competent.

The list of potentially naturally competent cyanobacteria might even be longer than suggested by this dataset due to the limited information available on the *com* genes. However, it becomes apparent that, especially in filamentous cyanobacteria, natural competence might be more prevalent than so far acknowledged.

5. Benefits and Drawbacks of Natural Competence in an Ecological Context

What benefits are derived for cyanobacteria from the uptake and integration of external DNA into their genome? Or, in other words, what are the evolutionary and ecological benefits of natural competence? The two most discussed hypotheses regarding the importance of competence are DNA-for-food and DNA-for-diversity (reviewed in [1,61]). In short, the DNA-for-food hypothesis favours the idea that competent bacteria take up foreign DNA as a nutrient, whilst the DNA-for-diversity explains the benefit of natural competence in acquiring new traits. It has been shown that *Synechocystis* can efficiently utilise external genomic DNA as a phosphate source and also as a carbon source under heterotrophic growth conditions [92], supporting the DNA-for-food hypothesis in cyanobacteria. Evidence can often be interpreted in favour of either of the hypotheses. For example, the induction of competence under nutrient-limiting conditions in some bacteria [1] could be directly linked to nutrient acquisition. Simultaneously, induction of natural competence as a stress response could also increase chances of acquiring new traits that may result in a selective advantage.

Cyanobacteria are present in a wide range of ecological niches and environments, including extreme habitats such as deserts and hot springs. Heterotrophic thermophilic bacteria, e.g., *Thermus thermophilus*, were shown to have high transformation efficiencies [1,93–95]. The same may likely apply to thermophilic cyanobacteria inhabiting such extreme habitats. In these environments natural competence can offer swift adaptation strategies in line with the DNA-for-diversity hypothesis.

Natural competence likely offers further ecological advantages beyond these more established hypotheses. For instance, it has previously been brought in context with sexual selection in bacteria, which is in line with the DNA-for-diversity hypothesis [96]. Under conditions unfavourable for replication, horizontal gene transfer may be the favoured way of transferring genetic material—similar to the concept of horizontal gene transfer promoting genetic mixing within a population [97]. Some bacteria secrete DNA into the environment, which is also the case for cyanobacteria [98]. No matter if environmental DNA stems from cell lysis or secretion, natural competence (as a mean of horizontal gene transfer) could promote genetic diversity, to some extent, analogous to sexual reproduction in animals.

Natural competence as a trait does not only offer benefits but also entails significant costs and potential drawbacks from an ecological point of view. Expressing the molecular machinery required for natural competence is costly. The cost of T4P formation can, however, be offset to some extent, given its functions across a wide range of important cellular mechanisms [34,35,44]. Sharing the cost for T4P formation means that the cost for natural competence itself is significantly reduced to mainly the Com proteins. Another risk comes with the uptake and potential incorporation of defective or harmful genes [1]. However, polyploidy is widespread amongst cyanobacteria (reviewed in [99]) and might efficiently compensate for this risk. Many cyanobacteria also have efficient restriction-modification systems that degrade foreign DNA without a matching methylation pattern [100], which is likely to offer additional protection. Additionally, CRISPR-Cas (clustered regularly interspaced short palindromic repeats/CRISPR associated proteins) systems are widespread amongst cyanobacteria (reviewed in [101,102]) and could also play a role in cellular protection from foreign DNA. Overall, this suggests that the benefits of natural competence, at least for cyanobacteria, clearly prevail the drawbacks.

6. Targeted Manipulation of Natural Competence

Cyanobacteria are of interest not only from a fundamental point of view given their immense ecological importance but also for applications as they are considered promising chassis to establish sustainable, light-driven biotechnological processes for a plethora of products from commodities to high-value compounds [103–106]. An important prerequisite for developing a target strain is the availability of reliable genetic manipulation tools and efficient transformation methods. Amongst the DNA delivery methods available for cyanobacteria, natural competence constitutes a simple and effective way of delivering DNA into a strain of interest. Thus, understanding underlying mechanisms and regulation of natural competence can open up new avenues of targeted manipulation and wider exploitation of these organisms.

Many cyanobacterial species contain all or almost all genes required for natural competence (Table 2 and Table S1), however experimental evidence of their transformability is still largely lacking. Restoring natural competence of some cyanobacterial species might be possible by replacing non-functional genes or adding functional copies from related cyanobacteria (Figure 3A). For instance, the fast-growing strain *Synechococcus elongatus* UTEX 2973 was found to not be naturally competent [107]. This is, at least partly, due to a mutation in the *pilN* gene [108]. Introducing a second, functional copy of *pilN* from *S. elongatus* into a neutral site on the genome lead to a restoration of natural competence [108]. However, the authors suggest the level of natural competence is lower than in *S. elongatus* [108]. Similar strategies could also be employed in other strains. It is important to note that restoring the target gene at its native locus might be beneficial to conserve regulatory elements and expression levels with other proteins of the competence machinery. In *Pseudomonas aeruginosa*, the stoichiometry of the PilM, PilN and PilO proteins was found to be important for stable PilM/N/O complex formation [109]. This is likely to also be the case in cyanobacteria, therefore introducing *pilN* into its native locus in *S. elongatus* UTEX 2973 might lead to an improved transformation efficiency.

Another interesting approach is to use on/off switches or even titratable systems, using defined external cues, to tightly control levels of natural competence (Figure 3B). Several inducible promoter systems have been characterised in cyanobacteria [110–112] and multiple types of genome editing tools have been developed (recently reviewed in [113]). These tools could also be deployed to engineer natural competence. The use of inducible systems with known effector molecules could potentially circumvent the need to depend on native (often unknown) regulators of natural competence.

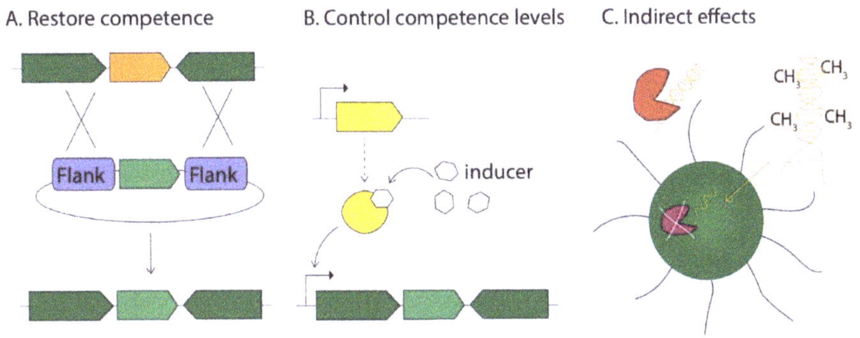

Figure 3. Examples of strategies for manipulating natural competence. (**A**) Restoration of natural competence by exchanging a non-functional gene (shown in orange) with a functional copy, e.g., by homologous recombination. (**B**) Controlling competence with a known (engineered) regulator, e.g., an external inducer molecule that binds to a transcription factor that activates a promoter for transcription of competence genes. (**C**) Control indirect effects on transformability by, e.g., knocking out nucleases (intracellular or excreted) or methylating DNA for transformation.

Difficulties with genetic manipulation by natural competence might, however, not only stem from a lack or disruption of the machinery itself but might be due to indirect effects. Many cyanobacteria, particularly filamentous species [82], have endogenous restriction-modification (RM) systems to protect the cell from incoming foreign DNA [100]. RM systems consist of methyltransferases, which methylate own DNA, and restriction endonucleases that digest DNA lacking the matching methylation pattern. However, restriction endonucleases only cut double-stranded DNA and should, therefore, not affect ssDNA as taken up by natural competence [96]. Nevertheless, DNA degradation by cellular protection mechanisms and extracellular nucleases might indirectly contribute to the lack of transformability in some cyanobacterial species. In *Synechocystis*, the deactivation of the exonuclease RecJ was shown to dramatically improve transformation efficiency [114]. Similarly, deletion of a type I restriction endonuclease in *T. elongatus* BP-1 also leads to higher transformation efficiencies [115]. Furthermore, pre-methylation of DNA for transformation could aid in overcoming transformability issues [116,117] (Figure 3C).

Ultimately, further research is needed to obtain a better understanding of natural competence in cyanobacteria. The development of strategies for controlling and manipulating natural competence is now enabled by advances in genetic engineering and genome editing tool development in cyanobacteria. This can not only contribute to understanding the molecular mechanisms of natural competence and its ecological context but also allows the wider exploitation of these organisms biotechnologically.

Supplementary Materials: The following are available online at http://www.mdpi.com/2075-1729/10/11/249/s1, Table S1: Prevalence of *pil* and *com* genes within cyanobacterial species.

Author Contributions: A.M.S. compiled the data of Table S1, A.M.S and J.A.Z.Z. wrote the manuscript with contributions from S.S.H. All authors have read and agreed to the published version of the manuscript.

Funding: This research received no external funding.

Acknowledgments: The authors thank David A. Russo for critical reading of the manuscript and Anthony W. Harris for technical support in data compilation.

Conflicts of Interest: The authors declare no conflict of interest.

References

1. Johnsborg, O.; Eldholm, V.; Håvarstein, L.S. Natural genetic transformation: Prevalence, mechanisms and function. *Res. Microbiol.* **2007**, *158*, 767–778. [CrossRef] [PubMed]
2. Dubnau, D.; Blokesch, M. Mechanisms of DNA Uptake by Naturally Competent Bacteria. *Annu. Rev. Genet.* **2019**, *53*, 217–237. [CrossRef] [PubMed]
3. Johnston, C.; Martin, B.; Fichant, G.; Polard, P.; Claverys, J.-P. Bacterial transformation: Distribution, shared mechanisms and divergent control. *Nat. Rev. Microbiol.* **2014**, *12*, 181–196. [CrossRef] [PubMed]
4. Haas, R.; Meyer, T.F.; Putten, J.P.M. Aflagellated mutants of *Helicobacter pylori* generated by genetic transformation of naturally competent strains using transposon shuttle mutagenesis. *Mol. Microbiol.* **1993**, *8*, 753–760. [CrossRef]
5. Bhaya, D.; Bianco, N.R.; Bryant, D.; Grossman, A. Type IV pilus biogenesis and motility in the cyanobacterium *Synechocystis* sp. PCC6803. *Mol. Microbiol.* **2000**, *37*, 941–951. [CrossRef]
6. Wolfgang, M.; Lauer, P.; Park, H.; Brossay, L.; Hébert, J.; Koomey, M. PilT mutations lead to simultaneous defects in competence for natural transformation and twitching motility in piliated *Neisseria gonorrhoeae*. *Mol. Microbiol.* **1998**, *29*, 321–330. [CrossRef]
7. Evans, K.J.; Lambert, C.; Sockett, R.E. Predation by *Bdellovibrio bacteriovorus* HD100 Requires Type IV Pili. *J. Bacteriol.* **2007**, *189*, 4850–4859. [CrossRef] [PubMed]
8. Pujol, C.; Eugene, E.; Marceau, M.; Nassif, X. The meningococcal PilT protein is required for induction of intimate attachment to epithelial cells following pilus-mediated adhesion. *Proc. Natl. Acad. Sci. USA* **1999**, *96*, 4017–4022. [CrossRef]
9. Klausen, M.; Aaes-Jørgensen, A.; Molin, S.; Tolker-Nielsen, T. Involvement of bacterial migration in the development of complex multicellular structures in *Pseudomonas aeruginosa* biofilms. *Mol. Microbiol.* **2003**, *50*, 61–68. [CrossRef]
10. Klausen, M.; Heydorn, A.; Ragas, P.; Lambertsen, L.; Aaes-Jørgensen, A.; Molin, S.; Tolker-Nielsen, T. Biofilm formation by *Pseudomonas aeruginosa* wild type, flagella and type IV pili mutants. *Mol. Microbiol.* **2003**, *48*, 1511–1524. [CrossRef]
11. O'Toole, G.A.; Kolter, R. Flagellar and twitching motility are necessary for *Pseudomonas aeruginosa* biofilm development. *Mol. Microbiol.* **1998**, *30*, 295–304. [CrossRef] [PubMed]
12. Bieber, D.; Ramer, S.W.; Wu, C.-Y.; Murray, W.J.; Tobe, T.; Fernandez, R.; Schoolnik, G.K. Type IV Pili, Transient Bacterial Aggregates, and Virulence of Enteropathogenic *Escherichia coli*. *Science* **1998**, *280*, 2114–2118. [CrossRef] [PubMed]
13. Lu, H.-M.; Motley, S.T.; Lory, S. Interactions of the components of the general secretion pathway: Role of *Pseudomonas aeruginosa* type IV pilin subunits in complex formation and extracellular protein secretion. *Mol. Microbiol.* **1997**, *25*, 247–259. [CrossRef]
14. Shestakov, S.V.; Khyen, N.T. Evidence for genetic transformation in blue-green alga *Anacystis nidulans*. *Mol. Gen. Genet.* **1970**, *107*, 372–375. [CrossRef] [PubMed]
15. Jacobsen, T.; Bardiaux, B.; Francetic, O.; Izadi-Pruneyre, N.; Nilges, M. Structure and function of minor pilins of type IV pili. *Med. Microbiol. Immunol.* **2020**, *209*, 301–308. [CrossRef]
16. Sun, D. Pull in and Push Out: Mechanisms of Horizontal Gene Transfer in Bacteria. *Front. Microbiol.* **2018**, *9*, 2154. [CrossRef]
17. Yoshihara, S.; Geng, X.X.; Okamoto, S.; Yura, K.; Murata, T.; Go, M.; Ohmori, M.; Ikeuchi, M. Mutational Analysis of Genes Involved in Pilus Structure, Motility and Transformation Competency in the Unicellular Motile Cyanobacterium *Synechocystis* sp. PCC 6803. *Plant Cell Physiol.* **2001**, *42*, 63–73. [CrossRef]
18. Yura, K.; Toh, H.; Go, M. Putative Mechanism of Natural Transformation as Deduced from Genome Data. *DNA Res.* **1999**, *6*, 75–82. [CrossRef]
19. Nakasugi, K.; Svenson, C.J.; Neilan, B.A. The competence gene, comF, from *Synechocystis* sp. strain PCC 6803 is involved in natural transformation, phototactic motility and piliation. *Microbiology* **2006**, *152*, 3623–3631. [CrossRef]
20. Bergé, M.; Mortier-Barrière, I.; Martin, B.; Claverys, J.-P. Transformation of *Streptococcus pneumoniae* relies on DprA- and RecA-dependent protection of incoming DNA single strands. *Mol. Microbiol.* **2003**, *50*, 527–536. [CrossRef]

21. Mortier-Barrière, I.; Velten, M.; Dupaigne, P.; Mirouze, N.; Piétrement, O.; McGovern, S.; Fichant, G.; Martin, B.; Noirot, P.; Le Cam, E.; et al. A Key Presynaptic Role in Transformation for a Widespread Bacterial Protein: DprA Conveys Incoming ssDNA to RecA. *Cell* **2007**, *130*, 824–836. [CrossRef] [PubMed]

22. Taton, A.; Erikson, C.; Yang, Y.; Rubin, B.E.; Rifkin, S.A.; Golden, J.W.; Golden, S.S. The circadian clock and darkness control natural competence in cyanobacteria. *Nat. Commun.* **2020**, *11*, 1688. [CrossRef] [PubMed]

23. Dienst, D.; Dühring, U.; Mollenkopf, H.-J.; Vogel, J.; Golecki, J.; Hess, W.R.; Wilde, A. The cyanobacterial homologue of the RNA chaperone Hfq is essential for motility of *Synechocystis* sp. PCC 6803. *Microbiology* **2008**, *154*, 3134–3143. [CrossRef] [PubMed]

24. Schuergers, N.; Ruppert, U.; Watanabe, S.; Nürnberg, D.J.; Lochnit, G.; Dienst, D.; Mullineaux, C.W.; Wilde, A. Binding of the RNA chaperone Hfq to the type IV pilus base is crucial for its function in *Synechocystis* sp. PCC 6803. *Mol. Microbiol.* **2014**, *92*, 840–852. [CrossRef]

25. Sergeyenko, T.V.; Los, D.A. Identification of secreted proteins of the cyanobacterium *Synechocystis* sp. strain PCC 6803. *FEMS Microbiol. Lett.* **2000**, *193*, 213–216. [CrossRef]

26. Yoshihara, S.; Geng, X.; Ikeuchi, M. *pilG* Gene Cluster and Split *pilL* Genes Involved in Pilus Biogenesis, Motility and Genetic Transformation in the Cyanobacterium *Synechocystis* sp. PCC 6803. *Plant Cell Physiol.* **2002**, *43*, 513–521. [CrossRef]

27. Okamoto, S.; Ohmori, M. The Cyanobacterial PilT Protein Responsible for Cell Motility and Transformation Hydrolyzes ATP. *Plant Cell Physiol.* **2002**, *43*, 1127–1136. [CrossRef]

28. Murphy, R.C.; Bryant, D.A.; Porter, R.D.; de Marsac, N.T. Molecular cloning and characterization of the *recA* gene from the cyanobacterium *Synechococcus* sp. strain PCC 7002. *J. Bacteriol.* **1987**, *169*, 2739–2747. [CrossRef]

29. Bhaya, D.; Watanabe, N.; Ogawa, T.; Grossman, A.R. The role of an alternative sigma factor in motility and pilus formation in the cyanobacterium *Synechocystis* sp. strain PCC6803. *Proc. Natl. Acad. Sci. USA* **1999**, *96*, 3188–3193. [CrossRef]

30. Lamb, J.J.; Hohmann-Marriott, M.F. Manganese acquisition is facilitated by PilA in the cyanobacterium *Synechocystis* sp. PCC 6803. *PLoS ONE* **2017**, *12*, e0184685. [CrossRef]

31. Lamb, J.J.; Hill, R.E.; Eaton-Rye, J.J.; Hohmann-Marriott, M.F. Functional Role of PilA in Iron Acquisition in the Cyanobacterium *Synechocystis* sp. PCC 6803. *PLoS ONE* **2014**, *9*, e105761. [CrossRef] [PubMed]

32. Allen, R.; Rittmann, B.E.; Curtiss, R. Axenic Biofilm Formation and Aggregation by *Synechocystis* sp. Strain PCC 6803 Are Induced by Changes in Nutrient Concentration and Require Cell Surface Structures. *Appl. Environ. Microbiol.* **2019**, *85*, e02192-18. [CrossRef]

33. Conradi, F.D.; Zhou, R.-Q.; Oeser, S.; Schuergers, N.; Wilde, A.; Mullineaux, C.W. Factors Controlling Floc Formation and Structure in the Cyanobacterium *Synechocystis* sp. Strain PCC 6803. *J. Bacteriol.* **2019**, *201*, e00344-19. [CrossRef] [PubMed]

34. Grigorieva, G.; Shestakov, S. Transformation in the cyanobacterium *Synechocystis* sp. 6803. *FEMS Microbiol. Lett.* **1982**, *13*, 367–370. [CrossRef]

35. Piepenbrink, K.H. DNA Uptake by Type IV Filaments. *Front. Mol. Biosci.* **2019**, *6*, 1. [CrossRef] [PubMed]

36. Pelicic, V. Monoderm bacteria: The new frontier for type IV pilus biology. *Mol. Microbiol.* **2019**, *112*, 1674–1683. [CrossRef]

37. Linhartová, M.; Bučinská, L.; Halada, P.; Ječmen, T.; Šetlík, J.; Komenda, J.; Sobotka, R. Accumulation of the Type IV prepilin triggers degradation of SecY and YidC and inhibits synthesis of Photosystem II proteins in the cyanobacterium *Synechocystis* PCC 6803. *Mol. Microbiol.* **2014**, *93*, 1207–1223. [CrossRef]

38. Nunn, D.; Bergman, S.; Lory, S. Products of three accessory genes, pilB, pilC, and pilD, are required for biogenesis of *Pseudomonas aeruginosa* pili. *J. Bacteriol.* **1990**, *172*, 2911–2919. [CrossRef]

39. Nunn, D.N.; Lory, S. Product of the *Pseudomonas aeruginosa* gene *pilD* is a prepilin leader peptidase. *Proc. Natl. Acad. Sci. USA* **1991**, *88*, 3281–3285. [CrossRef]

40. Strom, M.S.; Nunn, D.N.; Lory, S. A single bifunctional enzyme, PilD, catalyzes cleavage and N-methylation of proteins belonging to the type IV pilin family. *Proc. Natl. Acad. Sci. USA* **1993**, *90*, 2404–2408. [CrossRef]

41. Bhaya, D.; Takahashi, A.; Shahi, P.; Grossman, A.R. Novel Motility Mutants of *Synechocystis* Strain PCC 6803 Generated by In Vitro Transposon Mutagenesis. *J. Bacteriol.* **2001**, *183*, 6140–6143. [CrossRef]

42. Yoshihara, S.; Ikeuchi, M. Phototactic motility in the unicellular cyanobacterium *Synechocystis* sp. PCC 6803. *Photochem. Photobiol. Sci.* **2004**, *3*, 512–518. [CrossRef] [PubMed]

43. Ellison, C.K.; Dalia, T.N.; Ceballos, A.V.; Wang, J.C.-Y.; Biais, N.; Brun, Y.V.; Dalia, A.B. Retraction of DNA-bound type IV competence pili initiates DNA uptake during natural transformation in *Vibrio cholerae*. *Nat. Microbiol.* **2018**, *3*, 773–780. [CrossRef] [PubMed]

44. Chen, Z.; Li, X.; Tan, X.; Zhang, Y.; Wang, B. Recent Advances in Biological Functions of Thick Pili in the Cyanobacterium *Synechocystis* sp. PCC 6803. *Front. Plant Sci.* **2020**, *11*, 241. [CrossRef] [PubMed]

45. Diallo, A.; Foster, H.R.; Gromek, K.A.; Perry, T.N.; Dujeancourt, A.; Krasteva, P.V.; Gubellini, F.; Falbel, T.G.; Burton, B.M.; Fronzes, R. Bacterial transformation: ComFA is a DNA-dependent ATPase that forms complexes with ComFC and DprA. *Mol. Microbiol.* **2017**, *105*, 741–754. [CrossRef]

46. Johnsborg, O.; Håvarstein, L.S. Regulation of natural genetic transformation and acquisition of transforming DNA in *Streptococcus pneumoniae*. *FEMS Microbiol. Rev.* **2009**, *33*, 627–642. [CrossRef] [PubMed]

47. Meibom, K.L.; Blokesch, M.; Dolganov, N.A.; Wu, C.-Y.; Schoolnik, G.K. Chitin Induces Natural Competence in *Vibrio cholerae*. *Science* **2005**, *310*, 1824–1827. [CrossRef]

48. Lorenz, M.G.; Wackernagel, W. Natural genetic transformation of *Pseudomonas stutzeri* by sand-adsorbed DNA. *Arch. Microbiol.* **1990**, *154*, 380–385. [CrossRef]

49. Havarstein, L.S.; Coomaraswamy, G.; Morrison, D.A. An unmodified heptadecapeptide pheromone induces competence for genetic transformation in *Streptococcus pneumoniae*. *Proc. Natl. Acad. Sci. USA* **1995**, *92*, 11140–11144. [CrossRef]

50. Li, Y.-H.; Lau, P.C.Y.; Lee, J.H.; Ellen, R.P.; Cvitkovitch, D.G. Natural Genetic Transformation of *Streptococcus mutans* Growing in Biofilms. *J. Bacteriol.* **2001**, *183*, 897–908. [CrossRef]

51. Magnuson, R.; Solomon, J.; Grossman, A.D. Biochemical and genetic characterization of a competence pheromone from *B. subtilis*. *Cell* **1994**, *77*, 207–216. [CrossRef]

52. Kandel, P.P.; Lopez, S.M.; Almeida, R.P.P.; De La Fuente, L. Natural Competence of *Xylella fastidiosa* Occurs at a High Frequency Inside Microfluidic Chambers Mimicking the Bacterium's Natural Habitats. *Appl. Environ. Microbiol.* **2016**, *82*, 5269–5277. [CrossRef] [PubMed]

53. Merod, R.T.; Wuertz, S. Extracellular Polymeric Substance Architecture Influences Natural Genetic Transformation of *Acinetobacter baylyi* in Biofilms. *Appl. Environ. Microbiol.* **2014**, *80*, 7752–7757. [CrossRef] [PubMed]

54. Lorenz, M.G.; Wackernagel, W. High Frequency of Natural Genetic Transformation of *Pseudomonas stutzeri* in Soil Extract Supplemented with a Carbon/Energy and Phosphorus Source. *Appl. Environ. Microbiol.* **1991**, *57*, 1246–1251. [CrossRef] [PubMed]

55. Dorer, M.S.; Fero, J.; Salama, N.R. DNA Damage Triggers Genetic Exchange in *Helicobacter pylori*. *PLoS Pathog.* **2010**, *6*, e1001026. [CrossRef]

56. Charpentier, X.; Kay, E.; Schneider, D.; Shuman, H.A. Antibiotics and UV Radiation Induce Competence for Natural Transformation in *Legionella pneumophila*. *J. Bacteriol.* **2011**, *193*, 1114–1121. [CrossRef]

57. Prudhomme, M.; Attaiech, L.; Sanchez, G.; Martin, B.; Claverys, J.-P. Antibiotic Stress Induces Genetic Transformability in the Human Pathogen *Streptococcus pneumoniae*. *Science* **2006**, *313*, 89–92. [CrossRef]

58. Herriott, R.M.; Meyer, E.M.; Vogt, M. Defined Nongrowth Media for Stage II Development of Competence in *Haemophilus influenzae*. *J. Bacteriol.* **1970**, *101*, 517–524. [CrossRef]

59. Nielsen, K.M.; Bones, A.M.; Van Elsas, J.D. Induced Natural Transformation of *Acinetobacter calcoaceticus* in Soil Microcosms. *Appl. Environ. Microbiol.* **1997**, *63*, 3972–3977. [CrossRef]

60. Traglia, G.M.; Quinn, B.; Schramm, S.T.J.; Soler-Bistue, A.; Ramirez, M.S. Serum Albumin and Ca^{2+} Are Natural Competence Inducers in the Human Pathogen *Acinetobacter baumannii*. *Antimicrob. Agents Chemother.* **2016**, *60*, 4920–4929. [CrossRef]

61. Mell, J.C.; Redfield, R.J. Natural Competence and the Evolution of DNA Uptake Specificity. *J. Bacteriol.* **2014**, *196*, 1471–1483. [CrossRef] [PubMed]

62. Seitz, P.; Blokesch, M. Cues and regulatory pathways involved in natural competence and transformation in pathogenic and environmental Gram-negative bacteria. *FEMS Microbiol. Rev.* **2013**, *37*, 336–363. [CrossRef] [PubMed]

63. Porter, R.D. Transformation in Cyanobacteria. *CRC Crit. Rev. Microbiol.* **1986**, *13*, 111–132. [CrossRef] [PubMed]

64. Zang, X.; Liu, B.; Liu, S.; Arunakumara, K.K.I.U.; Zhang, X. Optimum conditions for transformation of *Synechocystis* sp. PCC 6803. *J. Microbiol.* **2007**, *45*, 241–245.

65. Golden, S.S.; Sherman, L.A. Optimal conditions for genetic transformation of the cyanobacterium *Anacystis nidulans* R2. *J. Bacteriol.* **1984**, *158*, 36–42. [CrossRef] [PubMed]

66. Wolk, C.P.; Kraus, J. Two approaches to obtaining low, extracellular deoxyribonuclease activity in cultures of heterocyst-forming cyanobacteria. *Arch. Microbiol.* **1982**, *131*, 302–307. [CrossRef]

67. Schuergers, N.; Wilde, A. Appendages of the Cyanobacterial Cell. *Life* **2015**, *5*, 700–715. [CrossRef]

68. Williams, J.G.K.; Szalay, A.A. Stable integration of foreign DNA into the chromosome of the cyanobacterium *Synechococcus* R2. *Gene* **1983**, *24*, 37–51. [CrossRef]

69. Jaiswal, D.; Sengupta, A.; Sohoni, S.; Sengupta, S.; Phadnavis, A.G.; Pakrasi, H.B.; Wangikar, P.P. Genome Features and Biochemical Characteristics of a Robust, Fast Growing and Naturally Transformable Cyanobacterium *Synechococcus elongatus* PCC 11801 Isolated from India. *Sci. Rep.* **2018**, *8*, 16632. [CrossRef]

70. Stevens, S.E.; Porter, R.D. Transformation in *Agmenellum quadruplicatum*. *Proc. Natl. Acad. Sci. USA* **1980**, *77*, 6052–6056. [CrossRef]

71. Herdman, M. Mutations arising during transformation in the blue-green alga *Anacystis nidulans*. *Mol. Gen. Genet. MGG* **1973**, *120*, 369–378. [CrossRef] [PubMed]

72. Orkwiszewski, K.G.; Kaney, A.R. Genetic transformation of the blue-green bacterium, *Anacystis nidulans*. *Arch. Microbiol.* **1974**, *98*, 31–37. [CrossRef] [PubMed]

73. Lightfoot, D.A.; Walters, D.E.; Wootton, J.C. Transformation of the Cyanobacterium *Synechococcus* PCC 6301 Using Cloned DNA. *J. Gen. Microbiol.* **1988**, *134*, 1509–1514. [CrossRef]

74. Takeshima, Y.; Takatsugu, N.; Sugiura, M.; Hagiwara, H. High-level expression of human superoxide dismutase in the cyanobacterium *Anacystis nidulans* 6301. *Proc. Natl. Acad. Sci. USA* **1994**, *91*, 9685–9689. [CrossRef] [PubMed]

75. Takeshima, Y.; Sugiura, M.; Hagiwara, H. A Novel Expression Vector for the Cyanobacterium, *Synechococcus* PCC 6301. *DNA Res.* **1994**, *1*, 181–189. [CrossRef] [PubMed]

76. Sugita, C.; Ogata, K.; Shikata, M.; Jikuya, H.; Takano, J.; Furumichi, M.; Kanehisa, M.; Omata, T.; Sugiura, M.; Sugita, M. Complete nucleotide sequence of the freshwater unicellular cyanobacterium *Synechococcus elongatus* PCC 6301 chromosome: Gene content and organization. *Photosynth. Res.* **2007**, *93*, 55–67. [CrossRef] [PubMed]

77. Tsinoremas, N.F.; Kutach, A.K.; Strayer, C.A.; Golden, S.S. Efficient Gene Transfer in *Synechococcus* sp. Strains PCC 7942 and PCC 6301 by Interspecies Conjugation and Chromosomal Recombination. *J. Bacteriol.* **1994**, *176*, 6764–6768. [CrossRef]

78. Onai, K.; Morishita, M.; Kaneko, T.; Tabata, S.; Ishiura, M. Natural transformation of the thermophilic cyanobacterium *Thermosynechococcus elongatus* BP-1: A simple and efficient method for gene transfer. *Mol. Genet. Genom.* **2004**, *271*, 50–59. [CrossRef]

79. Dittmann, E.; Neilan, B.A.; Erhard, M.; Von Döhren, H.; Börner, T. Insertional mutagenesis of a peptide synthetase gene that is responsible for hepatotoxin production in the cyanobacterium *Microcystis aeruginosa* PCC 7806. *Mol. Microbiol.* **1997**, *26*, 779–787. [CrossRef]

80. Koksharova, O.A.; Wolk, C.P. Genetic tools for cyanobacteria. *Appl. Microbiol. Biotechnol.* **2002**, *58*, 123–137. [CrossRef]

81. Al-Haj, L.; Lui, Y.; Abed, R.; Gomaa, M.; Purton, S. Cyanobacteria as Chassis for Industrial Biotechnology: Progress and Prospects. *Life* **2016**, *6*, 42. [CrossRef]

82. Stucken, K.; Ilhan, J.; Roettger, M.; Dagan, T.; Martin, W.F. Transformation and Conjugal Transfer of Foreign Genes into the Filamentous Multicellular Cyanobacteria (Subsection V) *Fischerella* and *Chlorogloeopsis*. *Curr. Microbiol.* **2012**, *65*, 552–560. [CrossRef]

83. Trehan, K.; Sinha, U. Genetic Transfer in a Nitrogen-fixing Filamentous Cyanobacterium. *Microbiology* **1981**, *124*, 349–352. [CrossRef]

84. Singh, D.T.; Bagchi, S.N.; Modi, D.R.; Singh, H.N. Evidence for Intergenetic Transformation in Filamentous, Diazotrophic Cyanobacteria. *New Phytol.* **1987**, *107*, 347–356. [CrossRef]

85. Verma, S.K.; Singh, A.K.; Katiyar, S.; Singh, H.N. Genetic transformation of glutamine auxotrophy to prototrophy in the cyanobacterium *Nostoc muscorum*. *Arch. Microbiol.* **1990**, *154*, 414–416. [CrossRef]

86. Nies, F.; Mielke, M.; Pochert, J.; Lamparter, T. Natural transformation of the filamentous cyanobacterium *Phormidium lacuna*. *PLoS ONE* **2020**, *15*, e0234440. [CrossRef]

87. Springstein, B.L.; Nies, F.; Dagan, T. Natural competence in *Chlorogloeopsis fritschii* PCC 6912 and other ramified cyanobacteria. *bioRxiv* **2020**, *2020*. [CrossRef]

88. Cassier-Chauvat, C.; Veaudor, T.; Chauvat, F. Comparative Genomics of DNA Recombination and Repair in Cyanobacteria: Biotechnological Implications. *Front. Microbiol.* **2016**, *7*, 1809. [CrossRef]

89. Wendt, K.E.; Pakrasi, H.B. Genomics Approaches to Deciphering Natural Transformation in Cyanobacteria. *Front. Microbiol.* **2019**, *10*, 1259. [CrossRef]

90. Ponce-Toledo, R.I.; Deschamps, P.; López-García, P.; Zivanovic, Y.; Benzerara, K.; Moreira, D. An Early-Branching Freshwater Cyanobacterium at the Origin of Plastids. *Curr. Biol.* **2017**, *27*, 386–391. [CrossRef]

91. Partensky, F.; Garczarek, L. *Prochlorococcus*: Advantages and Limits of Minimalism. *Ann. Rev. Mar. Sci.* **2010**, *2*, 305–331. [CrossRef]

92. Zerulla, K.; Ludt, K.; Soppa, J. The ploidy level of *Synechocystis* sp. PCC 6803 is highly variable and is influenced by growth phase and by chemical and physical external parameters. *Microbiology* **2016**, *162*, 730–739. [CrossRef]

93. Koyama, Y.; Hoshino, T.; Tomizuka, N.; Furukawa, K. Genetic transformation of the extreme thermophile *Thermus thermophilus* and of other *Thermus* spp. *J. Bacteriol.* **1986**, *166*, 338–340. [CrossRef]

94. Friedrich, A.; Hartsch, T.; Averhoff, B. Natural Transformation in Mesophilic and Thermophilic Bacteria: Identification and Characterization of Novel, Closely Related Competence Genes in *Acinetobacter* sp. Strain BD413 and *Thermus thermophilus* HB27. *Appl. Environ. Microbiol.* **2001**, *67*, 3140–3148. [CrossRef]

95. Friedrich, A.; Rumszauer, J.; Henne, A.; Averhoff, B. Pilin-Like Proteins in the Extremely Thermophilic Bacterium *Thermus thermophilus* HB27: Implication in Competence for Natural Transformation and Links to Type IV Pilus Biogenesis. *Appl. Environ. Microbiol.* **2003**, *69*, 3695–3700. [CrossRef]

96. Vos, M.; Buckling, A.; Kuijper, B. Sexual Selection in Bacteria? *Trends Microbiol.* **2019**, *27*, 972–981. [CrossRef]

97. Szöllősi, G.J.; Derényi, I.; Vellai, T. The Maintenance of Sex in Bacteria Is Ensured by Its Potential to Reload Genes. *Genetics* **2006**, *174*, 2173–2180. [CrossRef]

98. Russo, D.A.; Zedler, J.A.Z. Genomic insights into cyanobacterial protein translocation systems. *Biol. Chem.* **2020**, *2020*. [CrossRef]

99. Watanabe, S. Cyanobacterial multi-copy chromosomes and their replication. *Biosci. Biotechnol. Biochem.* **2020**, *84*, 1309–1321. [CrossRef]

100. Stucken, K.; Koch, R.; Dagan, T. Cyanobacterial defense mechanisms against foreign DNA transfer and their impact on genetic engineering. *Biol. Res.* **2013**, *46*, 373–382. [CrossRef]

101. Cai, F.; Axen, S.D.; Kerfeld, C.A. Evidence for the widespread distribution of CRISPR-Cas system in the Phylum Cyanobacteria. *RNA Biol.* **2013**, *10*, 687–693. [CrossRef] [PubMed]

102. Hou, S.; Brenes-Álvarez, M.; Reimann, V.; Alkhnbashi, O.S.; Backofen, R.; Muro-Pastor, A.M.; Hess, W.R. CRISPR-Cas systems in multicellular cyanobacteria. *RNA Biol.* **2019**, *16*, 518–529. [CrossRef] [PubMed]

103. Hitchcock, A.; Hunter, C.N.; Canniffe, D.P. Progress and challenges in engineering cyanobacteria as chassis for light-driven biotechnology. *Microb. Biotechnol.* **2020**, *13*, 363–367. [CrossRef] [PubMed]

104. Santos-Merino, M.; Singh, A.K.; Ducat, D.C. New Applications of Synthetic Biology Tools for Cyanobacterial Metabolic Engineering. *Front. Bioeng. Biotechnol.* **2019**, *7*, 33. [CrossRef]

105. Klemenčič, M.; Nielsen, A.Z.; Sakuragi, Y.; Frigaard, N.-U.; Čelešnik, H.; Jensen, P.E.; Dolinar, M. Synthetic biology of cyanobacteria for production of biofuels and high-value products. In *Microalgae-Based Biofuels and Bioproducts*; Gonzalez-Fernandez, C., Muñoz, R., Eds.; Woodhead Publishing: Sawston, UK, 2017; pp. 305–325.

106. Russo, D.A.; Zedler, J.A.Z.; Jensen, P.E. A force awakens: Exploiting solar energy beyond photosynthesis. *J. Exp. Bot.* **2019**, *70*, 1703–1710. [CrossRef]

107. Yu, J.; Liberton, M.; Cliften, P.F.; Head, R.D.; Jacobs, J.M.; Smith, R.D.; Koppenaal, D.W.; Brand, J.J.; Pakrasi, H.B. *Synechococcus elongatus* UTEX 2973, a fast growing cyanobacterial chassis for biosynthesis using light and CO_2. *Sci. Rep.* **2015**, *5*, 8132. [CrossRef]

108. Li, S.; Sun, T.; Xu, C.; Chen, L.; Zhang, W. Development and optimization of genetic toolboxes for a fast-growing cyanobacterium *Synechococcus elongatus* UTEX 2973. *Metab. Eng.* **2018**, *48*, 163–174. [CrossRef]

109. Ayers, M.; Sampaleanu, L.M.; Tammam, S.; Koo, J.; Harvey, H.; Howell, P.L.; Burrows, L.L. PilM/N/O/P Proteins Form an Inner Membrane Complex That Affects the Stability of the *Pseudomonas aeruginosa* Type IV Pilus Secretin. *J. Mol. Biol.* **2009**, *394*, 128–142. [CrossRef]

110. Behle, A.; Saake, P.; Axmann, I.M. Comparative dose-response analysis of inducible promoters in cyanobacteria. *ACS Synth. Biol.* **2020**, *9*, 843–855. [CrossRef]

111. Englund, E.; Liang, F.; Lindberg, P. Evaluation of promoters and ribosome binding sites for biotechnological applications in the unicellular cyanobacterium *Synechocystis* sp. PCC 6803. *Sci. Rep.* **2016**, *6*, 36640. [CrossRef]

112. Kelly, C.L.; Taylor, G.M.; Hitchcock, A.; Torres-Méndez, A.; Heap, J.T. A Rhamnose-Inducible System for Precise and Temporal Control of Gene Expression in Cyanobacteria. *ACS Synth. Biol.* **2018**, *7*, 1056–1066. [CrossRef]

113. Gale, G.A.R.; Schiavon Osorio, A.A.; Mills, L.A.; Wang, B.; Lea-Smith, D.J.; McCormick, A.J. Emerging Species and Genome Editing Tools: Future Prospects in Cyanobacterial Synthetic Biology. *Microorganisms* **2019**, *7*, 409. [CrossRef] [PubMed]

114. Kufryk, G.; Sachet, M.; Schmetterer, G.; Vermaas, W.F.J. Transformation of the cyanobacterium *Synechocystis* sp. PCC 6803 as a tool for genetic mapping: Optimization of efficiency. *FEMS Microbiol. Lett.* **2002**, *206*, 215–219. [CrossRef] [PubMed]

115. Iwai, M.; Katoh, H.; Katayama, M.; Ikeuchi, M. Improved Genetic Transformation of the Thermophilic Cyanobacterium, *Thermosynechococcus elongatus* BP-1. *Plant Cell Physiol.* **2004**, *45*, 171–175. [CrossRef]

116. Wang, B.; Yu, J.; Zhang, W.; Meldrum, D.R. Premethylation of Foreign DNA Improves Integrative Transformation Efficiency in *Synechocystis* sp. Strain PCC 6803. *Appl. Environ. Microbiol.* **2015**, *81*, 8500–8506. [CrossRef] [PubMed]

117. Elhai, J.; Vepritskiy, A.; Muro-Pastor, A.M.; Flores, E.; Wolk, C.P. Reduction of conjugal transfer efficiency by three restriction activities of *Anabaena* sp. strain PCC 7120. *J. Bacteriol.* **1997**, *179*, 1998–2005. [CrossRef]

Publisher's Note: MDPI stays neutral with regard to jurisdictional claims in published maps and institutional affiliations.

Review

Extracellular Vesicles: An Overlooked Secretion System in Cyanobacteria

Steeve Lima [1,2], **Jorge Matinha-Cardoso** [1,2], **Paula Tamagnini** [1,2,3] and **Paulo Oliveira** [1,2,*]

[1] i3S—Instituto de Investigação e Inovação em Saúde, Universidade do Porto, R. Alfredo Allen, 208, 4200-135 Porto, Portugal; steeve.lima@i3s.up.pt (S.L.); jorge.cardoso@i3s.up.pt (J.M.-C.); pmtamagn@ibmc.up.pt (P.T.)
[2] IBMC—Instituto de Biologia Molecular e Celular, Universidade do Porto, R. Alfredo Allen, 208, 4200-135 Porto, Portugal
[3] Departamento de Biologia, Faculdade de Ciências, Universidade do Porto, Rua do Campo Alegre, Edifício FC4, 4169-007 Porto, Portugal
* Correspondence: paulo.oliveira@ibmc.up.pt; Tel.: +351-22-607-4900

Received: 15 July 2020; Accepted: 29 July 2020; Published: 31 July 2020

Abstract: In bacteria, the active transport of material from the interior to the exterior of the cell, or secretion, represents a very important mechanism of adaptation to the surrounding environment. The secretion of various types of biomolecules is mediated by a series of multiprotein complexes that cross the bacterial membrane(s), each complex dedicated to the secretion of specific substrates. In addition, biological material may also be released from the bacterial cell in the form of vesicles. Extracellular vesicles (EVs) are bilayered, nanoscale structures, derived from the bacterial cell envelope, which contain membrane components as well as soluble products. In cyanobacteria, the knowledge regarding EVs is lagging far behind compared to what is known about, for example, other Gram-negative bacteria. Here, we present a summary of the most important findings regarding EVs in Gram-negative bacteria, discussing aspects of their composition, formation processes and biological roles, and highlighting a number of technological applications tested. This lays the groundwork to raise awareness that the release of EVs by cyanobacteria likely represents an important, and yet highly disregarded, survival strategy. Furthermore, we hope to motivate future studies that can further elucidate the role of EVs in cyanobacterial cell biology and physiology.

Keywords: cyanobacteria; extracellular vesicles; composition; biogenesis; biological roles; applications

---

1. Introduction

The phylum Cyanobacteria consists of a group of prokaryotes that typically carry out oxygenic photosynthesis with water as an electron donor and use carbon dioxide as a carbon source, or those secondarily evolved from such organisms [1]. Cyanobacteria are generally regarded as Gram-negative bacteria that can be found in a wide range of habitats [2], from aquatic to land-based ecosystems. In addition, these microorganisms can also be found in the harshest environments, such as hot springs, polar lakes, deserts and even polluted wastelands [3]. To colonize and thrive in such places, cyanobacteria evolved niche-specific adaptation strategies. They exhibit a great tolerance to environmental stresses, such as desiccation, radiation, nutrient deficiency and exposure to heavy metals [3]. Cyanobacteria adjust to these life threats through a series of adaptive physiological responses, resulting from the balance between competition for resources and intra- and interspecies cooperation [4]. This vital game of transmitting and receiving signals enable them to activate survival pathways and to maintain homeostasis. In this context, the extracellular milieu plays an important role, as it is a source of nutrients and structural components, but also of cell-damaging agents. Additionally, it can work as a sink for compounds, a channel for exchanging information, and a platform for interaction and for extending cellular functionalities [5].

In cyanobacteria, several uptake and secretion systems have been described, which mediate the transport of various substrates. However, the knowledge regarding some of the most fundamental aspects underlying transport across the cyanobacterial cell envelope remain unknown (details of the cyanobacterial cell envelope are shown in Figure 1). For example, only recently has the outer-membrane permeability of the unicellular cyanobacterium *Synechocystis* sp. PCC 6803 been investigated [6]. Remarkably, the permeability of this cyanobacterium outer membrane was found to be 20-fold lower than that of *Escherichia coli* [6], highlighting once more that what is known for the most well-documented Gram-negative bacteria should not be directly extrapolated to cyanobacteria.

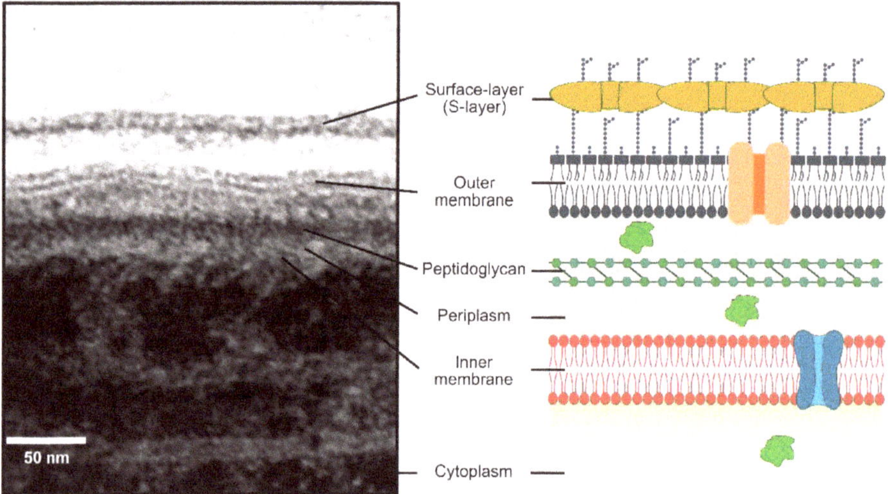

Figure 1. Cell envelope of the model, unicellular cyanobacterium *Synechocystis* sp. PCC 6803. Electron micrograph showing details of the *Synechocystis* sp. PCC 6803 cell envelope (left panel), which is schematically represented on the panel on the right-hand side (not to scale).

A number of biomolecules have been shown to be produced and secreted by cyanobacterial cells, including proteins (e.g., [7–10]), polysaccharides (e.g., [11–13]), glycolipids (e.g., [14–16]), and even fatty acids (e.g., [17–19]). Moreover, secretion is also vital to rid the cyanobacterial cell from harmful, exogenous compounds that reach the cell's interior, such as antibiotics (e.g., [9,20,21]) or excess of metals (e.g., [22,23]). Thus, product export represents an important survival strategy, playing crucial roles in a number of aspects of cyanobacterial physiology, like motility [24], adhesion [25], protection against desiccation [26], detoxification of extracellular reactive oxygen species [8,27], noxious compound efflux [9,20,21,28], etc. Throughout the years, several reports have focused on the identification and characterization of molecular machines mediating the translocation of products of metabolism, and others across the cyanobacterial inner and outer membranes (for reviews, see [29,30]). These systems, in which substrates are transported to the extracellular space via secretory portals located at the cell's inner and outer membranes, are classified as "classical secretion systems". Nevertheless, in terms of secretion strategies, one system has been highly overlooked in cyanobacteria—the release of extracellular vesicles (EVs), a mechanism of "non-classical secretion" (Figure 2).

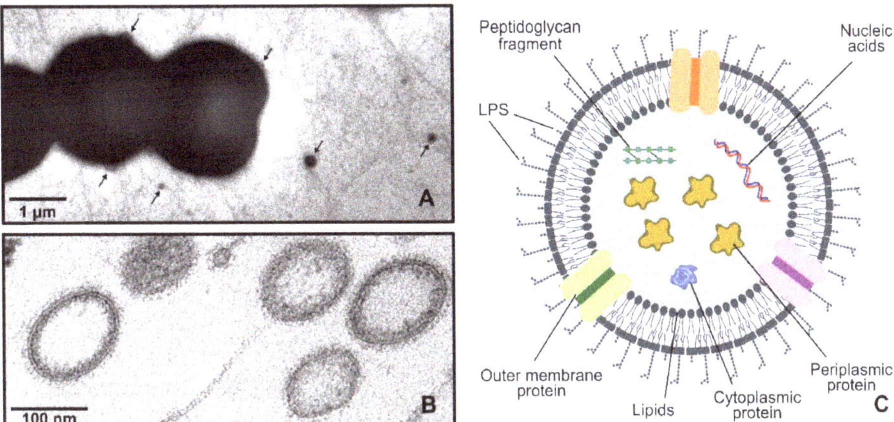

Figure 2. Cyanobacterial extracellular vesicles (EVs). (**A**) Transmission electron micrograph of negatively stained *Synechocystis* sp. PCC 6803 cells showing EVs (highlighted by arrows) on the cells' surfaces and in the extracellular medium. (**B**) Transmission electron micrograph showing ultrastructural details of EVs isolated from *Synechocystis* sp. PCC 6803 culture medium. (**C**) Diagram showing the general composition of EVs from Gram-negative bacteria, including those of the marine cyanobacterium *Prochlorococcus* MED4 [31]. These vesicles are composed mainly of outer-membrane components (lipopolysaccharides (LPS); outer-membrane proteins; and lipids), periplasmic content (proteins; peptidoglycan fragments; metabolites), and, in some instances, even cytoplasmic proteins and nucleic acids (DNA and/or RNA).

EVs have been detected in a number of cyanobacterial strains, including the unicellular *Synechococcus* sp. PCC 7002 [32], *Prochlorococcus* spp., *Synechococcus* sp WH8102 [31], *Synechocystis* sp. PCC 6803 [9,33], *Synechococcus elongatus* PCC 7942 [34] and *Cyanothece* sp. CCY 0110 [35], in the filamentous, non-heterocystous *Jaaginema litorale* LEGE 07176 [36], and in the filamentous, heterocystous *Anabaena* sp. PCC 7120 [8], *Cylindrospermopsis raciborskii* (CYRF-01) [37] and the *Azolla microphylla* symbiont [38]. However, the study of EVs in cyanobacteria has not received as much attention as in other bacteria. In this work, we focus on EVs released by Gram-negative bacteria, highlighting several aspects related to their composition, biogenesis and, most importantly, proposed biological roles. In addition, we gather the limited and fragmented information available on cyanobacterial EVs, and discuss this in light of what has been reported for other bacteria. As the biotechnological potential of cyanobacteria is becoming increasingly explored and established, we also draw special attention to several applications that have been tested and implemented with bacterial EVs. Altogether, our hopes are that researchers studying cyanobacteria become more acquainted with EVs, and that they decide to embark on the journey of elucidating further how relevant this unconventional secretion mechanism is for cyanobacterial cell survival.

2. EVs in Gram-Negative Bacteria

EVs were first reported more than 50 years ago [39]. Initially, they were regarded as mere artefacts of the electron microscopy technique or byproducts of bacterial growth [40]. However, subsequent studies definitively showed their production by different microorganisms [41]. Today, practically every microbial species tested, ranging from fungi to bacteria and archaea, has been shown to produce EVs, which suggests that EVs are a feature of every living cell [42]. EVs are discrete and non-replicable proteoliposomal nanoparticles [43], ranging in size between 20 and 500 nm in diameter [44]. Released by the bacterial cell envelope, EVs from Gram-negative bacteria naturally contain lipopolysaccharides, phospholipids and outer-membrane proteins; nevertheless, as EVs are not just small vesicles derived entirely from the outer membrane, periplasmic proteins, peptidoglycan fragments, and even cytoplasmic proteins and nucleic acids have been identified therein [40]. Due to the EVs' structure and composition, their content is strongly shielded against the activity of lytic enzymes present in the environment [45], preventing cargo degradation after leaving the cell. This way, the content of EVs can reach far longer distances than biomolecules released by the classical secretion systems. Moreover, EVs contain more than one cargo, all of which are simultaneously delivered to otherwise inaccessible targets. EVs' lumen, which resembles that of the periplasm in terms of its composition, contributes to the conservation of the cargos' features [45]. In particular, hydrophobic compounds, or those that demand targeted deliverance, can be transported in such nanostructures. Importantly, just as in the classical secretion systems, the formation and release of EVs can be tightly regulated by the cell, and so confer an adaptive advantage under stressful conditions.

3. EVs' Composition

One of the most well-characterized features of EVs is, perhaps, their biochemical composition. Several technical solutions are available to identify and quantify proteins, lipids and genetic material, which, combined with an ever increasing array of isolation methods (ultracentrifugation (including density gradient), ultrafiltration, hydrostatic dialysis, gel filtration (size-exclusion chromatography), precipitation, or even microfluidic devices and immunoprecipitation) [46], have helped clarifying many aspects of EVs' composition and biology. As briefly presented above, and despite cell envelope particularities found among Gram-negative bacteria, EVs are generally enriched in outer-membrane components [40]. Other cellular constituents have also been detected in EVs, namely periplasmic proteins and peptidoglycan fragments [47], and also metabolites, cytoplasmic proteins and nucleic acids [40,48] (Figure 2).

Increasing evidence supports the belief that bacteria modulate EVs' composition. For example, in *Salmonella* sp., EVs isolated from cultures grown under different conditions showed different compositions [49]. While proteins involved in translation and cellular metabolism were preferentially detected in EVs recovered from cultures grown with nutrient replete media, membrane proteins involved in nutrient acquisition were identified in EVs isolated from cultures maintained under limited nutritional conditions [49]. This study highlights the fact that EVs' content may vary according to the physiological state of the cell, and suggests that EVs' composition mirrors that of the parental cell at the moment of release. However, several others reported that the bacterial cell is capable of sorting what goes into EVs. The enrichment of a specific cargo that is scarce in the cell is illustrative of this fact. Conversely, the absence in EVs of highly abundant molecules in the cell and/or cell surface also indicate an EV-regulated packaging mechanism. In many bacterial species, whole-cell or outer-membrane and periplasmic proteome analyses showed differences when compared to isolated EVs [47,50]. An excellent showcase can be found in the bacteria of the genus *Bacteroides* [51–53]. These bacteria play a crucial role in decomposing complex polysaccharides in the human gut, being a typical resident of the intestinal microbiome. The ability is conferred by the expression of genes located in polysaccharide utilization loci (or PULs). A PUL represents a single genomic locus, which encodes proteins required to bind a specific polysaccharide at the cell surface, to perform an initial cleavage to oligosaccharides, to import these oligosaccharides into the periplasmic space, to complete the degradation into monosaccharides, and to regulate the PUL gene expression [54]. Comparative analyses between EVs and outer-membrane proteomes performed in *Bacteroides fragilis* and *B. thetaiotaomicron* revealed proteins exclusively detected either in EVs or in outer-membrane fractions [51,53]. Furthermore, EV proteomes were rich in PUL-encoded proteins, namely acidic lipoproteins with hydrolytic or carbohydrate-binding activities [51,53]. Thus, it appears that a selective packaging mechanism exists to load PUL-encoded hydrolytic enzymes into EVs [52]. Moreover, the regulated encapsulation of cargos into EVs is commonly observed in the context of pathogenesis, in which EVs work as a vehicle for virulence factors. Within the several cases reported, one highlight is the human oral *Porphyromonas gingivalis* that secretes EVs preferentially loaded with gingipains, a family of proteases that degrade host cytokines, leading to a decrease in the inflammatory response and facilitating bacterial invasion [55]. A three- to fivefold enrichment of gingipains was observed in EVs derived from *P. gingivalis* 33277 and W83, respectively, compared with levels in their parent bacterial strains [56], while major outer-membrane proteins seem to be excluded [56,57].

Because of the EVs' composition modulation capacity demonstrated by bacteria, the population of EVs isolated from a given bacterium at a specific time point can be highly heterogeneous. To illustrate this fact, one should note that within a determined population of recovered EVs, not all the vesicles are equal regarding their physical properties, namely in terms of their size and buoyant density. Thus, when studies report the detection of certain molecules in isolated EVs, it does not mean that all vesicles contain the same types of molecules. In a study that focused on a biological comparison of EVs from *E. coli*, the authors could separate EV subgroups from crude EV preparations [58]. The isolated EV subgroups presented molecularly distinct characteristics in terms of protein and endotoxin content, and amount of RNA [58], illustrating the natural biological heterogeneity of bacterial EVs.

Composition of Cyanobacterial EVs

Very limited information is available regarding the composition of cyanobacterial EVs. What is known is that EVs isolated from the marine cyanobacterium *Prochlorococcus* sp. have been shown to contain lipopolysaccharides, a number of typical cyanobacterial lipids (namely monoglycosyldiacylglycerol and sulfoquinovosyldiacylglycerol), and a diverse set of proteins, including periplasmic and membrane proteins (nutrient transporters, proteases, porins, hydrolases), but also cytoplasmic proteins (ribosome-associated proteins, and even RuBisCO) [31]. *Prochlorococcus* MED4 EVs were also reported to contain DNA and RNA: interestingly, DNA fragments detected in EVs measured at least 3000 bp, potentially encoding several genes. In addition, within the whole population of DNA fragments identified, an overabundance of reads was found, corresponding to a specific region of the chromosome [31]. As for RNA, sequences from as many as 95% of all open reading frames in the genome could be detected [31].

Altogether, the budding and detachment of EVs from the bacterial cell during active growth seem to constitute a regulated process rather than an accidental cell envelope disorganization event or a simple byproduct of cell lysis [40]. EVs' composition varies in agreement with whole-cell proteomic and lipidomic changes and adaptations, or as a result of specific sorting mechanisms, both of which are intimately dependent on how EVs are formed and released, i.e., on their biogenesis. While the presence of periplasmic and outer-membrane components in EVs can be easily perceived, the detection of cytoplasmic proteins and nucleic acids require an in-depth analysis of alternative routes for vesicle production. In the following section, we focus on the EVs' formation mechanisms.

4. EVs' Biogenesis

When addressing EVs' biogenesis, a point that requires clarification is their nomenclature. In the beginning, when EVs were first described, as the vesiculation process appeared to occur exclusively from the outer membrane, EVs were designated outer-membrane vesicles (OMVs). Even today, a great majority of the published articles refer to the vesicles isolated from the extracellular milieu as OMVs. Nevertheless, recent studies have shown additional routes for vesicle formation. Accordingly, the new types of vesicles were named outer–inner-membrane vesicles (OIMVs), explosive outer-membrane vesicles (EOMVs) and tube-shaped membranous structures (TSMSs) [41]. Hence, the term extracellular vesicles (EVs) is used here to refer to the whole population of vesicles released by cells, irrespective of their origin. However, it should be noted that Gram-negative EV studies predominantly discuss OMVs, and so most of the models describing vesicle formation and release address the classic concept of OMVs [59]. Here, we try to present a general overview of the various biogenesis hypotheses (Figure 3).

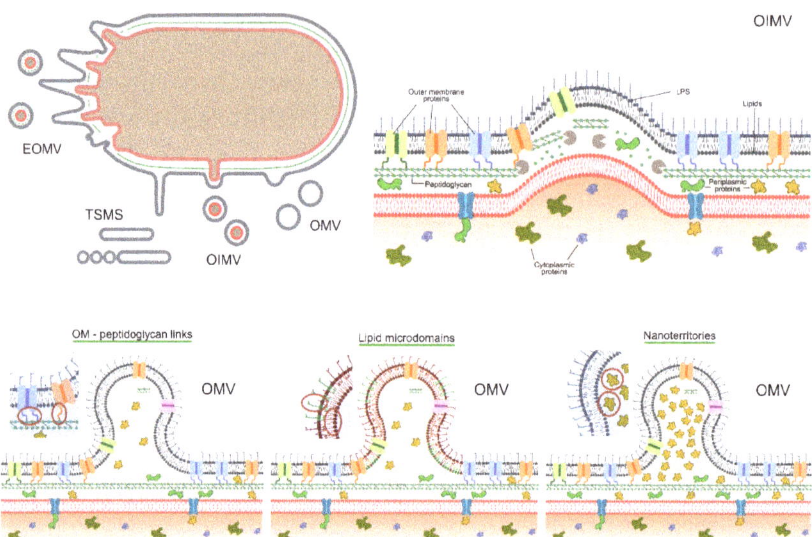

Figure 3. Gram-negative extracellular vesicles biogenesis models. The term "extracellular vesicle" is broadly used to describe various type of vesicles and structures released by bacterial cells. Four sub-groups of EVs have been described [41]: explosive outer-membrane vesicles (EOMV); tube-shaped membranous structures (TSMSs); outer–inner-membrane vesicles (OIMV), which have been proposed to form as a result of the action of peptidoglycan hydrolases in specific regions of the cell envelope, promoting the protrusion of the inner membrane into the periplasm and forcing the outer membrane to bulge [41,59,60]; and the classical outer-membrane vesicles (OMV) (lower panels). Different hypotheses have been formulated to explain the formation of OMVs [40,48,59], including modulation of the outer membrane–peptidoglycan links (left panel), lipid/LPS differential composition in specific regions of the outer membrane, termed lipid microdomains (middle panel), and accumulation of misfolded periplasmic proteins or envelope components in nanoterritories (right panel). Inset in each panel highlights different key components putatively involved in OMV formation, as per the respective model.

One aspect that is likely common for any EV formation model is the absence of linkages between the outer membrane and the underlying peptidoglycan in the blebbing region [61]. Envelope stability in most Gram-negative bacteria results from a number of interactions between the peptidoglycan and different components located in the outer and inner membranes [40]. The peptidoglycan-interacting outer-membrane components include the lipoprotein Braun (Lpp), the outer-membrane protein A (OmpA, which is a porin), and the Tol–Pal complex [40]. Thus, areas in the envelope devoid or depleted of attachments may represent hotspots for blebbing and EV formation. Accordingly, the overall number of Lpp–peptidoglycan crosslinks has been shown in some instances to inversely correlate with EV production. In *E. coli*, NlpI is an outer-membrane protein with a role in cell division, but it has also been shown to regulate the activity of a peptidoglycan endopeptidase (Spr, or MepS) that attacks peptide bonds in the peptidoglycan, modulating its crosslinks [62–64]. The *nlpI* mutant is a hypervesiculating strain, presenting approximately 40% less Lpp–peptidoglycan crosslinking than the wild-type strain [63]. Conversely, the loss of the diaminopimelic acid (a component of the peptidoglycan; DAP)–DAP peptide crosslinks resulted in an increase in Lpp–peptidoglycan crosslinking levels, which, in turn, led to hypovesiculation [65]. Furthermore, mutants lacking OmpA in several genera of Gram-negative bacteria resulted in EV overproduction, likely as a consequence of reduced outer membrane–peptidoglycan crosslinking [66–68]. Whether or not a specific mechanism exists to modulate these linkages remains unknown [61]. Consistent with the hypothesis of the outer membrane's dissociation from the underlying peptidoglycan, without compromising envelope integrity, portions of the peptidoglycan may be locally weakened by hydrolases in a targeted manner, allowing the invasion of the inner membrane into the periplasmic space. This protrusion, initiated by the inner membrane, ends up blebbing as an OIMV, whose envelope is similar to that of a cell. This model puts forward a possible explanation for the presence of nucleic acids and cytoplasmic proteins in EVs, as the OIMVs' lumen originates from the cytoplasm [41,60]. Nevertheless, it is not completely certain that these OIMVs are the only EVs to contain nucleic acids, and the mechanism that enables DNA fragments to be generated and incorporated therein is far from understood.

Despite the crucial role of outer-membrane components–peptidoglycan interactions in modulating EV formation, vesicle production has also been shown to occur independently of the total level of interactions and crosslinks. In fact, mutants with high "periplasmic pressure", resulting from an increase in misfolded periplasmic proteins, or accumulating high concentrations of envelope proteins, peptidoglycan fragments or LPS, were found to hypervesiculate [65]. This suggests an alternative route of EV biogenesis, in which these envelope components accumulate in nanoterritories, forcing the outer membrane to bulge outwards and to form a vesicle, relieving the cell from unwanted envelope components [40,59]. However, this model relies on a reduction in the crosslinking level in very specific regions, while the level of outer-membrane components–peptidoglycan crosslinks throughout the cell envelope remains constant [40,59,65].

As bacterial EVs are structurally composed of various lipid species and LPS (with rather different biophysical properties in terms of size, charge, and rigidity), determining membrane curvature and fluidity, some suggest their direct involvement in EV biogenesis. Thus, regions of the outer membrane may be enriched in particular types of lipids, LPS and/or specific LPS-associated molecules, representing lipid microdomains [40,59]. A combination of these lipid species and associated molecules may significantly increase the chance of producing outer-membrane blebs, resulting in EV production. To illustrate the role of lipid/LPS-associated molecules in the process of EVs biogenesis, one can highlight the major breakthroughs achieved in the study of the *Pseudomonas* quinolone signal (PQS). The PQS has been studied primarily in the context of its role as a quorum-sensing signaling molecule [69]. However, the exogenous addition of the hydrophobic PQS was found to affect the curvature of the outer membrane in *Pseudomonas* spp. as well as in other bacteria [70–72], promoting EV formation, as it incorporates into and/or fuses with LPS aggregates [70–72] and interacts with phospholipids [71].

Specific cellular responses may also lead to EV formation, which is particularly important in the context of environmental settings. This is the case for EOMVs, which are EVs formed as a consequence of cell lysis, triggered by phage-derived endolysins that degrade the peptidoglycan, after which the cell rounds up and explodes [41,73,74]. This mechanism of EV release has only been observed and reported in *Pseudomonas aeruginosa* [74], but is likely to occur in many other Gram-negative bacteria. EOMVs are formed due to the normal tendency of shattered membrane fragments to reorganize into vesicles, randomly capturing buoyant cellular material [41]. Hence, EOMVs' lumenal content may vary significantly in comparison to other EVs.

Finally, TSMSs represent a specialized type of EVs, comprising nanotubes, nanowires and nanopods. These structures are tube-like protrusions of the outer membrane, forming bridges between cells at the level of the periplasmic space, particularly produced in biofilms [41] but also detected in liquid media [75]. It has been suggested that TSMSs may facilitate the transfer between the cells of periplasmic proteins and metabolites, membrane proteins, and lipids [41,59]. Not much is known about how these structures are formed, and the simplest version detected was a chain of EVs in *Myxococcus xanthus* [76]. However, EVs have also been observed to pinch off from nanotubes in *Vibrio vulnificus* [77].

The models presented above elegantly address the contribution of various factors in EVs' biogenesis, and no vesicle formation process proposed thus far is widely accepted as the consensus. As the hypotheses discussed are not mutually exclusive, it is possible that a combination of factors and mechanisms operate to produce vesicles. Still, in the field of EV biogenesis, and in light of the presented models, a challenge that remains largely unsolved is to explain how soluble cargo is selected for EV packaging. It has been proposed that the sorting of soluble bacterial proteins into EVs may result from direct or indirect interactions with the periplasmic face of specific integral or auxiliary outer-membrane components, or even lipids, which are likely to be included in future vesicles. Conversely, components may be excluded from EVs by similar interactions with outer-membrane components that are not prone to budding [40]. Nevertheless, more experimental data are needed to validate this model.

EVs' Formation in Cyanobacteria

Several electron micrographs have shown EVs on the cell wall of different cyanobacterial strains (Figure 2; e.g., [9,31,36,78]). Nevertheless, the knowledge regarding EVs' biogenesis in these microorganisms is still very limited. Owing to the unique features presented by the cyanobacterial cell envelope, it can be anticipated that cyanobacterial EVs may have a distinct composition and rather specific formation mechanisms. On the outer-membrane level, structural differences in the lipid A moiety of LPS compared to that usually found in enterobacteria (in most cyanobacteria, the glucosamine disaccharide backbone is not phosphorylated, and the hydroxyl fatty acid chains are different [30]), the presence of species-specific carotenoids [79], the absence of the Braun lipoprotein Lpp (which establishes covalent interactions with the peptidoglycan; see above), are just a few of the factors that may determine specific mechanistic adaptations for EVs' biogenesis in cyanobacteria. Others may include peptidoglycan thickness (significantly higher when compared to other Gram-negative bacteria), the extent of peptidoglycan crosslinking (close to 60%, as opposed to the usual 20–30% found in most Gram-negative bacteria), the presence of specific polysaccharides complexed with the peptidoglycan [80], and unique lipid composition (while most bacteria contain phospholipids as major glycerolipids, cyanobacteria generally contain three glycolipids, monogalactosyldiacylglycerol, digalactosyldiacylglycerol and sulfoquinovosyldiacylglycerol, and a phospholipid, phosphatidylglycerol [81]). In addition, some cyanobacterial strains possess a surface layer (S-layer) (see Figure 1), the outermost structure of the cell envelope, composed of a glycoprotein with the capacity of self-assembly [82]. The S-layer in *Synechocystis* sp. PCC 6803 has been proposed to contribute to the integrity of the cell wall [83], conferring mechanical and osmotic cell stabilization, almost as an exoskeleton [84]. Using a number of *Synechocystis* sp. PCC 6803 mutant strains, impaired in several secretory functions, Gonçalves et al. observed that strains lacking the S-layer consistently presented a higher vesiculation capacity [28]. The authors hypothesized that the S-layer in *Synechocystis* sp. PCC 6803 represents a physical barrier for the biogenesis and release of EVs [28].

In summary, Gram-negative bacteria produce different types of EVs (OMVs, OIMVs, EOMVs and TSMSs). Remarkably, the production of EVs is apparently a vital mechanism as neither a bacterial species nor a mutant without the capacity to release EVs has been reported thus far [42]. As EVs' composition is intimately determined by the biogenesis process, an in-depth understanding of these mechanisms will continuously contribute to appreciate how bacteria selectively package EVs, and uncover their biological roles.

5. EVs' Biological Roles

Today, bacterial EVs are well established and are accepted to be involved in several biological functions. Intra- and interspecies communication, defense, the uptake of resources, the release of metabolic byproducts, detoxification, responsiveness to abrupt envelope stresses, horizontal gene transfer, acting as 'decoy' agents, as well as functioning as public goods, are among the proposed roles of bacterial EVs [40,48] (Figure 4). For instance, *Helicobacter pylori* releases EVs selectively decorated with catalase KatA on their surface, which contributes to neutralizing reactive oxygen species from the host, working as a mechanism to aid in evading host immune responses during infection [85]. Interestingly, *H. pylori* EVs showed a better performance in hydrogen peroxide detoxification than whole-cell lysates [85]. Moreover, Lekmeechai et al. demonstrated that these KatA-decorated EVs were able to protect *H. pylori* *katA*-null mutants from the hydrogen peroxide noxious effects, while EVs from these mutants or heat-inactivated KatA-decorated EVs could not [85]. With this example, it is possible to highlight that: (i) EVs can be enriched with a crucial biological component, which is active in the extracellular environment; (ii) EVs contribute to host infection;

and (iii) EVs released by specific bacteria may aid other bacteria to promote survival. Moreover, there are numerous cases in which antibiotic resistance enzymes have been detected in isolated EVs [86–88], resulting not only in a defense strategy, but also in large-spectrum community tactics. In this context, nucleic acid exchange between bacterial cells via EVs [89–91] represents a serious health-threatening issue, as antibiotic resistance determinants encoded in genetic material may be transferred by horizontal gene transfer. Another example to highlight EVs importance is their presence in biofilms [70,92,93], which are competitive niches that can comprise several bacterial species. It is thought that, due to the natural structure and secreting advantages of EVs, they could have a role in communication through the biofilm matrix to ease bacterial cooperation [70]. Biofilms may also function as well as a way to adhere to certain surfaces to facilitate the process of invasion, in which EVs may also play a critical role. That is the case described for EVs from the oral pathogen *P. gingivalis*, which enables the recruitment and adhesion of other bacterial players to the biofilm, increasing the chance of dental plaque formation [50,94]. The role of EVs in biofilms can be further illustrated by *Xylella fastidiosa*, a serious crop-threatening pathogen. During plant colonization, *X. fastidiosa* migrates and proliferates within xylem vessels. It has been shown that *X. fastidiosa* hypervesiculating mutants present a higher virulence capacity than the wild-type bacterium [95]. It was proposed that the production of EVs by *X. fastidiosa* represents a system that helps to control the state of biofilm formation: as the surface composition of EVs released by *X. fastidiosa* is very similar to that of the cells, EVs bind to the xylem vessels and block the attachment of *X. fastidiosa* to surfaces [95]. This allows the bacterial cell to modulate its adhesion capacity, in order to change from an "exploratory" lifestyle, for spreading within the plant host, to a more adhesive type, prone to insect transmission [95].

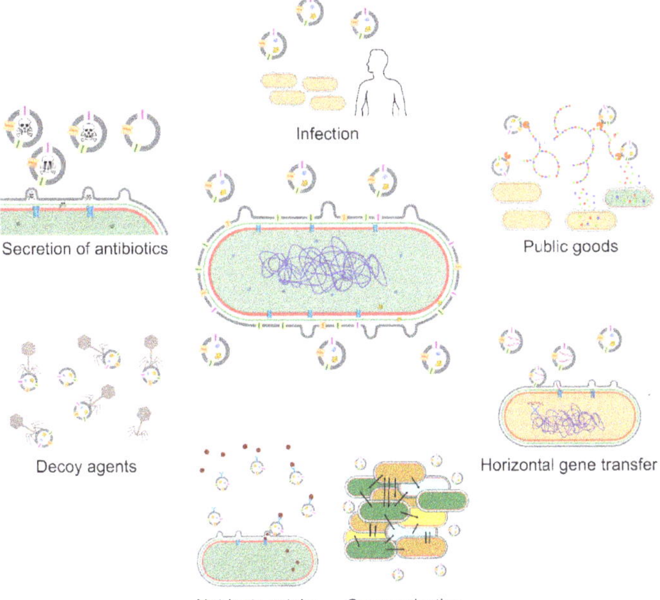

Figure 4. Diagram summarizing the biological roles proposed for extracellular vesicles from Gram-negative bacteria. EVs may have the following functions: in infection, by eluding the host immune response; as public goods, as they may carry catalytic enzymes that help deconstructing complex extracellular biomolecules, making them available for the EVs-releasing bacteria and for other organisms; in horizontal gene transfer, as they transport nucleic acids that can be transferred between organisms; in communication, e.g., in the context of biofilms; in the uptake of nutrients (e.g., [96,97]); as decoy, for example to evade phage infection; and in the secretion of antibiotics (e.g., [98,99]).

Biological Functions of EVs in Cyanobacteria

Cyanobacterial EVs have also been associated with key physiological roles. For instance, an increase in vesiculation was observed in *Cylindrospermopsis raciborskii* CYRF-01 in response to UV radiation [37]. A similar hypervesiculation response was detected when this cyanobacterium was co-cultured with *Microcystis aeruginosa* (MIRF-01). The authors suggested that hypervesiculation was an adaption response to the physical and biological environmental stressors [37]. Moreover, the work by Biller and co-workers on *Prochlorococcus* sp. EVs suggested that these spherical nanostructures can work as a "smoke-screen", preventing cyanophage infection [31]. Due to the similar composition between outer membranes from *Prochlorococcus* sp. and their EVs, cyanophage receptors are equally found on living cells and vesicles. This could explain the numerous accounts of phages bound to vesicles reported by Biller et al. [31], some of which seemed to have injected their genetic material into the EV. Thus, EVs may help the cell to escape phage infection, and so represent a crucial element of cyanobacterial population dynamics in the ocean. As *Prochlorococcus* sp. are the main primary producers in open oceans [100], this work represents also a major breakthrough in the understanding of the role of cyanobacterial EVs in other ecological aspects, namely as vectors for carbon fluxes in the ocean, as the metabolites and biomolecules packaged in *Prochlorococcus*-derived EVs were shown to be able to exclusively sustain the growth of heterotrophic bacteria [31]. Moreover, nucleic acids were found in EVs released by this highly abundant microorganism. Thus, vesicles from *Prochlorococcus* sp. truly represent a reservoir of genetic information, proposed to be active in horizontal gene transfer in marine ecosystems [31,101]. Another report supports the presence of DNA in cyanobacterial EVs: Zheng et al., showed that the cyanobiont present in the water fern *Azolla microphylla* sporocarp abundantly releases EVs [38]. Remarkably, up to 90% of the detected vesicles that were released by the cyanobiont filaments upon their initiation of proakinete/akinete differentiation contained DNA [38]. This observation suggests possible lateral gene transfer between the symbiotic partners. In addition, the released EVs may also contribute to cyanobacterial biofilm formation and outer envelope development during akinete differentiation and during sporocarp development, as proposed [38].

In line with reports from other Gram-negative bacteria, vesiculation in cyanobacteria also seems to be adjusted in response to specific genetic modifications, further suggesting that EVs may contribute to the homeostasis of the cyanobacterial cell. The deletion of the two glycogen synthase genes (*glgA*-I and *glgA*-II) in *Synechococcus* sp. PCC 7002 resulted not only in an accumulation of soluble sugars, but also in the spontaneous secretion of high levels of soluble sugars into the medium, without the need for additional transporters [32]. Moreover, numerous EVs budding from the outer membrane of the mutant could be observed, and the authors hypothesized that the release of EVs could mediate soluble sugar excretion [32]. In a different work, the deletion of the gene encoding the outer-membrane protein TolC in *Synechocystis* sp. PCC 6803 resulted in a strain with a high vesiculation capacity [9]. TolC is crucial for the export of several types of biomolecules, ranging from proteins [7,9,25,28] to fatty acids [17], and exogenous compounds that reach the cytoplasm [9,20,21,28]. It was proposed that the vesiculation was a response to the impaired classical secretion [9]. In a follow up study, deletion mutants in some inner-membrane and periplasmic components, parts of the TolC-dependent efflux pumps, showed as well-altered vesiculation [28]. Mutants with a higher vesiculation capacity than the wild-type strain were found to lack the S-layer (see above, EVs' biogenesis). Nevertheless, none of these mutant strains reached the high vesiculation level presented by the *tolC*-deletion mutant [28]. It remains to be fully determined why the *tolC*-deletion mutant hypervesiculates, but there is evidence that the strain is under some sort of envelope stress, as specific periplasmic chaperones and proteases are highly upregulated (*spy* and *degQ*) and overexpressed (Spy and DegP) compared to the wild-type [28]. Genetic modifications in cyanobacteria that affect vesiculation do not always result in hypervesiculation. In a recent work, a hypovesiculating *Synechocystis* sp. PCC 6803 mutant was reported, in which the gene encoding the alternative sigma factor SigF was deleted [102]. Nonetheless, SigF was found to be a broad player in the control of several metabolic and cellular processes, including different secretion mechanisms. The lower vesiculation capacity was therefore proposed to be part of a large adaptation process, rather than a specific target of SigF regulation [102].

As more bacterial species become characterized in terms of their EVs production capacities, more biological roles are attributed to EVs. The fact that many of these bacteria can be genetically modified and engineered has contributed towards the recognition that bacterial EVs may be used to fulfil specific technological goals.

6. Biotechnological Applications of Bacterial EVs

Fundamental research carried out in the past 20–30 years has significantly contributed to our better understanding of the mechanisms of EVs biogenesis in bacteria, their composition and properties, as well as their biological roles. The capacity of EVs to package multiple cargos, the fact that cargo properties are maintained in EVs, and the extraordinary advantage of EVs in protecting and trafficking cargo to otherwise inaccessible targets [45] are just a few of the EV attributes that have promoted their application in biotechnology. Furthermore, through the constant developments in bioengineering and genetic editing techniques, it is now possible to tailor EVs' exterior features, as well as modulate their cargo [103,104]. As a result, EV-based nanotechnology has been evolving quickly into a powerful and innovative toolkit for various fields (Figure 5), even though several challenges still need to be addressed, mainly regarding their scalability, commercial viability, and safe and efficient clinical use [103].

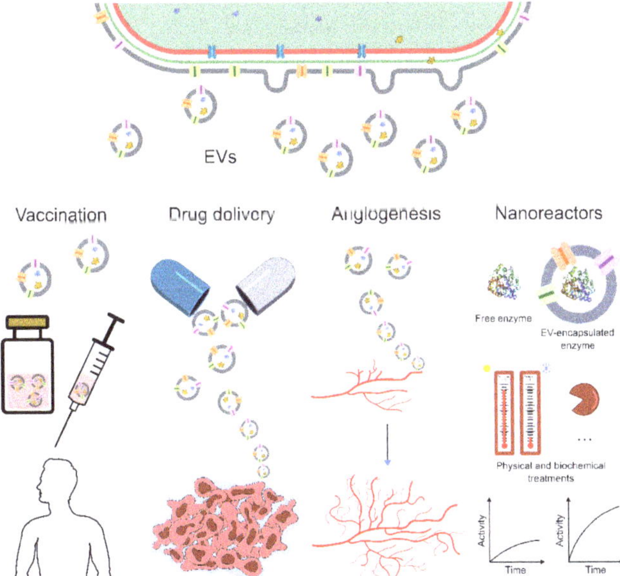

Figure 5. Illustration summarizing some applications that have been tested with bacterial EVs. One field that has benefited enormously from the potential of bacterial EVs is biomedicine. The best example is perhaps the use of EVs in vaccination, for which the commercial vaccine Bexsero® is a case of widely accepted success. Bacterial EVs have also been tested with promising results to deliver cargo in a cell-specific manner, in particular in the context of cancer research (e.g., [105,106]). EVs from the cyanobacterium *Synechococcus elongatus* PCC 7942 have been shown to positively modulate and even promote angiogenesis in the context of wound healing. Bacteria have also been genetically engineered as to release EVs packaged with enzymes, contributing to the protection of enzymatic functions (e.g., [107]), in what the EVs may be regarded as nanoreactors.

Inherent to the constant necessity of developing novel and more efficient vaccines against various pathogenic agents, EVs have been successfully implemented as potential vaccines/antigen-delivery platforms. Two major strategies have been followed for the development of EV-based vaccination. On one hand, naturally derived EVs from the target pathogenic bacterium can be isolated and directly administered to the subject. In this case, EV protection is mediated by the abundance of pathogen-associated molecular patterns, known to play a key role in stimulating innate immunity and promoting adaptive immune responses. The best example is the EV-containing meningococcal vaccine against *Neisseria meningitidis*, currently commercialized as Bexsero® [96,108–110]. On the other hand, through a genetic engineering approach, recombinant EVs harboring heterologous antigen(s) can be directly obtained from a non-pathogenic bacterial model. In this way, it is possible to develop safe and standardized EVs, capable of eliciting an adequate immune response with great versatility and flexibility [111–113]. The use of engineered *E. coli*-derived EVs loaded with heterologous proteins to work as antigen-delivery vehicles has been tried with very promising results [111]. All of these efforts towards the development of EV-based vaccine strategies offer the opportunity to accomplish highly effective, easy to produce and multi-valent vaccines.

EVs have also been used as specialized carriers for the delivery of therapeutic compounds [103,104]. For instance, bioengineered EVs produced by *E. coli*, presenting a human epidermal growth factor receptor 2-specific affibody molecule on their surface, could target cancer cells in a mouse model in a cell-specific manner [105]. As EVs were loaded with a small interfering RNA (siRNA), the specificity of the EVs' delivery resulted in targeted gene silencing, and thus in a reduction in the protein levels of a key enzyme involved in mitosis, ultimately inhibiting cancer cell proliferation [105]. Moreover, these nanostructures have also revealed their great potential in the delivery of chemotherapeutic agents, as demonstrated with EVs from an attenuated *Klebsiella pneumonia* strain that efficiently delivered the drug doxorubicin to lung cancer cells [106]. Moreover, in the context of cancer, EVs have also been reported as a promising tool for immunotherapy strategies, not only through EV-induced long-term antitumor immune responses [114], but also through their decoration with tumor antigens towards antigen-specific antitumor therapy [115]. Despite the fact that most efforts have been directed towards anti-cancer/anti-tumor therapies, other delivery applications have also been associated with EVs, including their development as antibiotic delivery vehicles. Recently, through the identification of a novel drug resistance mechanism mediated by EVs in *Acinetobacter baumannii*, antibiotic-loaded EVs were shown not only to efficiently deliver and facilitate the entry of antibiotics into pathogenic bacteria, but also to support the sustained bactericidal effect in the intestinal tract of a mouse model [99]. In an era of global concern regarding drug-resistant bacteria, this reported novel class of EVs mediating the delivery of antibiotics may turn out to be extremely useful in reducing antibiotic dosage, and in increasing the bactericidal effect.

Other applications are also being considered and developed, including the use of EVs as natural and versatile nanoreactors. The systematic organization of enzymes for the efficient operation of cascade reactions, resulting in cellulose degradation, has been demonstrated through the use of *E. coli*-derived EVs [116]. The functional assembly of multiple enzymes on the EV surface was accomplished by the construction of a system comprising three orthogonal cohesion domains and a cellulose-binding module, all assembled and anchored through OmpA [116]. By adding their species-specific dockerin parts, all cellulases could be activated, resulting in the amplification of cellulose hydrolysis by more than 20-fold over that of non-complexed enzymes [116]. In addition to performing complex biocatalysis, EVs have also been recently described to protect the native properties and enzymatic activity of phosphotriesterase (PTE), an ideal enzyme for bioremediation [107]. In light of this discovery, it is plausible to extend this EV-mediated enzymatic packaging/protection strategy to other biotechnological fields, including the health sciences context, for example, where the prevention of enzyme degradation on human serum would lead to a more efficient enzymatic delivery and bioactivity.

Cyanobacterial EVs in Biotechnology

Although most of the reports available on cyanobacterial EVs are fundamentally directed towards basic research on EV biology, these microorganisms represent an untapped tool with attractive features and promising applications. Indeed, EVs from *Synechococcus elongatus* PCC 7942 have already been linked to the promotion of angiogenesis and wound healing in vitro and in vivo through the promotion of interleukin 6 expression, highlighting their potential in the field of biomedicine [34]. As novel insights are gained in respect to cyanobacterial EVs, this can be the first of a broad range of prospective applications. In fact, cyanobacterial EVs might have a significant advantage over other bacteria in relation to a relevant issue concerning EVs' structural components, namely regarding LPS. In high dosages, LPS from some Gram-negative bacteria, including *E. coli*, *Salmonella* and *Pseudomonas*, can lead to pathological reactions, including systemic inflammatory responses [117]. Consequently, to avoid toxicity in biomedical applications, administration to the subjects of EVs derived from such organisms must follow rigorous and complex administration schemes, and/or demand vesicle bioprocessing approaches by implementing attenuated mutants or additional LPS detoxification steps [118,119]. Remarkably, evidence from different sources strongly suggests a significantly reduced toxicity in cyanobacterial LPS [120–122]. This low immune reactivity against cyanobacterial LPS can be explained by the missing phosphorylation at positions 1-4' of the glucosamine disaccharide in the cyanobacterial lipid A backbone [30]. The possibility that cyanobacterial LPS do not possess true endotoxic effects that can harm human health [30] underlines the potential of cyanobacteria-derived EVs as valid and safe alternatives to overcome this LPS-induced toxicity hurdle. With the current efforts regarding the development of more robust and reliable molecular tools to metabolically engineer cyanobacteria, EVs may become the target for packaging products of interest that can range from soluble and membrane-bound enzymes, to lipids, long-chain carbon molecules, and metabolites. With this in mind, an array of interesting applications may soon be associated with cyanobacterial EVs.

7. Conclusions

The EV research field is not novel, but there has always been some skepticism among researchers as to the existence and physiological function of EVs [42]. Some of the frequent arguments include the possibility of vesicles being artefacts of lipid self-aggregation, or debris from lysed cells, or even just ordinary waste products. Moreover, other argue that EV release represents a waste of energy, and thus serves no purpose, or that bacteria lack the complex machinery present in eukaryotic organisms determining the release of vesicular bodies, or even, as there are no mutants lacking vesiculation capacity, that EV release cannot be a regulated process [42]. Recent technical advances to isolate and analyze EVs have contributed significantly to dismissing these doubts. As the field develops and EVs become increasingly accepted, there is a demand for standardizing EV research through increased systematic reporting, which can already be followed on the Transparent Reporting and Centralizing Knowledge in Extracellular Vesicle Research (EV-TRACK) database [123]. In addition, the International Society for Extracellular Vesicles has also published guidelines for EV research, which are frequently updated.

In cyanobacteria, research into EVs is only just emerging, but the limited reports available already suggest that cyanobacterial EVs fulfil important ecological functions, such as evading phage infection [31]. In other instances, vesiculation was detected in response to the deletion of key metabolic enzymes [32] or transporters [9,28], but the full biological meaning of such observations remains to be elucidated. Nevertheless, it is clear that vesiculation in cyanobacteria can be modulated in response to environmental cues [37], thus pointing to the strong possibility of contributing to cell homeostasis and, ultimately, adaptation and survival. The detection of EVs in the context of symbiotic associations [38] opens the door to poorly addressed inter-kingdom communication, with large implications. In particular, the detection of genetic material therein may represent a route of gene transfer, maybe accounting for the presence of cyanobacterial genes in plant nuclear genomes [124].

In the past few years, the advances in the understanding of cyanobacterial biology and metabolism have been intimately related to rapid developments in synthetic biology and metabolic engineering. In addition, these contributions have also placed cyanobacteria in the spotlight in biotechnology, garnering increasing interest from several sectors. Combining cyanobacterial genetic tractability with metabolic plasticity, cyanobacteria can easily become attractive models for the production of EVs with tailored cargo. Some reports indicate that cyanobacterial lipopolysaccharides are less toxic to mammals compared to those isolated from other bacteria [121,122], which can be relevant for EV-mediated delivery.

In summary, the study of EVs in cyanobacteria definitely deserves more attention. Several open questions related to physiological responses, cellular modifications, differentiation, communication strategies and adaptability might be re-analyzed in light of the existence of EVs. Future advances in this field will surely contribute to a better understanding of how cyanobacteria have made their way through the last 2.1 Ga of the Earth's history, while shaping life as we know it.

Author Contributions: S.L., J.M.-C., and P.O. performed the literature search, compiled the relevant information and wrote the manuscript. P.T. gave suggestions regarding the outline of the manuscript and edited the text. All authors have read and agreed to the published version of the manuscript.

Funding: This work was financed by Fundo Europeu de Desenvolvimento Regional (FEDER) funds through the COMPETE 2020 Operacional Programme for Competitiveness and Internationalisation (POCI), Portugal, 2020, and by Portuguese funds through Fundação para a Ciência e a Tecnologia/Ministério da Ciência, Tecnologia e Ensino Superior in the framework of the project POCI-01-0145-FEDER-029540 (PTDC/BIA-OUT/29540/2017). Fundação para a Ciência e a Tecnologia is also greatly acknowledged for the PhD fellowship SFRH/BD/130478/2017 (SL) and FCT Investigator grant IF/00256/2015 (PO).

Acknowledgments: The authors acknowledge the support of the i3S Scientific Platform "Histology and Electron Microscopy", a member of the national infrastructure Portuguese Platform of Bioimaging (PPBI-POCI-01-0145-FEDER-022122).

Conflicts of Interest: The authors declare no conflict of interest.

References

1. Garcia-Pichel, F.; Zehr, J.P.; Bhattacharya, D.; Pakrasi, H.B. What's in a name? The case of cyanobacteria. *J. Phycol.* **2020**, *56*, 1–5. [CrossRef] [PubMed]
2. Carr, N.G.; Whitton, B.A. *Biol Blue-Green Algae*; University of California Press: Auckland, CA, USA, 1973.
3. Singh, H. Desiccation and radiation stress tolerance in cyanobacteria. *J. Basic Microbiol.* **2018**, *58*, 813–826. [CrossRef] [PubMed]
4. Agostoni, M.; Montgomery, B.L. Survival strategies in the aquatic and terrestrial world: The impact of second messengers on cyanobacterial processes. *Life* **2014**, *4*, 745–769. [CrossRef] [PubMed]
5. Giner-Lamia, J.; Pereira, S.B.; Bovea-Marco, M.; Futschik, M.E.; Tamagnini, P.; Oliveira, P. Extracellular Proteins: Novel Key Components of Metal Resistance in Cyanobacteria? *Front. Microbiol.* **2016**, *7*, 878. [CrossRef] [PubMed]
6. Kowata, H.; Tochigi, S.; Takahashi, H.; Kojima, S. Outer Membrane Permeability of Cyanobacterium *Synechocystis* sp. Strain PCC 6803: Studies of Passive Diffusion of Small Organic Nutrients Reveal the Absence of Classical Porins and Intrinsically Low Permeability. *J. Bacteriol* **2017**, *199*, e00371-17. [CrossRef] [PubMed]
7. Hahn, A.; Stevanovic, M.; Brouwer, E.; Bublak, D.; Tripp, J.; Schorge, T.; Karas, M.; Schleiff, E. Secretome analysis of *Anabaena* sp. PCC 7120 and the involvement of the TolC-homologue HgdD in protein secretion. *Environ. Microbiol.* **2015**, *17*, 767–780. [CrossRef]
8. Oliveira, P.; Martins, N.M.; Santos, M.; Couto, N.A.; Wright, P.C.; Tamagnini, P. The *Anabaena* sp. PCC 7120 Exoproteome: Taking a Peek outside the Box. *Life* **2015**, *5*, 130–163. [CrossRef]
9. Oliveira, P.; Martins, N.M.; Santos, M.; Pinto, F.; Büttel, Z.; Couto, N.A.S.; Wright, P.C.; Tamagnini, P. The versatile TolC-like Slr1270 in the cyanobacterium *Synechocystis* sp. PCC 6803. *Environ. Microbiol.* **2016**, *18*, 486–502. [CrossRef]
10. Vilhauer, L.; Jervis, J.; Ray, W.K.; Helm, R.F. The exo-proteome and exo-metabolome of *Nostoc punctiforme* (Cyanobacteria) in the presence and absence of nitrate. *Arch. Microbiol.* **2014**, *196*, 357–367. [CrossRef]

11. Fisher, M.L.; Allen, R.; Luo, Y.; Curtiss, R., III. Export of Extracellular Polysaccharides Modulates Adherence of the Cyanobacterium *Synechocystis*. *PLoS ONE* **2013**, *8*, e74514. [CrossRef]

12. Jittawuttipoka, T.; Planchon, M.; Spalla, O.; Benzerara, K.; Guyot, F.; Cassier-Chauvat, C.; Chauvat, F. Multidisciplinary Evidences that *Synechocystis* PCC6803 Exopolysaccharides Operate in Cell Sedimentation and Protection against Salt and Metal Stresses. *PLoS ONE* **2013**, *8*, e55564. [CrossRef] [PubMed]

13. Mota, R.; Guimarães, R.; Büttel, Z.; Rossi, F.; Colica, G.; Silva, C.J.; Santos, C.; Gales, L.; Zille, A.; De Philippis, R.; et al. Production and characterization of extracellular carbohydrate polymer from *Cyanothece* sp. CCY 0110. *Carbohydr. Polym.* **2013**, *92*, 1408–1415. [CrossRef] [PubMed]

14. Moslavac, S.; Nicolaisen, K.; Mirus, O.; Al Dehni, F.; Pernil, R.; Flores, E.; Maldener, I.; Schleiff, E. A TolC-Like Protein Is Required for Heterocyst Development in *Anabaena* sp. Strain PCC 7120. *J. Bacteriol.* **2007**, *189*, 7887. [CrossRef] [PubMed]

15. Staron, P.; Forchhammer, K.; Maldener, I. A novel ATP-driven pathway of glycolipid export involving TolC. *J. Biol. Chem.* **2011**, *286*, 38202–38210. [CrossRef]

16. Shvarev, D.; Nishi, C.N.; Maldener, I. Glycolipid composition of the heterocyst envelope of *Anabaena* sp. PCC 7120 is crucial for diazotrophic growth and relies on the UDP-galactose 4-epimerase HgdA. *MicrobiologyOpen* **2019**, *8*, e00811. [CrossRef] [PubMed]

17. Bellefleur, M.P.A.; Wanda, S.-Y.; Curtiss, R. Characterizing active transportation mechanisms for free fatty acids and antibiotics in *Synechocystis* sp. PCC 6803. *BMC Biotechnol.* **2019**, *19*, 5. [CrossRef]

18. Kato, A.; Use, K.; Takatani, N.; Ikeda, K.; Matsuura, M.; Kojima, K.; Aichi, M.; Maeda, S.-i.; Omata, T. Modulation of the balance of fatty acid production and secretion is crucial for enhancement of growth and productivity of the engineered mutant of the cyanobacterium *Synechococcus elongatus*. *Biotechnol. Biofuels* **2016**, *9*, 91. [CrossRef]

19. Liu, X.; Sheng, J.; Curtiss III, R. Fatty acid production in genetically modified cyanobacteria. *Proc. Natl. Acad. Sci. USA* **2011**, *108*, 6899. [CrossRef]

20. Hahn, A.; Stevanovic, M.; Mirus, O.; Lytvynenko, I.; Pos, K.M.; Schleiff, E. The Outer Membrane TolC-like Channel HgdD Is Part of Tripartite Resistance-Nodulation-Cell Division (RND) Efflux Systems Conferring Multiple-drug Resistance in the Cyanobacterium *Anabaena* sp. PCC7120. *J. Biol. Chem.* **2013**, *288*, 31192–31205. [CrossRef]

21. Hahn, A.; Stevanovic, M.; Mirus, O.; Schleiff, E. The TolC-like Protein HgdD of the Cyanobacterium *Anabaena* sp. PCC 7120 Is Involved in Secondary Metabolite Export and Antibiotic Resistance. *J. Biol. Chem.* **2012**, *287*, 41126–41138. [CrossRef]

22. Giner-Lamia, J.; López-Maury, L.; Reyes, J.C.; Florencio, F.J. The CopRS Two-Component System Is Responsible for Resistance to Copper in the Cyanobacterium *Synechocystis* sp. PCC 6803. *Plant Physiol.* **2012**, *159*, 1806. [CrossRef] [PubMed]

23. Stevanovic, M.; Hahn, A.; Nicolaisen, K.; Mirus, O.; Schleiff, E. The components of the putative iron transport system in the cyanobacterium *Anabaena* sp. PCC 7120. *Environ. Microbiol.* **2012**, *14*, 1655–1670. [CrossRef] [PubMed]

24. Bhaya, D.; Bianco, N.R.; Bryant, D.; Grossman, A. Type IV pilus biogenesis and motility in the cyanobacterium *Synechocystis* sp. PCC6803. *Mol. Microbiol.* **2000**, *37*, 941–951. [CrossRef] [PubMed]

25. Oliveira, P.; Pinto, F.; Pacheco, C.C.; Mota, R.; Tamagnini, P. HesF, an exoprotein required for filament adhesion and aggregation in *Anabaena* sp. PCC 7120. *Environ. Microbiol.* **2015**, *17*, 1631–1648. [CrossRef] [PubMed]

26. Tamaru, Y.; Takani, Y.; Yoshida, T.; Sakamoto, T. Crucial Role of Extracellular Polysaccharides in Desiccation and Freezing Tolerance in the Terrestrial Cyanobacterium *Nostoc commune*. *Appl. Environ. Microbiol.* **2005**, *71*, 7327. [CrossRef]

27. Shirkey, B.; Kovarcik, D.P.; Wright, D.J.; Wilmoth, G.; Prickett, T.F.; Helm, R.F.; Gregory, E.M.; Potts, M. Active Fe-Containing Superoxide Dismutase and Abundant *sodF* mRNA in *Nostoc commune* (Cyanobacteria) after Years of Desiccation. *J. Bacteriol.* **2000**, *182*, 189. [CrossRef] [PubMed]

28. Gonçalves, C.F.; Pacheco, C.C.; Tamagnini, P.; Oliveira, P. Identification of inner membrane translocase components of TolC-mediated secretion in the cyanobacterium *Synechocystis* sp. PCC 6803. *Environ. Microbiol.* **2018**, *20*, 2354–2369. [CrossRef]

29. Gonçalves, C.F.; Lima, S.; Tamagnini, P.; Oliveira, P. Chapter 18—Cyanobacterial Secretion Systems: Understanding Fundamental Mechanisms toward Technological Applications. In *Cyanobacteria*; Mishra, A.K., Tiwari, D.N., Rai, A.N., Eds.; Academic Press: Cambridge, MA, USA, 2019; pp. 359–381.

30. Hahn, A.; Schleiff, E. The Cell Envelope. In *The Cell Biology of Cyanobacteria*; Flores, E., Herrero, A., Eds.; Caister Academic Press: Norfolk, UK, 2014; pp. 29–87.

31. Biller, S.J.; Schubotz, F.; Roggensack, S.E.; Thompson, A.W.; Summons, R.E.; Chisholm, S.W. Bacterial Vesicles in Marine Ecosystems. *Science* **2014**, *343*, 183–186. [CrossRef]

32. Xu, Y.; Tiago Guerra, L.; Li, Z.; Ludwig, M.; Charles Dismukes, G.; Bryant, D.A. Altered carbohydrate metabolism in glycogen synthase mutants of *Synechococcus* sp. strain PCC 7002: Cell factories for soluble sugars. *Metab. Eng.* **2013**, *16*, 56–67. [CrossRef]

33. Pardo, Y.A.; Florez, C.; Baker, K.M.; Schertzer, J.W.; Mahler, G.J. Detection of outer membrane vesicles in *Synechocystis* PCC 6803. *FEMS Microbiol. Lett.* **2015**, *362*. [CrossRef]

34. Yin, H.; Chen, C.-Y.; Liu, Y.-W.; Tan, Y.-J.; Deng, Z.-L.; Yang, F.; Huang, F.-Y.; Wen, C.; Rao, S.-S.; Luo, M.-J.; et al. *Synechococcus elongatus* PCC7942 secretes extracellular vesicles to accelerate cutaneous wound healing by promoting angiogenesis. *Theranostics* **2019**, *9*, 2678–2693. [CrossRef] [PubMed]

35. Mota, R.; Vidal, R.; Pandeirada, C.; Flores, C.; Adessi, A.; De Philippis, R.; Nunes, C.; Coimbra, M.A.; Tamagnini, P. Cyanoflan: A cyanobacterial sulfated carbohydrate polymer with emulsifying properties. *Carbohydr. Polym.* **2020**, *229*, 115525. [CrossRef] [PubMed]

36. Brito, Â.; Ramos, V.; Mota, R.; Lima, S.; Santos, A.; Vieira, J.; Vieira, C.; Kaštovský, J.; Vasconcelos, V.M.; Tamagnini, P. Description of new genera and species of marine cyanobacteria from the Portuguese Atlantic coast. *Mol. Phylogenetics Evol.* **2017**, *111*, 18–34. [CrossRef]

37. Zarantonello, V.; Silva, T.P.; Noyma, N.P.; Gamalier, J.P.; Mello, M.M.; Marinho, M.M.; Melo, R.C.N. The Cyanobacterium *Cylindrospermopsis raciborskii* (CYRF-01) Responds to Environmental Stresses with Increased Vesiculation Detected at Single-Cell Resolution. *Front. Microbiol.* **2018**, *9*, 272. [CrossRef] [PubMed]

38. Zheng, W.; Bergman, B.; Chen, B.; Zheng, S.; Xiang, G.; Rasmussen, U. Cellular responses in the cyanobacterial symbiont during its vertical transfer between plant generations in the *Azolla microphylla* symbiosis. *New Phytol.* **2009**, *181*, 53–61. [CrossRef] [PubMed]

39. Work, E.; Knox, K.W.; Vesk, M. The chemistry and electron microscopy of an extracellular lipopolysaccharide from *Escherichia coli*. *Ann. N. Y. Acad. Sci.* **1966**, *133*, 438–449. [CrossRef] [PubMed]

40. Schwechheimer, C.; Kuehn, M.J. Outer-membrane vesicles from Gram-negative bacteria: Biogenesis and functions. *Nat. Rev. Microbiol.* **2015**, *13*, 605–619. [CrossRef]

41. Toyofuku, M.; Nomura, N.; Eberl, L. Types and origins of bacterial membrane vesicles. *Nat. Rev. Microbiol.* **2019**, *17*, 13–24. [CrossRef]

42. Coelho, C.; Casadevall, A. Answers to naysayers regarding microbial extracellular vesicles. *Biochem. Soc. Trans.* **2019**, *47*, 1005–1012. [CrossRef]

43. Caruana, J.C.; Walper, S.A. Bacterial Membrane Vesicles as Mediators of Microbe - Microbe and Microbe - Host Community Interactions. *Front. Microbiol* **2020**, *11*, 432. [CrossRef]

44. Zavan, L.; Bitto, N.J.; Kaparakis-Liaskos, M. Introduction, History, and Discovery of Bacterial Membrane Vesicles. In *Bacterial Membrane Vesicles: Biogenesis, Functions and Applications*; Kaparakis-Liaskos, M., Kufer, T.A., Eds.; Springer International Publishing: Cham, Switzerland, 2020; pp. 1–21.

45. Bonnington, K.E.; Kuehn, M.J. Protein selection and export via outer membrane vesicles. *Biochim. Biophys. Acta* **2014**, *1843*, 1612–1619. [CrossRef] [PubMed]

46. Konoshenko, M.Y.; Lekchnov, E.A.; Vlassov, A.V.; Laktionov, P.P. Isolation of Extracellular Vesicles: General Methodologies and Latest Trends. *BioMed Res. Int.* **2018**, *2018*, 8545347. [CrossRef] [PubMed]

47. Lee, J.; Kim, O.Y.; Gho, Y.S. Proteomic profiling of Gram-negative bacterial outer membrane vesicles: Current perspectives. *Proteom. Clin. Appl.* **2016**, *10*, 897–909. [CrossRef]

48. Kulp, A.; Kuehn, M.J. Biological functions and biogenesis of secreted bacterial outer membrane vesicles. *Annu. Rev. Microbiol.* **2010**, *64*, 163–184. [CrossRef] [PubMed]

49. Bai, J.; Kim, S.I.; Ryu, S.; Yoon, H. Identification and characterization of outer membrane vesicle-associated proteins in *Salmonella enterica* serovar Typhimurium. *Infect. Immun.* **2014**, *82*, 4001–4010. [CrossRef] [PubMed]

50. Nadeem, A.; Oscarsson, J.; Wai, S.N. Delivery of Virulence Factors by Bacterial Membrane Vesicles to Mammalian Host Cells. In *Bacterial Membrane Vesicles: Biogenesis, Functions and Applications*; Kaparakis-Liaskos, M., Kufer, T.A., Eds.; Springer International Publishing: Cham, Switzerland, 2020; pp. 131–158.

51. Elhenawy, W.; Debelyy, M.O.; Feldman, M.F. Preferential packing of acidic glycosidases and proteases into *Bacteroides* outer membrane vesicles. *mBio* **2014**, *5*, e00909-14. [CrossRef] [PubMed]

52. Rakoff-Nahoum, S.; Coyne, M.J.; Comstock, L.E. An ecological network of polysaccharide utilization among human intestinal symbionts. *Curr. Biol.* **2014**, *24*, 40–49. [CrossRef]

53. Valguarnera, E.; Scott, N.E.; Azimzadeh, P.; Feldman, M.F. Surface Exposure and Packing of Lipoproteins into Outer Membrane Vesicles Are Coupled Processes in *Bacteroides*. *mSphere* **2018**, *3*, e00559-18. [CrossRef]

54. Terrapon, N.; Lombard, V.; Drula, É.; Lapébie, P.; Al-Masaudi, S.; Gilbert, H.J.; Henrissat, B. PULDB: The expanded database of Polysaccharide Utilization Loci. *Nucleic Acids Res.* **2017**, *46*, D677–D683. [CrossRef]

55. Haurat, M.F.; Aduse-Opoku, J.; Rangarajan, M.; Dorobantu, L.; Gray, M.R.; Curtis, M.A.; Feldman, M.F. Selective sorting of cargo proteins into bacterial membrane vesicles. *J. Biol. Chem.* **2011**, *286*, 1269–1276. [CrossRef]

56. Mantri, C.K.; Chen, C.-H.; Dong, X.; Goodwin, J.S.; Pratap, S.; Paromov, V.; Xie, H. Fimbriae-mediated outer membrane vesicle production and invasion of *Porphyromonas gingivalis*. *Microbiologyopen* **2015**, *4*, 53–65. [CrossRef] [PubMed]

57. Xie, H. Biogenesis and function of *Porphyromonas gingivalis* outer membrane vesicles. *Future Microbiol.* **2015**, *10*, 1517–1527. [CrossRef] [PubMed]

58. Dauros Singorenko, P.; Chang, V.; Whitcombe, A.; Simonov, D.; Hong, J.; Phillips, A.; Swift, S.; Blenkiron, C. Isolation of membrane vesicles from prokaryotes: A technical and biological comparison reveals heterogeneity. *J. Extracell. Vesicles* **2017**, *6*, 1324731. [CrossRef] [PubMed]

59. Zingl, F.G.; Leitner, D.R.; Schild, S. Biogenesis of Gram-Negative OMVs. In *Bacterial Membrane Vesicles: Biogenesis, Functions and Applications*; Kaparakis-Liaskos, M., Kufer, T.A., Eds.; Springer International Publishing: Cham, Switzerland, 2020; pp. 23–46.

60. Pérez-Cruz, C.; Delgado, L.; López-Iglesias, C.; Mercade, E. Outer-inner membrane vesicles naturally secreted by Gram-negative pathogenic bacteria. *PLoS ONE* **2015**, *10*, e0116896. [CrossRef] [PubMed]

61. Mashburn-Warren, L.M.; Whiteley, M. Special delivery: Vesicle trafficking in prokaryotes. *Mol. Microbiol.* **2006**, *61*, 839–846. [CrossRef] [PubMed]

62. Ohara, M.; Wu, H.C.; Sankaran, K.; Rick, P.D. Identification and Characterization of a New Lipoprotein, NlpI, in *Escherichia coli* K-12. *J. Bacteriol.* **1999**, *181*, 4318. [CrossRef]

63. Schwechheimer, C.; Rodriguez, D.L.; Kuehn, M.J. NlpI-mediated modulation of outer membrane vesicle production through peptidoglycan dynamics in *Escherichia coli*. *Microbiologyopen* **2015**, *4*, 375–389. [CrossRef]

64. Singh, S.K.; SaiSree, L.; Amrutha, R.N.; Reddy, M. Three redundant murein endopeptidases catalyse an essential cleavage step in peptidoglycan synthesis of *Escherichia coli* K12. *Mol. Microbiol.* **2012**, *86*, 1036–1051. [CrossRef] [PubMed]

65. Schwechheimer, C.; Kulp, A.; Kuehn, M.J. Modulation of bacterial outer membrane vesicle production by envelope structure and content. *BMC Microbiol.* **2014**, *14*, 324. [CrossRef]

66. Deatherage, B.L.; Lara, J.C.; Bergsbaken, T.; Barrett, S.L.R.; Lara, S.; Cookson, B.T. Biogenesis of bacterial membrane vesicles. *Mol. Microbiol.* **2009**, *72*, 1395–1407. [CrossRef]

67. Moon, D.C.; Choi, C.H.; Lee, J.H.; Choi, C.-W.; Kim, H.-Y.; Park, J.S.; Kim, S.I.; Lee, J.C. *Acinetobacter baumannii* outer membrane protein a modulates the biogenesis of outer membrane vesicles. *J. Microbiol.* **2012**, *50*, 155–160. [CrossRef] [PubMed]

68. Song, T.; Mika, F.; Lindmark, B.; Liu, Z.; Schild, S.; Bishop, A.; Zhu, J.; Camilli, A.; Johansson, J.; Vogel, J.; et al. A new *Vibrio cholerae* sRNA modulates colonization and affects release of outer membrane vesicles. *Mol. Microbiol.* **2008**, *70*, 100–111. [CrossRef] [PubMed]

69. Lin, J.; Cheng, J.; Wang, Y.; Shen, X. The *Pseudomonas* Quinolone Signal (PQS): Not Just for Quorum Sensing Anymore. *Front. Cell. Infect. Microbiol.* **2018**, *8*, 230. [CrossRef] [PubMed]

70. Mashburn, L.M.; Whiteley, M. Membrane vesicles traffic signals and facilitate group activities in a prokaryote. *Nature* **2005**, *437*, 422–425. [CrossRef] [PubMed]

71. Mashburn-Warren, L.; Howe, J.; Garidel, P.; Richter, W.; Steiniger, F.; Roessle, M.; Brandenburg, K.; Whiteley, M. Interaction of quorum signals with outer membrane lipids: Insights into prokaryotic membrane vesicle formation. *Mol. Microbiol.* **2008**, *69*, 491–502. [CrossRef] [PubMed]

72. Tashiro, Y.; Ichikawa, S.; Nakajima-Kambe, T.; Uchiyama, H.; Nomura, N. *Pseudomonas* Quinolone Signal Affects Membrane Vesicle Production in not only Gram-Negative but also Gram-Positive Bacteria. *Microbes Environ.* **2010**, *25*, 120–125. [CrossRef]

73. Toyofuku, M.; Tashiro, Y.; Nomura, N.; Eberl, L. Functions of MVs in Inter-Bacterial Communication. In *Bacterial Membrane Vesicles: Biogenesis, Functions and Applications*; Kaparakis-Liaskos, M., Kufer, T.A., Eds.; Springer International Publishing: Cham, Switzerland, 2020; pp. 101–117.

74. Turnbull, L.; Toyofuku, M.; Hynen, A.L.; Kurosawa, M.; Pessi, G.; Petty, N.K.; Osvath, S.R.; Cárcamo-Oyarce, G.; Gloag, E.S.; Shimoni, R.; et al. Explosive cell lysis as a mechanism for the biogenesis of bacterial membrane vesicles and biofilms. *Nat. Commun.* **2016**, *7*, 11220. [CrossRef]

75. McCaig, W.D.; Koller, A.; Thanassi, D.G. Production of Outer Membrane Vesicles and Outer Membrane Tubes by *Francisella novicida*. *J. Bacteriol.* **2013**, *195*, 1120. [CrossRef]

76. Remis, J.P.; Wei, D.; Gorur, A.; Zemla, M.; Haraga, J.; Allen, S.; Witkowska, H.E.; Costerton, J.W.; Berleman, J.E.; Auer, M. Bacterial social networks: Structure and composition of *Myxococcus xanthus* outer membrane vesicle chains. *Environ. Microbiol.* **2014**, *16*, 598–610. [CrossRef]

77. Hampton, C.M.; Guerrero-Ferreira, R.C.; Storms, R.E.; Taylor, J.V.; Yi, H.; Gulig, P.A.; Wright, E.R. The Opportunistic Pathogen *Vibrio vulnificus* Produces Outer Membrane Vesicles in a Spatially Distinct Manner Related to Capsular Polysaccharide. *Front. Microbiol.* **2017**, *8*, 2177. [CrossRef]

78. Gorby, Y.A.; Yanina, S.; McLean, J.S.; Rosso, K.M.; Moyles, D.; Dohnalkova, A.; Beveridge, T.J.; Chang, I.S.; Kim, B.H.; Kim, K.S.; et al. Electrically conductive bacterial nanowires produced by *Shewanella oneidensis* strain MR-1 and other microorganisms. *Proc. Natl. Acad. Sci. USA* **2006**, *103*, 11358. [CrossRef] [PubMed]

79. Jürgens, U.J.; Weckesser, J. Carotenoid-containing outer membrane of *Synechocystis* sp. strain PCC6714. *J. Bacteriol.* **1985**, *164*, 384. [CrossRef] [PubMed]

80. Hoiczyk, E.; Hansel, A. Cyanobacterial Cell Walls: News from an Unusual Prokaryotic Envelope. *J. Bacteriol.* **2000**, *182*, 1191. [CrossRef]

81. Wada, H.; Murata, N. Membrane Lipids in Cyanobacteria. In *Lipids in Photosynthesis: Structure, Function and Genetics*; Siegenthaler, P.-A., Murata, N., Eds.; Springer: Dordrecht, The Netherlands; Berlin, Germany, 1998; Volume 6.

82. Fagan, R.P.; Fairweather, N.F. Biogenesis and functions of bacterial S-layers. *Nat. Rev. Microbiol.* **2014**, *12*, 211–222. [CrossRef] [PubMed]

83. Trautner, C.; Vermaas, W.F.J. The sll1951 Gene Encodes the Surface Layer Protein of *Synechocystis* sp. Strain PCC 6803. *J. Bacteriol.* **2013**, *195*, 5370–5380. [CrossRef] [PubMed]

84. Engelhardt, H. Are S-layers exoskeletons? The basic function of protein surface layers revisited. *J. Struct. Biol.* **2007**, *160*, 115–124. [CrossRef]

85. Lekmeechai, S.; Su, Y.C.; Brant, M.; Alvarado-Kristensson, M.; Vallström, A.; Obi, I.; Arnqvist, A.; Riesbeck, K. *Helicobacter pylori* Outer Membrane Vesicles Protect the Pathogen From Reactive Oxygen Species of the Respiratory Burst. *Front. Microbiol.* **2018**, *9*, 1837. [CrossRef]

86. Kulkarni, H.M.; Nagaraj, R.; Jagannadham, M.V. Protective role of *E. coli* outer membrane vesicles against antibiotics. *Microbiol. Res.* **2015**, *181*, 1–7. [CrossRef]

87. Liao, Y.T.; Kuo, S.C.; Chiang, M.H.; Lee, Y.T.; Sung, W.C.; Chen, Y.H.; Chen, T.L.; Fung, C.P. *Acinetobacter baumannii* Extracellular OXA-58 Is Primarily and Selectively Released via Outer Membrane Vesicles after Sec-Dependent Periplasmic Translocation. *Antimicrob. Agents Chemother.* **2015**, *59*, 7346–7354. [CrossRef]

88. Stentz, R.; Horn, N.; Cross, K.; Salt, L.; Brearley, C.; Livermore, D.M.; Carding, S.R. Cephalosporinases associated with outer membrane vesicles released by *Bacteroides* spp. protect gut pathogens and commensals against β-lactam antibiotics. *J. Antimicrob. Chemother.* **2015**, *70*, 701–709. [CrossRef]

89. Chatterjee, S.; Mondal, A.; Mitra, S.; Basu, S. *Acinetobacter baumannii* transfers the *blaNDM*-1 gene via outer membrane vesicles. *J. Antimicrob. Chemother.* **2017**, *72*, 2201–2207. [CrossRef] [PubMed]

90. Renelli, M.; Matias, V.; Lo, R.Y.; Beveridge, T.J. DNA-containing membrane vesicles of *Pseudomonas aeruginosa* PAO1 and their genetic transformation potential. *Microbiology* **2004**, *150*, 2161–2169. [CrossRef] [PubMed]

91. Rumbo, C.; Fernández-Moreira, E.; Merino, M.; Poza, M.; Mendez, J.A.; Soares, N.C.; Mosquera, A.; Chaves, F.; Bou, G. Horizontal transfer of the OXA-24 carbapenemase gene via outer membrane vesicles: A new mechanism of dissemination of carbapenem resistance genes in *Acinetobacter baumannii*. *Antimicrob. Agents Chemother.* **2011**, *55*, 3084–3090. [CrossRef]

92. Flemming, H.C.; Wingender, J.; Szewzyk, U.; Steinberg, P.; Rice, S.A.; Kjelleberg, S. Biofilms: An emergent form of bacterial life. *Nat. Rev. Microbiol.* **2016**, *14*, 563–575. [CrossRef] [PubMed]

93. Schooling, S.R.; Beveridge, T.J. Membrane vesicles: An overlooked component of the matrices of biofilms. *J. Bacteriol.* **2006**, *188*, 5945–5957. [CrossRef] [PubMed]

94. Kamaguchi, A.; Nakayama, K.; Ichiyama, S.; Nakamura, R.; Watanabe, T.; Ohta, M.; Baba, H.; Ohyama, T. Effect of *Porphyromonas gingivalis* vesicles on coaggregation of *Staphylococcus aureus* to oral microorganisms. *Curr. Microbiol.* **2003**, *47*, 485–491. [CrossRef]

95. Ionescu, M.; Zaini, P.A.; Baccari, C.; Tran, S.; da Silva, A.M.; Lindow, S.E. *Xylella fastidiosa* outer membrane vesicles modulate plant colonization by blocking attachment to surfaces. *Proc. Natl. Acad. Sci. USA* **2014**, *111*, E3910–E3918. [CrossRef]

96. Lappann, M.; Otto, A.; Becher, D.; Vogel, U. Comparative proteome analysis of spontaneous outer membrane vesicles and purified outer membranes of *Neisseria meningitidis*. *J. Bacteriol.* **2013**, *195*, 4425–4435. [CrossRef]

97. Prados-Rosales, R.; Weinrick, B.C.; Piqué, D.G.; Jacobs, W.R.; Casadevall, A.; Rodriguez, G.M. Role for *Mycobacterium tuberculosis* Membrane Vesicles in Iron Acquisition. *J. Bacteriol.* **2014**, *196*, 1250. [CrossRef]

98. Allan, N.D.; Beveridge, T.J. Gentamicin Delivery to Burkholderia cepacia Group IIIa Strains via Membrane Vesicles from *Pseudomonas aeruginosa* PAO1. *Antimicrob. Agents Chemother.* **2003**, *47*, 2962. [CrossRef]

99. Huang, W.; Zhang, Q.; Li, W.; Yuan, M.; Zhou, J.; Hua, L.; Chen, Y.; Ye, C.; Ma, Y. Development of novel nanoantibiotics using an outer membrane vesicle-based drug efflux mechanism. *J. Control. Release* **2020**, *317*, 1–22. [CrossRef] [PubMed]

100. Flombaum, P.; Gallegos, J.L.; Gordillo, R.A.; Rincón, J.; Zabala, L.L.; Jiao, N.; Karl, D.M.; Li, W.K.; Lomas, M.W.; Veneziano, D.; et al. Present and future global distributions of the marine Cyanobacteria *Prochlorococcus* and *Synechococcus*. *Proc. Natl. Acad. Sci. USA* **2013**, *110*, 9824–9829. [CrossRef] [PubMed]

101. Biller, S.J.; McDaniel, L.D.; Breitbart, M.; Rogers, E.; Paul, J.H.; Chisholm, S.W. Membrane vesicles in sea water: Heterogeneous DNA content and implications for viral abundance estimates. *ISME J.* **2017**, *11*, 394–404. [CrossRef] [PubMed]

102. Flores, C.; Santos, M.; Pereira, S.B.; Mota, R.; Rossi, F.; De Philippis, R.; Couto, N.; Karunakaran, E.; Wright, P.C.; Oliveira, P.; et al. The alternative sigma factor SigF is a key player in the control of secretion mechanisms in *Synechocystis* sp. PCC 6803. *Environ. Microbiol.* **2019**, *21*, 343–359. [CrossRef]

103. Jain, S.; Pillai, J. Bacterial membrane vesicles as novel nanosystems for drug delivery. *Int. J. Nanomed.* **2017**, *12*, 6329–6341. [CrossRef]

104. Wang, S.; Gao, J.; Wang, Z. Outer membrane vesicles for vaccination and targeted drug delivery. *WIREs Nanomed. Nanobiotechnology* **2019**, *11*, e1523. [CrossRef]

105. Gujrati, V.; Kim, S.; Kim, S.H.; Min, J.J.; Choy, H.E.; Kim, S.C.; Jon, S. Bioengineered bacterial outer membrane vesicles as cell-specific drug-delivery vehicles for cancer therapy. *ACS Nano* **2014**, *8*, 1525–1537. [CrossRef]

106. Kuerban, K.; Gao, X.; Zhang, H.; Liu, J.; Dong, M.; Wu, L.; Ye, R.; Feng, M.; Ye, L. Doxorubicin-loaded bacterial outer-membrane vesicles exert enhanced anti-tumor efficacy in non-small-cell lung cancer. *Acta Pharm. Sin. B* **2020**, in press. [CrossRef]

107. Alves, N.J.; Turner, K.B.; Medintz, I.L.; Walper, S.A. Protecting enzymatic function through directed packaging into bacterial outer membrane vesicles. *Sci. Rep.* **2016**, *6*, 24866. [CrossRef]

108. Holst, J.; Oster, P.; Arnold, R.; Tatley, M.V.; Næss, L.M.; Aaberge, I.S.; Galloway, Y.; McNicholas, A.; O'Hallahan, J.; Rosenqvist, E.; et al. Vaccines against meningococcal serogroup B disease containing outer membrane vesicles (OMV): Lessons from past programs and implications for the future. *Hum. Vaccines Immunother.* **2013**, *9*, 1241–1253. [CrossRef]

109. O'Ryan, M.; Stoddard, J.; Toneatto, D.; Wassil, J.; Dull, P.M. A multi-component meningococcal serogroup B vaccine (4CMenB): The clinical development program. *Drugs* **2014**, *74*, 15–30. [CrossRef] [PubMed]

110. Tani, C.; Stella, M.; Donnarumma, D.; Biagini, M.; Parente, P.; Vadi, A.; Magagnoli, C.; Costantino, P.; Rigat, F.; Norais, N. Quantification by LC-MS(E) of outer membrane vesicle proteins of the Bexsero® vaccine. *Vaccine* **2014**, *32*, 1273–1279. [CrossRef] [PubMed]

111. Fantappiè, L.; de Santis, M.; Chiarot, E.; Carboni, F.; Bensi, G.; Jousson, O.; Margarit, I.; Grandi, G. Antibody-mediated immunity induced by engineered *Escherichia coli* OMVs carrying heterologous antigens in their lumen. *J. Extracell Vesicles* **2014**, *3*. [CrossRef] [PubMed]

112. Van der Pol, E.; Böing, A.N.; Gool, E.L.; Nieuwland, R. Recent developments in the nomenclature, presence, isolation, detection and clinical impact of extracellular vesicles. *J. Thromb. Haemost.* **2016**, *14*, 48–56. [CrossRef] [PubMed]

113. Van der Pol, L.; Stork, M.; van der Ley, P. Outer membrane vesicles as platform vaccine technology. *Biotechnol. J.* **2015**, *10*, 1689–1706. [CrossRef] [PubMed]

114. Kim, O.Y.; Park, H.T.; Dinh, N.T.H.; Choi, S.J.; Lee, J.; Kim, J.H.; Lee, S.-W.; Gho, Y.S. Bacterial outer membrane vesicles suppress tumor by interferon-γ-mediated antitumor response. *Nat. Commun.* **2017**, *8*, 626. [CrossRef]

115. Grandi, A.; Tomasi, M.; Zanella, I.; Ganfini, L.; Caproni, E.; Fantappiè, L.; Irene, C.; Frattini, L.; Isaac, S.J.; König, E.; et al. Synergistic Protective Activity of Tumor-Specific Epitopes Engineered in Bacterial Outer Membrane Vesicles. *Front. Oncol.* **2017**, *7*, 253. [CrossRef]

116. Park, M.; Sun, Q.; Liu, F.; DeLisa, M.P.; Chen, W. Positional assembly of enzymes on bacterial outer membrane vesicles for cascade reactions. *PLoS ONE* **2014**, *9*, e97103. [CrossRef]

117. Park, K.S.; Choi, K.H.; Kim, Y.S.; Hong, B.S.; Kim, O.Y.; Kim, J.H.; Yoon, C.M.; Koh, G.Y.; Kim, Y.K.; Gho, Y.S. Outer membrane vesicles derived from *Escherichia coli* induce systemic inflammatory response syndrome. *PLoS ONE* **2010**, *5*, e11334. [CrossRef]

118. Baker, J.L.; Chen, L.; Rosenthal, J.A.; Putnam, D.; DeLisa, M.P. Microbial biosynthesis of designer outer membrane vesicles. *Curr. Opin. Biotechnol.* **2014**, *29*, 76–84. [CrossRef]

119. Tan, K.; Li, R.; Huang, X.; Liu, Q. Outer Membrane Vesicles: Current Status and Future Direction of These Novel Vaccine Adjuvants. *Front. Microbiol* **2018**, *9*, 783. [CrossRef] [PubMed]

120. Swanson-Mungerson, M.; Incrocci, R.; Subramaniam, V.; Williams, P.; Hall, M.L.; Mayer, A.M.S. Effects of cyanobacteria *Oscillatoria* sp. lipopolysaccharide on B cell activation and Toll-like receptor 4 signaling. *Toxicol. Lett.* **2017**, *275*, 101–107. [CrossRef] [PubMed]

121. Durai, P.; Batool, M.; Choi, S. Structure and Effects of Cyanobacterial Lipopolysaccharides. *Mar. Drugs* **2015**, *13*, 4217–4230. [CrossRef] [PubMed]

122. Stewart, I.; Schluter, P.J.; Shaw, G.R. Cyanobacterial lipopolysaccharides and human health—A review. *Environ. Health Glob. Access Sci. Source* **2006**, *5*, 7. [CrossRef] [PubMed]

123. The EV-TRACK Platform. Available online: http://evtrack.org/about.php (accessed on 14 July 2020).

124. Deusch, O.; Landan, G.; Roettger, M.; Gruenheit, N.; Kowallik, K.V.; Allen, J.F.; Martin, W.; Dagan, T. Genes of Cyanobacterial Origin in Plant Nuclear Genomes Point to a Heterocyst-Forming Plastid Ancestor. *Mol. Biol. Evol.* **2008**, *25*, 748–761. [CrossRef]